Limnology

Limnology

Charles R. Goldman
Division of Environmental Studies
University of California, Davis

Alexander J. Horne
Department of Sanitary and Environmental Engineering
University of California, Berkeley

McGraw-Hill Book Company

New York   St. Louis   San Francisco   Auckland   Bogotá   Hamburg
Johannesburg   London   Madrid   Mexico   Montreal   New Delhi
Panama   Paris   São Paulo   Singapore   Sydney   Tokyo   Toronto

This book was set in Times Roman by A Graphic Method Inc.
The editors were James E. Vastyan, Jay Ricci,
David T. Horvath, and Claudia Tantillo;
the production supervisor was Charles Hess.
The drawings were done by Fine Line Illustrations, Inc.
The cover was designed by Robin Hessel.
R. R. Donnelley & Sons Company was printer and binder.

The cover photograph was taken by Alexander J. Horne.
All color plate photographs were taken by Alexander J. Horne
with the exception of Plate 1a (George J. Malyj) and
Plate 6a (in cooperation with NASA-AMES).

LIMNOLOGY

3 4 5 6 7 8 9 0   DOCDOC   8 9 8 7 6 5 4

ISBN 0-07-023651-8

Library of Congress Cataloging in Publication Data

Goldman, Charles Remington, date
    Limnology.

    Bibliography: p.
    Includes indexes.
    1. Limnology.   I. Horne, Alexander J.   II. Title.
QH96.G63   574.5'26322   82.15356
ISBN 0-07-023651-8   AACR2

To the Students
Past, Present, and Future

Contents

Preface   xv

CHAPTER 1   Limnology: Past, Present, and Future   1

Overview   1
Introduction   1
    The Early Limnologists   2
    The Journals   6
Limnology Today   6
The Future   7
Further Readings   11

CHAPTER 2   The Structure of Aquatic Ecosystems   12

Overview   12
Lakes   12
    Morphometry   12
    Zonation   13
    Terminology of Physical Structure   15
    Chemical Structure   17
    Biological Structure   18
    The Watershed   19
    Ponds Versus Lakes   19

	Streams and Rivers	20
	Estuaries	22
	Further Readings	23

CHAPTER 3   Water and Light   24

	Overview	24
	Properties of Water	25
	Light	29
	Measurement	30
	Light and Heat	30
	Light in the Atmosphere	31
	Light under Water	32
	Lake Color	33
	Reflection (Albedo)	36
	Further Readings	37

CHAPTER 4   Heat   39

	Overview	39
	Measurements	40
	Introduction	40
	Thermal Stratification	41
	Thermocline Formation	43
	The Flow of Heat	45
	Heating and Cooling of the Epilimnion or Mixed Layer	47
	Upwelling	48
	Hypolimnetic Entrainment	48
	Establishment and Destruction of the Thermocline	50
	The Thermal Bar and Fall Overturn	51
	Heat Cycles	52
	Waste Heat Discharges	55
	Further Readings	56

CHAPTER 5   Water Movement   57

	Overview	57
	Introduction	58
	Measurement	59
	Laminar and Turbulent Flow	61
	The Kinetic Energy Spectrum	63
	Advective and Diffusive Transport of Energy	67
	Motion in the Epilimnion	69
	Surface Drift	69
	Surface Waves	71
	Langmuir Spirals	73
	Random	74
	Breaking Waves	75
	Motion in the Thermocline	75
	Internal Gravity Waves	76
	Short-Period Internal Gravity Waves	79
	Long-Period Internal Waves: Kelvin and Poincaré Waves	79
	Motion in the Hypolimnion	82

An Overall View   82
Effect of Rivers   83
Further Readings   84

CHAPTER 6   Chemicals and Growth Factors   87

Overview   87
Organic and Inorganic Compounds   88
The Influence of the Climate and Watershed   90
The Atmospheric Contribution   90
Chemical Pollution, Logging, and Erosion   92
Further Readings   94

CHAPTER 7   Oxygen and Carbon Dioxide   95

Overview   95
Measurement   96
Sources   96
    Diffusion from the Atmosphere   96
    Photosynthesis   97
pH   98
Effects of Temperature, Salinity, and Organisms   102
The Redox Potential   104
Diel and Seasonal Changes in Oxygen and Carbon Dioxide   105
    Diel Variations   106
    Seasonal Variations   106
    Types of Oxygen-Depth Curves   108
Oxygen and Carbon Dioxide in Streams   111
Further Readings   112

CHAPTER 8   Nitrogen   114

Overview   114
Introduction   115
Measurement   115
The Nitrogen Cycle   116
Forms of Nitrogen in Lakes   117
Nitrogen Fixation   118
Denitrification   121
Nitrate and Nitrite   122
    Seasonal Cycles of Nitrate   126
Ammonia and the Ammonium Ion   126
    Seasonal Cycles of Ammonia   127
Dissolved Organic Nitrogen (DON)   128
Further Readings   130

CHAPTER 9   Phosphorus   131

Overview   131
Introduction   132
Measurement   132
The Phosphorus Cycle   132
    In Pelagic Waters   133
    In the Sediment   135

Uptake of Phosphorus and Induction of Phosphatases   135
Recycling of Phosphorus   137
    Role of Rooted Macrophytes and Algal Decomposition   139
Sources of Phosphorus   140
Further Readings   140

CHAPTER 10   Other Nutrients   142

Overview   142
Elements Sometimes Required in Large Quantities:
Si, Ca, Mg, Na, K, S, Cl   143
                                                            143
    Silica (SiO$_2$) and Silicon (Si)   147
    Calcium   149
    Magnesium
    Sodium and Potassium   150
    Sulfur and Chlorine   150
    Iron, the "Trace" Element Needed in Moderate Quantities 151
Trace Elements: Mn, Zn, Cu, Mo, Co   158
    Manganese   159
    Zinc   160
    Copper   160
    Molybdenum   161
    Cobalt   163
    Other Trace Metals   163
Further Readings   163

CHAPTER 11   Organisms in Lakes, Streams, and Estuaries   165

Overview   165
Introduction   165
Functional Classification   166
The Major Groups of Organisms   166
    Viruses   166
    Bacteria   166
    Fungi and Fungi-Like Organisms   169
    Green Plants: Algae and Macrophytes   171
    Protozoans   180
    Rotifers   181
    Crustaceans   181
    Aquatic Insects   184
    Worms and Mollusks   185
    Fish   187
    Amphibians, Reptiles, Birds, and Mammals   192
Further Readings   196

CHAPTER 12   Phytoplankton   197

Overview   197
Introduction   198
Measurement   199
Algal Movements in Water   200
    Effect of Cell Shape   200
    Change in Density   201
    Effect of Size   201
    Flagella and Cilia   203

The Seasonal Variation of Phytoplankton 204
    The Spring Bloom of a Holoplanktonic Diatom:
      Asterionella formosa 204
    The Spring Bloom of a Meroplanktonic Diatom:
      Melosira italica 210
    Seasonal Cycles of Blue-Green Algae 212
Spatial Variations of Phytoplankton 216
Dinoflagellates and Red Tides 218
Patchiness 220
Further Readings 220

CHAPTER 13   Zooplankton and Zoobenthos 221

Overview 221
Introduction 222
Measurement 222
Zooplankton Population Structure 225
    Migrations 226
    Annual and Seasonal Variations 226
    Feeding 232
    Reproduction 236
    Population Dynamics 238
Zoobenthos 241
    The Benthic Environment 242
    Feeding 243
    Life Cycles 243
Further Readings 246

CHAPTER 14   Fish and Fisheries 248

Overview 248
Introduction 249
Measurement 250
Feeding 253
Population Changes 258
Fisheries Management and Conservation 260
    The Decline of the Great Lakes Fishery 261
    Exotic Predators, Fish Kills, and Pollution 265
    New Reservoirs 268
    Fish Stocking 270
Further Readings 271

CHAPTER 15   Food-Chain Dynamics 272

Overview 272
Introduction 273
Measurements 273
The Regulation of Nutrient and Energy Flow 273
Growth Kinetics and Isotopic Tracers 282
    Growth Kinetics 282
    Prediction of Phytoplankton Seasonal Cycles 283
    Isotopic Tracers 285
Primary Productivity 286
Further Readings 290

CHAPTER 16   Streams and Rivers   292

              Overview   292
              Introduction   293
              Measurement   294
              The Stream   295
              The River   296
              The Lotic Environment   297
                  Discharge   297
                  Temperature   298
                  Nutrients   302
                  The Food Chain   307
                  Carnivores and Herbivores   311
                  Drift   313
              Further Readings   316

CHAPTER 17   Estuaries   317

              Overview   317
              Introduction   318
              Measurements   318
              Variable Salinity and the Salt "Wedge"   318
              Eutrophic Estuaries   322
              The Biota   326
                  Marsh Plants, Seaweeds, and Phytoplankton   326
                  Zooplankton and Benthic Animals   331
                  Fisheries   336
              Tropical Estuaries: Mangrove Swamps   337
              Further Readings   338

CHAPTER 18   Origin of Lakes and Estuaries, Eutrophication,
              and Paleolimnology   341

              Overview   341
              Origin of Lakes   342
                  Tectonically Formed Lakes   343
                  Volcanically Formed Lakes   345
                  Glacially Formed Lakes   346
                  Lakes with Miscellaneous Origins   347
              Origins of Estuaries   348
              Lake Succession and Eutrophication   349
              Eutrophication in Flowing Waters   353
              Cultural Eutrophication   354
              Paleolimnology   358
              Further Readings   360

CHAPTER 19   Comparative and Regional Limnology   362

              Overview   362
              Introduction   363
              Measurement   363
              Lakes in the Temperate Zones   364
                  The English Lake District   364
                  The Laurentian Great Lakes   367

Lakes in Tropical or Arid Climates   372
    Lake George, Uganda   372
Polar and High Alpine Lakes   376
Salt Lakes   380
Special Stream Environments   383
Further Readings   385

CHAPTER 20   Applied Limnology   387

Overview   387
Introduction   388
Measurements   389
Case Studies   391
    Lake Washington: Waste Diversion   391
    Shagawa Lake: Advanced Wastewater Treatment   392
    Lake Tahoe: Preventative Measures   395
    Lake Trummen: Dredging   399
When Nutrient Diversion Is Impossible   400
    Flushing   400
    Aeration and Mixing   400
Plant Harvesting and Chemical Control   403
Streams, Rivers, and Estuaries   406
    Restoration of the River Thames, England   407
Further Readings   410

References   411

Indexes   435
    Name Index
    Subject Index

Preface

The purpose of this book is to present a balanced, comprehensive, and contemporary view of the science of limnology. This book deals with the variety of inland waters and is designed as a text for undergraduate and graduate students as well as those with a professional interest in the subject. We have attempted to write a text which will not date rapidly by using well-proven and generally applicable examples wherever possible. Traditionally the subject has been organized into the three divisions of biology, chemistry, and physics. Where possible, these have been integrated in our presentation of lake, river, and estuarine ecosystems. Although it is essential to have some background in the basic disciplines, an exact knowledge of them is no guide to understanding the workings of the system any more than knowledge of the parts of a watch gives us an ability to tell time.

Many of the examples used in this text are drawn from the lakes, streams, and estuaries we have personally studied in North, South, and Central America, Africa, New Guinea, New Zealand, Antarctic and Arctic regions, Japan, and Europe. These ecosystems differ markedly from one another in size, basin shape, and biota, yet illustrate a unity of principles so important to the science of limnology.

In attempting to unite the physical, chemical, and biological features into a comprehensive view of the subject, the most difficult problem has been to avoid redundancy. If the reader is to complete the book with the desired level of understanding, it is necessary to draw on the material already presented. It is hoped that the brief indications of what has come before or will follow will not greatly slow the advanced student and will be of considerable help to the beginner. An overview of the material is given at the beginning of every chapter. In this way a student may acquire a general idea of its contents before reading the chapter. Limnology is essentially a practical discipline, and there is general recognition both within and without the scientific community of a need for professionals more broadly trained to solve real-world problems. It is therefore logical that the reader should have some notion of the methods employed in securing the different kinds of limnological data. We have provided a short measurements section at the beginning of most chapters, but these brief summaries are no substitute for a complete analytical handbook.

The introduction of certain technical terms is essential. These terms are italicized as they first appear in the text, and they will usually be defined fully as they appear in context. Where this procedure becomes too complicated, the reader should refer to the index to find the chapter containing the full definition. Excessive numbers of literature references detract from the flow of the text and have been limited to key papers, major review articles or books, or the authors' own experience. Figures and tables contain further references. A short reading list of key or review works is provided at the ends of chapters. These will help locate more extensive literature.

Since limnology is a broad subject, we could not cover every aspect or example, even in the most general terms. We have neglected many good examples and interesting studies, for we believe a condensed treatment is justified. Nevertheless, we have attempted to give a sufficient range of topics so that the reader will be familiar with almost any aquatic ecosystem likely to be encountered.

ACKNOWLEDGMENTS

Many people have assisted us in the preparation of this text. We thank them all but are grateful especially to W. T. Edmondson for first encouraging the project, and to E. de Amezaga, R. Bachmann, M. L. Commins, D. C. Erman, P. Kilham, D. A. Livingstone, T. Powell, and J. C. Roth for reviewing all or parts of the manuscript and suggesting improvements. We also received help on various chapters from R. Axler, E. Byron, R. C. Cooper, M. S. Goldman, J. W. Hedgepeth, K. F. Lagler, R. W. Larimore, W. M. Lewis, B. A. Manny, M. D. Morgan, L. J. Paulson, G. W. Redfield, and P. J. Richerson. We thank our staff G. J. Malyj, P. A. Arneson, M. Smith, B. Jost, F. Orsi, and M. Eaton for their patience and assistance.

All unacknowledged black and white photographs were taken by Alexander J. Horne with the exception of Figures 2-4a, 2-4b, 4-3, 6-2, 18-1, and 18-5, taken by Charles R. Goldman.

Charles R. Goldman
Alexander J. Horne

Chapter 1

Limnology: Past, Present, and Future

OVERVIEW

Limnology is the study of fresh or saline waters which are contained within continental boundaries. Together with the closely related science of oceanography, they cover all aquatic ecosystems. Although limnological observations have a long history, they only evolved into a distinct science during the last two centuries, when the inventions of the microscope, the silk plankton net, and the thermometer began to provide data which showed that inland waters were microcosms of life with a distinctive structure.

At present, limnology plays a role in the decision-making process for problems of dam construction, pollution control, and fish and wildlife enhancement. Accurate predictions which are increasingly required in applied science are currently straining limnological knowledge, and this pressure has stimulated a variety of new experimental and theoretical developments.

An important goal of education in limnology is to increase the number of people who, although not full-time limnologists, can understand and apply its general concepts to a broad range of related disciplines. Further research in limnology will require a renewed dedication to working out more widely applicable principles. This can be accelerated by the wise selection of proven as well as new technologies which may need to be applied on a large scale.

INTRODUCTION

The science of limnology embraces lakes, streams, rivers, and estuaries as well as a host of microhabitats which are often overlooked by the casual observer. These microhabitats include springs, old watering troughs, tree holes, and even the unique environments formed in abandoned cans or in the water- and enzyme-filled cavities of insectivorous pitcher plants. Limnology applies to running, or lotic, as well as standing, or lentic, waters. This definition was given to limnology by the International Association of Theoretical and Applied Limnology (S.I.L.) in 1922. In this text limnology includes standing and running water, both salt and fresh, as long as the body of water is contained within continental boundaries. Brackish waters contained in estuaries constitute important areas of limnological investigation. The only remaining bodies of water are those dealt with in the science of oceanography, which shares basic principles with limnology.

The Early Limnologists

Originally the word *limnology* was used in reference to the study of a lake and is derived from the Greek word *limnos,* meaning "pool," "lake," or "swamp." It first appeared in the work of F. A. Forel, entitled *Le Léman: monographie limnologique,* on Lake Geneva, Switzerland. The first two volumes, published in 1892 and 1895, consist of the geology, physics,

and chemistry of the lake (Fig. 1-1), while the third volume, published in 1904, deals with the lake's biology. Earlier, in 1869, Forel published a paper on the bottom fauna of Lake Geneva entitled *Introduction à l'étude de la faune profonde du Lac Léman,* and in 1901 he published the first textbook on limnology, *Handbuch der Seenkunde: allgemeine Limnologie.* Forel, then a professor at the University of Lausanne, has long been considered the father of limnology.

The studies we now know as biological limnology originated in 1674 with the first microscopic description of the filamentous green algae *Spirogyra* from Berkelse-Lake, Netherlands, by Leeuwenhoek. Although Leeuwenhoek considered himself a microbiologist, his report contains the first account of the seasonal cycles of algae in lakes, hints about food-chain dynamics, and the influence of winds on algal ecology. He wrote:

About two Leagues from this Town (Delf) there lyes an Inland-Sea, called Berkelse-Lake, whose bottom in many places is very moorish. This water is in Winter very clear, but about the beginning or

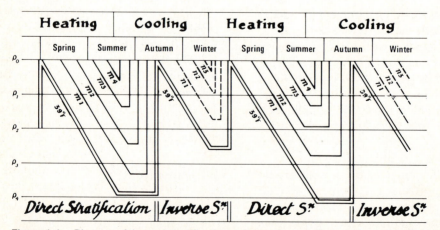

Figure 1-1. Diagram of thermal stratification patterns through the year for an idealized temperate lake as presented in Forel (1892). The double line at 39.2°F (4°C) is the maximum density of water. P_1, P_2 are depths, and m_1, m_2, n_1, n_2 are the isotherms showing the depth at which temperatures m_1, m_2, n_1, n_2, . . . are found at any period during the year. (For a more modern approach, see Fig. 4-2.)

in the midst of Summer it grows whitish, and there are then small green clouds permeating it, which the Country-men, dwelling near it, say is caused from the Dews when falling, and call it *Honey-dew*. This water is abounding in Fish, which is very good and savoury. Passing lately over this Sea at a time, when it blew a fresh gale of wind, and observing the water as above described, I took up some of it in a Glass-vessel which having view'd the next day, I found moving in it several Earthy particles, and some green streaks, spirally ranged, . . . and the compass of each of these streaks was about the thickness of a man's hair on his head. (Quoted in Fogg, 1968.)

Physical limnology began in Switzerland soon after Leeuwenhoek when the engineer F. de Duillier measured a seiche in 1730 and when Saussure made temperature observations in deep lakes from 1779 to 1796. The first description of how light, heat, water temperature, and wind mixing form the structure so important in lake ecosystems (see Chap. 2) was given by Sir John Leslie, who interpreted studies in Scottish lakes made by the civil engineer James Jardin in 1812 to 1814. Sir John could almost have composed the introduction to Chaps. 3 to 5 in this text when he wrote:

But the rays which fall on seas or lakes are not immediately arrested on their course; they penetrate always with diminishing energy till, at a certain depth, they are no longer visible. This depth depends without doubt on the clearness of the medium, though probably not one-tenth part of the incident light can advance five fathoms [10 m] in most translucid water. The surface of the ocean is not, therefore, like that of the land, heated by direct action of the sun during the day, since his rays are not intercepted at their entrance, but suffered partially to descend into the mass, and to waste their calorific power on a liquid stratum of ten or twelve feet in thickness. . . . But the surface of deep collections of water is kept always warmer than the ordinary standard of the place, by the operation of another cause, arising from the peculiar constitution of fluids. Although these are capable, like solids, of conducting heat slowly

through their mass, yet they transfer it principally in a copious flow by their internal mobility. The heated portion of the fluid being dilated, must continue to float on the surface; while the portions which are cooled, becoming consequently denser, will sink downwards by their superior gravity. Hence the bed of a very deep pool is always excessively cold, since the atmospheric influences are modified in their effects by the laws of statics. (Leslie, 1838, quoted in Murray and Pullar, 1910.)

The discovery of animal plankton was another important milestone in the development of aquatic biology. Although it is not known for certain who was the first to describe plankton, Johannes Müller conducted some of the earliest studies around 1845. A short time later another Müller, Peter Erasmus, observed microscopic crustacea for the first time in some Swiss lakes. These events launched a descriptive era for the classification of both freshwater and marine microorganisms.

The word *plankton*, which means "wandering," was first used by Hensen in 1887 to describe the suspended microscopic material at the mercy of the winds, currents, and tide. The meaning of the word was later expanded by the German biologist Ernst Haechel to include both large and small pelagic organisms. For a time only those organisms retained by a fine silk net were known, since those whose dimensions were less than 0.067 mm passed between the threads of the cloth and were not observed. Later these important smaller organisms were discovered and given the name *nannoplankton*.

The scientific study of the flowing waters of estuaries began with the discovery of the salt wedge by J. R. Lorenz in the Elbe in Germany in the 1860s (Fig. 1-2). At the same time, pollution research in the Thames in England (Figs. 20-13, 20-14) and the realization of the problems of survival in brackish waters initiated the biological approach (Meyer and Möbius, 1865, 1872).

Limnology in the United States began in the middle of the last century when Louis Agassiz

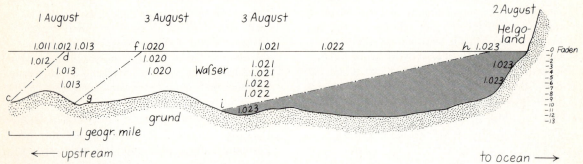

Figure 1-2 The salt wedge as discovered by J. R. Lorenz in the Elbe Estuary, Germany, in 1862. The island of Helgoland lies in the open sea (far right of figure), and North Sea water of specific gravity 1.023–1.025 forms a sharp wedge (see line *hi* and shaded area). The salt wedge becomes less pronounced upstream (lines *fg* and *cd*). Lorenz made these measurements in a drifting boat over 3 days in August using a homemade sampling device which frequently became jammed with sand or damaged by the swift estuarine current of 8 ft s⁻¹. He comments, "I spent many a bitter day in such exhausting exercises." Depth is shown in Faden (= fathom ≈ 2 m) on the right of the figure. (*Drawing much simplified from original published in Lorenz, 1863.*)

(1850) published *Lake Superior: Its Physical Character, Vegetation, and Animals,* which was primarily concerned with the biota of Lake Superior, especially the fishes. Perhaps the first man to consider lakes as functional ecological systems was Stephen A. Forbes, who in 1887 presented his *"The Lake as a Microcosm"* to the now-defunct Peoria, Illinois, Historical Society. The concept of an ecosystem has been particularly important in limnology. The term was first used by the English botanist Tansley but was put into popular usage in 1942 by G. E. Hutchinson and R. L. Lindeman in the latter's paper titled "The Trophic-Dynamic Aspect of Ecology."

At the turn of the century American limnology was dominated by four investigators: C. A. Kofoid, working on the Illinois River; James G. Needham, working on New York lakes; and E. A. Birge and C. Juday, studying Wisconsin lakes. Professor Birge at the University of Wisconsin is especially noteworthy for contributing a greater biological dimension to the field of limnology through his study of the plankton of Lake Mendota. Birge is also noted for studies of physical limnology which included light penetration, gases, currents, and the thermal characteristics of lakes. Birge's administrative duties (Dean, later University President, 1891–1925) left the execution of much of his limnological research to Chauncy Juday. Their first joint paper was published in 1908, and the two continued to make significant contributions for about 30 years. The life and contributions of Birge have been the subject of a biography by G. C. Sellery (1956), which includes a review of Birge's limnological contributions by C. H. Mortimer. The limnological laboratory established by A. D. Hasler on the shores of Lake Mendota continues this pioneering work.

In the United States limnology grew steadily through the early part of the twentieth century. Professor Paul S. Welch wrote the first American textbook on limnology for McGraw-Hill in 1935. This soon became a standard text. Franz Ruttner's *Fundamentals of Limnology,* which first appeared in 1940, in his own modest view "will in no way replace the introductions to limnology (Thienemann, Brehm, Lenz, and Welch) but complement them in certain respects." Many found the book as translated from the original German by D. G. Frey and F. E. J. Fry a useful introductory text well into the 1970s. At Yale University G. Evelyn Hutchinson began a comprehensive *Treatise on Limnology* (1957, 1967, 1975) which has become a standard reference work throughout the world.

In 1966 Bernard Dussart produced *Limnologie: L'étude des eaux continentales*, a text emphasizing biology and evolution. An individualistic treatment of the subject, *Physiological Limnology,* by Golterman, appeared in 1975, the same year that R. G. Wetzel, a distinguished former student of C. R. Goldman, published a scholarly text entitled *Limnology*. Two texts on streams appeared in the 1970s: Hynes' *The Ecology of Running Waters* (1972) and Whitton's edited text *River Ecology* (1975). Other introductory texts such as Cole (1979) and Reid and Wood (1976) cover the ecology of inland waters. A collection of essays on some aspects of physical and chemical limnology is given in the text edited by Lerman (1978), and a

review of Canadian limnology made up vol. 31 (1974) of the *Journal of the Fisheries Research Board of Canada*. Several reviews of estuarine limnology have been published over the last 20 years (e.g., Hedgepeth, 1957; Lauff, 1967; Perkins, 1974; Chapman, 1977).

At the same time that limnology was beginning in the United States, the science was well-developed in Europe. F. Simony studied thermal stratification in Austrian lakes about 1850, and Anton Fritsch began studying lakes in the Bohemian forests around 1888. It is now recognized that temperature change with depth in water bodies is one of the most important considerations in limnology. Simony made his investigations by lowering a crude insulated thermometer, allowing it to equilibrate, and then retrieving it rapidly for reading before it had had time to respond to the warmer surface temperatures. Similar physical limnological studies on temperature and water movement soon became common, particularly in Switzerland and Scotland (Figs. 1-1, 1-3). These studies, especially those concerning Forel's theory of seiches,

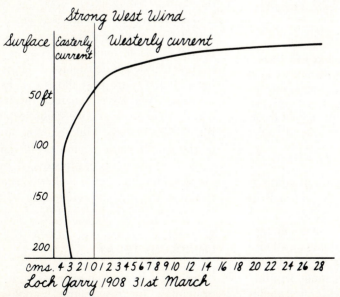

Figure 1-3 Measurements of the water currents in Loch Garry, Scotland, in 1908. This original figure shows rapid water movement downwind and slow return flows in the deeper water at about 50 ft (17 m). Water speed is given on the horizontal axis in cm s^{-1}, and water depth, in feet, on the vertical axis. (*From Murray and Pullar, 1910.*)

soon established the worldwide similarity of a variety of lake phenomena. Russian limnologists emphasized aquatic bacteriology, and S. N. Vinogradskiy, S. I. Kutznetsov, G. G. Vinberg, and V. I. Romanenko helped originate and sustain the interest in this often-neglected area of microbial limnology.

The Journals

In the nineteenth century the need arose to establish journals to pull together the increasing volume of information. Despite the existence today of limnological journals, the student must still go to the publications in a variety of other fields to search the literature to any real depth. This underscores the fact that limnology is a truly interdisciplinary science. On January 1, 1936, the Limnological Society of America was founded; in 1948 it was reorganized as the American Society of Limnology and Oceanography. This provided integration among limnologists, oceanographers, and marine biologists. The interrelationship of interest and activity was expressed by the first president of this new society in the first issue of its official publication, the journal *Limnology and Oceanography:*

> The American Society of Limnology and Oceanography was established in response to a need felt by its members for a common outlet for the publication of scientific papers on all aspects—physical, chemical, geological, and biological—of phenomena exhibited by natural bodies of water.

The year 1948 also marked the formation of the Freshwater Biological Association in Britain. Several decades of work distinguish this group from most others since it maintained a continuous record of physical, chemical, and biological information of 17 lakes in the famous Lake District of northwest England. In Italy the Istituto Italiano di Idrobiologia, first under the direction of Vittorio Tonolli and later Livia

Tonolli, furthered the science through intensive study of northern Italian lakes. The institute also became a Mecca for visiting scientists throughout the world who utilized the intellectual climate and the excellent library. Similar world-renowned institutes, for example, those at Plön, Germany; Uppsala and Lund, Sweden; Copenhagen, Denmark; on Lake Constance, Germany; and on several other lakes, estuaries, and rivers, were established about this time and many more have since joined them. In recent years what many have termed the environmental crisis has attracted engineers as well as a new generation of scientists who hope to cure the pollution of the world's lakes, streams, and estuaries by applying basic limnological principles.

Because lakes have been studied so extensively and have so much unity of physical, chemical, biological, and evolutionary structure, the field of limnology has largely been developed through the study of lakes. Although we have attempted to use examples from rivers and estuaries, it is often impossible to generalize for all aquatic ecosystems. The fact remains that most limnologists work on lakes, and frequently the comprehensive examples best-suited for a text of this type are lake-derived. However, some major limnological principles, such as the idea of detrital food chains or spatial heterogeneity, are best illustrated in rivers or estuaries. We have benefited from work on lakes, streams, rivers, wetlands, estuaries, and the world's oceans. Most existing limnological textbooks are built around the division of the subject on the basis of physics, chemistry, and biology, and the science makes its greatest contribution through the integration of the three. Our text is directed toward this integration.

LIMNOLOGY TODAY

Limnology, like other sciences, is a search for principles. In this search limnologists have found a number of generally applicable principles which provide a basis for comparison and prediction. For example, when a reservoir is being constructed the dam builder has several options as to where the water should be drawn off, how the basin should be prepared, and the location and height of the dam. Knowledge about water temperature at depth during the course of the year provides some of the information needed to reduce eutrophication and taste and odor problems. The experienced limnologist can often predict ecological problems in most areas of the world without even visiting the site. Other information on the physical shape or morphometry of the reservoir, quality of the water to be held, possible leakage through the basin, the burden of sediment, sources of pollution, and climate can greatly refine the information necessary to place the outlet(s) for optimum use of the water available. Cold bottom water might be desirable for a drinking-water supply or a trout stream below, but contains nutrients accumulated at depth, while warm surface water might better serve irrigation needs. Although higher in nutrients, the cold water slows germination and growth of crops. Adding the biota to the physical and chemical considerations of water quality, we can begin to appreciate the importance as well as the complexity of the structure and interaction of this system.

Environmental considerations have often been neglected, and resource developers frequently lacked both the knowledge and the interest to use this kind of information. With the recent rise of general environmental awareness and the recognition that reservoirs must often meet multipurpose objectives, recreational and aesthetic values now take their place in the decision-making process. As a result, more comprehensive planning studies are being undertaken in many parts of the world, and international funding is more likely where environmental problems have been considered (Horne, 1979a). Limnologists are playing an increasing role in decision making. Environ-mental analyses including recognition of alternative strategies were made in 1969–1970 at the Skippack watershed near Evansburg, Pennsylvania, where a large reservoir was proposed (McHarg and Clarke, 1973). As a result of the broad ecological study which included analyses of the limnological consequences of the large reservoir, the project was abandoned in favor of smaller reservoirs. In Papua, New Guinea, a huge hydroelectric scheme involving as many as 10 dams on the Purari River was reviewed and modified by the United Nations Development Program (Goldman and Hoffman, 1977). Reassessment of overall costs often allows more time for the development of less costly and less environmentally damaging alternative programs and mitigating measures.

THE FUTURE

The future of limnology is closely tied to the general advance of science and technology. Six areas which offer good prospects are long runs of data, replicated studies in large enclosures, integrated laboratory and field studies, some aspects of mathematical modeling, whole-lake manipulation, and comparative limnology, especially in very large lakes and rivers. Some of the most interesting prospects are being sought in the analyses of fairly long runs of reasonably spaced limnological sampling. Data sets, collected specifically for time-series analysis, cover more than 50 years in the marine sciences and over 30 years in limnology. Analyses of these data should provide better insights into the little-understood interactions between climate, watersheds, and the long-term succession of aquatic organisms (e.g., Figs. 12-2, 13-4, 14-7, 14-9, 16-11, 18-10, 20-6).

Acid rain, by no means a recent discovery (Smith, 1852; Barret and Brodin, 1955; Gorham, 1955), is a long-range global concern. Increasing scientific effort is going into better quantifying its effect on both terrestrial and aquatic ecosystems. Although the industrial

(a)   (b)

Figure 1-4 Analog models of aquatic ecosystems. (*a*) A medium-scale analog model of an estuary. Electric pumps provide tidal motion and allow use of any desired hydraulic residence time. Replication is possible in these systems, but the cost of construction and maintenance is high. The figure shows part of 20 tanks of 3 m^3 used for flow-through tests in the San Francisco Bay Estuary. Precise amounts of waste can be added continually via Plexiglas header tanks (center right), and plants or animals can be held under precise experimental conditions for months or even years. This facility is part of the University of California, Berkeley, Sanitary Engineering and Environmental Health Research Laboratory's extensive analog systems. (*b*) An in situ lake analog experiment showing 5 of 12 separate 0.7-m^3 containers made of 4-mil polyethylene (2.5 m deep by 0.7 m wide, tied at the bottom). Triplicate measurements of a control and three experimental treatments can be made. Although mixing occurs easily and oxygen or temperature are identical to those in the surrounding lake, heavier phytoplankton settle out. They can easily be resuspended using a Secchi disk on a pole as illustrated. The bags are suspended on springs to enable experiments to be carried out in the center of large Clear Lake, California (*A* = 17,000 ha), where wave action is often great. (*Analogs designed and built by C. J. W. Carmiggelt and J. C. Roth.*)

sources of acid rain in northern Europe are known, the origins of the variety of acids contained in the rains of North America require further study. Again, long runs of intercalibrated synoptic data should prove of great importance in future analyses.

Sometimes data collected over 10 or 20 years are unavailable. For example, studies in high mountains or in polar regions are physically difficult. Integrated laboratory and field studies which simultaneously use several different limnological approaches over a short period may then be the best solution. We have used this procedure in rivers, lakes, estuaries, and oceans, but it requires a major organizational effort. The methods can be divided into three sections: field measurement, pure culture studies, and analog experiments. Field measurements have inherently high variability, but results are usually directly applicable to the problem concerned. Pure culture studies using a single species of algae, invertebrate, or fish have the advantage of low variability but the disadvantage of limited applicability to real systems. The benefits and drawbacks of the third approach, using analog or actual scale models of the aquatic ecosystem (Fig. 1-4), lie between the other two. There are two advantages of the analog method. First, whole-model ecosystems can be easily adjusted experimentally to enable prediction of future effects. Second, the experimental design can be changed to give the degree of precision required by the investigator. Manipulation, while decreasing variability, has the inherent danger of departing from the reality of the system. For example, in a planktonic or riverine analog model, zooplankton or benthic invertebrates can be physically excluded to establish more precisely the relationships between algae and nutrients without the complicating effects of grazing.

The manipulation of whole lakes is an excellent method for limnological research. The entire food web can be explored, and the long-term changes found cannot be confounded by effects inherent in even the largest bottles or bags. Since all lakes are individuals, adjacent "control" lakes are not adequate, and several years of data, at all trophic levels, must be collected prior to manipulation. One of the first examples of whole-lake manipulation was Castle Lake, California, where the effects of molybdenum addition were studied (Fig. 10-12). Similar whole-lake manipulations have been undertaken, although the difficulties in undertaking and interpreting the results should not be underestimated (Schindler, 1980; Fee, 1980).

Comparative limnology, the study of lakes within a lake district or between similar lakes in different regions, is likely to produce new limnological principles (Chap. 19). The greatest challenge is the comparative limnology of large lakes, rivers, and estuaries. The comprehensive comparative limnology of the Laurentian Great Lakes, the African Rift lakes, and the Amazon, Nile, and Rhine rivers, all pose formidable logistic and financial problems. Helicopters, satellites, and remote data-analyzing devices are new methods which must be further developed for these large ecosystems.

Studies in large enclosures such as those used in the English Lake District (Fig. 1-5) or the giant CEPEX bag experiments in the Pacific Ocean near Vancouver, Canada, give some idea of what might be done in the future. These experiments lacked the replication needed to determine between-bag variation and thus a level of reliability for future predictions. The multiple ponds at Cornell, New York (Hall et al., 1970), or the multiple-model estuary system at the University of California's Berkeley Field Station are examples of well-replicated analog systems (Fig. 1-4a).

In recent years a popular fashion in biology has been mathematical modeling. The idea of applying mathematics to the solution of complex biological problems has enormous appeal and launched a whole generation of model builders. Much of the North American International Biological Programme (IBP) was in fact directed toward this panacea for the afflictions of modern biology. Looking back now at over a decade of frenzied modeling efforts, we can make some assessments and some generalizations on progress to date and hopes for the future.

The inescapable conclusion is that most major modeling efforts have either failed to adequately describe real-world situations or lack the predictive capability that biologists and mathematicians have strived to achieve. The excesses of model building have been well-

Figure 1-5   Large-scale model systems. Three 46-m-diameter, 12-m-deep enclosures in Blelham Tarn in the English Lake District. These experiments are too large to replicate easily but can simulate lake conditions for a whole season and have been run continually for 11 years. The enclosures are made of butyl rubber and are dug about 1 m into the mud. In this picture the central tube has a spring bloom of the blue-green algae *Oscillatoria agardhii* which appears white in this picture. The effect is due to a complex several-month interaction between fungal parasitism, zooplankton grazing, and fertilization with phosphate and silicate in the previous fall. (*Photograph by J. W. G. Lund, F. R. S.*)

described in Joel Hedgepeth's paper on "Models and Muddles" (1977). To date their major value appears to be in the conceptualization and compartmentalization of ecosystems. In contrast, smaller, simpler models, referred to as *submodels,* with fewer compartments, are often used in the decision-making process. It is a major temptation for the young mathematically inclined biologist to construct a simplistic model of a problem when it would be more effective to collect the empirical data necessary to solve it.

Of considerable importance to future advances in the aquatic sciences is the continued interaction between marine and freshwater workers. The existence of the American Society of Limnology and Oceanography is important in promoting their cross-fertilization. Limnol-

ogists often maintain close working relationships with marine scientists. Both groups stand to benefit from this interaction, and there is no logical reason for restricting oneself to one or the other ecosystem.

The next chapter considers in detail the structure of lakes, streams, and estuaries. The physical, chemical, and biological elements of aquatic systems change with time and depth, and study of these variations provides the basis of limnology. Additional understanding can be gained when different bodies of water are compared. Limnology provides a unique opportunity to combine observations with both laboratory and field experimentation. Much of the information derived from fundamental investigations of aquatic ecosystems can and should be applied to problems of water quality.

FURTHER READINGS

Limnological Journals

Limnology and Oceanography
Freshwater Biology
Journal of Plankton Research
Canadian Journal of Fisheries and Aquatic Science
    (formerly *Journal of the Fisheries Research
    Board of Canada*)
Hydrobiologia
Archiv für Hydrobiologie
Swiss Journal of Hydrobiology (abbreviation
    Schweiz. Z. Hydrol.)
*Proceedings of the International Association of
    Theoretical and Applied Limnology* (abbreviation
    Verh. Int. Ver. Limnol.)
Journal of Great Lakes Research
Memorie dell'Istituto Italiano di Idrobiologia

Hydrobiological Journal (USSR)

Journals That Publish Some Limnological Articles

Ecology
Journal of Ecology
Oikos

Other Readings

Elster, H-J, 1974. "History of Limnology." *Mitt.
    Int. Ver. Theor. Angew. Limnol.*, **20**:7–30.
Giesy, J. P. 1980. "Microcosms in Ecological Re-
    search." Technical Information Center, U.S. De-
    partment of Energy, Springfield, Va. (DOE Sym-
    posium Series 52). Conf.-781101. 1110 pp.
Rodhe, W. 1979. "The Life of Lakes." *Arch.
    Hydrobiol. Beih.*, **3**:5–9.

Chapter 2

The Structure of Aquatic Ecosystems

OVERVIEW

Lakes, streams, and estuaries have discernible structures based on their morphometry as well as that of their drainage basins. The distribution of other physical properties such as light, heat, waves, and currents produces a physically distinct structure which varies by day and season. Under certain circumstances the distribution of chemicals gives additional form to the system. Superimposed upon these structural components is still another type of organization provided by the distribution of the biota.

LAKES

Morphometry

The geologic origin of a lake sets the limits for the morphometry or shape of its basin. Once the lake basin is formed, a variety of physical,

chemical, and biological factors interact to produce discernible structure within the water which persists despite the continual motion characteristic of the aquatic ecosystem. The relatively still waters of lakes have given them the general term *lentic* environment.

The lake basin, as opposed to the drainage basin, is that portion which actually holds the water. Clues to its morphometry may be gained from examining the topography of the surrounding area, but morphometric details, i.e., depth and contour of bottom, must be gained from sounding with a weighted line or an echo sounder.

Measures of lake morphometry provide a very useful means for rapid description which we will use frequently throughout this book. Lake surface area A varies with season and is most constant where a natural dam and outflow stream maintain a fairly uniform water level.

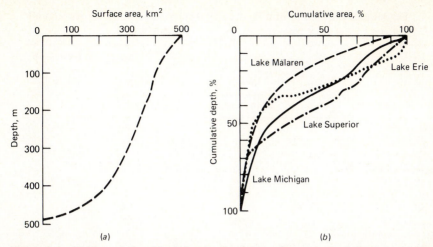

Figure 2-1 (*a*) The relationship between absolute surface area and depth in the deep-sided graben of Lake Tahoe as indicated by cumulative hypsographic curves derived from measurement by planimetry. (*b*) Relative hypsographic curves for three of the glacially excavated Laurentian Great Lakes and the pan-shaped Lake Mälaren, Scandinavia. (*Modified from Håkanson, 1981.*)

The area may be measured with a planimeter from a good map or aerial photograph which includes an elevation datum. Volume V can be calculated from underwater contour lines by summing the volume of the various layers of water contained between all depth contours. It is most conveniently expressed in cubic meters or cubic kilometers, although the term acre-foot (one acre, one foot deep) is still commonly used in engineering and agriculture in the United States. The mean depth $\bar{z}$ is obtained by dividing the volume V of the lake by its surface area A. Other frequently used morphometric terms are maximum depth z_{max} and length L of the shoreline. The hypsographic curve is a plot of depth along the vertical axis and area along the horizontal axis (Fig. 2-1), from which the area of the lake at any desired depth can be obtained.

Shoreline development D_L reflects the degree of irregularity and is expressed as the ratio of the length L to the circumference of a circle of area equal to that of the lake surface. The more irregular the shoreline, the greater D_L will be. Some of these measurements are given for selected lakes throughout the world in Table 2-1 and for those found in two lake districts in Tables 19-1 and 19-3.

The time required for all the water in the lake to pass through the outflow is termed the *hydraulic retention time*. This is an important parameter in lake pollution studies or for nutrient-dynamics calculations. It is mainly determined by the interplay between lake inflow and basin morphology. The retention time of a nutrient may be somewhat different from the hydraulic residence time since sedimentation and recycling are taking place at the same time as outflow. Table 2-1 gives hydraulic retention times for selected lakes and reservoirs. Table 19-3 shows similar data for the Laurentian Great Lakes. Lake Tahoe, with a retention time of about 700 years, is almost a permanent sink for nutrients. In contrast, Marion Lake in Canada has a hydraulic retention time of only a few days, and most nutrients are removed almost as fast as they enter.

Zonation

Considering the structure within the lake basin, two major depth zones are generally recognized

Table 2-1   Physical Dimensions and Hydraulic Residence Times (Where Available) for Selected Lakes and Reservoirs Throughout the World.

	V, km³	A, km²	$\bar{z}$, m	z_{max}, m	Retention time, years
Americas					
Tahoe, California	156	499	313	501	~700
Castle, California	0.0023	0.201	11.4	35	3–5
Clear, California	1.4	176	~8	14	~3
Superior, North America	12,000	83,300	144	307	184
Pyramid, Nevada	~25	~500	~50	~100	Sink
Okeechobee, Florida	~6	1,880	~3	~4.5	
Crater, Oregon	20	55	364	608	
Winnipeg, Canada	311	24,530	13	19	
Managua, Nicaragua	~8	~1,000	5	19	Sink
Titicaca, Andes	866	8,100	107	281	70
Europe					
Esrom, Denmark	0.21	17.3	12.3	22	8.5
Windermere, England	0.35	14.8	24	67	0.75
Balaton, Hungary	~1.8	596	~3	~4	
Constance, Germany	49.3	540	91	252	
Tjeukemeer, Netherlands	~0.03	~20	~1.5	~3	
Africa					
Victoria, East Africa	2,700	68,800	40	79	
George, Uganda	0.63	250	2.4	3	0.34
Tanganyika, East Africa	18,940	34,000	557	1,470	Sink
Chad, Chad	20–40	16,000	1.5–4	12	Sink
Kariba (reservoir), Africa	130	4,300	30	93	~3
Kainji (reservoir), Africa	15.6	1,280	12.3	50	0.25
Asia					
Biwa, Japan	28	685	41	104	5.4
Baikal, Siberia	23,000	31,500	730	1,741	
Caspian Sea, U.S.S.R.	79,319	436,400	182	946	Sink
Kinneret, Israel	4.301	168	26	43	7.32
Antarctica					
Vanda, Dry Valley	0.15	5.2	29	66	Sink
Lake 2, Signy Island	0.00009	0.04	2	6	

which have exact analogies in the marine environment (Fig. 2-2). The *littoral zone* extends from the shore just above the influence of waves and spray to a depth where the light is barely sufficient for rooted aquatic plants to grow. For practical purposes the littoral habitat extends from the shore to a depth where the well-mixed warm surface waters still reach the lake bed in summer. Away from the shore, the lighted and usually well-mixed portion is

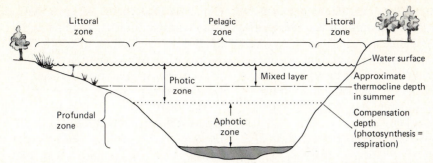

Figure 2-2 How the physical structure of a lake is influenced by the distribution of light and temperature (see also Fig. 2-3). The situation shown is typical of a small temperate lake in summer. The zones are not exact, and in turbid lakes or near dawn and dusk in any lake the photic zone may only extend partway down the mixed layer. In winter the mixed layer may extend to the lake bed. The strong wave action along the shore as well as favorable light establish the limits of the littoral zone. For stratified lakes, the mixed layer is the same as the epilimnion (see Fig. 2-3).

termed the *photic* or *euphotic zone* which extends from the lake surface down to where light is about 1 percent of that at the surface. In transparent lakes the photic zone may extend below the region of thermal stratification. The *aphotic* zone extends below the littoral and photic zones to the bottom of the lake. Here light levels are too low for photosynthesis. Respiration, however, proceeds at all depths, so that the aphotic zone is a region of oxygen consumption.

Light zonation establishes a major element of lake structure. It should be noted that the lower boundaries vary daily and seasonally with changing solar intensity and transparency of the water. Similarly the upper zones are not exact since inhibition by too much light, wave action, or rocky substrate frequently prevents plant growth at the upper boundary. Only a very small amount of light penetrates to great depths even in the most transparent lakes. In Lake Tahoe, for example, a few specialized primitive plants receive enough light to permit their growth at depths of over 150 m. This deepest area of plant growth is a transition between littoral and profundal zones and may be referred to as the *sublittoral* zone. Many shallow lakes with relatively transparent water have no sublittoral or profundal zones. Plant growth may

cover the entire lake bed, and all the open lake water is sufficiently illuminated to permit photosynthesis.

Lake zones may change rapidly, however, if transparency is reduced by algae blooms or sediment inflows. In contrast to shallow transparent lakes, shallow lakes with low transparency often have an aphotic zone where no plant growth can occur. In Clear Lake, California, turbidity from algae and suspended sediments shades out the growth of attached plants within a few meters of shore despite the shallow depth. Terminology based on light levels alone is thus insufficient for a complete description of lakes. For example, very deep ($\bar{z} = 313$ m) Lake Tahoe may be described as optically clear since much of its water column is illuminated. Shallow Clear Lake ($\bar{z} = 8$ m) is, by contrast, optically turbid since only the small upper fraction of its waters is illuminated.

Terminology of Physical Structure

The elements of lake structure which involve water movement and the distribution of heat are often used to describe the offshore conditions. In deep lakes the area beyond the influence of the shore or bottom is termed the *pelagic* or *limnetic* zone (Fig. 2-2). The organisms which inhabit this zone must be adapted for swimming, suspension, or flotation. The water mass itself has a characteristic vertical temperature structure which is independent of the shape of the basin. Swimmers in lakes find that in summer colder water is only a short distance below the surface, and fishermen can cool their bottled drinks on a hot summer day by suspending them below the surface.

During *thermal stratification,* three vertical zones are found in temperate regions. The upper warmer water of a stratified lake is called the *epilimnion;* the middle portion where the rate of temperature change with depth is greatest, the *thermocline*, is called the *metalimnion;* and the deeper portion is called the *hypolimnion*

(Fig. 2-3). Sometimes the term thermocline has been restricted to that zone having a change of at least 1°C per meter of depth. This unnecessarily restrictive definition has limited value, particularly in the tropics, and will not be adhered to in this text.

The distribution of water as measured by temperature is a reflection of the differences in its density. The colder denser water is on the bottom, the zone of rapid change above, and the warmer less dense water is isolated near the surface by the density change which occurs in the thermocline. Although storms may stir the warm waters of the epilimnion into furious motion, little energy is transmitted through the

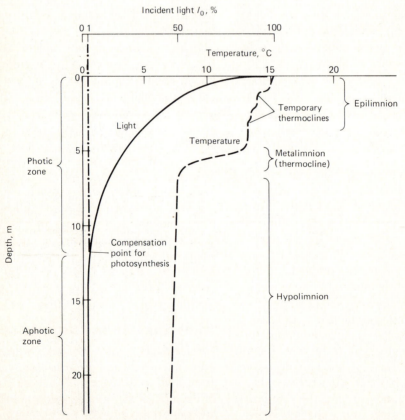

Figure 2-3   The thermal and optical structure of a lake with depth during the period of summer thermal stratification.

thermocline to the cool quiescent hypolimnion. The epilimnion is often called the *mixed layer*. This applies equally to both stratified and unstratified lakes. In shallow lakes which never stratify for more than short periods, the mixed layer is the more appropriate term. The great resistance to mixing from density differences that establishes and maintains the stratified structure in some lakes is very important to the distribution of dissolved chemicals, gases, and the biota.

In the fall with less solar radiation reaching the water and greater heat loss at night, convection and wind mixing begin to erode the thermocline. The epilimnion then increases in depth as it decreases in temperature. Eventually the temperature and density difference between the overlying water and that beneath is so slight that a strong wind in late autumn overcomes the remaining resistance to mixing and the lake undergoes the *fall overturn.* The stratified structure is lost and the lake becomes homothermous; i.e., it has a uniform temperature with depth. In cold climates mixing and further cooling continue until the surface of the lake freezes. Freezing occurs only when surface waters reach 0°C on a windless, cold winter night. Some slightly warmer water is left below.

Ice forms on the top of the lake rather than on the bottom due to a unique property of water, namely, that it attains its maximum density at 4°C, not at its freezing point (Fig. 3-3). The presence of ice cover prevents further wind mixing and actually conserves the heat remaining in the lake. The ice increases in thickness as the winter progresses and may acquire a layer of insulating snow. Winter heat gains under ice contribute to a characteristic thermal structure known as *inverse stratification*. Here water at 0°C is in contact with the ice but is warmer only a few centimeters below. The water is warmest near the bottom where it is often 4°C. In warm climates or where the ocean moderates winter's cold, some lakes never freeze over and mix continually throughout the winter.

In spring the lake ice melts and wind mixes the cold water during the *spring overturn* until lake temperature increases enough for thermal stratification to be reestablished. Other patterns of circulation are established in tropical regions and at high latitudes or high altitudes (Chap. 4). It will suffice at this point to note from Fig. 4-4 that there is a greater density change per degree of temperature change in warm water than in cold. This is illustrated by the fact that it takes about 30 times as much energy to completely mix equal volumes of 24 and 25°C water as it takes to mix the same volumes of water at 4 and 5 °C! Herein lies the explanation for thermal stratification and the remarkable stability of the epilimnion. A useful classification of lakes has been developed on the basis of how often and how completely or incompletely their waters mix (Chaps. 4 and 5).

Chemical Structure

The distribution of chemicals, especially nutrients, through lake waters provides the second major element of lake structure. After the onset of thermal stratification, nutrients often become depleted in the epilimnion or photic zone while at the same time accumulating in the hypolimnion or aphotic zone (Figs. 7-4, 8-4, and 10-10). By analogy with temperature, the depth at which rapid change in a substance occurs is called the *chemocline*. In a few lakes the chemocline is a permanent feature (Chap. 4), but in the usual case chemical stratification is determined by temperature and density distribution. There tends to be little vertical stratification of chemical constituents in the well-mixed littoral zone. If the shoreline in any lake is well-developed (D_L is large), with numerous bays and points, there will be more contact of water with shore and bottom. In addition, a shallow lake has a greater percentage of its water mass in direct contact with the bottom than does a deep lake. This contact usually permits nutrients such as phosphorus, nitrogen, iron, and trace elements to dissolve more effectively from the basin and sediments into the lake waters.

The littoral zone and the bed of a shallow lake provide a good habitat for bottom-dwelling organisms whose foraging, burrowing, and excretion aid in recycling nutrients from the bottom. The chemical structure of the lake has a vertical and a horizontal component. The vertical component is seasonal and depends on the presence of layers of density-stabilized water. The horizontal element may occur year-round and results from the influence of the lake's edge.

Biological Structure

The third major element of lake structure is biological. Although some organisms may spend part of their lives in different zones of the lake, for the most part they may be classified on the basis of their most important habitat. Except for fish and for floating and emergent plants, many lake organisms are more or less invisible to the casual observer (Plates 1, 4b). If we look carefully, however, there are myriads of tiny organisms living beneath the lake surface which have no permanent association with the bottom. These organisms of the pelagic zone are further classified by their mode of transport or their association with the water surface. *Plankton* are the floating or weakly swimming organisms at the mercy of the waves and currents. The word is derived from the Greek term meaning "wanderer." The animals of this group are called *zooplankton* and the plants *phytoplankton*. Planktonic bacteria are common, and planktonic fungi and viruses, although present, have been little studied. Bacteria and fungi are often most abundant in association with particles, phytoplankton, or suspended sediments.

The stronger swimming animals such as fish that inhabit the pelagic zone and are masters of their position in the water column are called *nekton*. There is also a special community of organisms, the *neuston,* which inhabit the surface of the water. The *pleuston* comprises large floating assemblages blown about by the wind. In African lakes the aquatic macrophyte *Pistia* (water cabbage) can be an abundant component of the pleuston (Plate 4a).

Among the biota of the littoral zone the *attached algae* are important where rocks or higher plants provide a firm substrate, while higher plants, the *aquatic macrophytes,* are likely to dominate the sandy or muddy littoral zones of lakes if wave action is not great enough to uproot them. The whole community of microscopic attached organisms composed of algae, bacteria, fungi, protozoa, and small metazoa are called *aufwuchs*. Such communities are responsive to environmental change and are excellent biological indicators of the degree of eutrophication.

The organisms associated with the lake's bottom are called *benthic organisms* and are referred to collectively as *benthos*. These include all forms found in or upon submerged substrates regardless of whether they are in the littoral, sublittoral, or profundal zone. Those which live and move about on the lake bottom, such as crayfish and dragonfly larvae, are called *epibenthic organisms,* while those which burrow beneath the mud surface, such as aquatic worms and insect larvae, are known as *infauna*.

In lakes with sand or gravel beaches there is another element of the fauna which is hidden from view. This is the *psammon* which occupies a unique aquatic habitat between the grains of sand. Water flow in this *psammolittoral zone* is provided by waves and the capillary action which draws the water up between the sand particles. Microorganisms such as bacteria, algae, rotifers, and copepods occupy this restricted zone. Moving as they do through this sand and gravel filter, they depend in part on what the waves bring in for food and what the algae present can produce from the limited penetration of light. Waves produce a foam high in organic matter along some beach areas, which may provide food for the psammolittoral zone. Around the margins of Mono Lake, California, fly larvae develop in this accumulation of foam.

The Watershed

The lake's watershed constitutes a fourth element in lake structure, as important as the physical, chemical, or biological elements. The size, slope, geological composition, and climate of the lake's drainage basin influence the identity and quantity of minerals dissolved in the lake and what sediments are deposited there (Chap. 6). The granite basins of the Precambrian shield of Canada, Scandinavia, and Scotland are well known for their low fertility. In contrast, lakes farther south in areas of glacial drift and sedimentary rock are more productive and able to produce large algal crops (Figs. 19-1, 19-2).

The importance of drainage-area size in relation to the surface area is important in many lakes since there is often a higher fertility in lakes with larger drainage areas. Eutrophication is also strongly influenced by the lake surface/watershed ratio (Table 18-1).

The general climate of the watershed influences sediment and nutrient transport in a manner which can be generalized as follows. In temperate climates rainfall is spread out over much of the year and does not often fall torrentially. Such rain patterns produce a continuous vegetative cover in both forests and grasslands which show little natural soil erosion. By contrast, regions with semiarid climates have a few severe rainstorms, and ground cover is not continuous. In these hot or Mediterranean climates soil erosion is frequently extensive, and sediments usually move more easily from their watersheds to the lakes.

The nutrient phosphorus, which is generally transported as phosphate adsorbed to soil particles, will also move more easily in semiarid climates (Plate 5a). By contrast, nitrogen, as highly soluble nitrate, is easily transported by either clear or muddy water (Plate 5b). Nitrate

passes most easily from the land to water in the high-rainfall temperate zones. Rivers and lakes in semiarid climates tend to have excess phosphate and be nitrogen-limited, while temperate climates have excess nitrate and may be more limited in their fertility by lack of phosphate. Other limitations imposed by trace elements or silica are controlled more by the geology of the basin than by the climate. Another effect of climate is the existence or nonexistence of outflow. Freshwater lakes without an outflow eventually become salt lakes through evaporation and may dry up completely. The great Bonneville Salt Flats, now used as a track for developing land-speed records, were once an enormous freshwater lake. As discussed later, there is almost as much water contained in saline lakes, including the inland "seas" ($> 3^0/_{00}$* salinity), as there is in freshwater ones (Table 3-1, Chap. 19).

Beyond the natural sources of chemicals from sediment or erosion and leaching of the watershed, there are agricultural, forest, and urban sources. These, in addition to slightly altering the morphometry of lake basins by filling, also modify the chemical environment.

Ponds versus Lakes

The problem of where a pond ends and a lake begins has been the subject of considerable definition and redefinition. P. S. Welch (1952) concluded that to be considered a lake, a body of water must have a barren wave-swept shore. Forel (1892) considered ponds to be lakes of slight depth. This definition seems inappropriate for lake waters such as Lake Chad in Africa or Lake Winnipeg, Canada, which are very large when full but very shallow. The presence of higher aquatic plants is considered important in Welch's definition of ponds. He would classify as lakes all the shallow coastal waters of the Antarctic regardless of their

---

*$^0/_{00}$ represents parts per thousand.

shallow depth and small size. We feel that a precise definition is not often essential but that one may use the type of lake mixing to aid in definition. Water bodies may be considered lakes when the wind plays the dominant role in mixing (Chap. 5). In ponds gentler convective mixing predominates (Chap. 4). By definition ponds are shallow, but often thermally stratified waters, with abundant growths of rooted and floating aquatic macrophytes. Employing this definition, lakes Chad and Winnipeg remain lakes while the familiar tree-shaded ponds would not be considered lakes. The abundant shallow-water bodies of the tundra which are subject to extensive wind mixing are probably more similar to the littoral zone of lakes than to ponds.

STREAMS AND RIVERS

Viewing the *lotic* or running-water environment from a stream bank, one is struck by the variety of habitat, largely determined by *gradient* and substrata, over a relatively short distance. Here are shallow gravel-covered *riffles,* a little farther down a quiet *pool,* and below, a small waterfall. Chapter 16 deals with rivers and streams, and some special features of food-chain dynamics, cold-water streams, and river pollution are discussed in Chaps. 15, 19, and 20. As with lakes and ponds the distinction between streams and rivers is vague, but rivers are larger, faster moving, and often warmer. We use the terms river and stream interchangeably in this text.

The dominant feature which structures the lotic environment is the swift unidirectional water flow. The *discharge* (volume per time) and *current* (distance per time) interact with the substrate to determine if the streambed will be stony or composed of mud and detritus (Chap. 16). Most streams possess a series of distinct physical structures which have a regular vertical and horizontal periodicity. Familiar to all are the horizontal *meanders* which occur in the flatter portions of the watercourse,

no matter if the stream is large or small, mountain or lowland. These meanders are due to the water seeking the least energetic path. Meanders produce areas of deeper, swifter flows near the eroding, outer edge of the meander circle and depositing, shallow areas on the opposite bank (Fig. 2-4). In the broad valleys of major rivers, extensive meanders create *oxbow lakes* in abandoned channels and *scroll lakes* through deposition at existing bends (Fig. 2-4).

In the vertical plane, shallow stony riffles alternate regularly with deeper muddy-bottomed pools (Fig. 16-2). The periodicity of riffles and pools has long fascinated limnologists. One effect of this periodicity in the streams may be a regular periodic chemical recycling which Elwood et al. (1981) have called *nutrient spiraling*. This spiraling may produce the same general kind of power spectrum effect now known for waves in lakes and oceans (Fig. 5-7). Streams have a great biotic variability, even over short distances, some of which may be due to this effect.

Such aspects of the lentic environment as thermal or chemical-density stratification are usually unimportant in the turbulent lotic environment. However, in very large rivers the mixed layers may not reach the bottom. There the stream resembles an elongated lake and often shows a definite thermal and chemical stratification.

The physical structure of streams provides an abundance of specialized biological niches. For example, there is a fast current on the upstream face of a rock and an eddy formed behind where little downstream flow occurs. In addition, beneath the rock lies a well-protected dark hiding place for small animals, while the upper surface provides a well-lighted site for attached algal growth. Much of the biological structure of streams is dependent on the spatial patterns of *drift* and *detritus* which reflect the dominant effect of currents. Drift consists of living benthic invertebrates and algae which have released or lost their attachment to the substrate. Swept

(a)

(b)

Figure 2-4 (a) Meanders which occur in flat areas in almost every river and stream. These meanders are on the Salado River in Argentina. Oxbow lakes (see text) are formed when the meander becomes cut and blocked off at the ends. (b) Meanders may form a series of multiple oxbows or scroll lakes as deposition moves the river channel to a new bed.

downstream by the current, they may find another favorable site or source of food. Fish and invertebrates which feed on drift are distributed to make optimum use of this food sup-

ply. Detritus, also important in lake sediments and estuarine environments, consists of dead organic fragments coated with bacteria and fungi, small protozoans, and rotifers (see Chap.

16). These biological components provide further structure to the stream ecosystem by virtue of their distribution in relation to current speed, substrate, and food supply.

ESTUARIES

An estuary is the place where the river meets the sea. Estuarine structure is modified by the shape of the estuary, the tide, and the amount of inflowing fresh water. Estuaries are discussed in more detail in Chap. 17. The density difference provided by the salt and fresh water results in a *salt wedge* which provides a structure to the water mass not unlike that caused by thermal stratification in lakes. However, the salinity effect dominates over thermal stratification, which is usually unimportant in estuaries. The salt wedge, because of its high salinity, is heavier than the overlying fresh water and therefore extends upstream beneath it (Fig. 2-5). The wedge advances upstream at high tide and retreats at low tide. In most regions freshwater inflow varies considerably during the year, and this modifies the degree of separation between the fresh and salt water as well as the upstream extension of the salt wedge.

At the interface between the salt- and freshwater masses, small particles and some dissolved organic material "salt out" or flocculate into larger, heavier detrital aggregates. The saline bottom water tends to collect particles settling into it which may return upstream with the high tide. The area covered by the salt wedge as it moves up and down the estuary is often the zone of greatest phytoplankton and zooplankton abundance.

There are more distinct physical elements to estuarine structure than are found in lakes and rivers. In addition to pelagic and littoral zones, estuaries frequently have extensive mud flats exposed at each low tide, and large tidal marshes. The tide mixes nutrients and food between these structural elements. This interaction produces high plant and animal growth, and estuaries are important nursery areas for both fish and shellfish. The tide controls the size of the structural elements in an estuary. Its influence varies greatly according to its height and the estuary's morphometry. The daily tidal variation is over 10 m in some funnel-shaped estuaries like the Severn estuary in Britain. In contrast, in the Mediterranean Sea or San Francisco Bay tides are only about 1 m in height.

Large human populations are often situated near estuaries which have traditionally been of great importance for food, water transport, and waste disposal. As populations and industries have grown, even some of the world's largest estuaries have been polluted and their value as a food source for human beings has declined.

The degree of structure imparted by the

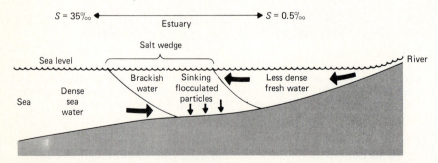

Figure 2-5 Diagrammatic representation of a typical estuary showing the salt wedge. Outflowing freshwater overlies the denser seawater to create the wedge. This is also the site of "salting out" or flocculating of small silt particles (see also Chap. 17).

various physical, chemical, and biological components discussed in this chapter provides a common conceptual thread that ties together lakes, streams, and estuaries. In the following chapters on the characteristics of water, the distribution of light, heat, and water motion, it should be increasingly evident that aquatic ecosystems, despite their great diversity of type, maintain a remarkable continuity of structure that makes them ideal subjects for intensive ecological study.

FURTHER READINGS

Hakanson, L. 1981. *A Manual of Lake Morphometry*. Springer-Verlag, New York. 78 pp.

Hutchinson, G. E. 1957. *A Treatise on Limnology*, vol. I, chap. 2. Wiley, New York.

Rawson, D. S. 1955. "Morphometry as a Dominant Factor in the Productivity of Large Lakes." *Verh. Int. Ver. Limnol.*, **12:**164–175.

Chapter 3

Water and Light

OVERVIEW

Most of the structural elements of aquatic ecosystems are a result of the unusual properties of water and its interaction with light. Water covers seven-tenths of the earth's surface. Over 10^{18} m³ is in the oceans, and there is also a great amount of water in the atmosphere, the ground, and in the polar ice fields (Table 3-1). The molecular structure of water is responsible for many unique properties such as its high melting and vaporization points, or its maximum density at 4°C. This structure allows weak *hydrogen bonding* between hydrogen and oxygen atoms in adjacent molecules, producing a matrix known as a *liquid crystal*.

Because the maximum density of water is at 4°C rather than at its freezing point, ice forms only at the lake surface, leaving warmer, denser water below. The ice formed insulates the liquid

water and prevents the lake from freezing solid. Few organisms can survive in solid ice. At the other extreme the high specific heat of water allows it to absorb large amounts of heat without a large temperature increase. This enables aquatic organisms to survive even the intense solar radiation at the equator which produces only a small daily increase in lake temperature. Water changes rapidly in density with only a small change in temperature. This is responsible for the remarkable resistance to mixing found in stratified lakes. The viscosity or drag of water on objects moving through it is not unique to water but plays an important role in determining the shape of most planktonic organisms from minute bacteria and algae to streamlined fish.

The intensity, color, direction, and distribution of light in lakes are major components in the structure of lake ecosystems. Primary

Table 3-1 Distribution of Fresh and Saline, Liquid and Frozen Water in the World

Most surface freshwater is contained in the Antarctic ice caps, and there is roughly as much saline ($>3^0/_{00}$ salinity) lake water as fresh.

Site	Volume, thousands of km³
Freshwater lakes	125
Saline lakes and inland seas	104
Rivers and streams	1.3
Soil water	67
Groundwater	8,350
Polar ice caps and all glaciers	29,200
Total for land	≈37,800
Total for atmosphere	13
Total for sea	1,320,000

Annual inputs	
Surface runoff to sea	30
Groundwater runoff to sea	1.6

Precipitation	
On ocean	320
On land	100

production in larger deeper lakes is entirely dependent on light transmitted through the water to provide energy. Small shallow lakes may receive contributions from emergent plants or shoreline vegetation. Sunlight is modified both in intensity and color as it passes through the atmosphere. The air as well as clouds and dust particles change the spectral composition and intensity of light. More importantly for limnologists, similar processes of reflection, refraction, scattering, and selective absorption of certain wavelengths occur in water. Light absorption is measured by the absorption coefficient ϵ_λ, which is fairly specific for each lake water and wavelength of light. In general, transparent lakes have a low ϵ_λ, while eutrophic or muddy lakes have a high ϵ_λ. In the visible range, green light penetrates farthest and red is usually most rapidly absorbed. Blue light is strongly backscattered by water molecules, which pro-

duces the characteristic blue color of transparent mountain lakes. All water rapidly absorbs invisible ultraviolet and infrared light. *Albedo*, the ratio of light reflected to that received, is high at low solar angles or over snow-covered ice. Throughout the water column, water molecules and suspended particles reflect a variety of colors back to the observer. This *apparent color* is further modified by the *true color* of dissolved materials. True and apparent colors combine with topography, nearby vegetation, the lake bed, and a changing solar spectral pattern to produce the lake color actually seen by an observer.

PROPERTIES OF WATER

The physiological importance of water to organisms can scarcely be overemphasized. *All* organisms are largely composed of water. In fact, it was necessary for animals to develop the ability to transport water with them before they could inhabit the terrestrial environment. All basic life processes involve the use of water. Often referred to as a universal solvent, it transports through biological systems the gases, minerals, and dissolved organic components that drive life's machinery. Plants depend upon hydrolysis by water for the simple conversion of starches to sugars. Water serves as an electron source in photosynthesis. A typical plant consumes 100 times its weight in water during the growing process; most of this is lost through transpiration. Where organisms have developed circulatory systems, water is the essential transporting medium.

Many of the characteristics of lakes, rivers, and estuaries result from the unusual properties of water. When compared with other similar compounds, water shows a very anomalous behavior. If water (H_2O) behaved at natural environmental temperatures as do H_2S, NH_3, or HF, it would be present only as vapor. In fact, water and mercury are the only inorganic liquids that can exist at the earth's surface under ordinary pressures and temperatures. Many of the unique properties of water result from its molecular structure. Oxygen is highly electronegative and, in water, shares electrons with its two associated hydrogens. The covalent chemical bonding has electrons skewed in the direction of the strongly negative oxygens. This gives oxygen a slight negative charge, while the hydrogens retain a small positive charge. The asymmetry of the charge in water molecules allows the oxygen in one molecule to form a weak *hydrogen bond* with the oppositely charged hydrogen atom of two adjacent molecules (Fig. 3-1). The hydrogen bond is strongly directional, and the covalent chemical bond —O—H of one molecule must point almost directly at the oxygen nucleus of another to form a bond with the hydrogen atom (Fig. 3-1). Liquid water consists of a continuous network of randomly connected hydrogen bonds which form a *liquid crystal* rather than a true fluid within which individual molecules move freely. The water in a lake, for example, has an uninterrupted network of hydrogen bonds running in all directions throughout its entire volume.

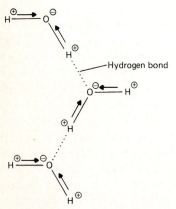

Figure 3-1   Hydrogen bonding (dotted lines) in water. The small electronegative oxygen atom attracts part of the valence-electron cloud, leaving a positively charged hydrogen atom. The hydrogen bond forms between the hydrogen of one molecule and the oxygen of another. The bond is weak and contains only about one-sixteenth the energy of a normal covalent bond. This hydrogen bond is more flexible than other bonds.

The covalent bonds are too strong to be significantly deformed by hydrogen bonding. As H_2O changes from ice to liquid water and then to vapor, it is the hydrogen bonds which break or are strained. The change in arrangement of hydrogen bonds from the hexagonal crystal lattice of ice to the bulkier and variable polyhedral crystalline lattices of water is shown in Fig. 3-2. The crystalline matrix of liquid water is not static but undergoes continual topological reformation as the hydrogen bonds move between the different atoms. The anomalous properties of water result from competition between the strongly hydrogen-bonded, bulky crystalline-lattice form and the more compact structures which have many strained or broken hydrogen bonds. As the covalent molecular bonds twist and turn, the hydrogen bonds continuously switch their attachment sites in the matrix. This complex bonding holds the water together as a liquid to a much higher temperature than similar compounds such as H_2S, HF, or NH_3, all of which exist as vapors at room temperature. As ice melts the water molecules can move more freely and assume different hydrogen-bond angles than are formed in the more rigid ice crystals. This allows the molecules to pack together, and water density continues to increase up to 4°C. Heating ruptures the hydrogen bonds which increases the distance between adjacent molecules. Over 4°C rupture predominates over closer packing and density decreases as temperature increases.

Water expands when it freezes so that the ice formed requires more volume than it did in an unfrozen state. The expanding frozen surface of a lake may groan and crack as it expands and can exert enormous pressure around the lake's margin. This may force up a ridge of soil. From 0 to 4°C, thermal expansion of water is negative, and it contracts with increasing temperature to a maximum density at 3.94°C. Because water becomes lighter below 4°C, ice forms on the surface of lakes. Ice contains water molecules in several crystal-lattice forms (Fig. 3-2*a* to *e*). The hydrogen bonds in ice which hold

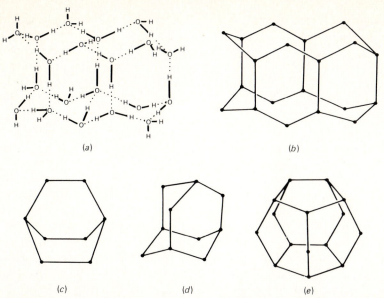

Figure 3-2 Crystal structure of the various forms of ice. (*a*) Details of the compact hydrogen bonding of water in ordinary hexagonal ice. Each water molecule has four nearest neighbors and is hydrogen-bonded to all four with a tetrahedral symmetry. The oxygen atoms of the neighbors occupy the vertices of a regular tetrahedron surrounding the oxygen of the central molecule. The bond angle of water (104.5°) is less than the ideal tetrahedron (109°), so the hydrogen bonds are slightly bent and strained. (*b*) A simplified version of (*a*); the black dots are the oxygen atoms. (*c*), (*d*), and (*e*) The bulky, unstrained polyhedra also found in ice are drawn in the simplified fashion of (*b*). The maintenance of this liquid crystal gives water most of its unusual properties. (*Modified from Stillinger, 1980.*)

water molecules in such open networks are geometrically positioned in an almost ideal thermodynamic arrangement with little bending or straining of either hydrogen or covalent bonds. As ice melts the rigid ice crystal becomes a fluid and each water molecule is free to move. The water molecules collapse into a denser, more compact polygonal or polyhedral crystalline structure which strains the hydrogen bonds out of ideal alignment (not shown). The closer packing increases up to 4°C. Above this temperature rupture of hydrogen bonds from a heat-induced *vibrational motion* successfully competes with the *configurational* close packing. When hydrogen bonds are ruptured, water molecules move farther apart than when bonded. Between 0 and 4°C configurational packing exceeds vibrational expansion and the

compact strained polyhedra previously mentioned are maintained. Above 4°C the vibrational component exceeds the tendency to pack, and the water assumes a normal, positive thermal-expansion coefficient (Fig. 3-3). These density anomalies are entirely due to the large role played by hydrogen bonding in water.

One important consideration of heating or cooling water is that its heat capacity, or *specific heat,* is higher than all elements with the exception of liquid lithium, hydrogen, and helium. The specific heat of a substance is the ratio of its thermal capacity to that of water at 15°C. The specific heat of water is defined as unity since it takes one calorie to heat one gram of water by 1°C. Water also has a high latent heat of fusion, i.e., the heat required to melt ice.

The presence of lakes everywhere has, at the

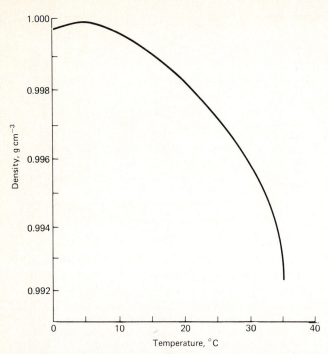

Figure 3-3 Change of the density of freshwater with temperature. Note the maximum density at about 4°C and the almost exponential decrease in density at higher temperatures (see also Fig. 4-4).

very least, micrometeorological effects on the surrounding land. Because water stores a lot of heat per unit volume, the presence of large volumes of water has the ability to alter climate. In the U.S. Great Lakes region, for example, orchards may be protected from frost in the spring by the proximity of the lakes. The region north of the Arctic circle in Swedish Lapland also has a much milder climate than the tundra at similar latitudes in Alaska. In this case an extension of the tropical Gulf Stream current warms the coast of Sweden and Norway.

The difference in density is responsible for the great resistance to mixing of water masses of different temperatures. The rate of change of the density of water is not constant with increasing temperature. The higher the temperature, the greater the density change per degree change (Chap. 4, especially Fig 4-4).

Water's density may be determined by factors other than temperature. The presence of dissolved salts increases the density of water and gives stability to the bottom waters of some meromictic lakes (see Chap. 4 for definition of meromictic). Salinity-induced density effects dominate over those of temperature in estuaries and the ocean. The density of pure water is 1.000, and that of normal seawater at $35^0/_{00}$ (35 parts per thousand) is 1.02822. This decreases seawater's temperature of maximum density to $-3.52°C$. The freezing point of a solution is lowered as salinity increases, although on Mono Lake, California (salinity $\sim70^0/_{00}$), ice forms at $-2.5°$ C.

Other properties of water which are important in limnology include its viscosity and surface tension. *Viscosity*, a measure of a liquid's resistance to flow, provides considerable resis-

Table 3-2   Decrease in Viscosity of Water with Temperature

Temperature, °C	Viscosity, cP
0	1.79
5	1.52
10	1.31
15	1.14
20	1.00
25	0.89
30	0.80
35	0.72
40	0.65

tance, or *viscous drag,* to organisms moving through it and serves to regulate the rate of sinking of planktonic organisms. Viscosity is much higher at lower water temperatures (Table 3-2).

Surface tension, like viscosity, results from cohesive hydrogen bonding in the liquid-crystal lattice. Certain plants and animals maintain their position by surface tension. Water striders (*Gerridae*) take advantage of surface tension and skate rapidly about on the surface. Dissolved salts, in addition to increasing the density and viscosity of water, also increase surface tension. Organic surfactants, which are foaming or wetting agents produced by aquatic plants or animals, reduce surface tension. Lines of foam which accompany some water motions are often caused by natural surfactants (Fig. 5-14).

LIGHT

Solar radiation provides the heat which drives the world's wind patterns. Wind energy propagates waves which provide the mixing forces in marine and fresh waters (Chap. 5). Similarly the light transmitted directly to the aquatic environment through solar radiation influences the dis-

tribution of both organisms and heat (Chap. 4) in the lake as well as powering the photosynthetic mechanism of plants as far down as the light can penetrate.

Much of the structure of aquatic ecosystems described in Chap. 2 is directly related to the solar radiation which heats and lights the water. The photosynthetic base of the aquatic food chain is represented by large and small attached or free-floating plants which owe their existence in water to solar radiation. The distribution of all aquatic plants and most animals is also strongly influenced by sunlight. Direct solar radiation is most important although indirect light from the sky and clouds also contributes significantly. Energy from the moon and stars ($1/30,000$ to $1/50,000$ of that from the sun), except for aiding the night feeding and migration of animals, is insignificant as an energy source. Light which has been polarized into a unidirectional plane by the water may serve as guideposts for the distribution of aquatic animals. The compound eyes of arthropods can discriminate between normal and polarized light, and they may use this ability for orientation.

The behavior of both higher and lower organisms is strongly influenced by the strength of the underwater light field. This is exemplified by the diurnal migration of lake animal plankton in response to changing light (Figs. 13-17, 14-1). The epibenthic forms in lakes and estuaries avoid light by burrowing into the bottom mud or hiding under stones during the day and emerging at night. In streams a drift of small invertebrates to new habitats occurs at night (Fig. 16-16). Night scuba dives with underwater lights are an instructive experience. Fish which may inhabit only the deep waters by day move into the shallows to feed by night. Crayfish and other benthic predators, which hide under rocks or vegetation during the daylight hours, may also be observed foraging after sundown.

Nowhere is the importance of light so clearly illustrated as in the Antarctic. The inland, permanently frozen lakes such as Lake Vanda (Figs. 4-3, 18-7) receive enough light energy through their ice cover to support a simple food chain of algae, bacteria, and protozoans.

Measurement

Estimates of total solar radiation at the lake's surface can be made by using continuously recording *pyrheliometers*. These instruments record changes in light by simple mechanical tension resulting from differential expansion of an absorbing black surface and a reflecting silver surface. Other types of pyrheliometers convert light energy to electricity by using solar cells in combination with electronic circuitry to measure light energy in langleys (gram-calories per square centimeter), einsteins, or watts.

Underwater light measurement is usually made with waterproofed photocells and compared with simultaneous readings taken at the surface with a deck cell. Measurements can be refined by adding selective absorptive filters to the photocell so that the different wavelengths of light can be measured. In this way the limnologist can compute the percentage of various wavelengths of light penetrating to any particular depth. Some modern instruments are full-fledged spectroradiometers which may be equipped with submersible quartz-fiber optical probes. These can scan the spectral response at almost nanometer (nm) intervals over the whole spectrum from 400 to 800 nm every 20 μs. Subtle changes in light, color, and intensity due to different types of plants and animals are now amenable to analysis.

A simple, inexpensive device which quickly measures the transparency of a lake is the *Secchi disk*. This white disk, usually 20 cm in diameter, is among the oldest of limnological devices. It was first used by the Italian Prof. P. A. Secchi aboard the *SS L'Immacolata Concezione* in the employ of the Papal Navy (Tyler, 1968). The disk is lowered into the lake until it disappears from sight and is then raised slowly until it is just visible again. The distance halfway between the points of disappearance and reappearance of the disk is taken as the Secchi depth. The Secchi depth of eutrophic and muddy lakes or estuaries ranges between 0 and 2 m but may be as great as 40 m in highly oligotrophic lakes or the open ocean (see Table 18-1). In many lakes the photic zone is equal to one-half to one-third the Secchi depth. Such simple, inexpensive measurements as temperature and Secchi depth can tell the experienced limnologist a great deal about a lake's trophic state (see Chap. 18).

Light and Heat

As the light penetrates and is absorbed, it is converted to heat, and if it were not for the unique nature of water and the frequent mixing of the lake, the same vertical distribution would be observed for temperature as for light (Fig. 2-3). The density differences and resistance to mixing mentioned previously and in Chap. 4 provide the explanation.

Light reaching the earth can be thought of as a continuous flow of electromagnetic waves or, alternately, as photons, or *quanta,* which are discrete packets of energy. Wavelength is a useful measure of light color, while quanta most conveniently define energy. An important characteristic of a light beam is its *intensity,* which is the number of quanta passing through a certain unit area. Light energy is now measured in einsteins but formerly was defined in langleys or footcandles. In general, more light means more photosynthetic production, but in almost all lakes on sunny days the intense light around noon inhibits plant photosynthesis near the surface. In polar regions partial inhibition is likely to occur for most of the cloudless days (see Figs. 15-9, 18-7, 19-9).

A second important feature of a beam of light is its wavelength (λ), or color, which is a qualitative measure of light energy. Light from the sun is a mixture of many different wavelengths which are attenuated as they pass through the earth's atmosphere. A third characteristic of a light beam is the direction in which it is traveling. This can be of considerable importance to the many phototactic lake organisms which migrate toward or away from the source of light.

Light in the Atmosphere

Light leaving the sun has a wide and uneven spectral distribution ranging from very short ultraviolet to very long infrared wavelengths. The penetration of light through the earth's atmosphere and through water results in the selective absorption and scattering of light, especially at the ends of the spectrum. The spectral composition and percentage of light at various wavelengths arriving at the lake surface is crucial since the physical properties of photons (scattering, absorption, reflection) and their suitability for photosynthesis depend on the wavelengths actually reaching the plant pigments. Figure 3-4 illustrates the spectral distribution of light outside the atmosphere, after it passes

through the earth's atmosphere, and the portion of the spectrum visible to the human eye. About half the total energy occurs in the visible part of the spectrum, and the peak energy input lies near 380 nm (Fig. 3-4).

The spectrum and intensity of direct solar radiation reaching the surface of a lake depends upon various transient phenomena in the atmosphere since clouds, dust, and fog determine which wavelength of light traversing a given distance will be absorbed or scattered. In Fig. 3-4 there are dips in the curve of light as it penetrates the earth's atmosphere. These are due to selective absorption in the atmosphere of sunlight of various wavelengths by such things as carbon dioxide, water, and ozone.

Of major importance for solar intensity is the angle of incidence of sunlight to the lake surface, which determines the actual distance sunlight must travel through a given atmospheric thickness before reaching the lake surface. Since the sun's angle varies with different

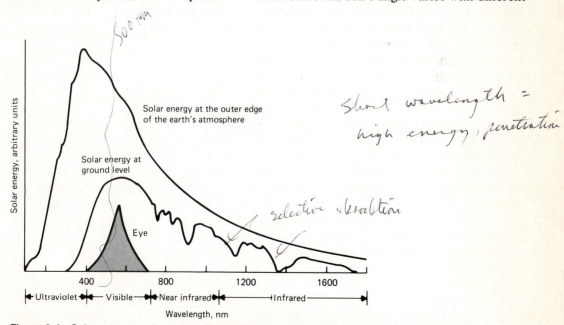

Figure 3-4 Solar energy as it reaches the earth's atmosphere and at ground level. The black area shows the sensitivity of the human eye (400–700 nm). Thermal radiation (not shown) is at about 5000 to 14,000 nm and together with infrared radiation accounts for about half the solar energy at the lake surface. (*Modified from Hutchinson, 1957.*)

latitudes, seasons, and time of day, the distance, and thus the amount and quality of light penetrating the atmosphere, will also vary. This variation in color and intensity is the main reason to use in situ measurements of primary productivity (Chap. 15). In addition to latitude, season, and time of day, the distance traveled by sunlight through the atmosphere is less at high altitudes than it is at sea level.

The solar radiation incident on the lake surface is both *direct* sunlight and *indirect, or diffuse*, light which is scattered and reflected off mountains, clouds, and atmospheric particles. The ratio of indirect to direct light depends on (1) the amount of atmosphere to be traversed, (2) the amount of scattering constituents, and (3) the local topography. Indirect radiation accounts for about 20 percent of the total but is extremely variable. The land and vegetation which surrounds a lake also influences solar radiation reaching the lake surface since mountains, for example, will shade the lake early and late in the day. Tall trees around a small lake will have the same effect. Cloud cover or haze generally reduces light intensity and filters out some portions of the spectrum. Besides the direct shading effect of clouds and mountains at low sun angles, scattered clouds present above the lake or snow-covered slopes adjacent to the lake may serve to reflect light toward the water surface and actually increase the intensity of solar radiation.

In summary, light arriving from the sun has a characteristic wavelength, intensity, and direction. Light enters the earth's atmosphere and is selectively absorbed, scattered, and bent or refracted before it finally reaches the surface of a lake to be further modified by the denser liquid medium.

Light under Water

Light penetrating water is first refracted, then either absorbed, reflected, scattered, or transmitted. Light that is absorbed is changed to heat. As light moves between media of differing optical densities, as from air to water, it is bent by refraction. Refraction is why submerged objects appear to be slightly out of place to an above-water observer reaching beneath the water to grasp them. Diving birds such as kingfishers must correct for this displacement when they dive for fish. The refractive index is higher for shorter wavelengths and lower for longer wavelengths, which separates light into its constituent colors. The refraction of sunlight through raindrops to form a rainbow illustrates this effect.

As light penetrates the waters of a lake, it is rapidly absorbed and its light intensity decreases exponentially. This loss of light is designated by the extinction coefficient ϵ_λ of the solution in question and is simply a value of the fraction of light held back per meter of depth by absorption and diffusion. The higher the value of ϵ_λ, the lower will be the transmission of light through the water. For parallel beams of monochromatic light (single-wavelength), the intensity I at a depth when the sun is directly overhead is given by the formula

$$I_z = I_0 e^{-\epsilon_\lambda z}$$

where I_0 = intensity penetrating the surface
z = path length
ϵ_λ = extinction coefficient for the wavelength in question

The percentile transmission is 100 times the fraction of light of a given type which is transmitted through one meter. For distilled water and daylight, transmission is approximately 50 percent. The extinction coefficient can also be expressed as

$$\epsilon_\lambda = \epsilon_w + \epsilon_d + \epsilon_p$$

where ϵ_w = extinction due to water molecules
ϵ_d = extinction due to dissolved materials
ϵ_p = extinction due to particulate matter

For pure water, $\epsilon_d = \epsilon_p = 0$, and extinction $\epsilon_\lambda = \epsilon_w$.

Lake Color

As the light penetrates the lake there is a selective absorption which is most pronounced at both ends of the spectrum (Fig. 3-5). Absorption differs slightly between clear and cloudy water since the scattering effect of turbidity permits a greater relative penetration of long-wavelength red light than is encountered in clear water (Figs. 3-5a, 3-6). In all cases the ultraviolet and the infrared ends of the spectrum are absorbed first and therefore penetrate least. As we go deeper into the water we find a progressive narrowing of the spectrum so that in a deep, transparent lake like Tahoe, virtually monochromatic blue-green colored light (475 nm) is found at 100 m. If we divide up the spectrum into the categories ultraviolet (UV), blue (B), green (G), red (R), and infrared (IR), we discover that for pure water the order of increasing extinction, which is decreasing light transmission, is B, G, UV, R, and IR. However, in water with dissolved materials, ϵ_d will be highest for short wavelengths and will decrease for longer wavelengths. Consequently, B and UV transmissions are decreased most, and the resulting order of increasing extinction is G, B, R, UV, and IR for most lakes. The presence of highly colored particulate matter (in which case ϵ_p becomes selectively important) can alter this basic relationship since water may assume the color of the suspended particles. For example, the presence of a "red tide" of photosynthetic microorganisms will absorb green strongly, decreasing its transmission. Perhaps the most extreme example is shown in muddy lakes or estuaries or those with very dense algal crops. Here, although red light may penetrate farthest, it will not be very far in absolute terms (Fig. 3-6a).

Organisms may be especially adapted with particular pigments to utilize the light available at the depths at which they are living. This is called *chromatic adaptation* and is important for deep-dwelling plants which must capture as much solar radiation as possible. Other shade-adapted forms may have unusually high pigment/cell-volume ratios for the same purpose. The converse is also true in that some organisms living in shallow water where solar irradiance is high often possess extra pigments to protect them from light damage to cellular DNA and chloroplasts. Benthic blue-green algae which form feltlike growths in shallow Antarctic lakes have a layer of red carotenoid-rich cells above the photosynthetically active ones. The absence of this protection in phytoplankton is probably the reason for their scarcity in such lakes (see also Fogg and Horne, 1970; Goldman, Mason, and Wood, 1963).

Another factor reducing the penetration of light is scattering by water molecules. The shortest wavelengths are scattered the most. Since the scatter of light is proportional to $(1/\lambda)^4$, blue light is the dominant wavelength scattered back to the surface in transparent lakes. White sunlight is effectively filtered through the blue filter of the lake water and the direction reversed by scatter. It is this scattering of blue light by water molecules back to the observer's eyes that gives highly oligotrophic lakes a deep cobalt blue color (Plate 1) and a clear sky its blueness. In less transparent lakes, dissolved and particulate matter normally obscures this molecular scattering. Absorption of light is due to water molecules, suspended particles, and colored dissolved substances. Bog lakes, for example, usually have a brown color from the humic substances present.

The color of lakes to the human eye is one of their most appealing features. Who has not been cheered by the blue of a lake glimpsed down in a mountain valley or seen between the trees in a woodland? Likewise, some people may be disappointed if the lake is a muddy brown or green color. Lake color is composed of reflected light consisting of two parts, *real* and *apparent* color. The interpretation of what the human eye receives is often complicated by *bottom color* in shallow lakes or at the edges of deeper ones. Changes in daily and seasonal

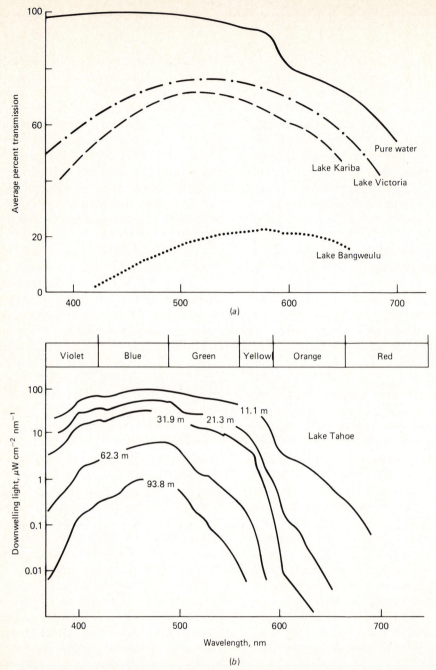

Figure 3-5 The color of light transmitted by different lakes, and changes with depth in one lake. (*a*) The average transmission of various colors (wavelengths) in pure water and the relatively transparent lakes Kariba and Victoria, Africa. Also shown is the shallow, less transparent, Lake Bangweulu, Africa, which has more suspended particles and humic acids from papyrus-swamp drainage. The maximum transmission shifts from blue in pure water through green in normal lakes to the orange-red in the muddy productive lake. The spectrum for Lake Victoria is similar to those of many subalpine European lakes. (*Modified from Balon and Coche, 1974.*)(*b*) A shift toward longer wavelength also occurs with increase in depth, even in very transparent lakes such as Lake Tahoe. Here the shift only occurs between blue and green. (*Modified from Smith, Tyler, and Goldman, 1973.*)

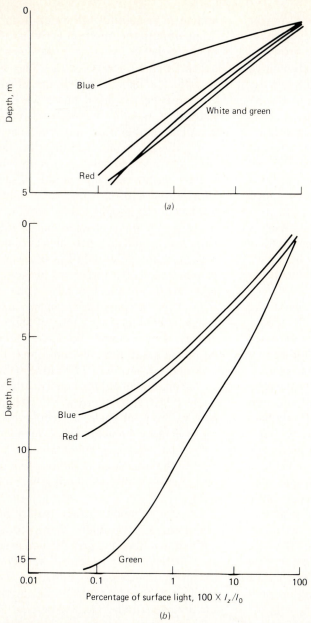

Figure 3-6 Absorbance of different colors of light by (a) a highly productive lake with dense phytoplankton and (b) a relatively oligotrophic lake. Note that in the unproductive lake, green penetrates farthest. In very muddy water, red may penetrate farther than other colors.

spectral distribution of incident radiation, cloud cover, reflection of vegetation, and hills around the margin of the lake also modify the apparent water color. Real as opposed to apparent color depends on the interaction between the wave-

lengths that are scattered back toward the eye and the absorption of these wavelengths in the water between the depth of scattering and the surface. We can, however, make the following generalizations. If there is little or no dissolved or suspended matter, a deep lake will appear blue. However, under storm clouds even these ultraclear waters may appear grey or brown to the observer. If there is a moderate amount of dissolved matter, the lake will probably appear green. A green color is also produced by the presence of some suspended particles such as fine clay. For this reason lakes in chalk or limestone quarries have a green color. In a series of glacial valley lakes, the lake nearest the end of the glacier is usually white in color due to a dense suspension of fine particles often called "glacial flour." The brilliant green characteristic of the second glacial lake in the chain is often due to a thinner suspension of glacial flour as the heavier particles have settled out. Finally, with more settling, transparent blue lakes terminate the chain (Plate 2).

If there are large quantities of dissolved material, especially organics, a lake will appear yellow or brown. Large amounts of particulates can produce a variety of colors from red-brown to yellow and green. Phytoplankton blooms produce a green or chartreuse color; suspended iron hydroxide or a dinoflagellate bloom, red or reddish brown water. On exposed mud flats pools of sulfur particles from the activity of hydrogen sulfide–oxidizing bacteria can produce a yellow appearance. Some algae which become filled with oil and float in the autumn also produce a yellow color. Spectacular lake colors may be produced by dense blooms of near-surface blue-green algae (Plates 6a and 7b) and by a mixture of algae and bacteria (Plate 3b).

In shallow lakes sunlight striking the bottom will permit the growth of attached plants and be reflected upward to boost phytoplankton photosynthesis. It may also be absorbed to heat the sediments and perhaps to develop convection currents which may aid in the transfer of nutrients from the sediments to the overlying water.

Reflection (Albedo)

Although some light striking a lake surface is absorbed, a significant portion of it is reflected and never penetrates the water. This reflection is apparent to the observer early in the morning or late in the day when water assumes a metallic sheen from surface reflection. The percentage of reflected light is called the *albedo*. Light coming from overhead has a lower reflection than slanted light. Reflection from a mirror-calm lake is therefore at a minimum when the sun is directly overhead since the light strikes the water perpendicularly. It is maximal during seasons of low sun angle and is always high at sunrise and sunset. Since real water surfaces are frequently roughened by waves, there will be a continuous variation in reflection even from overhead light since the waves themselves present a certain percentage of nearly perpendicular and nearly horizontal surfaces. In many lakes there are more winds in the afternoon than in the morning which causes an asymmetrical underwater distribution of photosynthetic energy. For example, in Lake Tahoe about noon, the onset of waves can change the Secchi disk depth from 30 to 18 m in less than 2 h.

The amount of light penetrating the surface of a lake depends on a variety of water properties besides the solar angle. Molecular backscattering of light by the water and suspended and dissolved particles can reflect about 5 to 10 percent of the total solar radiation. Any dust or oil on the surface film will both reflect and absorb light.

Ice is a common surface feature of lakes at high latitude or altitude. If the ice is covered with snow, albedo is greatly increased and approximately 90 percent of sunlight will be reflected. By contrast, snow-free ice, if free from air bubbles, is almost transparent. In Antarctic Lake Vanda, 14 to 20 percent of incident solar radiation penetrates the nearly 4 m of dense clear ice covering this permanently frozen lake (Fig. 18-7 and Goldman, Mason, and Hobbie, 1967).

In summer blue-green algae and submerged macrophytes often form dense layers just beneath the water surface which absorb strongly in the red and blue regions but reflect near-

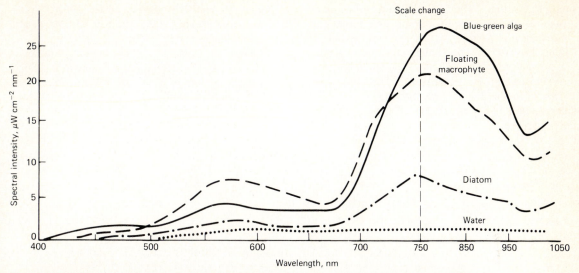

Figure 3-7   Reflection of light by algae and higher aquatic plants. Note the large near-infrared (NIR) reflectance of the blue-green alga (*Aphanizomenon*) and the floating higher plant, duckweed (*Lemna*). The gas vacuoles of the blue-green alga and the air spaces inside *Lemna* are probably the reason for NIR reflectance. The diatom (*Cymbella*) and other green plants reflect more light in the visible spectrum, giving them a green color. (*Modified from Anderson and Horne, 1975.*)

infrared (NIR) and green light. Other algae reflect much less NIR but similar amounts of green light (Fig. 3-7). Since NIR is invisible to the human eye, lakes with dense blue-green algae or macrophyte crops seem to reflect mostly green light. Diatoms and green algae rarely form dense near-surface layers, and their green reflectance is greatly modified by absorption due to dissolved and nonliving materials in the water column. Despite its invisibility to humans and the absorption of NIR by water, the strong reflectance by some plants is of value to limnologists. Modern false-color infrared film accurately detects NIR reflectance (Plate 6*a*). Photographs made using this method demonstrate the complicated distributional patterns made by blue-green algae or dust and debris such as floating dead grasses, leaves, and submerged macrophytes (Figs. 3-7, 5-3). Muddy water, such as from winter river inflows or from some forms of pollution, also produces distinct reflectance patterns on lakes but at shorter wavelengths than green plants. Pictures made

from satellites or airplanes can tell the limnologist a great deal about the instantaneous distribution of highly variable phenomena, especially on large lakes where conventional limnology covers too small an area.

This chapter has stressed the important aspects of the quantity and quality of light in the aquatic ecosystem. Light which is not reflected or backscattered is absorbed as heat by the water or the organisms, particles, and pigments it contains. This heat produces the extremely important thermal structure of lakes that is described next in Chap. 4.

FURTHER READINGS

Hutchinson, G. E. 1957. *A Treatise on Limnology*, vol. I, chap. 6. Wiley, New York.

James, H. R., and E. A. Birge. 1938. "A Laboratory Study of the Adsorption of Light by Lake Waters." *Trans. Wis. Acad. Sci. Arts Lett.,* **31**:1–154.

Jerlov. N. G. 1968. *Optical Oceanography*. Elsevier Press, London. 194 pp.

Smith, R. C. 1968. "The Optical Characterization of Natural Waters by Means of an 'Extinction Coefficient'." *Limnol. Oceanogr.*, **13**:423–429.

Talling, J. F. 1971. "The Underwater Light Climate as a Controlling Factor in the Production Ecology of Freshwater Phytoplankton." *Mitt. Int. Ver. Theor. Angew. Limnol.*, **19**:214–243.

Tyler, J. 1968. "The Secchi Disc." *Limnol. Oceanogr.*, **13**:1–6.

Tyler, J. E., and R. W. Priesendorfer. 1962. "Transmission of Energy within the Sea." In M. N. Hill (ed.), *The Sea*, **1**:397–451.

Vollenweider, R. A. 1961. "Photometric Studies in Inland Waters. 1. Relations Existing in the Spectral Extinction of Light in Water." *Mem. Ist. Ital. Idrobiol.*, **13**:87–113.

Chapter 4

Heat

OVERVIEW

Thermal stratification, which contributes so much to lake structure, is a direct result of heating by the sun. The only other important sources of cooling or heating are evaporative cooling from wind, sensible heat loss or conduction, inflow of streams, and, in some cases, geothermal heating of sediments or tributary hot springs. Thermal stratification, rather than any absolute temperature, is the most important. Even in rivers the absolute temperature is less important than daily or spatial changes in temperature. In estuaries stratification is largely controlled by salinity, not temperature. Heating decreases the density of the upper water which in combination with the wind results in a three-layered system. Thermal stratification is the most important physical event in the lake's annual cycle. In deeper lakes in summer, stratification results in an upper warm, lighter layer—the *epilimnion;* a cool, dense deep layer—the *hypolimnion;* and a transitional zone between them—the *thermocline,* or *metalimnion.*

The thermocline is not fixed in depth. It gradually descends during the summer until the lake turns over in fall. Due to the increased density change per degree rise in temperature, at higher temperatures less temperature difference is needed between epilimnion and hypolimnion for a stable stratification. Lakes which mix from top to bottom are termed *holomictic;* deep or chemically stratified ones may only partially mix and are called *meromictic.* The rate of winter cooling and spring heating controls the duration of unstratified mixing. *Dimictic* lakes mix twice, once in the fall and once in the spring; they are covered with ice in winter. *Monomictic* lakes do not freeze; they have one long mixing period all through the winter.

Polymictic lakes are shallow; they mix every few days or even daily all year round. *Amictic* lakes have year-round ice cover and never mix.

The heat cycle of a lake, the balance between heat gained and lost, is mainly a function of mean depth and geography and is useful in estimating the onset of thermal stratification. In deep lakes with a large maximum heat content, the water takes a long time to heat. Shallow lakes stratify, destratify, freeze, and thaw more rapidly than deeper lakes in the same vicinity. Maximum heat contents are greater in hot climates, but evaporative cooling may almost equal the solar input. The maximum temperatures in tropical lakes are similar to those on the warmest days in some temperate lakes. If it were not for evaporative cooling, they would be too hot for most organisms to survive. Overall, the heat changes and distribution in lakes provide one of the most important elements of structure influencing a host of chemical and biological processes.

MEASUREMENTS

An ordinary mercury-in-glass thermometer or electric thermistor are adequate for most limnological purposes. Recently great progress has been made in increasing the accuracy of thermistors, which measure the change in resistance with temperature of a metal probe in a tiny gas bubble under a thin glass cover that is in close contact with the water. Accuracies greater than $\pm 0.1°C$ with thermistors having a remote readout require more effort and cost, but $\pm 0.01°C$ is achievable with some salinity or conductivity-temperature-depth instruments often used in lakes or at sea.

A third device used is the bathythermograph, a rocket-shaped instrument which is lowered rapidly to the lake bed on a hydrographic wire. On the way down it records both temperature and depth by producing a continuous trace which is etched on a piece of smoked or gold-plated glass. Bathythermographs may be of a recoverable or expendable type. The expendable variety, frequently used at sea, transmits temperature and depth information back to the ship as it free-falls to the bottom.

Incoming radiation, which is needed for heat cycles, can be measured directly or calculated from existing meteorological stations in the vicinity. Recording thermopiles or pyrheliometers give solar radiation directly in gram-calories per square centimeter per minute which can be integrated over any time period. More detailed work may require a net radiometer which simultaneously measures incoming and back radiation reflected from the surface.

INTRODUCTION

In Chap. 3 we discussed light in terms of its quality and quantity and how it is modified by passage through both the atmosphere and water. Most light entering water is converted directly to heat. Water has an enormous capacity for heat storage and has the highest specific heat of any naturally occurring substance. Major heat losses occur by evaporation and conduction. The slow but continuously changing heat content of temperate lakes, rivers, and estuaries provides an important part of the structure of aquatic systems already described. In tropical regions the annual change in the heat content closely reflects the seasonal pattern of temperature and as expected varies much less during the year than in temperate regions. Lakes and streams at higher latitudes and elevations, although often subject to more intense radiation, tend to be cooler because of lower ambient air temperatures. Winter ice in these areas may persist well into summer, and the open-water period may be very brief.

Although heat and temperature in streams and estuaries show many similarities with lakes, they do not show significant thermal stratification. Shading by vegetation is very important in streams but not in most lakes. Solar heating of mud flats at low tide is unique to estuaries, and for these reasons features of heat and temperature peculiar to rivers and estuaries are discussed in Chaps. 16 and 17.

THERMAL STRATIFICATION

It is fairly obvious that sunlight falling on the surface of a lake will heat it more near the surface and thus form a layer of less dense warmer water overlying a denser cooler zone. Since the absorption of light in lakes is nearly exponential with depth, one might expect a similar distribu-

tion of heat. Because of convection and wind stirring, however, a layer of water is formed which has a more defined boundary than the illuminated zone (Fig. 4-1). The lake is then stratified by being divided into separate strata. The regions formed are the *epilimnion*—the warmer less dense upper layer; a cooler denser lower layer—the *hypolimnion;* and the *metalimnion* with *thermocline* between them. The definition of exactly what constitutes the boundaries of the metalimnion is not fixed, but if temperature is plotted against depth the thermocline lies in the area of greatest inflection of the curve. In

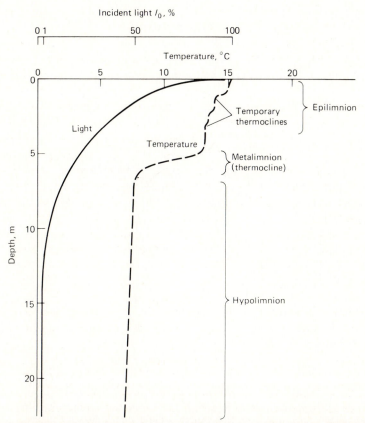

Figure 4-1 Idealized summer distribution of light and heat in a stratified lake. Note the irregular distribution of heat and sharp thermoclines and the smooth exponential curve for light distribution. An actual example is given in Fig. 5-18 where complex temporary thermoclines can be seen. The temporary thermoclines in the epilimnion are caused by heating on calm days and are usually destroyed each night by convective cooling or by afternoon winds (Fig. 5-15).

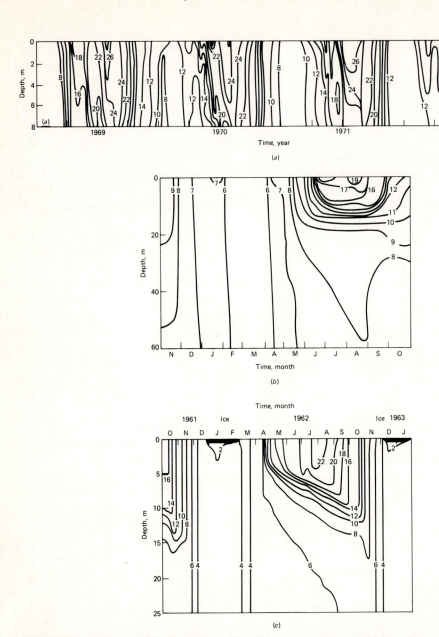

Figure 4-2 Seasonal depth distribution of isotherms for three contrasting lakes. (*a*) Four years for a warm, shallow polymictic lake, Clear Lake, California. (*Redrawn from Horne, 1975.*) (*b*) A deep monomictic lake, Windermere, England. (*Redrawn from Jenkin, 1942.*) (*c*) A dimictic lake, Mountain Lake, Virginia. (*Redrawn from Roth and Neff, 1964.*) Note the regularity of the seasonal thermal stratification (horizontal lines close together) and winter destratification (vertical lines far apart). Uniform temperatures exist from surface to lake bed in winter for the monomictic lakes, indicating good mixing. Isotherms every 2°C in (*a*) and (*c*), every degree for (*b*).

lakes the epilimnion may range from 2 to more than 20 m, the thermocline is usually several meters thick, and, depending on lake depth, the hypolimnion may be large or small.

Lakes at midlatitudes form a thermocline in spring, if they are deep enough, and lose it in the fall. Many have thick ice cover in winter. In tropical climates the rainy season is often the time of mixing. Lakes which have winter ice cover are called *dimictic* since they mix twice a year. Mixing occurs between ice breakup and the onset of thermal stratification in spring and between the breakdown of thermal stratification in fall and the onset of winter ice cover (Fig. 4-2c). The ice cover prevents wind energy from mixing the cold water beneath the ice. Most temperate lakes are dimictic. Typical examples of dimictic lakes are Castle Lake, California, and Lake Mendota, Wisconsin. *Monomictic* lakes are never ice-covered, and winter is a single continuous wind-stirred event (Fig. 4-2a, b). Typical monomictic lakes are the Great Lakes except Lake Erie, Lake Tahoe, California, and Lake Windermere, England.

If a lake is shallow, exposed to the wind, and located where winters are mild, thermal stratification may last for a week or two, be destroyed by a storm, reestablish for a few days, be disrupted again, and so on. This rather common type of lake is called *polymictic* since it mixes many times per year (Fig. 4-2a). Examples are Clear Lake, California, and Lake George, Uganda.

Two other sets of definitions are based on the degree of lake mixing and its seasonal variations. If, during the annual mixing cycle, a lake mixes from top to bottom it is said to be *holomictic* (holo = "whole"; mixis = "mixing"). If, on the other hand, it is so deep that there is insufficient energy to stir it from top to bottom, it is called *meromictic*. Some of the world's deep lakes in tropical climates, such as Lake Tanganyika, a very deep African Rift lake, are meromictic with a permanent thermocline at about 400 m and seasonal thermoclines between 20 and 100 m. Lakes with significant accumulations of salts near the bottom may not mix. These lakes are chemically meromictic and may have a warm bottom layer or *monimolimnion* where the additional buoyancy from heat is more than counterbalanced by an increased density from dissolved salts. Even quite shallow lakes with high salt content at the bottom may be chemically meromictic.

Occasionally, meromictic lakes are mixed by severe storms. This mixing may be a dramatic event since hydrogen sulfide accumulated in the bottom water is mixed with surface water, oxygen is depleted, and massive fish kills may occur. The nutrients released usually produce a series of algal blooms until equilibrium is reestablished. In the vicinity of stagnant coastal lagoons a similar meromixis may cause extensive fish kills when severe storms flush the oxygen-deficient waters into the sea.

The four types, meromictic, dimictic, monomictic, and polymictic, encompass almost all the world's lakes, but there are a few curious examples which show different mixing patterns. Lakes that are always covered with ice never mix and are called *amictic*. Lakes Vanda and Bonney in the Antarctic (Figs. 4-3, 19-12) are examples of permanently ice-covered large lakes, but many smaller lakes at high altitudes or latitudes are amictic. Other lakes which are found in less cold climates may thaw only once every few years. Such lakes are called *oligomictic*.

THERMOCLINE FORMATION

The formation of the thermocline in spring with consequent restriction of nutrient circulation is the single-most important physical event for lake biota. The wax and wane of blooms of diatoms is an obvious biological change directly associated with thermocline formation (Chap. 12).

Figure 4-3   An example of an amictic lake, Lake Vanda, Antarctica. This large lake (A = 5.2 km², $\bar{z}$ = 29 m) is permanently covered with a thick layer of ice. The lake lies in a dry valley, but adjacent hills were covered by snow in summer when this picture was taken.

An approach to thermocline formation, useful to biologists and chemists, considers energy as a unifying concept. Boyce (1974) suggests a mechanical energy budget for the water column and relates this to a critical wind speed and rate of surface heating or cooling [see Eqs. (2) and (3)]. During thermocline formation the sun heats the water surface. The wind stirs this warmer lighter water down to a depth where the turbulence is eventually dissipated. This depth becomes the thermocline. The thermocline is established at a depth that is shallow in relation to the depth to which sunlight penetrates. Most heat is absorbed in the first few meters and to extend farther down must be physically stirred by wind or convection-induced turbulence. The downmixing water is warmer and positively buoyant. It resists mixing in proportion to the density difference between the warm and cold water. The density of water changes rapidly with temperature (Fig. 4-4), and a large effect can be expected with a few days of sunshine and calm weather.

Why then, when sunshine and calm weather occur in winter, does not a thermocline appear,

at least temporarily, when the winds resume? The answer lies in the relative strengths of the turbulent wind mixing and the warmer water's buoyant resistance to mixing. These forces can be compared by using the Richardson number, which is expressed by the equation

$$R_i = \frac{g \; d\rho/dz}{\rho \; (dU/dz)^2} \tag{1}$$

where R_i = the Richardson number, a dimensionless quantity

g = acceleration of gravity

ρ = density of the fluid

U = horizontal water current

$g \dfrac{d\rho}{dz}$ = a measure of the water's buoyancy

z = depth

$\rho \left(\dfrac{dU}{dz}\right)^2$ = a measure of the stirring due to shearing of water currents

Richardson's number defines the ratio of work done by stirring to overcome the stabilizing buoyancy forces relative to the available kinetic energy. Where R_i becomes critical, there is adequate kinetic energy available as wind stirring to produce shearing stress sufficient to overcome the buoyancy forces. This R_i (critical) is close to $1/_4$. When R_i is below $1/_4$, stirring increases and stratification is destroyed. Above this value buoyancy dominates and thermal stratification is preserved. In winter R_i is low since there is just too much kinetic energy available compared to the small buoyancy effects

caused by heating from winter sunshine. Under this circumstance mixing prevails.

THE FLOW OF HEAT

An important feature of lake heating is the average summer temperature of the epilimnion and hypolimnion. The epilimnion heat balance is complicated since heat losses at night and back radiation or evaporative cooling vary

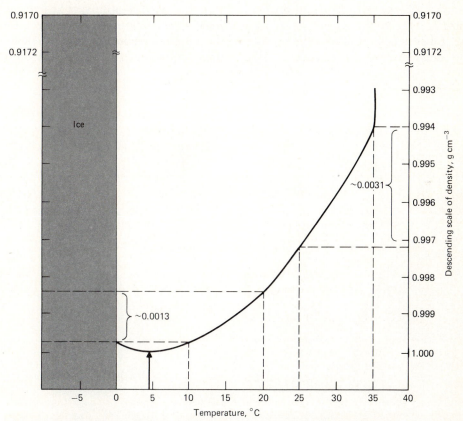

Figure 4-4 Changes in freshwater density with temperature. Note that the maximum density (arrow) of 1.000 g cm^{-3} occurs at $+ 4°$C. There is a great increase in density per degree rise in temperature at higher temperatures, and much of the curve is parabolic. For example, from 10 to 20°C density changes about 0.0031 g cm^{-3}, but for a similar 10° rise starting at 25°C, the density rise is almost 3 times as much ($\sim$0.0031 g cm^{-3}) A few degrees temperature difference can form stable thermoclines in tropical lakes but not in temperate waters. Wind velocity, and thus wind stirring of the water, is similar the world over but warm water is more resistant to mixing. Ice is very much less dense than water ($\sim$0.9100 g cm^{-3}) and thus floats.

with day-to-day weather patterns. The situation in the hypolimnion is much simpler since heat flows are buffered by the lake bed and the overlying blanket of the epilimnion (Fig. 4-5). The main factor influencing the hypolimnetic temperature is the onset of stratification. If the lake stratifies early in the season when the water is cold, the hypolimnion will be cooler than when stratification occurs later. The first permanent thermocline is usually established by a storm, and these do not occur at the same time each year. Therefore variation from year to year is likely in the starting temperature of lake hypolimnia. The activity of many organisms including bacteria is strongly influenced by temperature. Since the most lake bacteria are found in the sediments, the rate of decomposi-

tion and recycling will be partially dependent on hypolimnetic temperature.

Once stratification is established, *direct heating* is the only important heat source for the hypolimnion. This occurs when the lake is sufficiently transparent to allow significant amounts of light to penetrate below the thermocline (Fig. 4-5). As described earlier, longwave radiation, including thermal radiation, is absorbed in the first few centimeters of lake water. Visible light penetrates deeper and is converted to heat in the hypolimnion. In transparent lakes with dark, rocky or muddy bottoms the lake bed is directly heated and releases heat to the overlying water. In deep transparent lakes some or even most of the heating of the upper hypolimnion can be accounted for by

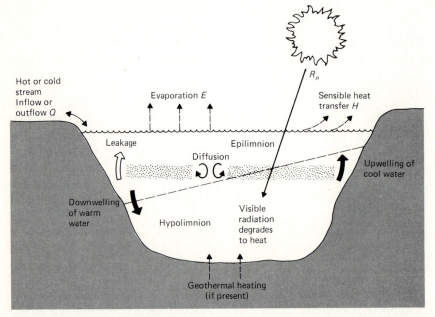

Figure 4-5 Diagram of the major heat and water flows across or at the edges of the thermocline during summer. Bold lines are advective flows of water (currents); thin lines are diffusive transfer. Upwelling is most obvious in large lakes when the thermocline tilts significantly. Leakage probably occurs continually as a flow around the edge of the thermocline where it meets the lake bed. Heat diffusion is always present but very difficult to measure directly. In transparent lakes some sunlight will penetrate below the thermocline and directly heat the hypolimnion or sediments.

Figure 4-6 Volcanic heating of Lake Managua, Nicaragua. The bare areas contain hot springs with water at near-boiling temperatures which runs into the lake. The heat from the slopes of Momatombo Volcano (to the right) extends below the water's edge. Geothermal heating is, however, insignificant in most of the world's lakes.

direct conversion of visible radiation into heat (e.g., Bachmann and Goldman, 1965). The thermocline depth and that of the photic zone are roughly equal in many lakes, but in transparent lakes this is not so.

In Lake Vanda, Antarctica, significant geothermal heating of the lake sediment and overlying water occurs in the salt-stabilized monimolimnion at 25°C, which exists beneath a layer of 8°C water. Geothermal heating can also occur in deep lakes in areas of volcanic activity when heat transfer may be direct or result from hot springs entering the lake. Lake Managua in Nicaragua has hot springs entering the lake which are being tapped for geothermal steam to be used for electric power generation (Fig. 4-6).

Heating and Cooling of the Epilimnion or Mixed Layer

The discussion that follows is also applicable to polymictic and unstratified lakes and even rivers since their single mixed layer is similar to an epilimnion. It is important to be able to predict changes in the epilimnetic temperatures of lakes. Examples are prediction of the onset of ice, the effect of thermal discharges on lakes, and the effect of dry cloudless summers on reservoir levels. The mixed-layer temperature also controls the rate of egg development in zooplankton and stream invertebrates, as well as the success of some fish-year classes. These predictions are just a few of the concerns of limnologists. Unfortunately, because of the variability of the weather, predictions are often difficult. A basic problem is that lakes do not heat up in direct proportion to the applied light and heat inputs. Evaporative cooling and back radiation also occur, and the balance between these two losses and solar heating determines the resulting lake temperature.

The fraction of incoming radiation absorbed by the epilimnion is calculated from pyrheliometer measurements, lake extinction coefficients, and knowledge of the lake's altitude and latitude. Evaporation is much more difficult to measure since it depends largely on the relative humidity, lake surface temperature, and wind speed. Further, the strength of the wind may vary considerably over the lake surface.

In lakes in semiarid zones there is often little surplus of inflow over outflow, and the evaporation eventually produces salt lakes (Chap. 19). Small changes in weather patterns can also produce salt lakes. A good example is Lake Managua, Nicaragua, where inflows and evaporation are about equal and the former outflow channel is now filled for a road. In other wetter climates the main effect of evaporation is to slightly increase the hydrologic residence time and decrease the maximum surface temperature. In tropical lakes evaporative cooling is essential for the survival of many organisms since these lakes often show noontime temperatures as high as 35°C, which is close to the thermal death point for many aquatic organisms.

Upwelling

Upwelling in the ocean along the western shores of the major continents produces rich fisheries. Upwelling also occurs in lakes when high winds drive the warm less dense epilimnion to the lee shore and expose the underlying thermocline to the wind (Figs. 4-5, 5-17). Under certain conditions cold nutrient-rich hypolimnetic water upwells to the surface where the wind mixes it into the epilimnion. Only larger lakes like Baikal, Michigan, and Tahoe exhibit significant upwelling, but this results mostly from resonant internal waves moving around the lake edges (see Chap. 5 for details of wave motion).

Hypolimnetic Entrainment

One might expect indirect heating of the hypolimnion by conduction from the warmer adjacent epilimnion. Heating would thus be regulated by the thermal conductivity of the metalimnion or thermocline in the same way as a window controls heat flow from a warm room to the outside air. Since the thermocline is a dynamic mobile layer, this uniform heat flow does not in fact occur as expected. What actually happens is that the thermocline in lakes gradually descends from spring to fall and the epilimnion becomes deeper at the expense of the hypolimnion. This process is not uniform, and the thermocline may fall and rise over some weeks as it descends (Fig. 4-7). Parts of the hypolimnion are incorporated into the epilimnion throughout the summer by storm-induced turbulence near the top of the thermocline, by upwelling, or by hypolimnetic entrainment (Fig. 4-7).

The very existence of a thermocline is testimony to its strong resistance to vertical mixing. Diffusion (Fig. 5-15) across the metalimnion is very low, but some transfer of nutrients has been observed. An important question is whether the hypolimnetic water moves, carrying nutrients with it, or whether vertical migrations of fish, zooplankton, and flagellate algae account for the transfer. This is a difficult problem since direct measurement is not practical for either the water currents or the heat changes that occur near the thermocline. In many cases physical transfer of water accounts for nutrient transport. Movement of hypolimnetic water across the thermocline from the hypolimnion occurs in several lakes including Lake Perris, California, a reservoir with an anoxic hypolimnion in late summer. When nutrients in the epilimnion of this lake are monitored on a diel cycle, there is a sudden rise in epilimnetic iron and hydrogen sulfide during the period of maximum wind in midafternoon (Table 4-1). This upward movement of iron and hydrogen sulfide indicates entrainment of hypolimnetic water into the epilimnion.

In some cases measurements have been made in large lakes where edge leakage is sufficiently small relative to the epilimnetic volume to allow measurement of the transfer of heat and nutrients across the thermocline. This pro-

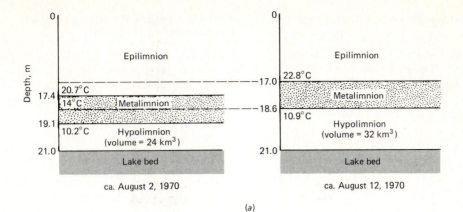

ca. August 2, 1970   ca. August 12, 1970

(a)

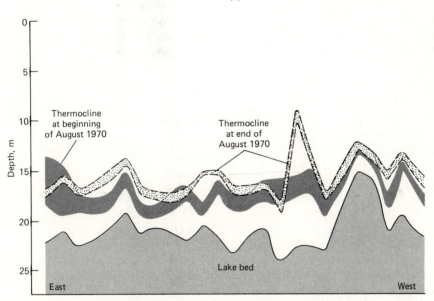

Distance along lake (stations start near the east and proceed to west)

(b)

Figure 4-7 Spatial and temporal changes in the thermocline of Lake Erie in summer. Short bursts of high wind energy after periods of calm cause the observed changes. Heat, water, and nutrients are transferred indirectly between hypolimnion and epilimnion via the thermocline. The effect on the otherwise insulated hypolimnion is more obvious than in the epilimnion where heat loss or gain to the atmosphere occurs continually. The effect is exaggerated for Lake Erie as its hypolimnion is unusually thin for such a large lake. (a) In 10 days, 10^{15} kcal of heat and 8 km³ of water were transferred down into the hypolimnion. During relatively calm conditions (early August) the metalimnion or thermocline was lower and thicker (1.7 m) than after some windy days when it was 1.58 m thick. The metalimnion incorporates water from the hypolimnion and epilimnion in calm weather but is reduced in size by winds. (b) How these effects varied from place to place in the lake over a longer time period. Note that thermocline and hypolimnion thickness can vary considerably from the average values shown in (a). Large internal gravity waves on the thermocline may cause the up and down motion obvious in (b). (*Modified from Burns and Ross, 1972.*)

Table 4-1   Diel Variations in Concentrations of Total Dissolved Iron (μg liter^{-1}) in Perris Reservoir, California, on August 17, 1976, due to Transfer of Iron from the Anoxic Hypolimnion through or around the Thermocline*

Note that in the lower epilimnion at 10 m iron, as well as a noticeable smell of H_2S, increased following afternoon winds.

	Hour		
Depth, m	0900	1200	1500
0	7.5	6.0	9.0
10	7.5	17.3	20.3
16	47.1	53.4	49.9

*From Elder and Horne, 1977.

cess is *hypolimnetic entrainment* and involves an initial transfer of water from hypolimnion and epilimnion to the metalimnion during calm weather (Burns and Ross, 1972; Blanton, 1973). During a storm the metalimnion decreases in thickness, transferring water back to the epilimnion and sometimes to the hypolimnion (Fig. 4-7). Relatively small changes in wind speed affect the thermocline depth (Boyce, 1974). During calm weather the thermocline depth decreases and the metalimnion volume increases at the expense of both the epilimnion and the hypolimnion. Thus hypolimnetic water is incorporated into the metalimnion. When strong winds resume, the thermocline descends and the epilimnion reincorporates some of the metalimnion and indirectly some of the hypolimnion water. Figure 4-7 indicates changes in volumes of these three layers of Lake Erie. The volume of water transferred occurs over several days and is quite small compared to that of the entire epilimnion. The effect on the temperature of the epilimnion will thus be minimal. The most significant effect of hypolimnetic entrainment is the transport of nutrients upward since relatively large amounts of growth-limiting elements may be transferred.

Establishment and Destruction of the Thermocline

Students studying small lakes are often surprised at the rapidity with which the thermocline forms in spring. This can occur in a matter of only a few days. Such speed would not be expected from a uniform diffusion process based on steadily increasing solar radiation. A good explanation employs a nonuniform event such as a storm. Above a critical wind speed ($U_{a,\,crit}$) the rate of change of potential energy (dP/dt) supplies sufficient surface turbulence to entrain deeper water layers. To do this the resistance of any density gradients must be overcome. At thermocline onset the sun warms the lake surface and wind blows across it (Boyce, 1974). Then

$$\frac{dP}{dt} = 0 = \gamma_w U_{a,\,crit}^3 + \gamma_h g \frac{\alpha_{vs}}{C_p} q_s D \qquad (2)$$

where γ_w = a constant dependent on environmental conditions encompassing wind drag and vertical wind-mixing efficiency

γ_h = a constant also dependent on environmental conditions which represents the fraction of heat energy used in mixing of the water column

α_{vs} = volume coefficient of thermal expansion of the surface water

C_p = specific heat of water

q_s = net heat flux across the water surface

D = depth over which heat is distributed, i.e., depth to which uniform vertical mixing proceeds for $U_{a,\,crit}$

Equation (2) says that a lake's potential energy may change in two ways: The first term ($\gamma_w U_{a,crit}^3$) is wind stirring which redistributes layers of different densities; the second term ($\gamma_h g \dfrac{\alpha_{vs}}{C_p} q_s D$) shows that heat loss or gain can change the density distribution.

When the wind blows at speeds faster than $U_{a,\,crit}$, as in a mild storm, dP/dt is positive, and active turbulent entrainment proceeds down from the surface.

Rearranging Eq. (2),

$$U_{a,\,crit} = \left(\frac{\gamma_h g \alpha_{vs}}{C_p \gamma_w} \right)^{1/3} (-q_s D)^{1/3}$$
$$= \beta(|q_s|D)^{1/3} \qquad (3)$$

This has been related to values in a lake by Boyce (1974), where in summer $\gamma_h = 1$, $g = 980 \text{ cm s}^{-2}$, $\alpha_{vs} = 10^{-4}\,°C^{-1}$, $C_p = 1 \text{ cal g}^{-1}\,°C^{-1}$, $\gamma_w = 1.9 \times 10^{-9}$, and thus $\beta = 372$ (cgs units). For $D = 10$ m and heating rate q_s of 400 cal cm^{-2} day^{-1} (4.6 × 10^{-3} cal cm^{-2} s^{-1}),

$$U_{a,\,crit} = 618 \text{ cm s}^{-1} \qquad (\approx 12 \text{ knots})$$

This is a typical spring-storm wind velocity and is correlated with both the onset of Langmuir spirals in many lakes and the whitecapping of breaking waves (Chap. 5). These extra inputs of turbulent energy increase entrainment, overcoming the previous slight temperature gradients, and rapidly establish the permanent thermocline.

The Thermal Bar and Fall Overturn

In the special case of very large cold lakes such as the Great Lakes, the thermocline does not form all over the lake at the same time. In spring the Great Lakes' waters are divided in two sections, offshore unstratified water at less than 4°C and a warmer weakly stratified mass near shore. The 4° water inbetween the two is the densest (Fig. 4-4) and sinks. The zone of dense sinking water is called the *thermal bar* (Fig. 4-8); it gradually moves offshore until the entire lake stratifies. The nearshore water becomes warmer than the main water mass because it is relatively shallow and the heat is contained in a small volume. In smaller lakes the nearshore water would quickly be wind-mixed horizontally into the offshore water. The thermal bar enhances early algal growth by ef-

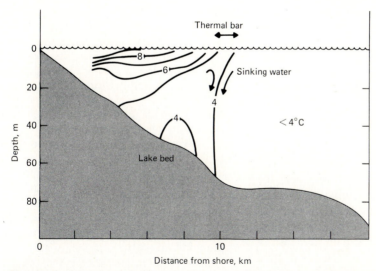

Figure 4-8 The thermal bar in Lake Ontario resulting from the differential effect of solar heating in the open lake and at the margin of the lake. The warmer water is confined to shallows by a sinking mass of dense 4°C water derived from lighter, cooler water offshore which must be less than 4°C. (*Redrawn from Boyce, 1974.*)

fectively trapping both heat and nutrients from spring meltwater.

The prime reason for the loss of thermal stratification in autumn is the reduction in heat input from the sun's radiation which occurs while evaporative cooling continues at high summer rates. Most lakes experience a gradual lowering of the thermocline in late summer (Fig. 4-2c). The eventual total mixing or *overturn* is of great importance in dimictic lakes since only then can nutrients and gases such as oxygen be uniformly distributed. In summer the lighter epilimnion overlies the denser hypolimnion to produce a low overall center of gravity. The large turbulent energy input due to storms raises the center of gravity of the lake water to the winter position. From Eqs. (2) and (3) a decrease in heat inflow q_s amplifies the effect of wind speed U_a^3 on the thermocline depth D. The net effect is a series of declining thermoclines. This incorporation from below of hypolimnetic water into the epilimnion provides nutrients for the autumn algal bloom (Fig. 12-3).

Temporary patches of downwelling several meters across appear and vanish on the surface of lakes in autumn. These turbulent sinking patches are analogous to the air thermals so effectively utilized by soaring birds and gliders. They help to maintain the underwater microclimate and high species diversity often found in the autumn plankton.

Heat Cycles

The effects of the mechanisms already described can be summarized in the concept of the heat cycle. In contrast to the difficult measurements required for upwelling and entrainment, *heat cycles* are relatively easily measured. They give a total accounting of the gain and loss of heat by the system during a specified time period. A square centimeter of lake surface is a convenient unit since heat transfer is expressed in terms of area. Thus units are calories per square centimeter. The length of the water column considered is taken as the average depth of the lake. The heat income and losses from a lake surface vary greatly with time of day, latitude, altitude, and exposure to the sun as well as any shading from surrounding mountains. In general, there is greater variation in heat input and output at higher elevations and latitudes than there is for the more uniform climate of the equator (Fig. 4-9a, b).

The components of a lake's heat cycle are summarized in the following equation:

$$S = R_n - E - H - Q \tag{4}$$

where S = storage rate of heat in lake
R_n = net radiation
E = evaporation
H = sensible heat transfer, which is roughly equal to conduction
Q = advective heat inputs and outputs due to water currents or inflow and outflow of streams

In detail, S is the rate of heat entering a lake (cal cm^{-2} min^{-1}) and is due to the following processes:

R_n (net radiation) is the direct and indirect solar radiation absorbed at the lake surface (see Chap. 3), plus the long-wave radiation given off by the air minus long-wave radiation back-radiated from the water surface. Net radiation depends on air temperature, air vapor pressure, and cloud cover.

E (evaporation) is the amount of heat lost by evaporation and depends on the wind, surface temperature, and air vapor pressure.

H (sensible heat transfer) is the amount of heat lost by conduction from the water to air and depends on the difference between the air and the water-surface temperature as well as on the wind velocity.

Q (advective heat inputs and outputs) is the heat loss due to an outflow of warm surface water being replaced by an inflow of cold stream water (may be reversed in special circumstances, e.g., hot springs).

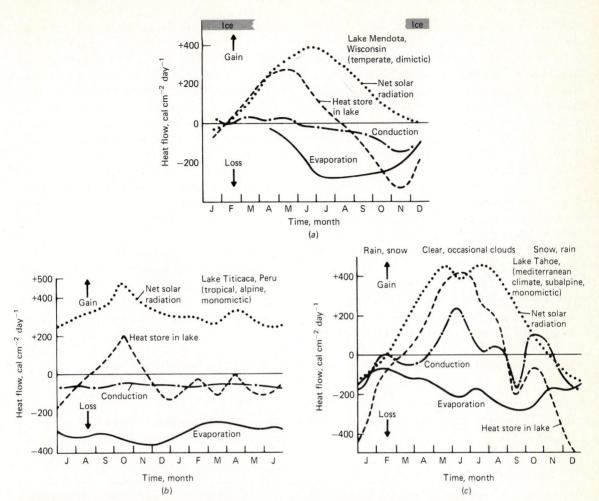

Figure 4-9 Heat cycles in three contrasting lakes. (a) In a typical temperate lake (Lake Mendota, Wisconsin) spring and summer radiation R_n heat the lake. There are no significant heat losses at this time. At midsummer the warm epilimnion begins to lose substantial amounts of heat due to evaporation E. This loss continues at much the same rate through autumn when sensible heat loss H, akin to conduction, amplifies the evaporative losses. (*Redrawn from Ragotzkie, 1978.*) (b) In a large high-altitude monomictic tropical lake (Lake Titicaca), both net radiation and evaporation are more uniform and show little seasonal change. (*Redrawn from Kittel and Richerson, 1978.*) (c) In the subalpine monomictic Lake Tahoe, summers are relatively cloudless and the net radiation input varies seasonally. Evaporation in temperate lakes is relatively low compared with that in tropical lakes. Heat storage S in temperate lakes shows a single large regular cycle each year, while in tropical lakes cycles are smaller and irregular. (*Redrawn from Myrup et al., 1979.*)

We neglect here geothermal heating, loss by conduction to the basin, or solar heating of sediments.

Thus the storage rate S depends on solar radiation, cloud cover, surface and air tempera- tures, humidity, and wind speed and is the net amount of heat entering a lake. As already noted, lakes at high altitudes or low latitudes re- ceive more heat from the sun than those at other sites (Fig. 4-9). A deep lake with a large water volume can contain more heat than a shallow lake at the same location merely because it has more water (Table 4-2). Local effects such as exposure to sunlight, protection from the wind, and warm submerged springs are most important in smaller lakes. Large lakes such as the Great Lakes actually modify the surrounding area by virtue of their great heat content. The microclimate created may be locally important in preventing early frosts.

The *maximum heat content,* once called the *annual heat budget,* of a lake is the total heat per unit surface area which enters the lake between the time of lowest and highest heat content of the lake's water. It is not a true budget since it only contains heat gains. The term maximum heat content, as used in this text, is a more appropriate term. It reflects the total heat input from the coldest winter to the warmest summer conditions without considering daily fluctuations. Not all the heat will be in the water, as some will have been absorbed by the

sediments. The maximum heat content of a lake is the product of the weighted averages of mean lake temperatures and lake depth. Mean temperature is measured by summing the product of temperature and volume for a series of layers of the lake. The mean temperature is the content of heat per unit volume (cal cm^{-3}), assuming density and specific heat are equal to unity. Total heat is heat per unit area (cal cm^{-2}) above 0°C (Birge, 1915; Stewart, 1973).

The maximum heat contents of dimictic Lake Mendota and adjacent lakes near the center of the continental United States may be considered rather typical of inland temperate lakes near sea level (Table 4-2). Those of the monomictic lakes of the English Lake District and Loch Ness, Scotland, are typical of coastal waters (Table 4-2). Table 4-2 also shows that mean depth is the dominating factor in maximum heat content (Gorham, 1964). Shallow lakes stratify, destratify, freeze, and thaw more rapidly than adjacent deep lakes.

Table 4-2 Physical Characteristics, Maximum Heat Contents, Stratification Period, and Length of Isothermal Mixing for Three Adjacent Wisconsin Dimictic Lakes and Three Monomictic Lakes in Britain*

Note that the deeper the lake, the greater the heat stored and the longer the period of isothermal mixing.

Lake	$\bar{z}$, m	Mean max. temp., °C	Heat contents, cal cm^{-2}	Area, km^2	Volume, m$^3 \times 10^6$	Isothermal mixing, days Fall	Spring
Mendota	12.4	20.0	24,100	38.7	480	50 or 66	73
Monona	7.7	23.3	17,600	14.0	108	66	63
Waubesa	4.6	25.2	11,400	8.3	38	Polymictic in summer	
Windermere: North Basin	27	~20	17,500	8.2	221	All winter (Nov–May)	
South Basin	19.4	~20	15,700	6.1	118	All winter (Oct–May)	
Loch Ness	133	~20	37,200	56.4	7500	All winter (Jan–June)	

*Modified from Stewart, 1973; Mortimer and Worthington, 1942; and Mill, 1895.

The effect of increasing the maximum heat content is to delay the freezing of dimictic lakes. Inflowing sediment and nutrients as well as thermal power-plant discharges may also affect heat cycles. Pollution may change freezing and thawing dates or the onset of thermal stratification, which will affect the lake's chemistry and biota.

Variations in the times of maximum blooms of the diatom *Asterionella* in various lakes of the English Lake District are due to the mean depth of the lakes and associated maximum heat contents since the shallower lakes stratify before the deep ones. The deep lakes with their larger heat contents require more solar heating in spring to raise the temperature of the epilimnion to the level where density differences prevent mixing.

The onset of freezing or the time of thermal stratification is important. It produces major physical and, later, chemical changes to which almost all lake biota respond. The most obvious effect of stratification is nutrient depletion in the epilimnion.

Dimictic lakes have only a short period in which to satisfy any summer oxygen debt in the hypolimnion. In some high-latitude lakes overturn in autumn is followed so rapidly by ice formation that there is insufficient time to transfer all the needed oxygen from air to sediments. This is most obvious in eutrophic lakes. Similar conditions may also occur in spring, but due to the greater solubility of oxygen in cold water and photosynthetic oxygen production from spring algae blooms, an oxygen deficit is less likely.

"Winter kills" can occur where under-ice oxygen is depleted; large numbers of fish may die. Since most fish would not overwinter in lakes without oxygen, the variable period of autumn mixing prior to ice cover may be a deciding factor in the winter kill. In addition, the problem may be aggravated by snow cover which prevents light from providing photosynthetically produced oxygen during winter.

Strong winter winds and cold air temperatures will produce a water column temperature below 4°C prior to ice formation. After freezing the lake gains some heat from solar radiation and a little heating from the sediments. Any heat loss occurs via the ice layer which thickens through midwinter. When the ice melts there is a slight dip in the spring temperature rise produced by the latent heat required to melt the ice. Such a small dip is surprising since in Lake Mendota, for example, the heat required to freeze and melt the ice cover is about a quarter of the heat content (Stewart, 1973). In many lakes the daily heat flux is surprisingly large in comparison to the maximum heat content.

Waste Heat Discharges

The effects of changes in the heat content due to thermal discharges are of major concern to applied limnologists and environmental engineers. The example of Lake Monona, Wisconsin, illustrates this. On Lake Monona, a power station uses sufficient cooling water to theoretically heat all the lake water by 10°C each year, or 0.03°C day^{-1}. The amount of heat represents 50 percent of the total heat content for the lake. Despite all this heating, the lake temperature is apparently unaltered, possibly because 0.03°C is a very small change compared to normal daily variations. Daily heat losses and gains mask the effects of the thermal pollution. The natural heat-buffering capacity of the lake prevents a rapid change in temperature from unusually sunny or cloudy days. One noticeable effect of the thermal pollution of Lake Monona is an accelerated ice breakup in spring. The 115-year average for ice breakup in lakes Mendota and Monona differs by only 1 day, but for the last 15 years (since the start of the power station) Lake Monona has opened 19 days before Lake Mendota.

Only by influence during overturn or by changing hypolimnial heating can waste heat discharges produce large shifts in lake heat content and cause biotic changes. The lake mean depth is very important for the heat content but is not easily altered. The same can be said for all other variables in Eq. (4). The absolute amounts of daily heating and cooling in the mixed layer are large compared with the amount retained or lost in the epilimnion. The effect of small, slow changes in temperature is normally minimal for living communities because temperature adaptation occurs. There are few, if any, cases where dramatic shifts in populations could be induced by a change in temperature of less than 1°C. The very high specific heat of water (Chap. 3) buffers the thermal environment against change.

The heat transmitted with light is responsible for establishing the thermal stratification in water bodies, and at the same time it influences the respiration of both plants and animals. Phytoplankton exhibit higher respiration and hence slightly lower efficiency at elevated temperatures. As temperatures rise, bacteria, which are responsible for the important recycling of dissolved organic matter and minerals in the system, make an increasing demand on the available oxygen. We will discuss this oxygen depletion in greater detail in Chap. 7. As summer temperatures increase, the filtration and excretion rates of zooplankton also increase, and with this rise in feeding there are in turn more fecal pellets for the bacteria to regenerate. This accelerates the growth of the phytoplankton population upon which the zooplankton feed. Thus an increase in heat during summer turns the metabolic wheels at a faster rate. Even the *poikilothermal,* or cold-blooded, fish require more food and exhibit a slightly lowered efficiency of food conversion at elevated temperatures.

The stratification of heat and light imposes on lakes a fundamental structure within which other forces must operate. Waves and currents weave a chaotic pattern on this stratified water column, modifying the distribution of plankton within and among the strata. This motion is described next in Chap. 5. A study of water movement is perhaps less intuitive than the topics discussed so far, in that it requires a mathematical approach. However, the study of water in motion reveals an important aspect of lake structure.

FURTHER READINGS

Birge, E. A. 1915. "The Heat Budgets of American and European Lakes." *Trans. Wis. Acad. Sci.,* **18:**166–213.

Boyce, F. M. 1974. "Some Aspects of Great Lakes Physics of Importance to Biological and Chemical Processes." *J. Fish. Res. Board Can.,* **31:**689–730.

Denison, P. J., and F. C. Elder. 1970. "Thermal Inputs to the Great Lakes 1968–2000." *Proc. 13th Conf. Great Lakes Res.,* pp. 811–828.

Gorham, E. 1964. "Morphometric Control of Annual Heat Budgets in Temperate Lakes." *Limnol. Oceanogr.,* **9:**525–529.

Hutchinson, G. E. 1957. *A Treatise on Limnology,* vol. I, chap. 7, Wiley, New York.

Lewis, W. M. 1973. "The Thermal Regime of Lake Lanao (Philippines) and Its Theoretical Implications for Tropical Lakes." *Limnol. Oceanogr.,* **18:**200–217.

Likens, G. E., and N. M. Johnson. 1969. "Measurement and Analysis of the Annual Heat Budget for the Sediments in Two Wisconsin Lakes." *Limnol. Oceanogr.,* **14:**115–135.

Mason, D. T. 1967. "Limnology of Mono Lake, California." *Univ. Calif. Berkeley Publ. Zool.,* **83.** 102 pp.

Stewart, K. M. 1972. "Isotherms under Ice." *Verh. Int. Ver. Limnol.,* **18:**303–311.

Talling, J. F. 1969. "The Incidence of Vertical Mixing, and Some Biological and Chemical Consequences, in Tropical African Lakes." *Verh. Int. Ver. Limnol.,* **17:**998–1012.

Chapter 5

Water Movement

OVERVIEW

The motion of water has profound consequences for the chemistry and biology of lakes. Wind is the primary force moving lake water at all depths, although the mixing due to evaporative cooling discussed in Chap. 4 is important. The combination of heat and wind produces a lake structure consisting of many short-lived layers or patches which are more turbulent and chemically and biologically distinct from the surrounding water. The kinetic energy of the wind is transferred to the water to produce an irregular *turbulent* cascade of kinetic and potential energy from that contained in large lake-sized eddies to the smaller more rapid ones. The distribution of energy over various wave lengths, called an *energy spectrum,* can be used as a unifying concept for understanding the complexity of features that characterize movement in an enclosed body of water.

There are two kinds of water motion, *periodic* or *rhythmic waves* and *nonperiodic* or *arrhythmic currents*. Waves consist of the rise and fall of water particles, involving some oscillation but no net flow. Currents consist of net unidirectional flows of water. Currents and waves normally occur together. Part of the wind's kinetic energy goes into the continuous formation of surface waves which lose their form and dissipate their energy as they break on the downwind shore. Some of the wind energy is transferred indirectly via breaking waves to currents. Currents build up much more slowly than waves but eventually contain most of the lake's kinetic energy. In addition, the wind induces *internal waves* in the thermocline and the hypolimnion.

Water currents at the lake surface are called

surface drift. In large lakes and estuaries surface currents eventually flow at approximately 45° to the direction of the prevailing wind. This flow results from the effect of the earth's rotation. Water currents below the surface move at progressively greater angles to the wind the greater the distance from the surface. Finally, the deepest currents flow in the opposite direction to the wind. This spiral-staircase-like flow of currents is called an *Ekman spiral.*

Lakes are mixed most vigorously by storm winds which produce surface and internal waves as well as strong horizontal currents. Although surface waves are obvious, internal waves occur at the thermocline. The short-wavelength low-amplitude surface gravity wind waves are familiar to all. Less familiar are the very long surface waves which may resonate and reflect back and forth from shore to shore. These are termed *surface seiches.* The natural resonant frequencies of the basin select among the large-amplitude *internal gravity waves* for certain wavelengths. These are the *internal seiches* which occur at the thermocline. Short-wavelength internal gravity waves may become unstable and break in midlake, causing considerable local turbulent mixing and even transfer of hypolimnion water into the epilimnion. Formation of these *turbulent billows* is most pronounced near the base of the thermocline.

Vertical mixing as well as horizontal flow is caused by the surface wind. The currents of the Ekman spiral may be thought of as slabs of water moving at different speeds and directions. The contact between these layers imparts a vertical displacement or shear stress to the water in each layer, producing vertical mixing between them. *Surface wind-shear* effects are mostly confined to the epilimnion. *Breaking waves* contribute to vertical mixing, and *Langmuir spirals* provide a more-organized vertical mixing energy with a wavelength about equal to the depth of the thermocline. At certain times of year evaporative cooling at the lake surface is the major vertical mixing force. Nocturnal sinking of cooler waters to the bottom of the epilimnion stirs the mixed layer by *convective mixing.* This process is more efficient than wind mixing and is sometimes the main source of mixing in stratified lakes in temperate and tropical climates. The epilimnion usually mixes daily unless prevented by thermal barriers which occur during sunny, windless periods. Turbulent energy is finally lost as heat. Movement in deep water below the thermocline is slow, and substances near the lake bed are only gradually circulated throughout the hypolimnion. In large lakes the main energy for deepwater mixing may come from large-amplitude long-period internal gravity waves at the thermocline. The resulting turbulence at the lake bed moves dissolved substances up from the sediment-water interface. In large exposed lakes *upwelling* and *downwelling* produced at the lake edges by storms are a major mixing force.

Inflowing rivers are usually more important sources of biostimulants and toxicants than of kinetic energy. River plumes extending into lakes spread more horizontally than vertically and sink if their density is greater than the lake's. However, mixing of discrete plumes into a lake is quite small, and there is a tendency for such inflows to cling to one shoreline. Water movement in estuaries, rivers, and streams has many features in common with lakes, but important differences are discussed in Chaps. 16 and 17 on flowing waters and estuaries.

INTRODUCTION

The continual motion of the aquatic environment provides its unique character among ecosystems (Fig. 5-1). Water movement transports phytoplankton from a high-light, low-nutrient, low-predator environment near the surface to the dark nutrient-rich deeper waters which contain numerous predators. The one-way rapid water currents dominate the ecology of plants and animals in rivers, streams, and estuaries (Chaps. 16 and 17). On land large physi-

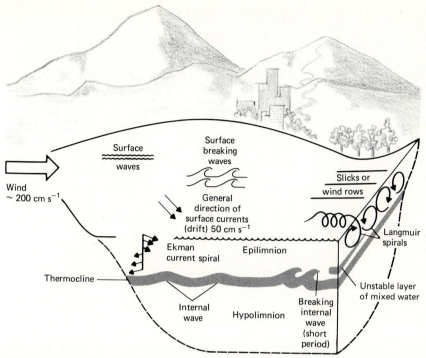

Figure 5-1 Composite diagram of forces (wind, gravity, evaporation and the earth's rotation) and the resultant water currents and waves. Wind moves the water, gravity makes horizontal flow easier than vertical, evaporation cools surface water which then sinks, and the earth's rotation moves surface flows to the right (northern hemisphere), and to the left (southern hemisphere).

cal changes in the environment tend to take place over a period greater than the life span of most terrestrial organisms, which promotes specific adaptations. Aquatic organisms must adapt to a wide range of conditions, and even then are truly at the mercy of their everchanging environment.

Water movements are critical to the distribution of all forms of energy, momentum, nutrients, dissolved gases, algae, some zooplankton, and sedimentary material. The distribution of solar and wind energy produces some form of thermal stratification in all lakes and stable summer stratification in most (Chap. 4). The lack of water mixing between the warm epilimnion and cooler hypolimnion is a major factor determining lake productivity. Similarly, the rapid top-to-bottom stirring in shallow lakes

during the summer is extremely important in recycling the nutrients which often support high levels of productivity.

MEASUREMENT

Lake currents range from almost zero to more than 30 cm s⁻¹. The measurement of rapid water currents is relatively straightforward for horizontal flows, but weak and vertical currents are very difficult to estimate accurately. The three principal devices used are current meters, drifters, and dyes. The basic instrument for measuring water flow usually consists of a propeller that is turned by the water current. A common choice is the Savonius rotor. Some instruments swivel horizontally and vertically and thus always face into the current (Fig. 5-2).

Figure 5-2 A modern profiling current meter. The current velocities are measured by a freely suspended vane (white can-shaped object). The direction of the current flows are measured by first finding the position of the vane (which has an electrode) relative to the two other pairs of electrodes (stick-like objects) that are also freely suspended. The vane electrode picks up signals at different frequencies from the two pairs of electrodes, indicating its orientation and hence the direction of the water current. Results from this apparatus are shown in Figs. 5-21 and 5-22. (*Photograph courtesy of S. A. Thorpe, Institute of Oceanographic Sciences, England.*)

A number of current meters are available but most are too insensitive or fragile for routine use. For example, for rapidly changing currents a meter with speedy response, such as the delicate hot-film probe, may be necessary.

Alternatives to current meters are *drifters* and *drogues* which present a large surface area to the current. Except at the water surface, a marker such as a small flag must be attached so that movements can be followed. The drag of the surface marker should be slight compared to that of the main device. Usually several drogues are placed at different depths. Motion can be plotted over time using photography or observations from a ship's crow's nest or the nearest hillside. This method is normally used over a few hours on calm days.

Fluorescent dyes have been used in studies of horizontal and vertical water motion. This method requires many measurements of a mov-

ing and expanding dye patch but is not recommended for measuring strong currents.

An inexpensive but effective method of using an orange attached to a fine fishing line was demonstrated by I. R. Smith in Lake George, Africa. The neutrally buoyant orange is carried by the surface currents. Keeping the line taut and placing knots at 50-m intervals gives an accurate short-term current measurement. A similar technique on the ocean prompted the seamen of old to call the velocity term a *knot*.

Remote sensing from aircraft or satellites measures surface-water masses over periods of days provided that a sufficiently obvious feature is available, for example, a sediment plume or temperature difference. The method can be extended to shallow-surface films by using the high reflectance of blue-green algae in the near-infrared (NIR) wave bands. Repeated photography of distinct features shows movement

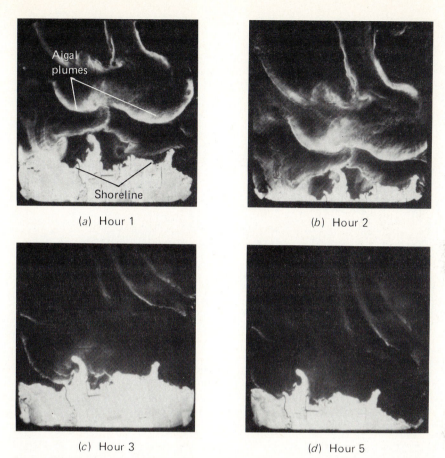

(a) Hour 1

(b) Hour 2

(c) Hour 3

(d) Hour 5

Figure 5-3 Changes in surface patterns over a 5-h period in Clear Lake, California, beginning at 7 A.M. The large turbulent eddy is outlined by the high reflectance of blue-green algae in the near-infrared (Fig. 3-7). Note how the eddy vanished in a few hours probably due to the setting up of a temporary thermocline. The areas shown are 5 km across. There was no inflow to the lake at this time, and there was also no wind. Thus motion was mostly generated by convection currents due to evaporation. (*From Wrigley and Horne, 1975.*)

relative to a fixed point, normally the shoreline (Fig. 5-3). The effects of a typical daily wind pattern as it affects near-surface organisms are given in Fig. 5-4.

LAMINAR AND TURBULENT FLOW

We experience laminar and turbulent flows every day. *Laminar flow* is the smooth slipping of water particles past each other or an obstruction and has little drag on moving objects.

Dolphins, most fish, and submarines possess body designs which maximize laminar flow past them, causing a minimal energy expenditure. In contrast, *turbulent flow* is the random, chaotic tumbling of the water particles around each other or any object passing through the water. These tumbling motions are described as *eddies*. Most lake, river, or estuarine flow motions are turbulent. A good example of the two types of motion is a burning cigarette: the initial smoke plume is laminar while the more distant

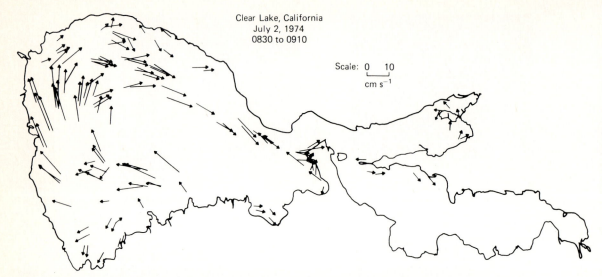

Clear Lake, California
July 2, 1974
0830 to 0910

Scale: 0   10
         cm s⁻¹

Figure 5-4   Vector diagram of wind-driven resuspended sediment movements on July 2, 1974, as recorded by remote sensing. The length of the lines is proportional to the water velocity. Most sustained currents are in the range 2 to 15 cm s⁻¹. (*From Wrigley and Horne, 1975.*)

plumes show turbulent flow. A drop of dye in a beaker of water or the pouring of milk into a cup of coffee produces three-dimensional turbulence.

The onset of turbulence in a lake can be predicted from the *Reynolds number R_e*, which is the ratio of inertial forces ($\sim U^2/d$) to viscous forces ($\sim \nu U/d^2$).

Thus

$$R_e = \frac{Ud}{\nu} \tag{1}$$

where U = water-current velocity
d = depth or thickness of water layer concerned
ν = kinematic viscosity of water, a property of the water molecules (Table 3-2)

A large R_e value implies high water velocities, a thick water layer, low viscosity, or a combination of these three factors. A value can be assigned to R_e to distinguish turbulent from laminar flows. When the Reynolds number is less than a critical value ($R_e \sim 500$), the flow will be laminar, and at $R_e \sim 2000$ the flow will be turbulent (Fig. 5-5). It is now accepted that most lake motion is turbulent and that if laminar flows are present they are in transition to a turbulent state. For a typical lake, values might be $U = 10$ cm s⁻¹, $d = 10$ m (depth of lake or thermocline), $\nu = 0.01$ cm² s⁻¹. Thus from Eq. (1), $R_e = 10^6$, much larger than 2000, and the flow is turbulent. In a stratified fluid, such as a lake, the balance between the buoyancy of a water mass and the mixing power of the wind must also be considered. The ratio of these two forces, the *Richardson number*, was considered in Chap. 4. Detailed analysis of turbulent flows has been described as "one of the last unsolved problems of classical physics." However, some progress can be made by considering the typical scales or dimensions of various lake motions which are usually measurable as waves. For example, the common surface waves, surface gravity waves, have approxi-

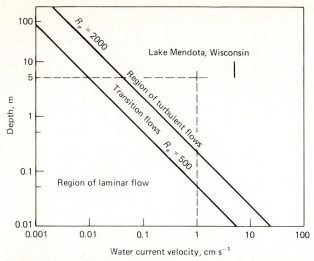

Figure 5-5 Idealized diagram to illustrate the conditions for turbulence in lakes. Lake surface currents are rarely slow enough to allow laminar flow. Lake water motion is almost always turbulent. (*Modified from Smith, 1975.*)

mately the following dimensions:

Length L:
   Horizontal, 10 m
   Vertical, 1 m
Time T, 1 s
Velocity V, 10 m s^{-1}

Another common lake surface phenomenon, Langmuir spirals (Figs. 5-13, 5-14), has very different dimensions: L (horizontal) = m to km, L (vertical) = 10 m, $T = 10^3$ s, $V = 1$ cm s^{-1}. Using these scales, a kind of taxonomic order can be given to these apparently chaotic motions (Table 5-1).

THE KINETIC ENERGY SPECTRUM

Energy of motion is designated kinetic energy (KE) and manifests itself in two main phenomena: periodic wavelike motion and aperiodic currents. Each type of wave and current has captured a characteristic proportion of the applied wind energy. Waves are characterized by dis-

tinctive lengths, heights, periods, and frequencies (Fig. 5-6). *Length* is measured between two adjacent crests or troughs. *Height* is the vertical distance between the crest and trough, and the *amplitude* is half this distance. *Period* is the time required for the passage of two crests or two troughs past a fixed point. The inverse of the period is called the *frequency*. The major types of motion occurring in lakes are shown in Table 5-1 and Fig. 5-1.

After KE is transferred from wind to the water surface, the energy is distributed vertically through the water column. Energy is not imparted to the whole lake but is transferred unevenly depending on the lake's morphometry, latitude, and on the irregular nature of the wind. As mentioned previously, free wind-driven surface waves have wavelengths up to about 10 m and regular periodicity of approximately 1 s. Wavelength and period are mainly determined by the *fetch*, or distance of open water over which the wind blows. *Surface waves* retain little of the wind's energy and merely oscillate water particles in a small

Table 5-1 Size, Frequency, Velocity, and Importance of Waves, Currents, and Other Lake Water Motions*

Type of Motion	Length scale		Time scale	Velocity scale	Importance to kinetic energy spectrum	Importance for plankton (P_t) or nutrient recycling (R)
	Horizontal	Vertical				
Horizontal						
Surface systems:						
Wind-driven surface gravity waves	1–10 m	1 m	1 s	10 m s^{-1}	Small	P_t, R: small
Standing surface gravity waves (surface seiches)	1 km–100 km	10 cm	2–10 h	2 cm s^{-1}	Small	P_t, R: small
Surface wind drift and whole-lake gyres	1 km up	1–25 m	Days	1–30 cm s^{-1}	Large	P_t, R: large
Deepwater systems:						
Short freely propagating internal waves	100 m	2–10 m	2–10 min	2 cm s^{-1}	Major mixing energy at the thermocline	R: summer moderate
Long freely propagating internal waves steered by lake shape (including internal seiches)	To 10 km	2–20 m	1 day	50 cm s^{-1}	Major source of motion in hypolimnion of large lakes	P_t, R: moderate
Vertical (in epilimnion)						
Random flows:						
Vertical diffusion of momentum	1 cm–100 m	1 cm–10 m	1 min	1 cm s^{-1}	A major vertical force	P_t, R: important
Breaking waves	1 m	1 m	Mins	50–500 cm s^{-1}	Moderate to small	P_t: Moderate
Organized flows, Langmuir spirals	50 m–100 m	2–20 m	5 min	0–8 cm s^{-1}	Moderate to small	P_t: important
Hypolimnial	1 km up	Up to 200 m	long	slow	Small	P_t: important in clear lakes R: small

*Modified from Boyce, 1974.

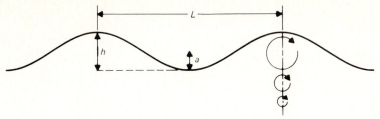

Figure 5-6 Diagram of the motion of water parcels and the various linear definitions of a rhythmic wave. The water parcels oscillate elliptically but show no net lateral motion. The motion of the water parcels is not exactly to scale. The amount of movement decreases exponentially with depth. L = wavelength, h = wave height, a = wave amplitude ($=\frac{1}{2}h$).

ellipse (Fig. 5-6, Table 5-1). In contrast, wind-driven horizontal lake *currents* are basinwide, and the eddies or gyres produced may be 10^5 m in diameter in large lakes. They have an irregular "periodicity" of about 10^5 s or even days. These large-scale currents contain the major part of the applied wind energy and cause much of the vigorous horizontal and vertical mixing in lakes (Table 5-1, Figs. 5-1, 5-7).

Measurement of lake waves shows that motions are restricted to certain time and space scales. We call $KE(\lambda_1, \lambda_2)$ the kinetic energy in lake water motion between two wavelengths λ_1 and λ_2. For small intervals,

$$KE(\lambda_1, \lambda_2) = \bar{E}_{(\lambda)} (\lambda_1 - \lambda_2) \qquad (2)$$

or more generally,

$$KE(\lambda_1, \lambda_2) = \int_{\lambda_1}^{\lambda_2} E_{(\lambda)} \, d\lambda$$

where $\bar{E}_{(\lambda)}$ is the average value of $E_{(\lambda)}$ over the interval (λ_1, λ_2). $E_{(\lambda)}$ is called the KE spectral density. It is also called the *energy* or *power spectrum*. More precisely, it is the KE per unit mass per unit wavelength interval.

The energy spectrum shows that KE cascades from large lake-sized eddies driven by storms down to smaller eddies and is eventually lost as heat. It is often convenient to plot the log of the KE density on the vertical axis and the log of the wavelength on the horizontal. Such a

plot produces peaks of energy connected by smooth curves. The energy spectrum is shown for an idealized large lake in Fig. 5-7, where the sizes of the various types of motion are shown. For a smaller lake the spectrum would be similar in shape but with different absolute values. Familiar lake phenomena such as waves and currents show up as peaks in the density of KE and are well-defined in both time and space. The smooth curve connecting the energy peaks indicates its redistribution into turbulent, chaotic eddies as energy cascades down the spectrum.

An important feature of lake structure is the space-time scale of turbulent eddies. The size and energy of these motions can be seen in Fig. 5-7 and Table 5-1. The lake can be thought of as containing a series of patches of water, each with slightly different physical and chemical constituents (Fig. 5-22a). Whether plankton inside a patch will be broken into new patches depends on the size of the patch, the size and strength of nearby turbulent eddies, and the organisms' swimming ability. As can be seen from Fig. 5-7 the most energetic water turbulence is in long wavelengths, greater than 100 or 1000 m. This motion also has a period of many hours. Thus organisms in a patch 50 m across will only be moved (i.e., *advected*) along with the patch by the most energetic eddies. Eddies of a smaller size and more rapid period will penetrate the patch and increase the separation or

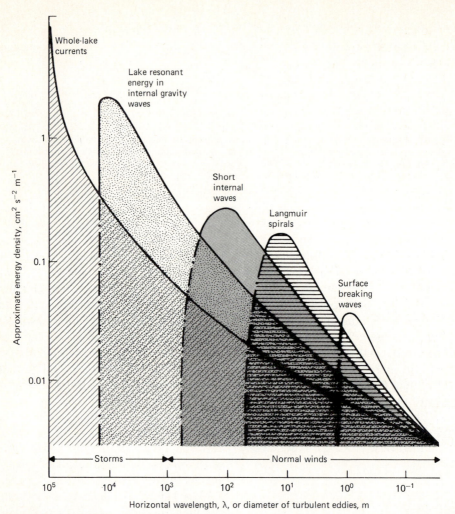

Figure 5-7 Diagrammatic representation of wind-energy distribution in a hypothetical large lake. Surface wind waves ($\lambda = 10$ m) have little energy and do not show on this figure. An energy or power spectrum can be measured in lakes by using current meters. $E_{(\lambda)} =$ KE spectral-density function, where $E_{(\lambda)}\, d\lambda$ is proportional to the KE in the wavelength range $\lambda = -d\lambda/z$ to $+d\lambda/z$.

diffusion of individuals within the patch. Mobile plankton such as *Daphnia* or buoyant blue-green algae can move several meters in an hour. They can thus counteract the smaller eddies and either maintain or disperse from the patch. Other nonmotile plankton like many of the algae depend entirely on small low-energy turbulence to diffuse them out of their nutrient-depleted patch into richer waters.

Limnologists using the concept of energy density, or power spectrum, usually know the scales or wavelengths of motion of common lake phenomena. Reference to the energy spectrum will show approximately how energetic the water motion is at that wavelength. For example, consider the difference between two types of lake water movement, *mean flow* and *turbulent flow*. Mean flow has a specified direction and *advects* substances around the lake, while turbulence is random in direction and causes *diffusion* and spreading of a patch. The relative amounts of energy in the advective and diffusive motions can be found in the appropriate segments of the energy spectrum (Fig. 5-7).

Kinetic energy in lakes is finally dissipated as heat in small eddies of about 10^{-2} m. Lakes do not heat up despite the enormous input of wind energy. Similarly, white-water streams on steep mountain slopes do not heat up as the water loses potential energy in its descent. The reason why water in motion does not heat significantly is the high heat capacity of water (Chap. 3). The calorie, the energy needed to heat one cubic centimeter of water by one degree Celsius, is very large in comparison to the energy generated by environmental scales of motion. For example, water must fall 418 m to gain a KE equivalent of 1 cal. In the heyday of Victorian physical sciences Lord Kelvin, known for the absolute temperature scale and Kelvin waves on the thermocline, met Mr. Joule, who had defined an energy unit bearing his name. Joule was in the Swiss Alps carrying an enormous thermometer in search of a high waterfall to measure the amount of heating produced. Unfortunately, Joule was unsuccessful in measuring a temperature difference, largely due to the ability of water to absorb heat with little temperature increase.

ADVECTIVE AND DIFFUSIVE TRANSPORT OF ENERGY

Mean horizontal movement or advection as, for example, of a dissolved substance in water, can be described by

$$\frac{\partial s}{\partial t} \text{ (advective)} = -\frac{\partial(us)}{\partial x} - \frac{\partial(vs)}{\partial y} - \frac{\partial(ws)}{\partial z} \quad (3)$$

where s = concentration of substance
u, v, w = current velocities in the three dimensions x, y, z (width, breadth, depth of water layer), respectively
Here we visualize motion as a simple directional movement of water particles.

Diffusion is random, chaotic motion. The flux of a substance due to diffusive transport is proportional to the spatial gradient of concentrations between the substance and the surrounding water. The proportionality constant is an *eddy diffusion constant* K. The rate of change of a substance due to diffusive transport of a substance S in water is described as

$$\frac{\partial S}{\partial t} \text{ (diffusive)} = \frac{\partial}{\partial x}\left(K_x \frac{\partial S}{\partial x}\right) \quad (4)$$
$$+ \frac{\partial}{\partial y}\left(K_y \frac{\partial S}{\partial y}\right) + \frac{\partial}{\partial z}\left(K_z \frac{\partial S}{\partial z}\right)$$

where K_x, K_y, and K_z are the eddy diffusion coefficients in the three directions x, y, and z described above. Note the formal similarity between, us, vs, ws, and $K_x \partial S/\partial x$, $K_y \partial S/\partial y$, $K_z \partial S/\partial z$. The values of the various diffusion coefficients are given in Fig. 5-8.

It should be noted that by including the pressure gradient and Coriolis force due to the rotation of the earth, the "substance" can become momentum in Eq. (4). The equations which

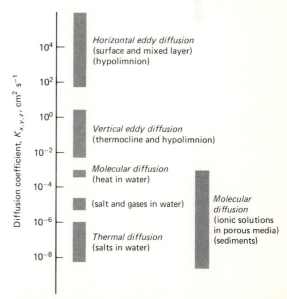

Figure 5-8 Various diffusion coefficients in lakes. (*Modified from Murphy, 1972.*)

result are called *equations of motion*. More-advanced discussions of water movement begin with these equations, and the reader is referred to Pond and Pickard (1978) for a recent example. Other discussions with a biological or chemical orientation are the reviews by Boyce (1974), Mortimer (1974), Csanady (1975), and Smith (1975).

Horizontal eddy diffusion is several orders of magnitude faster than that in the vertical plane (Fig. 5-8) since no buoyancy forces have to be overcome. Although less easily seen in water than in the atmosphere, advection usually involves wavelengths greater than 1000 m, while

diffusion or small-scale turbulence occurs at wavelengths less than 100 m (Figs. 5-7 and 5-9). The average velocity of the turbulent flows in all directions is zero due to their random eddying characters. However, the energy involved in random motion does not vanish and can be described as the sum of the mean square values (or the variance) for each dimension averaged over a given time period. Thus for energy of turbulent flow,

Turbulent KE per unit mass:

$$E' = \frac{1}{2}\left[\overline{(u')^2} + \overline{(v')^2} + \overline{(w')^2}\right] \tag{5}$$

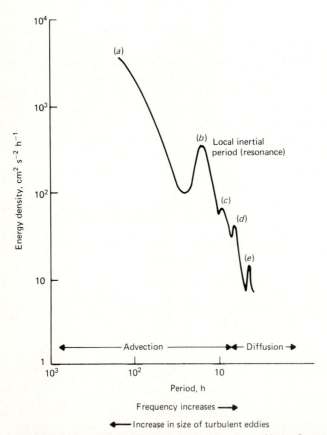

Figure 5-9 Input of the KE of wind into the waters of Lake Ontario. Note advective energy peaks for (*a*) whole-lake motion, (*b*) lake resonance, and (*c*), (*d*), and (*e*) other unidentified peaks, which may be Langmuir spirals or gravity waves. The values were measured 6 km offshore, $z = 8$ m; total variance of the measurements = 107 cm^2 s^{-2}. (*Modified from Boyce, 1974.*)

where u', v', and w' are fluctuating water-velocity components and are averaged over a suitable time scale (denoted by an overbar). For energy of the mean flow,

Mean KE per unit mass:

$$E = \frac{1}{2} (\bar{u}^2 + \bar{v}^2 + \bar{w}^2) \qquad (6)$$

where u, v, and w are as in Eq. (3). Total energy in the water motion is the sum of both E and E', which are only separated for convenience. Generally, for the surface of the lake in windy conditions, E (associated with advection) is greater than E' (associated with diffusion), but at depth or on calm days the opposite may be true.

Kinetic energy in the water may be estimated directly from long-term measurements of horizontal lake currents. This is how Fig. 5-8 was derived for Lake Ontario. Storms commonly produce large horizontal water currents (Fig. 5-21), but such events occur infrequently. If whole-lake horizontal current speeds are high, overall lake energy content is high. This is shown graphically by values at the upper left-hand section of Figs. 5-7 and 5-9. By contrast rapid, common surface waves represent a low-energy component for the lake.

Another peak of energy density is often found at the wavelength of breaking waves. Effects due to surface waves are indicated in Fig. 5-7 and in detail in Fig. 5-16. The breaking of both surface and internal waves imparts turbulence to lake waters. It is an important way in which vigorous mixing can occur between layers of water with different density that form below the surface in stratified lakes.

Waves break when the slope of the leading edge becomes too steep. The increased slope occurs because the velocity of water at the base of the slope is held back by friction more than the upper part. Photographic and schematic representations of turbulent billows produced by such mixing is shown for an internal wave in Fig. 5-10a, b. Such mixing is called Kelvin-Helmholtz mixing or Kelvin-Helmholtz instability and is most important at the lake surface or in the layers of slightly different density within the thermocline (Figs. 5-15, 5-18, and 5-22).

MOTION IN THE EPILIMNION

The main sources of water currents are (1) the wind, which is the major force, (2) pressure gradients, caused by nonequilibrium distribution of water masses, (3) buoyant forces, caused by heating and cooling from evaporation, which can lead to vertical motion (Chap. 4), and (4) inflowing and outflowing rivers. These are modified by the Coriolis effect, especially in large lakes, and by bottom and side friction, which is most important in small lakes. Wind is usually the major energy-supplying agent but may not always be the most significant from the view of lake biology or chemistry, since convection currents are also important (Chap. 4). Friction develops when wind blows over the surface of a lake, resulting in a shear stress at the wind-water interface. The air movement sets up a corresponding motion in the upper water layer which will reduce the stress; i.e., a current develops in the same direction as the wind at a reduced velocity.

Surface Drift

Under most conditions the wind, if allowed to reach a steady state, would impart a fairly constant fraction of its speed (roughly 3 percent) to the surface layers of the water over which it moves. This is largely attributable to the differences in density between the two fluids, air and water. When the stress of the air on the water (τ_{air}) is equal and opposite to the stress of the water on the air (τ_{H_2O}), we can use the fact that the stress on a fluid should be proportional to fluid density $\rho \times$ (velocity u_*)^2 to find

$$\tau_{air} = \rho_{air}\, u^2_{*air} = \rho_{H_2O}\, u^2_{*H_2O} = \tau_{H_2O} \qquad (7)$$

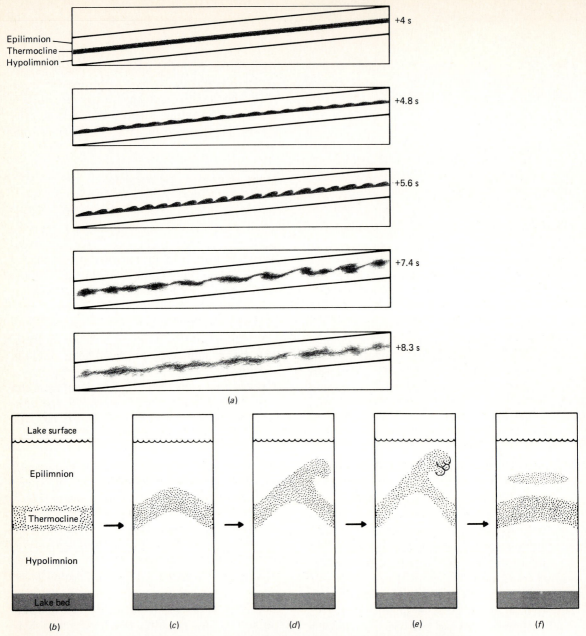

Figure 5-10 Kelvin-Helmholtz mixing (or instability). (*a*) Onset of Kelvin-Helmholtz shear instability in a three-layer experiment with densities 1.000 (epilimnion), 1.092 (thermocline), and 1.172 g cm^{-3} (hypolimnion). The central dyed layer had an initial thickness of 1.3 cm. Initially horizontal the tube was tilted 5.6°. The first diagram, taken 4 s after tilting, shows the smooth three-layered system. Kelvin-Helmholtz instability occurred soon after. The values to the right show the elapsed time after tilting occurred. Billows formed by mixing represent conditions similar to those found in lakes where internal waves occur. (*Redrawn from Thorpe, 1971.*) (*b*)-(*f*) Diagrammatic representation of changes shown in (*a*). (*b*) Condition prior to wave initiation. (*c*) An internal wave forms, (*d*) becomes unstable, (*e*) breaks and incorporates different water layers at the crest, and finally (*f*) produces a layer of water of intermediate density. See also Fig. 5-22.

so that

$$\frac{u_{*H_2O}}{u_{*air}} = \left(\frac{\rho_{air}}{\rho_{H_2O}}\right)^{1/2} \approx \left(\frac{10^{-3}}{1}\right)^{1/2} \approx 0.03$$

over real lakes. That is, water speeds equal about 3 percent of wind speeds at equilibrium. Under most lake conditions this value is only an approximation. Lake water speeds generally range up to 30 cm s^{-1} and are much slower than the water velocity of a typical stream.

The earth's rotation is important for lake currents. In large lakes the water does not move in exactly the same direction as the wind. It moves at an angle to the wind due to the rotation of the earth. The apparent force produced is called the *Coriolis force,* and the surface drift to the right in the northern hemisphere and to the left in the southern hemisphere is referred to as *Ekman drift.* The deflection of the surface current is less-pronounced in shallow lakes but may be as great as 45° to the direction of the wind in deep lakes and in the ocean.

Water blown to one end of a lake must return eventually, and this occurs either at the lake surface or, more usually, by return flows deeper down in the water column. Most studies have been carried out in rectangular-shaped lakes, and in these, return flows are often just above the thermocline under stratified conditions (Fig. 5-21) or near the bottom when the lake is unstratified. The maximum current at the surface decreases with depth to a point of zero net flow. Below this depth there is a fairly uniform return flow which is much slower than the near-surface flow. In addition, due to the Coriolis force, movement down the water column is twisted into a coiled spiral known as an *Ekman spiral* (Fig. 5-11). Large deviations from a perfect spiral may form during high or changeable winds or where basin bathymetry is irregular. The Ekman spiral is a classical but useful simplification found in large deep lakes and the

ocean, particularly when steady, strong winds blow in one direction (Fig. 5-21).

Surface drift can be a nuisance in eutrophic or polluted lakes because concentrations of algae, bottles, cans, or dead fish are rapidly moved by wind. Movements of floating algae are shown in productive Clear Lake, California (Fig. 5-3*a* to *d*), and in eutrophic, tropical Lake George, Uganda (Fig. 5-12).

Another type of whole-lake surface drift occurs in most lakes in the northern hemisphere —a slow mean flow in a counterclockwise direction (Emery and Csanady, 1973). This circular motion with a relatively stagnant center is familiar to oceanographers and is called a *gyre.* The importance of gyres is that lake organisms can be transported without energy expenditure. For example, it may be advantageous for plankton to pass near a nutrient-laden inflow. In rectangular lakes, especially in those with a flat bottom, the wind produces topographic gyres. Usually two gyres form, a cyclonic one to the right of the wind and an anticyclonic one to the left (Csanady, 1975).

Surface Waves

Wind-driven surface gravity waves are the normal waves observed on the lake surface. Although less impressive on lakes than on the ocean, there is little difference between the two except that in lakes they are smaller. Their properties are described in Table 5-1 and Fig. 5-7, and they contain little of the lake's total KE.

Standing surface gravity waves or surface seiches are the regular, resonating waves which pass from one edge to the other and back. Surface seiches were among the first physical events noted by early European limnologists. In other parts of the world their existence did not go unnoticed. In beautiful Lake Wakitipu in New Zealand the slow seiche rhythm gave rise to a Maori legend that a sleeping giant lay breathing on the lake bed. The word *seiche* may

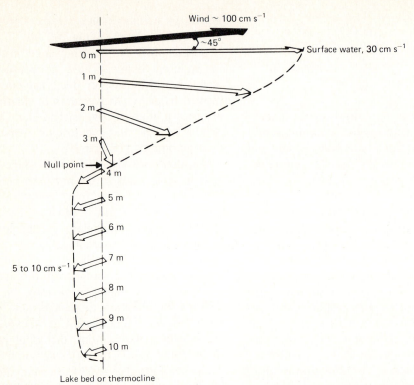

Figure 5-11   An idealized diagram of the Ekman spiral. An actual example is shown in Fig. 5-21. Surface water moves most quickly at the surface at about 45° to wind direction due to the Coriolis force of the earth's rotation. Motion is to the right of the wind in the northern hemisphere and to the left in the southern. In this diagram the water is considered to be layers of equal thickness moving in various directions. Note the reversal of direction in the deeper water. To conserve mass the volume of water moving downwind should approximately equal that returning; i.e., the total area under the two curves should be about equal. At the null point no flow occurs in any direction. (*Modified from Gross, 1977.*)

be derived from the French word *seche,* which means "dry," and was originally used because one shore of the lake becomes exposed and dry while the other becomes flooded.

Seiches are commonly generated when winds blow fairly constantly from one direction, driving the surface water downwind. The wind piles up water on the lee shore, holding it there until the wind drops, at which time the driving force is released and the accumulated water mass flows back under the influence of gravity. This produces a standing wave which rocks back and forth with gradually decreasing mo-

tion. A series of waves is produced which are called standing surface gravity waves, surface seiches, or simply seiches. It is a common property of mechanical systems that inputs of energy at any of a wide range of frequencies will excite the resonant or harmonic frequencies more than others. Thus the sloshing back and forth of the water produces a standing wave at the resonant frequencies of the lake in the same way that a plucked guitar string produces one main note irrespective of the plucking frequency. In lakes, surface seiches have a period given by

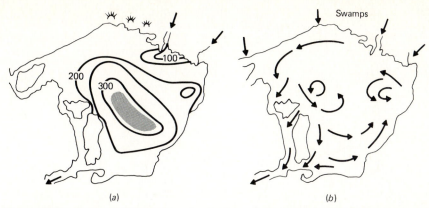

Figure 5-12 Surface-drift patterns in shallow tropical Lake George, Uganda, Africa. (*a*) Chlorophyll *a* at the surface in milligrams per cubic meter. Isopleths at intervals of 50 μg liter⁻¹. (*Redrawn from Burgis, 1971.*) (*b*) Water currents derived from changes in chlorophyll. (*Redrawn from Ganf, 1974.*)

$$T = \frac{1}{n} \frac{2L}{\sqrt{gZ}} \qquad (8)$$

where T = resonance period

Z = lake depth where the lake is a rectangular basin

n = number of nodes of the standing wave

L = basin length

Surface seiches are apparent on lakeshores as regular but small (≤ 10 cm) up-and-down motions of the lake surface. They are of little importance for lake biology or chemistry as their energy is low. They dissipate their energy by frictional shear at the air-water interface and at the lake edges.

Langmuir Spirals

Langmuir spirals are often as obvious as breaking waves. On a windy day it may be possible to see lines of foam called *windrows* oriented in the same direction as the wind and at right angles to the waves (Figs. 5-13, 5-14). These lines mark the boundaries of pairs of *Langmuir spirals,* a series of adjacent vertical clockwise and counterclockwise rotating cells of water.

They produce alternate areas of upwelling and downwelling, and foam accumulates above the downwelling zone. Windrows, also called *slicks,* contain algae and zooplankton as well as oily substances or natural foaming agents from the death and decay of plankton or shoreline vegetation (Fig. 5-14). Langmuir-type circulation is due to an interaction between surface waves and wind-driven drift currents. A detailed explanation of this circulation can be found in Craik and Leibovich (1976). Langmuir spirals produce measured downwelling flows of between 2 and 8 cm s⁻¹. This is considerably faster than the swimming of most zooplankton or algae. Langmuir spirals may rapidly mix plankton, heat, or dissolved gases throughout the epilimnion. Upwelling is much less concentrated than downwelling, and currents may be slower. In many cases the spirals have a diameter approximately equal to the depth of the thermocline or to the total depth of shallow lakes.

An important feature of these rapid vertical movements is their effect on algal photosynthesis and zooplankton. Zooplankters such as *Daphnia* may concentrate in the slicks (Chap. 13), presumably offsetting the disadvantages of being eaten there against the advan-

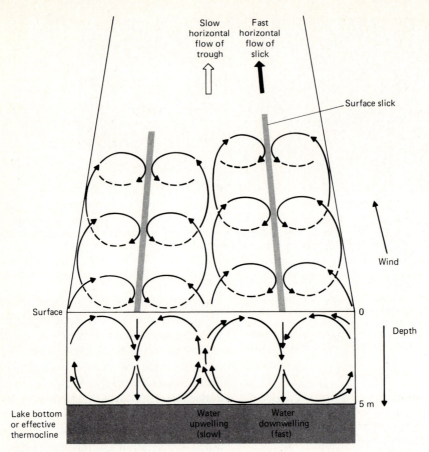

Slow horizontal flow of trough

Fast horizontal flow of slick

Surface slick

Wind

Surface

0

Depth

5 m

Lake bottom or effective thermocline

Water upwelling (slow)

Water downwelling (fast)

Figure 5-13   As viewed downwind, wind-produced Langmuir spirals show surface water mixed down between two adjacent left- and right-handed spirals.

tages of concentrations of food in the form of algae and bacteria. The inhibition of algal photosynthesis by exposure to surface-light intensities is well known (Goldman et al., 1963; Watt, 1966), and rapid mixing in Langmuir spirals may increase light inhibition and reduce primary production. Thus the system of bottles suspended at fixed depth normally used to estimate photosynthesis (Chap. 7) may not reproduce the natural-light regime.

Random

A combination of the diffusion of momentum from the surface with additional energy provided by breaking waves accounts for random mixing of the lake's epilimnion. Wind blowing over lakes causes fluctuating turbulent movement in the vertical as well as horizontal direction. Because surface water is almost always warmer and thus more buoyant than deeper water, more energy is needed to mix vertically than horizontally. Since turbulence is essentially a random three-dimensional motion, its effect is to mix or diffuse adjacent water masses. Such mixing is important in the exchange of nutrients and heat between top and bottom water.

The vertical transfer of the wind's energy is similar to that described for a substance in Eq.

Figure 5-14 Langmuir spirals in Clear Lake, California, as seen from an aircraft 2000 m above the lake. Wind is blowing from the west (left to right). Boundaries of Langmuir spirals are indicated by white streaks parallel to the wind. Wave fronts are obvious as ripples at right angles to the wind. Wind speed was approximately 6 m s^{-1}. The Langmuir spirals are about 2 to 3 m across per cell in this photograph. The picture shows an area approximately 2 km across.

(4) but is further modified by pressure gradients and the Coriolis force. This gives a slightly larger diffusion coefficient for momentum than for diffusion of a substance. A 5- to 20-m-thick layer of epilimnion in a temperate lake is well-mixed vertically each day by the shear stress imparted by moderate winds (Fig. 5-15).

Breaking Waves

Extra energy is available for both vertical and horizontal transport via breaking waves or whitecaps when organized wave motion is degraded into turbulence. The energy concentrated in breaking waves can be measured directly as part of the whole-lake energy-frequency spectrum (Fig. 5-7). Measurements of this spectrum in a lake with and without breaking waves demonstrate the formation and decay of energy at a particular wavelength due to breaking waves (Fig. 5-16).

The thermocline is stably stratified and thus resists mixing from the cascade of random, chaotic turbulence. Most motion at the thermocline appears as organized waves.

MOTION IN THE THERMOCLINE

The fairly gentle turbulence introduced by internal waves at the thermocline (Figs. 5-9, 5-10) helps prevent the stagnation which would otherwise prevail due to the slowness of molecular diffusion (Fig. 5-8). This may be insufficient if detritus or zooplankton cause metalimnetic oxygen depletion. Gentle mixing may be one reason for the abundant growth of algae and bacteria around the thermocline. The danger of being swept into unfavorable zones is minimal, but stirring provides fresh supplies of nutrients from the hypolimnion. Internal waves are usually detected by the periodic rise and fall of the thermocline (Fig. 5-17e). Some types of

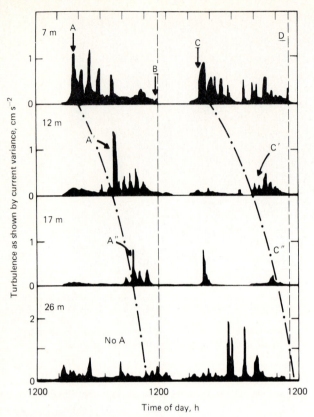

Figure 5-15 Mixing in the epilmnion of Lake Tahoe. The afternoon wind mixes the lake down to the top of the thermocline (20 m). Turbulence starts with the afternoon wind near the surface (*A*, *C*) and drops down to the top of the thermocline (dotted lines *A'*, *A''*, *C'*, *C''*). By midmorning, solar heating restratifies the epilmnion and extinguishes turbulence (*B*, *D*, dashed line) until the onset of the next afternoon wind. (*Redrawn from Dillon and Powell, 1979.*)

waves such as the Kelvin and Poincaré waves discussed later in this chapter are important only in large lakes, while others like short freely propagating internal gravity waves occur in almost all stratified lakes.

Internal Gravity Waves

Internal gravity waves (internal seiches), which can be formed only under stratified conditions, are not apparent from the lake surface. They are usually much larger than surface waves and may be as much as 10 m high. In a rectangular basin the period of an internal seiche is given by

$$T = \frac{2L}{\sqrt{g\left(\frac{\rho_h - \rho_e}{\rho_h}\right)\Big/(z_h^{-1} + z_e^{-1})}} \qquad (9)$$

where ρ_h, ρ_e = densities of hypolimnion and epilimnion, ≈ 1 g cm^{-3}, respectively

z_e, z_h = respective thicknesses of epilimnion and hypolimnion

L = basin length

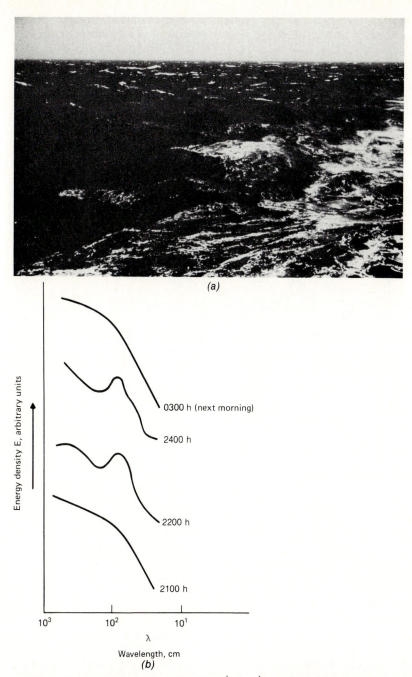

(a)

(b)

Figure 5-16 (*a*) Large whitecapped waves in the South Atlantic Ocean. (*b*) Transfer of energy due to whitecapping (breaking waves) into energy of vertical mixing. This occurred after a windy afternoon was followed by a calm evening on Lake Ontario. Note the rise of a turbulent energy peak with a wavelength of about 100 cm, which is similar to the size of a breaking wave crest. (*Modified from Lemmin et al., 1974.*)

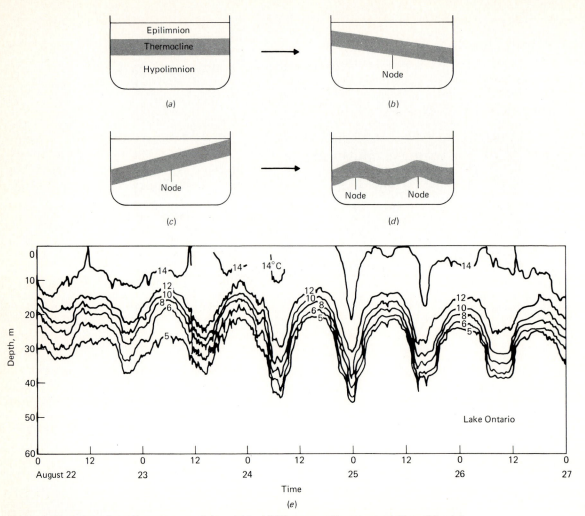

Figure 5-17 Internal gravity waves or seiches. (a) to (d) Diagrammatic representation of the initiation of internal gravity waves. The single central node is usually replaced by several nodes.
(a) Calm three-layered structure. (b) After strong winds the epilimnion becomes displaced
downwind, depressing the thermocline. (c) After the wind stops, the epilimnion tilts back due
to gravity. (d) After a few hours or days of oscillation, internal gravity waves are set up on the
thermocline. These waves may persist for several days. (e) Internal gravity waves on Lake Ontario, Canada (Station 6, off Oshawa, northwest of lake), as recorded by thermocline movement. (*Redrawn from Boyce, 1974.*)

Figure 5-17 *a* to *d* shows the growth of internal
gravity waves on the thermocline. Figure 5-17*e*
shows the dramatic changes in the horizontal
position of the thermocline after the passage of
Hurricane Agnes over Lake Ontario in 1972.

Small lakes are little affected by internal
waves. The height of all waves increases with
fetch, so waves on small lakes have a small
amplitude and wavelength. Kettle lakes, for example (Chap. 18), have extremely stable stratification, and motile or buoyant algae dominate
the phytoplankton (Chap. 12).

Short-Period Internal Gravity Waves

These freely propagating waves within the thermocline may cause the intricate patterning indicated by taking very accurate depth-temperature profiles. An example from Lake Tahoe is shown in Fig. 5-18. These could be due to the breaking of the waves from a Kelvin-Helmholtz instability. These short-period waves have a maximum frequency $N/2\pi$, where N, called the *Brunt-Vaisala frequency* is:

$$N^2 = -\frac{g}{\rho}\frac{d\rho}{dz} \qquad (10)$$

where ρ = water density
z = depth
g = acceleration of gravity

For Lake Tahoe these waves would pass an observer on the thermocline every 100 s during late-summer stratification.

Planktonic organisms in the grip of turbulence from the breaking of these short-period internal waves are vertically mixed toward or away from light over periods of several hours. Such turbulence may assist metalimnion populations of the blue-green algae *Oscillatoria rubescens*, which grew so abundantly in Lake Zurich during periods of enrichment.

Long-Period Internal Waves: Kelvin and Poincaré Waves

These two long propagating internal waves are probably the only types which produce significant currents in the hypolimnia of large lakes.

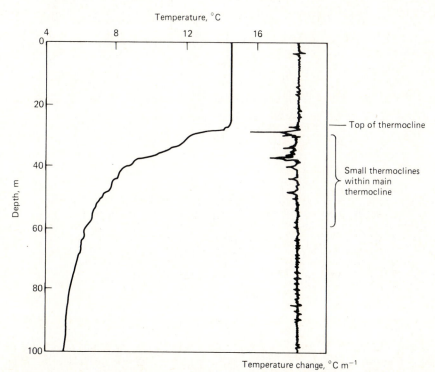

Figure 5-18 Temperature microstructure and temperature-gradient profiles for September 29, 1973, Lake Tahoe, midlake. Note the fine detail of thermal microstructure and several small thermoclines. Some of these may be only temporary and will be modified by the next storm. (*Redrawn from Dillon, Powell, and Myrup, 1975.*)

They are internal gravity waves which differ from previously discussed internal seiches because they are modified by basin morphometry and latitude. They normally reach large size on the thermoclines of stratified lakes. The rotational influence of the earth, the Coriolis force, is vital for their maintenance so they occur only in large lakes. Kelvin waves are low-frequency waves whose frequency depends on the size of the lake basin and its resonant frequency. Kelvin waves are important only in lakes with a minimum fetch in any direction of about 5 km. Poincaré waves have a frequency close to the inertial frequency (or period), which depends on the latitude of the lake. The *inertial period* is the period of a wave only acted on by Coriolis force. It is about 17 h for the North American Great Lakes (45°N) and is much longer for an equatorial lake such as Lake Tanganyika (6°S) (Fig. 5-19). Poincaré waves occur only in the very largest lakes.

Kelvin waves have a pronounced effect in nearshore waters less than 5 km from the lake edge. When an internal seiche forms in a large lake, the Coriolis force induces a counterclockwise rotation (northern hemisphere) on the single plane whose node is at the center of the lake (Fig. 5-20). Kelvin waves may displace the thermocline several meters near the lake edges and will pass around the lake at its inertial period. The motion imparted may be important in causing "leaks" around the edges of the ther-

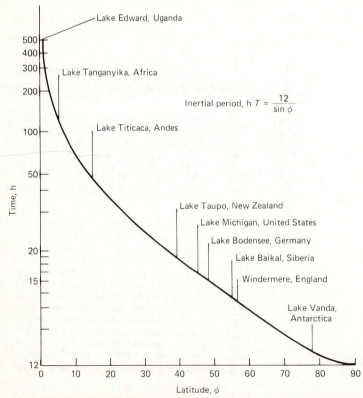

Figure 5-19   Inertial period as a function of latitude. This is the period of a wave acted upon only by the Coriolis force of the earth's rotation. (*Courtesy of C. P. Duncan and G. Schladow.*)

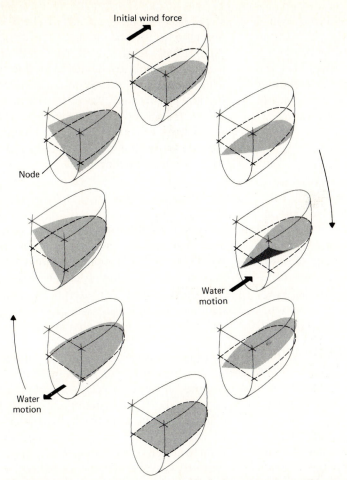

Figure 5-20   One cycle of a Kelvin wave (an internal gravity wave) on the thermocline of half of a large northern hemisphere lake. The wave has a counterclockwise rotation (due to the Coriolis force) about a stationary center-lake node. Note the large wave near the lake edge which characterizes a Kelvin wave. Also note the reversal of net water flow in the hypolimnion during the cycle. This flow reversal also applies to any internal gravity wave in large lakes. (*Modified from Mortimer, 1974.*)

mocline which allow nutrients from the hypolimnion to reach the epilimnion and increase algal growth. Kelvin waves may be important in the shoreline areas where fisheries are concentrated since this region of the lake is the major source of sediment-nutrient recycling during stratification.

Poincaré waves influence the hypolimnion in the open waters away from the shores of large lakes. They are very large waves, and their effects on water currents may extend 50 m or more below the thermocline. Poincaré waves can be visualized by imagining the thermocline to be divided into huge squares of alternating black and white colors. Initially, all the black squares will be depressed toward their centers like a pit while the white ones will be elevated like a hill. After a period near the inertial frequency, the positions will be reversed with black hills and alternating white pits. This will continue producing alternating water mounds and depressions. In addition, horizontal water currents will be caused by Poincaré waves.

MOTION IN THE HYPOLIMNION

Sources of organized motion in this layer are generally absent and most mixing is diffusive and weak. Some motion is due to Poincaré and Kelvin waves. These waves have large vertical displacements (Table 5-1 and Fig. 5-17e) and produce a back-and-forth movement of water over the sediment. This motion sets up turbulence, especially in lakes with irregular beds, and mixes substances dissolved in interstitial water into the open water (Powell and Jassby, 1974). Wave turbulence is negligible at a depth equal to half the wavelength. Thus even a high wave must have a long wavelength to affect deep water. Since such waves are generally stable and do not break, there is no source of vigorous mixing in the hypolimnion.

The amount of nutrient release depends on the composition of the lake bed, which may range from soft muds to hard stony areas. Some lakes have oozes which are protected from mixing by a rigid crust formed by insect activities.

In eutrophic lakes where the bottom sediments become anoxic, large quantities of nutrients diffuse out of the interstitial mud water. This outflow is much less in the oxygenated muds of very oligotrophic lakes but is still significant since important algal growth occurs in the hypolimnion of very clear lakes (Goldman et al., 1973; Fee, 1976). It is important to know how this layer of nutrient-rich water overlying the sediments is circulated throughout the hypolimnion. First, it should be noted that such mixing does not always occur. In small well-stratified lakes such as Castle Lake, California, mixing over the sediment surface is slight, and progressive depletion of near-bottom nitrate occurs, probably by denitrification in the mud. This chemical stratification indicates that mixing is weak at best. Because of the relatively weak currents, bottom hardness, biotic activity, and temperature remain the most important factors in establishing the absolute quantity of nutrients in the hypolimnion.

AN OVERALL VIEW

A good example of overall current patterns for a short period during a storm is that for Loch Ness (Fig. 5-21). This figure shows strong surface flows and slower deeper flows in the reverse direction. An alternative view of this data is estimation of changes in turbulence. Turbulence is essentially random and cannot be measured directly. If current or temperature measurements are known accurately, the proportion of the water which is unstably stratified relative to the surrounding water can be calculated. After mixing, some parts of the epilimnion have a greater or lesser density than the water around them since they originated from warmer or cooler layers. This produces unstable stratified water since heavier water overlies lighter warm water. Eventually, turbulent diffusion will eliminate these density differences. Until that time, the proportion of unstably stratified water, χ, called an *intermittency index*, measures the amount of turbulence in the water. When greatly above zero, χ indicates downmixing of surface water, internal waves, and possible Kelvin-Helmholtz mixing (Fig. 5-10a to f). In Fig. 5-22 the intermittency index χ is plotted for Loch Ness in September. The resultant layered structure of turbulence is similar to that found in Lake Tahoe (Fig. 5-18). Most turbulence ($\chi > 30$ percent) is found near the surface. Layers of lesser turbulence are in-

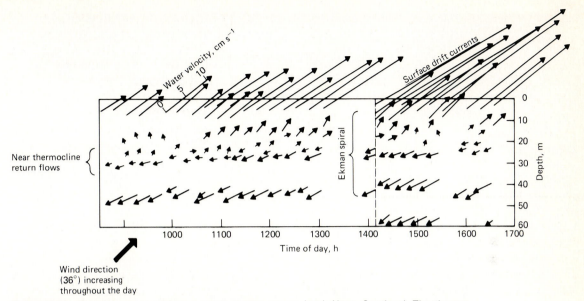

Figure 5-21 Currents measured as a storm moves across Loch Ness, Scotland. The thermo-cline starts at about 20 m. The long axis of the lake is downwind (at 36°). Note the very rapid rise in surface flows as the day progresses and as winds increase. Slower return flow is indicated near the thermocline. Ekman spirals are also shown. (*Modified from Thorpe, 1977.*)

terleaved between 10 and 30 m; below this there is little turbulence. As would be expected, turbulence is low in the hypolimnion.

EFFECT OF RIVERS

Most rivers flowing into large lakes have only a local effect on water movement. There are some smaller lakes, such as Marion Lake in British Columbia, which are essentially riverine. Lakes of this sort have very short retention times of only a few days. Since rivers transport most of the annual supply of nutrients, pollutants, and toxicants, the dispersion of river inflow in lakes is particularly important. Plumes of inflowing muddy water are easily seen (Fig. 5-23). The horizontal spread of a plume can be approximated where the horizontal eddy diffusivity coefficient is K_y [see also Eq. (4)] as

$$K_y = \text{const. } U \frac{\partial w}{\partial x} = \text{const. } w^{4/3} \qquad (11)$$

where w = width of a plume at distance x from the source (river mouth)

U = horizontal water-current speed.

The vertical spread of the plumes is usually negligible compared with horizontal spread (refer to Csanady, 1969, for further details).

Near the river mouth, in the initial phases of plume spreading,

$$w = \text{const. } x^{3/2} \qquad (12)$$

That is, there is an initial rapid and nonlinear horizontal spread of the plume which is due to incorporation of larger and larger eddies up to 50 m in diameter. Soon after this, plume dissipation slows. A few kilometers farther out the

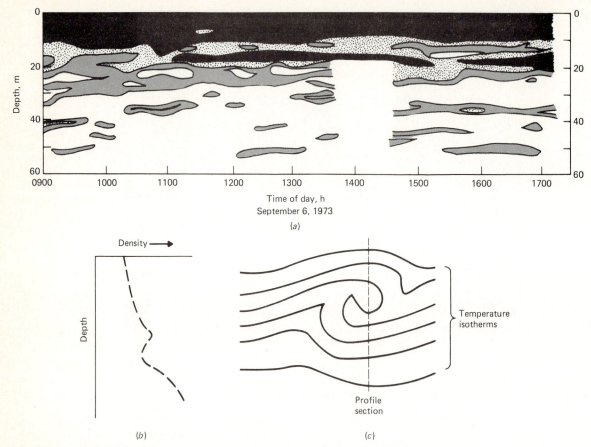

Figure 5-22 Estimated turbulent mixing in temperate Loch Ness, Scotland. (*a*) The amount of turbulence over one day as measured by the function χ which increases as instability and mixing increase. χ represents the frequency of the occurrence of density inversions, i.e., regions of gravitational instability in which the density increases upward. Such inversions are caused by the Kelvin-Helmholtz mixing shown in Fig. 5-10 and in parts (*b*) and (*c*) below (see text for further details). Solid black = high turbulence ($\chi > 30$ percent), dotted = lower turbulence ($\chi = 15$–30 percent), grey = low turbulence ($\chi < 15$ percent). Clear areas of no density inversions have little turbulence. (*b*) the density measured by an instrument as it profiles the billow. The instrument normally measures temperature which is converted to density before plotting (*b*). (*c*) The buildup of a Kelvin-Helmholtz billow. (*Redrawn and modified from Thorpe, 1977.*)

plume meanders and often follows the shoreline, especially in calm weather.

Because of their visual impact, sediment plumes are likely to arouse public concern and are particularly good subjects for monitoring by remote sensing (Fig. 5-23). Plumes provide direct evidence of the dispersion into lakes of nutrients, particularly nitrogen and phosphorus

from the watershed. Chapters 6 to 10 deal with nutrients and their transformations in aquatic ecosystems.

FURTHER READINGS

Boyce, F. M. 1974. "Some Aspects of Great Lakes Physics of Importance to Biological and Chemical

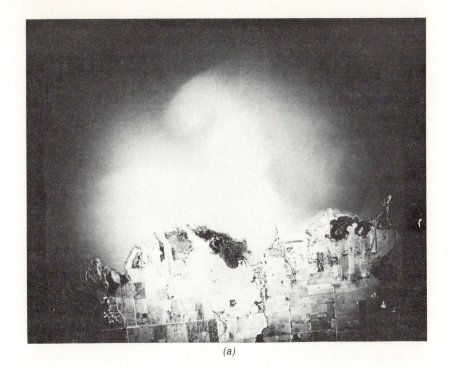

(a)

(b)

Figure 5-23 (*a*) Black-and-white rendering of a color infrared photograph of the muddy plume of the inflowing river at flood stage at Clear Lake, California. Note the large eddy forming at the end of the sediment plume. The view shown is about 5 km across. (*Photograph courtesy of R. C. Wrigley, NASA-Ames.*) (*b*) Normal spectral range photograph of the muddy plume of the Flathead River as it enters Flathead Lake, Montana, during a high flow period. (*Photographed by T. J. Stuart.*)

Processes." *J. Fish. Res. Board Can.,* **31**:689–730.

Burns, N. W., and C. Ross. 1972. "Project Hypo —An Intensive Study of the Lake Erie Central Basin Hypolimnion and Related Surface Water Phenomena." Canadian Centre for Inland Waters, Paper 6. 182 pp.

Csanady, G. T. 1969. "Dispersal of Effluents in the Great Lakes." *Water Res.,* **3**:835–972.

Mortimer, C. H. 1974. "Lake Hydrodynamics." *Mitt. Int. Ver. Theor. Angew. Limnol.,* **20**:124–197.

Smith, I. R. 1975. "Turbulence in Lakes and Rivers." *Freshwater Biol. Assoc., U.K. Publ.* 29. 79 pp.

Thorpe, S. A. 1977. "Turbulence and Mixing in a Scottish Loch." *Philos. Trans. R. Soc. London, Ser. A.,* **286**:125–181.

Chapter 6

Chemicals and Growth Factors

OVERVIEW

The major driving forces controlling the chemical composition of natural waters are the age-old processes of rainfall, erosion and solution, evaporation, and sedimentation. The daily, seasonal, and long-term cycles of the major elements are also influenced by the biological components of the watershed, stream, and lake waters. A detailed discussion of each biologically important element is given in the four chapters which follow. Some specialized chemical effects on physiological processes in various habitats are illustrated in Chaps. 16 to 20. In this chapter we introduce the broad spectrum of chemicals present in natural waters in relationship with each other and to the watershed from whence most of them originate. We also consider human modification of water chemistry throughout the world.

There is no pure water in nature, and since water is the universal solvent, it contains a vast array of inorganic and organic compounds which are present as dissolved solids and gases. Chemicals in lakes, streams, and estuaries may exist as simple molecules but are often complex combinations of organic and inorganic molecules. Also included is any material adsorbed onto the exterior of particles. Water is constantly picking up impurities of all kinds from the air as well as from the land, and some of these provide the growth factors needed to support the aquatic food chain. Organisms already present in the water excrete ammonia, phosphate, carbon dioxide, and a large percentage of everything they have eaten. Even distilled water usually carries enough trace quantities of nutrients to grow bacteria and algae. If one places an open container of sterile, highly purified water outdoors, life will quickly colonize it. First the water would come into equilibrium with such dissolved gases as carbon dioxide, oxygen, and nitrogen. Some nutrients will dissolve from the walls of the container and some enter as aerosols or from atmospheric dust. Spores or resting stages of bacteria, algae, fungi, protozoans, and some small higher animals then colonize the water, and evaporation concentrates the dilute nutrients until growth is possible.

Methods for determining the concentration of important nutrients are discussed in general terms, but the details should be sought in such references as *Standard Methods for the Examination of Water and Wastewater* (American Public Health Association, or APHA, 1981) and "A Practical Handbook of Seawater Analysis" (Strickland and Parsons, 1972), which are applicable to freshwater. Most biologically important chemicals can be detected in the waters of lakes and streams, although sometimes advanced analytical methods may be needed. Chemical methods change, and the current literature should be examined when low concentrations are investigated.

ORGANIC AND INORGANIC COMPOUNDS

Inorganic nutrients provide the chemical constituents upon which the entire food chain is based. They may be converted by microorganisms to organic growth factors such as vitamins. The multitude of inorganic salts and dissolved gases in natural waters can be conveniently divided into three groups: (1) major and minor nutrients (see also Chaps. 7 to 10), (2) those whose effect is largely ionic (Chaps. 10, 17), and (3) toxicants (Chap. 10). The nutrients that are important in lakes are those which are often in short supply and which limit growth of plants and animals. Common nutrients needed in large quantities for cell development include CO_2, O_2, NH_4, NO_3, PO_4, SiO_2, SO_4, and Fe. Important minor nutrients which may occasionally be in short supply include Mn, Co,

Mo, Cu, and Zn. Sodium, potassium, and chloride ions are often abundant in lakes, but their effect is largely ionic and occurs at the surface of delicate membranes through which ions are exchanged.

Some common inorganic substances such as copper or zinc can act either as toxicants or as growth stimulators (Fig. 10-12). Although copper is an essential trace element for algae, high levels of copper sulfate (called bluestone) have been used for many years to poison algae in reservoirs, recreational pools, and lakes. Normally these metals are present at low levels or in a nontoxic form but may become toxic under unusual, acid conditions such as in some volcanic springs or in the leachate from coal and metal mining operations. Some organisms can adapt to toxic metals if given sufficient time for physiological adaptation. We have observed streams below 20-year-old copper and zinc mines which were totally devoid of life, but have found organisms below mines abandoned in Roman times.

Small quantities of organic compounds are common in natural waters, especially in estuaries and eutrophic lakes. The organics can be divided into five classes by their source or function in aquatic ecosystems: (1) refractory compounds, (2) those providing food for microbes, (3) chelating compounds, (4) odiferous substances which are notable because they cause problems in water supplies for human beings, and (5) those used by animals for communication or defense.

Refractory compounds are defined as those which are decomposed very slowly, if at all. Thus they tend to accumulate even in oligotrophic lakes. They include solutions of the genetic material DNA together with various organic nitrogen and phosphorus compounds. The most common refractory compounds are the family of brownish *humic acids* leached out of the soil mantle. It is these compounds which give the characteristic yellow-brown color to most bogs and some dystrophic lakes.

Bacteria, fungi, and some protozoans are dependent for their growth on a supply of readily decomposable organic compounds. Acetate, glucose, and glycolate are consumed by lake bacteria which have been shown to keep the level of such desirable organic foods to a level too low for most algae to utilize.

An interesting set of organic compounds are those which alter the chemical state of the natural waters. These compounds are produced by algae and some animals. One very important group acts as *chelating agents* which change the ionic state of metals that might otherwise be toxic. They may even make some metals available for assimilation which are normally chemically inert. Humic acids and citrate are perhaps the most common natural chelating agents in lakes and streams. The word *chelator* is derived from the word *chela,* or claw, of crustaceans such as crabs. The metal is held by a weak chemical bond between parts of the organic molecule in the same way as a food particle is held between the two pincers of a crab. Chelation complexes play an important role in blood pigments and in the structure of the chlorophyll molecule (Fig. 10-4).

A group of enzymes with a function similar to chelating agents but which acts on nonmetallic nutrients are the *alkaline phosphatases*. These enzymes are excreted by algae when phosphate is scarce and function by splitting off phosphate originally bound to an organic molecule (Chap. 9). Half the total phosphorus that zooplankton excrete may be organic phosphorus, but this fraction would be unavailable for plant growth if it were not for the existence of alkaline phosphatases. A similar enzyme which makes iron available is called a *siderochrome* (Chap. 10).

Plants, like animals, excrete organic compounds which stimulate or inhibit their own growth or that of their competitors. The process is often called *allelochemistry,* and the substances responsible are collectively called *metabolites* when they stimulate growth or *antimetabolites* when they inhibit growth. Organisms which are not able to synthesize such growth factors as vitamins must depend on their production by other organisms. Although important in terrestrial plant ecology and known to occur in dense cultures of algae, allelochemical effects are unlikely to occur in aquatic ecosystems due to the enormous dilution of any excreted material. Although a few apparent examples of allelochemistry such as suppression of green algae by blue-green algae are known in lakes, this problem requires careful study since other factors can also control the seasonal cycles of phytoplankton (Chap. 12).

Some organic compounds produced by freshwater blue-green algae or red tides of marine and brackish-water dinoflagellates are among the most toxic substances known. Paralytic shellfish poisoning of human beings is due to saxitoxins concentrated in shellfish which have fed on a toxic dinoflagellate. Geosmin, an organic compound secreted by some algae and fungi, is one of several organic compounds which impart a musty odor to water. The presence of these is a nuisance in municipal water supply systems which are dependent upon eutrophic reservoirs as their source. Odor removal requires expensive chemical treatment such as oxidation at the treatment plant.

Animals communicate chemically in water as they do on land (Hasler, 1966). Very dilute quantities of organic compounds play a vital role in the return of salmonid fish from the ocean to the freshwater streams where they were originally spawned. Other aquatic animals key their reproductive behavior to dissolved organic excretions. Interference in this chemical communication is one of the more subtle effects of pollution, and dechlorination of sewage effluent after disinfection is now mandated in California to help prevent salmon from being blocked in their upstream migration.

THE INFLUENCE OF THE CLIMATE AND WATERSHED

If we are to understand the chemical composition of the waters of a particular drainage system, we must consider the nature of the vegetation, the weathering of the parent rock, and climatic factors. The dissolution of rock is dependent upon the solvent properties of water, the extent of time that water has contact with the substrata, and to some degree the temperature.

The four factors of parent rock, climate, topography, and vegetation cover are important in determining the chemistry of water draining the lake's watershed. The gradual weathering and decay of the parent rock provides elements directly to the runoff water. A general rule is that the soft waters of mountain lakes and streams in hard, old rocks such as granite, metamorphosed grits, or volcanic "tuffs" are like rainwater in their chemical composition, while sedimentary rocks dissolve more easily to produce lakes and streams with hard water (Chap. 10). Climate strongly influences the rate of weathering and dissolution of minerals. Extreme heat and cold speed the weathering process by cracking rocks and altering soil conditions (Fig. 16-11). Temperate climates with year-round rainfall usually produce well-vegetated soils with little erosion. Lakes and streams in these areas often have ample nitrogen but may be deficient in phosphorus or silica. In arid climates where a few heavy storms provide most of the water and erode the sparsely vegetated soils, nitrogen is often in short supply (Chaps. 8, 9).

Basin topography also controls the stability of the site and regulates both erosion and the exposure of the soil and rock to the water passing over it. The snow and rain falling in a vol-

canic caldera is little altered by the surrounding volcanic rim which intercepts only a small portion of the precipitation (Plate 1a). However, alteration of water chemistry may result from sulfurous fumaroles or hot springs from the old volcano. In contrast, a lake at lower elevations may collect the water and nutrients dissolved from thousands of square kilometers of drainage area. The ratio of lake surface area to drainage basin area is a major factor influencing the lake's trophic state (Table 18-1).

Vegetation protects the soil from erosion and provides dissolved organic material when leaf litter decomposes. Leaves may be particularly high in combined nitrogen if symbiotic nitrogen-fixing bacteria are associated with the roots and soil. Nitrogen-fixing alder trees, for example, increase the nitrogen content of water filtering through the high-nitrogen humus layer that develops beneath them. In Castle Lake, California, one-third of the lake's nitrogen is derived directly from the nitrogen-fixing mountain alders (*Alnus tenuifolia*). The distribution of nitrogen between trees, soil, spring water, and lake is illustrated in Fig. 6-1.

THE ATMOSPHERIC CONTRIBUTION

The atmosphere is a source of chemicals for the aquatic ecosystem. Even rainwater, most of which has been distilled from the surface of the ocean, brings with it measurable quantities of sodium and halogens (chlorine, bromine, iodine, and fluorine), not to mention the host of aerosol pollutants from the world's industrial activities. The farther one gets from the seacoast, the smaller the proportion of sodium and halogens relative to such continental contaminates as sulfate. Iodine is essential for the functioning of the thyroid gland, and deficiencies of this element in food produce a swelling of the neck called *goiter,* once known as *Derbyshire neck* in that English county which has no seacoast. The central United States, far from the sea, has long

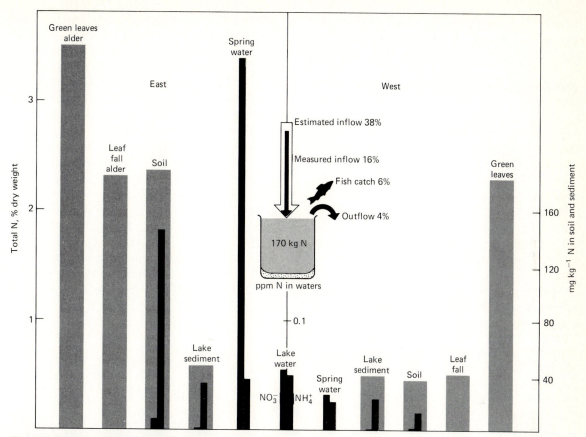

Figure 6-1 The contribution of nitrogen by alder trees (*Alnus tenuifolia*) to Castle Lake, California, is evident from a comparison of the nitrogen distribution on the east and west sides of the lake. Alders are common on the east side. The crosshatched bars represent the total Kjeldahl nitrogen as percent dry weight of the alder and other deciduous plant leaves, the soil, and lake sediment. The solid bars are the micrograms per liter NO_3 and NH_4^+ levels in the soil and sediment in accordance with the scale on the right side of the figure, while the lake-water and spring-water values are indicated by the more exaggerated scale in the center of the figure. The average nitrogen content of the 19 ha lake with inflow, outflow, and fish removal as percentages of this average value is also given. (Data from Goldman, 1961; Figure re-drawn from Frey, 1963).

been referred to as the *goiter belt*, and iodized salt has become the normal table salt in these areas.

The great affinity that water has for picking up ammonia from the atmosphere is appreciated by anyone who has attempted to prepare ammonia-free blanks in the laboratory. There is an interesting study of a pigsty near, but not draining into, a farm pond where the farmer had carefully sloped the drainage so that there would be no direct pollution of the pond from the pigs. Still, the pond was nitrogen-fertilized by aerial transport of ammonia from pig excreta to the pond. Aerosols containing ammonia and some other chemicals are frequently produced by animals and can travel great distances (Chap. 8). Dust contains adsorbed nitrogen and phosphorus as well as trace metals. Preliminary studies show airborne dust to be a major source of trace metals in some areas but apparently insignificant in others.

CHEMICAL POLLUTION, LOGGING, AND EROSION

On a worldwide scale the natural processes dominate the chemistry of natural waters. Only in a few cases such as increases in atmospheric carbon dioxide, lead, and DDT has human pollution been truly global. Most limnologists will be concerned, however, with the specific effects of local human pollution and water diversion which tip the natural chemical balance toward eutrophication, salt lake formation, or a poisoned biota.

The greatest cause of altered water chemistry in natural waters is now pollution from human waste, agriculture, and industry. Many wa-

terways receive industrial and domestic waste in various stages of treatment, and lumbering and agricultural activities supply sediment and dissolved salts. Together this chemical pollution is responsible for the increasing eutrophication of many lakes and reservoirs and the poisoning of some groundwater supplies.

Forest fires, particularly the crown or wildfire variety, provide several inches of ash. Together with the erosion of unprotected soil, ash may represent an important loss of fertility from the land and an increase in nutrient loading of receiving waters. Controlled burns of forests are an emerging management strategy designed to reduce the fuel level of the forest floor. Although they yield some ash and a temporary increase in erosion, they can prevent the catastrophic effects of wildfires (Richter et al., 1982).

Nutrient input to the aquatic ecosystem as-

Figure 6-2 Extensive sediment deposits in a stream valley resulting from a mud flow in the hills of this tropical watershed. The erosion was caused by poor land management. The original stream valley is now buried by 10 m of soil.

sociated with agricultural practices is greatly influenced by sediment production and the use of fertilizers. Good agricultural practices can reduce erosion and sediment production, though erosion is less of a problem in flat irrigated lands. Where swidden (slash-and-burn) agriculture is practiced in the tropics or where steep slopes are plowed in temperate regions, the loss of topsoil threatens future agricultural production while fertilizing receiving waters (Fig. 6-2).

Forest practices associated with timber harvest may yield great quantities of sediment and dissolved nutrients. Forest cutting is frequently done on steep slopes and currently involves extensive use of bulldozers and other heavy equipment as well as the construction of roads

into wild areas. The importance of the forest cover is twofold. First, it protects the soil from physical erosion and, second, it recycles the nutrients from the forest floor. With the vegetation removed, the loss of plant nutrients from the land is greatly accelerated. Likens et al. (1970) have demonstrated how a forest will release large quantities of soil nutrients to the drainage system if recycling is interrupted by simply killing the vegetation. Improved forest practices are essential, particularly in the tropics, if we are to reduce the excessive nutrient input into aquatic ecosystems while at the same time maintaining the fertility and hence productivity of the surrounding land.

The sediment particles which cause turbidity in the streams and lakes they enter also serve as

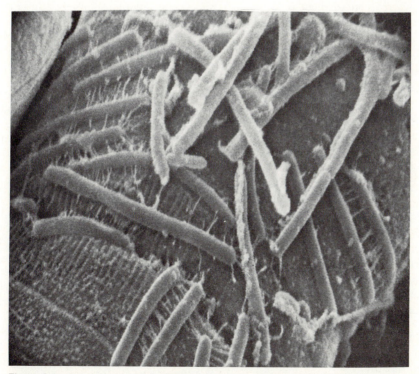

Figure 6-3 Detritus magnified 4000 times. In this case dead stream algae have become covered with long rods of bacteria whose pililike appendages hold them to the algae in the fast-flowing waters. The bacteria also bind many small algae together to form larger clumps of detritus which become food for animal suspension feeders. (*SEM photograph by H. W. Paerl.*)

sites for ion exchange between waters and sediment (e.g., Fig. 10-7). If nutrients, trace elements, or biocides are in high concentration in the water, the sediments may remove them by adsorption. If their concentration is low, they tend to be released from the sediment into the water. Thus a dynamic equilibrium exists between the nutrients adsorbed to sediment particles and the surrounding water. Bacteria, utilizing sediment and detritus for attachment, benefit from concentration of dissolved organic matter at the particle-water interface (Fig. 6-3).

In addition to the problem of direct sediment production, some nutrients in fertilizers leach out of the fields. Nitrates, in particular, move readily through most soils and end up in aquatic ecosystems, while phosphates, which are adsorbed onto soil particles, are less mobile. Heavy rainfall in the tropics makes the effect of inorganic fertilizers short-lived as most of the chemicals added are rapidly washed into the watercourse.

Once point sources of pollution such as sewage outfalls are controlled, the diffuse pollution from poor land use tends to dominate the degradation of lakes and streams. Urbanization has become an ever-increasing diffuse source of nutrients to the aquatic environment. Roads invariably increase runoff and promote erosion along their margins. Fertilizers used for lawns and golf courses contribute further to eutrophication. Studies are badly needed to determine what strategies of land development are least harmful and how best to regulate the use of heavy earth-moving equipment to reduce vegetation loss and soil disturbance. This is particularly important in mountain habitats or at high latitudes where regrowth is extremely slow as a result of the cold winter and dry summer conditions.

FURTHER READINGS

Droop, M. 1957. "Auxotrophy and Organic Compounds in Nutrition of Marine Phytoplankton." *J. Gen. Microbiol.,* **16**:286–293.

Fogg, G. E. 1971. "Extracellular Products of Algae in Freshwater." *Arch. Hydrobiol. Beih.,* **5**:1–25.

Folt, C., and C. R. Goldman. 1981. "Allelopathy between Zooplankton: A Mechanism for Interference Competition." *Science,* **213**:1133–1135.

Frey, D.G. 1963. *Limnology in North America,* The University of Wisconsin Press, Madison. 734 pp.

Gold, K. 1964. "Aspects of Marine Dinoflagellate Nutrition Measured by ^{14}C Assimilation." *J. Protozool.,* **11**:85–89.

Goldman, C. R. 1961. "The Contribution of Alder Trees (*Alnus Tenuifolia*) to the Primary Productivity of Castle Lake, California." *Ecology,* **42**:282–288.

Golterman, H. L., and R. S. Clymo (eds.). 1967. *Chemical Environment in the Aquatic Habitat.* Noord-Hollandsche Uitgevers-Mij., Amsterdam. 322 pp.

Grant, J. W. G., and I. A. E. Bayly. 1981. "Predator Induction of Crests in Morphs of the *Daphnia Carinata* King Complex." *Limnol. Oceanogr.,* **26**:201–218.

Kaushik, N. K., and H. B. N. Hynes. 1971. "The Fate of Dead Leaves That Fall into Streams." *Arch. Hydrobiol.,* **68**:465–515.

Krueger, D. A., and S. I. Dodson. 1981. "Embryological Induction and Predation Ecology in *Daphnia Pulex*." *Limnol. Oceanogr.,* **26**:219–223.

Paerl, H. 1973. "Detritus in Lake Tahoe: Structural Modification by Attached Microflora." *Science,* **180**:496–498.

Provasoli, L. 1963. "Organic Regulation of Phytoplankton Fertility." In M. N. Hill (ed.), *The Sea,* **2**:165–219.

Saunders, G. W. 1971. "Carbon Flow in the Aquatic System." In J. Cairns (ed.), *Structure and Function of Freshwater Microbial Communities.* Research Division Monograph 3, pp. 31–45. Virginia Polytechnic Institute, Blacksburg.

Wetzel, R. G. 1979. "The Role of the Littoral Zone and Detritus in Lake Metabolism." *Arch. Hydrobiol. Beih.,* **13**:145–161.

Chapter 7

Oxygen and Carbon Dioxide

OVERVIEW

Oxygen participates in many important chemical and biological reactions and has become the most widely studied chemical in the aquatic environment. It is continually consumed in respiration by both plants and animals and is produced by plant photosynthesis only when sufficient light and nutrients are available. Very cold water contains less than 5 percent of the oxygen contained in a similar volume of air. The amount rapidly decreases as the water temperature increases. Water contains little oxygen due to the relatively low partial pressure of oxygen in the atmosphere and its quite low solubility. The lack of oxygen in water relative to air means that it is easily depleted by respiration and decomposition unless continually replenished from the air. The short- and long-term variations in dissolved oxygen of lakes and rivers give a good measure of their trophic

state. Oligotrophic waters show little variation from saturation, while eutrophic ones may range from virtual anoxia to 250 percent saturation. Organic matter from natural sources or domestic and industrial sewage may result in serious depletion of dissolved oxygen. When this occurs for a long enough time, most aquatic organisms perish or are replaced by a few specialized organisms tolerant of low oxygen.

Carbon dioxide is a product of respiration by both plants and animals, provides the major carbon source for photosynthesis, and in most ways shows an inverse relationship to oxygen. Although only a minor component of air, carbon dioxide is quite abundant in water because its solubility is about 200 times that of oxygen. Carbon dioxide dissolves in water to produce carbonic acid (H_2CO_3), which dissociates into various fractions (HCO_3^-, CO_3^{2-}) depending upon the hydrogen-ion concentration (pH). At typical pH levels of 6 to 8, bicarbonate is the most abundant of the carbon fractions. The rate-limiting step in the solution of carbon dioxide in water is the hydration reaction which produces the carbonic acid. This reaction may limit photosynthetic rates for a short time during calm days in some productive waters where free carbon dioxide levels are low relative to plant demand. Some primitive plants seem to require carbon dioxide at unusually high concentrations and are restricted to well-aerated sites such as waterfalls. When the demand for carbon dioxide for photosynthesis is high, precipitation of calcium carbonate occurs, especially in hard-water lakes. This results in two common lake phenomena: benches of limestone or marl are deposited around the edges of *marl lakes* and colloidal suspensions of calcium carbonate produce a *lake whitening* which greatly increases water turbidity. Although the single-most important environmental factor regulating the concentration of oxygen and carbon dioxide is temperature, the level also depends upon (1) photosynthesis of plants, (2) respiration of all organisms, (3) aeration of the water, (4) presence of other gases, and (5) any chemical oxidations that may occur. Further, oxygen and carbon dioxide can enter or leave an aquatic ecosystem chemically combined with other elements as well as dissolved in water or as gas.

MEASUREMENT

The distribution of oxygen and carbon dioxide in natural waters provides a convenient measure of organic production and decomposition and forms the basis of most methods of measuring primary productivity (Chap. 15). The first extensive measurements of dissolved oxygen were probably in the estuary of the River Thames, England, in 1882 (Fig. 20-14). S. A. Forel studied oxygen in Lake Geneva, Switzerland, in 1885, and modern studies of oxygen distribution began with Birge and Juday in 1911. Oxygen measurements, together with temperature, have become routine limnological

measurements. The Winkler method for measuring oxygen concentration is a simple oxidation-reduction reaction routinely performed by aquatic biologists. Physiological studies of plants and animals now use a modification of the classic Warburg system of manometric measures of oxygen consumption. In recent years oxygen-sensitive electrodes, which should be regularly calibrated with the Winkler method, have made continuous measurement easier and have helped improve our knowledge of the distribution of oxygen in streams, lakes, and estuaries. The main problem with most oxygen probes is that the delicate membrane over the electrode must be replaced frequently. Probes now exist which use longer-lasting gel membranes.

Despite a decade of effort devoted to mathematical modeling of many nutrient and physical variables in lakes, a simple model using oxygen is usually the most successful for pollution control (Chaps. 14 and 20). Carbon dioxide is less frequently measured but is needed to estimate organic production and decomposition. Total inorganic carbon ($CO_2 + HCO_3^- + CO_3^{2-}$) is easily measured by titration or modern instruments. A common technique involves the sealing of a small water sample in a glass ampule which is then either acidified and flushed with inert carrier gas or heated. Acidification converts all the inorganic carbon to carbon dioxide, which can be measured by its adsorption spectrum in an infrared gas analyzer. From knowledge of the original pH and temperature of the lake, the free carbon dioxide present can be determined from the total inorganic carbon present through the use of standard tables of dissociation constants.

SOURCES

Diffusion from the Atmosphere

Both oxygen and carbon dioxide are important constituents of air. Air contains about 21 percent oxygen by volume or 300 mg oxygen per liter of air. Carbon dioxide in the atmosphere is about 0.033 percent by volume. Water at 0°C at sea level contains only 14.6 mg liter^{-1} O_2 and 1.10 mg liter^{-1} CO_2. At 1 atm pressure both oxygen and carbon dioxide diffuse slowly into static liquids. Unless there is turbulent surface contact with the air, as occurs in waves on lakes or in streams, the process remains slow. The gas laws of Henry and Dalton quantify the solution of gases. Henry's law, dealing with the mixture of gases, states that the concentration of a solution of a gas which has reached equilibrium is proportional to the partial pressure at which the gas is supplied. Nitrogen (79 percent of air) has a partial pressure of 0.79 atm, oxygen (21 percent), a partial pressure of 0.21 atm, and carbon dioxide (0.033 percent), a partial pressure of 0.00033 atm. Since air is a mixture of gases including oxygen, carbon dioxide, and nitrogen, gases dissolved in water tend to be in equilibrium with the atmosphere. Their different solubilities, however, determine how much of a given gas water can hold. A combination of partial pressure and solubility determine exactly how much of each gas is dissolved. Oxygen is about 2 times as soluble as nitrogen, but its partial pressure is one fourth that of nitrogen. Thus the ratio of both gases dissolved in water is roughly 1:2. Carbon dioxide is about 30 times more soluble than oxygen and has a partial pressure about 1/700 that of oxygen. Unlike nitrogen and oxygen, carbon dioxide is more abundant in water than in the air. Changes in barometric pressure will slightly alter the concentrations of dissolved gases since all are more soluble at higher pressures. Lakes at high elevations contain slightly less dissolved gas per unit volume than lakes of equal temperatures at sea level (Fig. 7-5).

Dalton's law also applies to a mixture of gases in solution and states that the pressure of each component of gas is proportional to its concentration in the mixture, the total pressure being equal to the sum of the components. This is important in aquatic systems because bubbles of a particular gas rising through the fluid medium will exchange gas with the water and alter its concentration. At the surface a similar exchange will occur, tending to establish equilibrium with the atmosphere. Bubbling a gas, such as nitrogen, through a solution will lower the concentration of all other gases present.

Wind plays an important role in the distribution of dissolved gases by providing the energy required to move them through the water column. Rapid unpolluted streams are naturally aerated and are usually saturated with oxygen. As a result streams provide an effective system of self-purification which unfortunately has long been overexploited by streamside civilizations. Waves breaking on shore are also exposing a large surface area for diffusion and, like the water in the rapids of a stream, are saturated with dissolved gases.

Artificial reoxygenation by bubbling air or pure oxygen into polluted waters is a technique with increasing application for lake and river restoration. This is further discussed in Chap. 20. The method is in particular demand for maintaining a cold-water salmonid fishery in otherwise anoxic hypolimnia. Large reservoir operations, such as those of the Tennessee Valley Authority, have been forced to consider an expensive method involving the injection of pure oxygen into the hypolimnetic outflows from their reservoirs. Without augmentation, the water does not gain sufficient oxygen in the river below the dam to meet municipal water standards.

Photosynthesis

The photosynthetic activities of various organisms are a major source of oxygen in the aquatic environment. Since light provides the energy for photosynthesis, it can only occur in the lighted zone of a water body called the *euphotic* or *photic zone* (Fig. 2-2). The depth where oxygen production by algae is in equilibrium with algal respiration is termed the *compensation depth.* The dark waters below the compensation depth are called the *aphotic zone,* where respiration exceeds oxygen production.

Obviously the penetration of light influences the distribution of both oxygen and carbon dioxide in the aquatic system. When photosynthesis is occurring at a high rate, oxygen production may exceed the diffusion of oxygen out of the system and bubbles from supersaturation may result. If filamentous green algae are common, these bubbles may serve to float them to the surface, forming scumlike mats which blow up on shore to the annoyance of lakeside dwellers. The very presence of this high production has the opposite effect at night when respiration by the same organisms may lower oxygen levels appreciably.

Both pH and dissolved oxygen increase when vigorous photosynthesis occurs in productive waters. Carbon dioxide levels decrease but are buffered to some extent by the reserve of HCO_3^- and CO_3^{2-} present. Photosynthesis per unit area in eutrophic waters does not follow the changes in light intensity directly but frequently shows a peak in the morning and relative depression in the early afternoon (Fig. 7-1*a*). Because photosynthesis is directly dependent on light, some chemical factor must be influencing the afternoon depression. Almost all nutrients show some diel variability but none as much as CO_2 and O_2. Over the course of any one day, either of these gases may limit primary productivity. In contrast, overall seasonal production is limited by other nutrients such as nitrogen, phosphorus, or silica or by the turbidity or turbulence of the water. Oxygen at the high levels reached in some eutrophic lakes may actually inhibit algal photosynthesis. Naturally productive unpolluted lakes like Lake George, Uganda, may reach 250 percent O_2 saturation, while lakes like Clear Lake, California, may reach 150 percent saturation. Current opinion is that a CO_2 deficit rather than an O_2 toxicity is the most important factor.

Despite the reservoir of carbonate in water,

experimental additions of carbon dioxide or carbonate to phytoplankton from eutrophic lakes usually result in an increase in photosynthesis if two conditions are met. First, the increase in photosynthesis must be measured in a sealed bottle which restricts atmospheric input, and second, additions must be made when natural CO_2 levels are low relative to the algal demand. The role of CO_2 as a limiting nutrient in algal production is disputed. Much of the controversy was engendered by the rapid cultural eutrophication of some lakes caused by pollution. Some limnologists blamed phosphorus from detergents; others believed nitrogen from sewage and agriculture was more important. Some representatives of the detergent industry took the position that N or P pollution was irrelevant to the eutrophication of lakes and that CO_2 limitation was the main factor controlling phytoplankton growth. These arguments all have some elements of truth. Most limnologists now agree that there will be a number of exceptions to any postulation that states "only one nutrient controls eutrophication in lakes." Discharges of sewage and many industrial wastes contain both nitrogen and phosphorus along with carbonaceous matter which can reduce dissolved oxygen and thus release more nutrients from the sediments. In some cases (see Chap. 20), removal of phosphorus or nitrogen alone can relieve the symptoms of cultural eutrophication. It is to be hoped that such removals, which leave toxic metals and organic matter untreated, will not become a substitute for total waste management.

pH

The acidity or alkalinity of lakes is measured in units called pH, an exponential scale of 1 to 14. If one remembers the French origin of the unit's name, *puissance d'Hydrogène* (strength of the hydrogen), pH makes more sense, since the hydrogen ion H^+ controls the acidity. pH is defined as the negative log of the hydrogen-ion

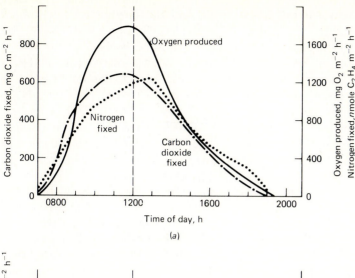

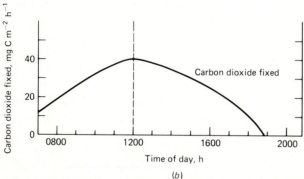

Figure 7-1 (a) Diurnal production of oxygen and fixation of carbon dioxide and nitrogen in very eutrophic Lake George, Uganda. Note that none of the three curves is symmetrical around noon and that there is a relative depression in the afternoon. Oxygen production is more symmetrical because the CO_2 to O_2 ratio was brought to equilibrium with the atmosphere before incubation was started. This is not necessary with $^{14}CO_2$ and nitrogen fixation, which may represent more realistic lake conditions. (*From Ganf and Horne, 1975.*) (b) Diurnal fixation of carbon dioxide in unproductive Lake Tahoe, California. Note the more symmetrical curve around noon indicating that CO_2 or O_2 depletion does not occur. Also note that the values for carbon fixed (up to 40 mg C m^{-2} h^{-1}) are much lower than for Lake George, Uganda, despite the very deep photic zone of Lake Tahoe (100 m versus around 1 m for Lake George). (*Redrawn from Tilzer and Horne, 1979.*)

concentration. Acidity is denoted from 0 to 7 and alkalinity from 7 to 14. Most lakes have a pH of 6 to 9; more acid waters show values down to 2, and some eutrophic or soda lakes have pH values of 10 to 11.5.

When the pH of a lake falls below 4 or 5, the species diversity is severely restricted. Low-pH waters due to the acid rain of industrial air pollution are blamed for the disappearance of some plants and animals in a few areas of the eastern United States and northern Europe. The drainage of these areas is mostly granitic or grits and is particularly susceptible to acid pollution since the rocks are deficient in alkaline materials. They also lack the buffering capacity to minimize the pH change resulting from the addition of acid rain. In contrast, many small central African lakes and other desert lakes like the

Great Salt Lake in Utah have less inflow than outflow and become very alkaline. These also support a limited variety of plant and animal life able to tolerate the high-pH and salt concentration.

The hydrogen-ion concentration also controls the chemical state of many lake nutrients, including carbon dioxide. Changes in pH influence other important plant nutrients such as phosphate, ammonia, iron, and trace metals (Chaps. 8 to 10). Carbon dioxide gas dissolves in water to form soluble carbon dioxide. This reacts with water to produce undissociated carbonic acid (H_2CO_3), which dissociates and equilibrates as bicarbonate (HCO_3^-) and carbonate (CO_3^{2-}) according to the equation

$$CO_2 \rightleftharpoons CO_2 \overset{H_2O}{\rightleftharpoons} H_2CO_3 \rightleftharpoons HCO_3^- + H^+$$

Gas   Dissolved   Undissociated   Bicarbonate
          gas   carbonic acid

$$\rightleftharpoons CO_3^{2-} + 2H^+$$

Carbonate   (1)

The relative abundance of each ionic or molecular state at various pH levels is shown in Figs. 7-2 and 7-3. The precise levels of each component phase will vary with both the temperature and the ionic strength of the lake water. The free CO_2 necessary to maintain HCO_3^- in solution is called *equilibrium CO_2*.

Most plants can only utilize CO_2 for photosynthesis. Since CO_2 can be supplied by diffusion from the air or from HCO_3^- or CO_3^{2-} [Eq. (1)], there is a potentially inexhaustable supply of carbon for plant growth. Some plants may also use HCO_3^- directly after converting it to CO_2 by using the enzyme *carbonic anhydrase*. However, for all practical purposes, the amount available will depend on the rate at which the various reactions in Eq. (1) reach equilibrium. We already showed that pH is not a major factor in carbon availability unless it exceeds pH 8 (Fig. 7-2).

In approximate diagrammatic form the relative rates of reaction and relative quantities of the various carbon components available in typical lake water are shown in Fig. 7-3. The rate-limiting step is obviously the hydration-dehydration of carbon dioxide to carbonic acid. CO_2 is used during photosynthesis from a relatively small dissolved CO_2 reservoir. Away

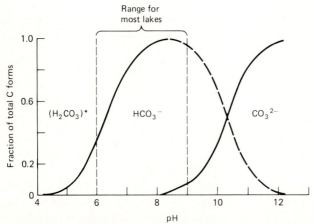

Figure 7-2 Distribution of the various forms of inorganic carbon in a lake with changes in pH. $(H_2CO_3)^* = H_2CO_3 + CO_2$. Note that at the usual pH in a lake, bicarbonate (HCO_3^-) is most abundant. CO_2 is most abundant at low pH, while CO_3^{2-} dominates the high-pH conditions.

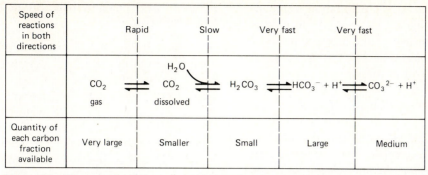

Speed of reactions in both directions	Rapid	Slow	Very fast	Very fast
Quantity of each carbon fraction available	Very large	Smaller	Small	Large	Medium

Figure 7-3 Diagram of rates between and amounts of carbon fractions in a typical lake at about pH 7 and 15°C. Note the small amount of dissolved CO_2 available for photosynthesis relative to the large HCO_3^- and atmospheric CO_2 pools.

from the lake surface where atmospheric CO_2 is available, HCO_3^- and CO_3^{2-} represent the main reserve of CO_2.

Since the H_2CO_3, HCO_3^-, CO_3^{2-} reactions are very fast, the equation can be simplified to

$$\underset{\text{Gas}}{CO_2} \underset{\text{rapid}}{\overset{\text{rapid}}{\rightleftharpoons}} \underset{\text{Dissolved}}{CO_2 + H_2O} \underset{\text{slow}}{\overset{\text{slow}}{\rightleftharpoons}} \begin{pmatrix} H_2CO_3 \\ HCO_3^- \\ CO_3^{2-} \end{pmatrix} \quad (2)$$

or away from the lake surface,

$$\underset{\text{Dissolved}}{CO_2 + H_2O} \underset{\text{slow}}{\overset{\text{slow}}{\rightleftharpoons}} \begin{pmatrix} H_2CO_3 \\ HCO_3^- \\ CO_3^{2-} \end{pmatrix} \quad (3)$$

In the simplified form the rate-limiting step for photosynthetic carbon uptake is the dehydration of carbonic acid. In most lakes mixing by wind provides additional CO_2. During short periods of intensive photosynthesis even this source may be insufficient to meet the maximum demands of the plants for CO_2 (Talling, 1976).

Carbon, often as CO_2 gas, can be used to stimulate photosynthesis in algal cultures or sewage oxidation ponds. The question is at which level does the natural influx of CO_2 from the atmosphere balance the demand for CO_2 at

peak photosynthesis. Since gas diffusion also depends on the degree of water mixing, the weather is important. The general conclusion from several studies in the carbon dioxide–deficient Canadian lakes, the English Lake District, and the deep oceans is that carbon dioxide availability may restrict daily production in some waters. The result is that the crop may be slightly delayed in reaching its seasonal maximum, which is then controlled by other factors such as the supply of nitrogen, phosphorus, or silica. Some aspects of algal physiology, such as the rising and sinking of blue-green algal colonies (see Chap. 12) or depressed photosynthesis in eutrophic lakes (Fig. 7-1a), are related to carbon dioxide metabolism.

A further complication in some lakes is the precipitation of solid calcium carbonate during periods of high photosynthesis. Such lakes are called *marl lakes,* and layers of whitish precipitate are formed around the lake edges below the surface. As deposits build up they form a characteristic marl bench in the euphotic zone. Where higher aquatic plants are present, marl may precipitate on them and slowly encapsulate the leaves and stems in calcium carbonate. Obviously, the process can only occur in hardwater areas where there is a good supply of calcium.

In these lakes the precipitation removes much of the available CO_2 reservoir in a form which is not returned to the lake waters. The reaction of carbonic acid (H_2CO_3) in rainwater as it flows through the soil is

$$H_2CO_3 + CaCO_3 \rightarrow Ca^{2+} + 2HCO_3^- \quad (4)$$

Thus inflowing streams often contain both calcium ions and bicarbonate. These calcium ions enter the lake from the inflowing rivers and react with the most abundant inorganic carbon ion, bicarbonate, as follows:

$$Ca^{2+} + 2HCO_3^- \rightleftharpoons Ca(HCO_3)_2 \rightarrow$$
$$\text{Soluble} \qquad \text{Soluble}$$
$$CaCO_3 \downarrow + H_2O + CO_2 \rightarrow \quad (5)$$
$$\text{Insoluble}$$

The reaction to $CaCO_3$ is essentially irreversible in lakes since the weak carbonic acid (H_2CO_3) which could redissolve the $CaCO_3$ is soon lost after CO_2 uptake by plants. The photosynthetic uptake of equilibrium CO_2 continually pulls the reaction toward $CaCO_3$ until photosynthesis is stopped by darkness, lack of CO_2, or other limiting nutrients.

The $CaCO_3$ precipitates may form solid masses like limestone or remain as colloidal suspensions. Either of these forms can coprecipitate other chemicals or become coated with a layer of organic compounds. This coating can be of great importance to the allochthonous carbon budget of marl lakes (Wetzel and Otsuki, 1974). The fine suspension of $CaCO_3$ particles can rapidly reduce the light penetration in some lakes. This occurs even in large lakes such as Pyramid Lake, Nevada, or Lake Michigan where the Secchi depth can change from 9 to 3 m. The phenomenon is called *lake whitening* and may be widespread, although the chemistry causing the precipitation is not fully understood (Kelts and Hsü, 1978).

EFFECTS OF TEMPERATURE, SALINITY, AND ORGANISMS

The concentration of oxygen in solution is inversely proportional to temperature. If one wishes to produce an oxygen-free solution of water, boiling will quickly achieve this result. At any given pressure, cold waters contain a much higher oxygen concentration when saturated than warm waters. Of course, photosynthesis and respiration may alter this relationship. In unpolluted lakes and streams, the greatest loss of oxygen from the system occurs during the summer warming period. Temperature increase during spring and summer alone can account for a loss of 50 percent of the dissolved-oxygen content of the coldest water. Figure 7-4 illustrates this important relationship between temperature and dissolved-oxygen concentration. As mentioned previously, there is less dissolved oxygen in high-altitude lakes (Fig. 7-5).

The inflow of oxygen-rich waters from tributaries may be important for small lakes in spring if respiration under winter ice cover leaves the deep waters oxygen-depleted. After thermal stratification, cold oxygen-rich meltwater of high density can flow directly into the hypolimnion or into some intermediate level. This inflow displaces the warmer surface waters containing less dissolved oxygen that flow out of the lake in proportion to the inflow. This lowers the overall heat content of the lake while at the same time increases its oxygen content.

Increasing salinity has only a minor effect upon dissolved-oxygen concentration, namely, that salts dissolved in water reduce the intermolecular space available for oxygen. Unless salinities are very high, such as in some Texas salt ponds (Simpson and Gunther, 1956), the Great Salt Lake, Utah, and Mono Lake, California, salinity influences are very small indeed.

As already noted in the discussion of pho-

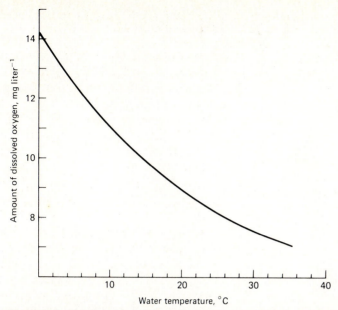

Figure 7-4 Relationship between dissolved oxygen and water temperature in a well-mixed system of pure water at sea level and at 760-mm Hg pressure. There is less dissolved oxygen at higher altitudes or in very saline water. The percent saturation is often used to illustrate oxygen deficits (Figs. 20-13, 20-14).

tosynthesis, respiration consumes oxygen and produces carbon dioxide. In highly eutrophic situations where algae grow rapidly during the daytime, respiration and the decomposition of their products occur just as rapidly and dominate during the dark hours. On calm nights this imbalance may deplete the oxygen concentration to levels harmful or even fatal to many aquatic organisms. A considerable amount of organic matter, whether it be detritus suspended in the water column or organic accumulation on the bottom, undergoes decomposition by bacteria with a concomitant loss of oxygen and gain in carbon dioxide by the system.

It is most convenient to consider photosynthesis and respiration together, as in Eq. (6):

$$CO_2 + H_2O \rightleftharpoons (CH_2O)_n + O_2 \qquad (6)$$

where n is usually 3, 6, or 12 (e.g., pyruvate, glucose, or sucrose). Relative concentrations of carbon dioxide and oxygen are dependent upon whether the process is going from left to right or from right to left. In productive waters photosynthesis and respiration are dominant in establishing the environmental level of oxygen. However, in more sterile environments these processes only slightly influence the concentration of oxygen and carbon dioxide. Respiration is more temperature-dependent than photosynthesis, and lower temperatures slow the process. This is of some benefit to plants during low-light winter conditions.

Wherever bubbles of gas are released by decomposition in the bottom of a lake, stream, or estuary, they will tend to reach equilibrium with the surrounding water as they rise through the fluid medium and will effectively purge

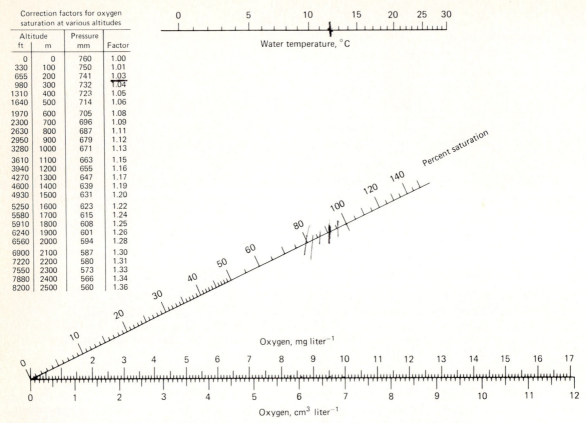

Correction factors for oxygen
saturation at various altitudes

| Altitude | | Pressure | |
ft	m	mm	Factor
0	0	760	1.00
330	100	750	1.01
655	200	741	1.03
980	300	732	1.04
1310	400	723	1.05
1640	500	714	1.06
1970	600	705	1.08
2300	700	696	1.09
2630	800	687	1.11
2950	900	679	1.12
3280	1000	671	1.13
3610	1100	663	1.15
3940	1200	655	1.16
4270	1300	647	1.17
4600	1400	639	1.19
4930	1500	631	1.20
5250	1600	623	1.22
5580	1700	615	1.24
5910	1800	608	1.25
6240	1900	601	1.26
6560	2000	594	1.28
6900	2100	587	1.30
7220	2200	580	1.31
7550	2300	573	1.33
7880	2400	566	1.34
8200	2500	560	1.36

Water temperature, °C

Percent saturation

Oxygen, mg liter^{-1}

Oxygen, cm^3 liter^{-1}

Figure 7-5   Nomograph for calculating the percent dissolved oxygen at various temperatures, pressures, and altitudes. The values derived from this nomograph may be as much as 2 percent in error due to the assumptions of water-vapor pressure used in the original calculations. (*Drawn from data of Truesdale, Downing, and Lowden, 1955.*) A new, more precise nomograph has recently been prepared (Mortimer, 1981).

other gases from the system. As mentioned earlier, this procedure may be used with nitrogen gas to remove oxygen from solution.

THE REDOX POTENTIAL

Oxygen combines readily with other elements to form oxides. If sufficient soluble iron is present, ferric oxide will form an orange *ocher* layer on the bottom of lakes and streams. A similar iron stain occurs in a washbasin where a faucet has leaked slowly for a long time. The change in oxidation state of many metal ions and some nutrient compounds is defined by the *redox*, or *oxidation-reduction*, *potential*. The

redox potential E_h is the electrical voltage which exists between two electrodes, one made of hydrogen and the other made of the material under study. At neutral pH and 25°C, most oxygenated lake water has a redox potential of about +500mV (millivolts). Under these conditions most common metals and nutrients are thermodynamically stable in their most oxidized forms. Iron will be present as Fe^{3+}, not Fe^{2+}, for example. As oxygen falls to zero, a series of substances undergo chemical reductions, each at a specific redox potential. For example, between an E_h of +450 and +300 mV, ammonia becomes favored over nitrate, and between +300 and +200 mV, Fe^{2+} is favored over Fe^{3+}. A *reduced microzone* is formed at the sediment-water interface where most of these reactions occur (Chaps. 9, 10). The reverse will occur if oxygen is reintroduced and E_h rises again to the +500-mV level of fully oxygenated water. As mentioned in Chap. 11, most of these oxidation-reduction reactions are carried out by bacteria which gain energy from converting the substance to the thermodynamically favored state as oxygen and redox potentials change. Although in situ estimation of redox potentials in natural waters is possible (Mortimer, 1942), for most purposes measurement of dissolved oxygen is sufficient since oxygen controls redox. Low oxygen concentrations near the mud-water interface lower the nut redox potential and release dissolved nutrients dvai such as PO_4^{3-} and reduced ferrous iron, Fe^{2+}. Under high oxygen conditions ferric iron, Fe^{3+}, is produced which binds with PO_4^{3-} as an insol-

uble oxide. Changes in the redox potential are important only in eutrophic conditions or in special environments such as highly reducing acid bogs.

DIEL AND SEASONAL CHANGES IN OXYGEN AND CARBON DIOXIDE

Measurements of the distribution of oxygen and carbon dioxide within stratified lake waters provide the limnologist with a great deal of information about the nature of the lake, including its trophic status. The concentration of oxygen in an aquatic environment is a function of biological processes such as photosynthesis or respiration and physical processes such as water movement or temperature. Typical summer distribution of CO_2, O_2, and temperature with depth is shown for a eutrophic lake in Fig. 7-6. Thus *diel* (that is, 24 h) as well as seasonal

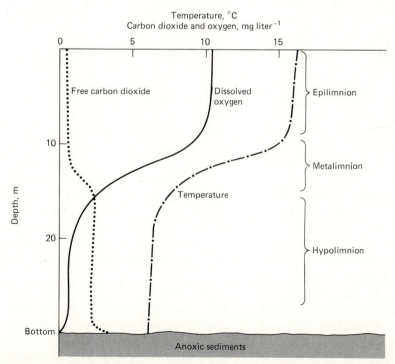

Figure 7-6 Diagrammatic representation of dissolved oxygen, carbon dioxide, and temperature in midsummer in a typical productive lake. None of the dissolved gases are in perfect equilibrium with the respective amounts contained in air.

variations in oxygen distribution can be expected to occur under natural conditions. Diel means day plus night. *Nocturnal* is nighttime, in contrast to *diurnal* which spans only the lighted period of the day.

Diel Variations

It is not uncommon for a productive lake to pass from below-oxygen saturation to supersaturation in the course of a 24-h period. Carbon dioxide will typically follow an inverse relationship to oxygen if the system is photosynthetically controlled. Because of biological modification of the oxygen and carbon dioxide regime, the more productive the environment, the greater will be the fluctuation in concentration of these two gases during the day. Figure 7-7*a* illustrates the net oxygen gain during the daylight hours, with a low during the dark hours in a highly productive lake. Even some unproductive lakes show small changes (Fig. 7-7*b*). The most dramatic situation is to be found in a sewage oxidation lagoon with a high concentration of nutrients supporting such high algal growth that oxygen levels may reach supersaturation during the day and drop to zero during the night. The magnitude of variations in oxygen concentration within all lakes is influenced by the distribution of plants. Quite large variations in oxygen concentration may occur near shore in some lakes where higher aquatic plants are more abundant than in deeper or more open water.

The sediments themselves often contain much decaying organic material and a coating of photosynthetic organisms if there is sufficient light. The respiratory activity of benthic invertebrates in the sediments creates a continuous exchange of carbon dioxide and oxygen with the overlying water. At night the combination of respiration by plants, bacterial decomposition, and invertebrate respiration can remove significant amounts of oxygen from the overlying water, with an exchange of carbon dioxide and dissolved nutrients in the process.

This benthic contribution to the oxygen deficit is particularly important in shallow, thermally stratified lakes which have a small hypolimnetic volume. For example, in Lake Erie the epilimnion is about 17 to 20 m thick, but in some places the underlying hypolimnion is only 2 to 3 m thick (Fig. 4-7).

Seasonal Variations

Summer Stratification As previously noted in Chap. 2 thermal distribution allows a distinct chemical structure in lakes. At the beginning of summer in temperate lakes, there is usually a high oxygen concentration in the hypolimnion. However, as summer progresses, oxygen in the hypolimnion of eutrophic and mesotrophic lakes decreases and carbon dioxide may increase due to decomposition of organic matter (Figs. 7-6, 7-9). In deep lakes like Lake Superior, Lake Baikal, or the North Basin of Lake Windermere, this change is negligible (Fig. 7-10). The oxygen decline in the hypolimnion is due to the productivity of the epilimnion since the more organisms of all types in the epilimnion, the more dead and decomposing material that will sink through the hypolimnion to the lake bottom. The hypolimnetic oxygen depletion can therefore be used to estimate the productivity of the lake. Figure 7-6 illustrates the distribution of oxygen, carbon dioxide, and temperature during summer stagnation.

Fluctuations in oxygen concentration have a marked effect on the biota. "Summer kills" are a result of the depletion of oxygen, particularly in warm unstratified waters where oxygen-depleted waters accumulate overnight during windless periods. Large numbers of dead fish may litter beaches when this occurs.

Fall, Winter, and Spring Overturn As autumn approaches, the waters cool and the resistance to mixing decreases; eventually, a storm will mix the waters from top to bottom (Chap. 4), and the oxygen and carbon dioxide levels will reach equilibrium with the atmosphere

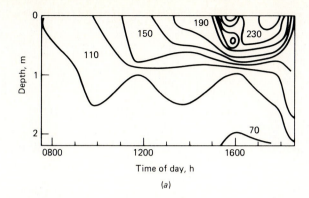

(a)

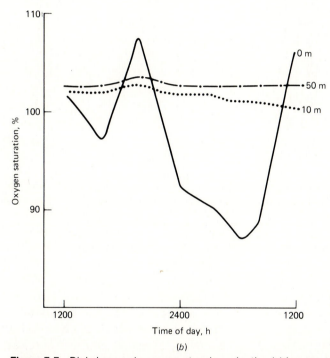

(b)

Figure 7-7 Diel changes in oxygen at various depths. (a) In a very productive lake, Lake George, Uganda; isopleths of oxygen saturation are at 20 percent intervals. Numbers are percent oxygen saturation. (*From Ganf and Horne, 1975.*) (b) In Lake Baikal, an unproductive system. (*Redrawn from Kozhov, 1963.*) Both are shown at periods of maximum algal photosynthesis. Note the very small changes in the oligotrophic lake (Secchi depth 10–25 m) and large ones for the eutrophic lake (Secchi depth less than 0.5 m).

(Figs. 7-8, 7-9). If the lake reaches 4°C, additional cooling floats the colder water to the surface, and at 0°C the lake surface freezes. When clear ice persists, oxygen is often produced by algae immediately under the ice cover through-

out the winter. If snow covers the ice, or during winter darkness at polar latitudes, the lake waters are without light for much of the winter. Then photosynthesis ceases or is limited to a small twilight zone immediately beneath the ice,

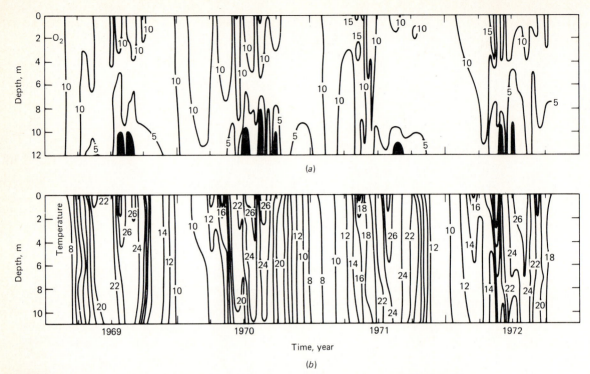

Figure 7-8   Seasonal changes in (*a*) oxygen and (*b*) temperature in a productive polymictic lake, the Lower Arm of Clear Lake, California. Note the large changes in dissolved oxygen relative to those in unproductive Lake Baikal (Fig. 7-10). Note also the intermittent anoxic benthic water masses (black areas) which indicate a low redox potential and large releases of otherwise bound nutrients from the sediments. The high temperatures during summer, when combined with moderate to high concentrations of dissolved oxygen, produce supersaturations of oxygen as great as 50 percent (see also Fig. 7-7*a*). Oxygen isopleths are every 5 mg liter^{-1} and temperature isotherms every 2°C. (*From Horne, 1975.*)

and the respiration occurring even at low temperatures slowly depletes the oxygen level, particularly near the bottom of deep lakes. In shallow ice-covered lakes an abundance of aquatic plants may result in a winter kill as these plants die and consume oxygen during their decomposition.

With the radiation increase in early spring, photosynthetic oxygen production under the ice may produce the highest dissolved-oxygen values of the year (Fig. 7-10). When the ice melts, a spring storm will usually again mix the lake from top to bottom. As discussed in Chap. 4, lakes protected from winds, as in some volcanic

cones or in cirque basins like Castle Lake (Fig. 18-5), a calm spring period with high solar radiation may establish summer stratification without allowing spring overturn. This carries the winter oxygen deficit into the summer stagnation period and prevents nutrients released from the sediment from being mixed into the entire water body.

Types of Oxygen-Depth Curves

Four general types of oxygen distribution in lakes are recognized. In waters of low productivity, oxygen distribution will largely be a function of temperature, resulting in fairly uniform

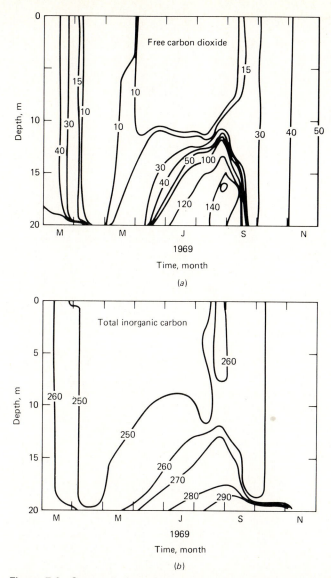

Figure 7-9 Seasonal changes in (*a*) free carbon dioxide and (*b*) total inorganic carbon in productive Lake Estrom, Denmark. Free carbon dioxide levels are reduced in surface water in summer by photosynthetic uptake. Carbon dioxide is abundant near the sediments due to the respiration of benthic organisms. Total inorganic carbon, which can buffer photosynthetic uptake of free carbon dioxide, shows much less variation. Units are in mM liter^{-1}. (*Redrawn from Jonasson et al., 1974.*)

orthograde distributions (Fig. 7-11*a*). This condition is characterized by no appreciable decrease in oxygen with depth. If productivity is high and thermal stratification occurs, oxygen depletion is likely to occur in the hypolimnion in summer and during winter stagnation under ice. This oxygen distribution is called *clinograde* and is characterized by relatively

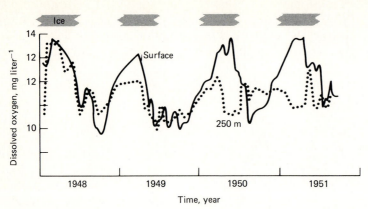

Figure 7-10 Seasonal changes in dissolved oxygen in oligotrophic Lake Baikal (solid line = surface, dotted line = 250 m, crosshatching indicates ice cover). Note that there is little difference between the two depths except under ice cover in spring during the main phytoplankton bloom. (*Redrawn from Kozhov, 1963.*)

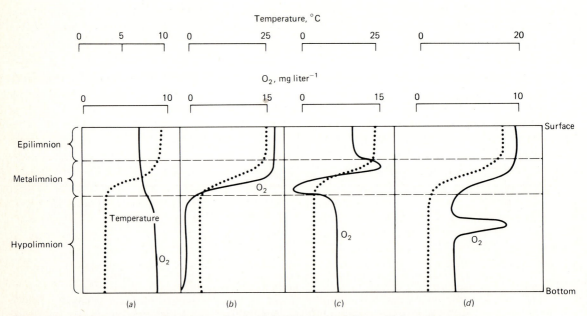

Figure 7-11   Types of oxygen distribution with depth in diagrammatic form. (*a*) An orthograde curve typical of an unproductive lake. (*b*) A clinograde curve from a productive lake. (*c*) Positive and negative heterograde curves. Here photosynthesis from a layer of algae just above the thermocline raises oxygen in the upper part of the metalimnion. Respiration occuring during decomposition lowers oxygen levels just below the thermocline (metalimnion minimum). (*d*) Anomalous curves due to the inflow of dense, cool, oxygen-rich stream inflows which form a discrete layer. In this example, the oxygen-rich stream inflow is in midhypolimnion.

higher oxygen content near the surface where photosynthesis replenishes the supply (Fig. 7-11*b*). A *heterograde* oxygen curve (Fig. 7-11*c*) exhibits an irregular slope from the lake surface to the depths. Concentration of animals may produce a *negative heterograde* distribution, if respiration (oxygen consumption) dominates at some middepth, or a *positive heterograde* distribution, if photosynthetic organisms (oxygen producers) are concentrated in the same fashion. Other anomalous oxygen distributions may result from the settling of cooled high-oxygen surface waters or by layering at intermediate depths of inflow waters (see Fig. 7-11*d*) that have a different oxygen concentration than the lake in question.

The actual *oxygen deficit* in a lake is equal to the amount of oxygen necessary to reach saturation minus the amount of oxygen actually present. For example, for a given temperature, say, 11.5°C, the amount of oxygen to reach saturation (10.6 ppm) minus the amount of oxygen actually present (9.0 ppm) is equal to the actual deficit (1.6 ppm).

Arctic and tropical lakes present different problems in terms of their oxygen distribution. The high temperatures of tropical lakes may preclude mixing if they are protected from wind and are not at high elevation. These lakes may have permanent hypolimnetic oxygen deficits together with high concentrations of methane or hydrogen sulfide. Some tropical lakes like Lake Tanganyika in Africa may never mix or mix on very rare occasions during unusual tropical storms. They are characterized by permanent clinograde oxygen curves frequently reaching zero below the density discontinuity. In the Arctic and Antarctic, lakes may pass from a frozen condition to a continuously mixed condition without ever warming sufficiently to stratify. Some lakes are permanently ice-covered as in the case of the lakes Vanda and Bonney in the Antarctic. If more productive, these lakes would be expected to show severe oxygen depletion near the bottom. However, because of the clear ice, low productivity, and very low respiration, they do not show severe oxygen depletion despite the fact that they are permanently frozen over.

OXYGEN AND CARBON DIOXIDE IN STREAMS

When unpolluted streams contain a large population of algae and benthic invertebrates, they frequently show considerable diel variation in oxygen in the warmer months (Fig. 7-12). This diel variation may be used to estimate primary production. Most rivers and streams in inhabited areas receive organic pollution, and oxygen levels may fluctuate more drastically. Modern sewage treatment plants greatly reduce the biological oxygen demand of their effluent, but where no treatment or only primary treatment is provided a permanent oxygen depletion, called an *oxygen sag*, occurs below the outfalls. This downstream region has a characteristic flora and fauna. Diffuse sources of organic materials can dramatically reduce oxygen levels in streams. For example, in the Chicago area, a great deal of material in the form of dead leaves, dog excreta, and other debris washes into the main waste diversion canal with each summer storm. The oxygen demand of this diffuse matter is so great that it alone reduces the stream oxygen to zero even without the addition of any sewage effluent. Unpolluted streams also become clogged with dead leaves in fall, and during low flow this produces a major oxygen depletion.

Large rivers have much less ability to reoxygenate their waters since rapids are few and the water is deeper than in streams. Many rivers have had all rapids, waterfalls, and shallows removed to facilitate navigation. This has the unfortunate effect of removing almost all natural rapid reaeration. In the River Thames in England, artificial devices have now been installed to replace natural aeration. Su-

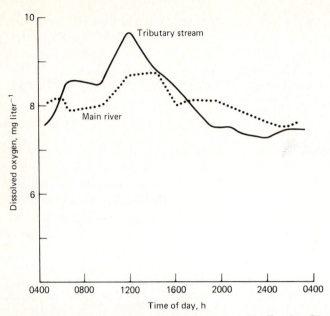

Figure 7-12   Diel changes in dissolved oxygen in the Truckee River, California, and a small tributary, Martis Creek, in August 1979. Note that both show changes but the river shows the smaller variation due to its larger volume and depth. The changes are not related to temperature but are due to respiration by the extensive benthic invertebrate population which exceeds river reaeration at night. During the day reaeration and diatom photosynthesis increase oxygen levels. (*Data from F. R. McLaren.*)

persaturation of oxygen and other gases like nitrogen to values of 105 to 140 percent saturation occurs beneath the spillways of high dams like those on the Columbia River in the United States. This supersaturation causes the grotesque *gas bubble* disease which kills salmon during migration. When mixing is vigorous, CO_2 is usually at saturation levels in streams. Some plants, particularly aquatic mosses and the common stream blue-green alga *Nostoc,* require such a high level of CO_2 that they can only grow well in very turbulent water. Consequently, such plants are usually found only in rapids or small waterfalls.

In the limnological cycles of oxygen and carbon dioxide the gaseous form plays a major role. Next, in Chap. 8, we will discuss nitrogen, whose gaseous phase is less important, despite the abundance of nitrogen gas in the atmos-

phere and in the water. The chemical inertness of N_2 relative to the O_2 and CO_2 molecules causes this difference. As we will see in Chap. 8, the $N \equiv N$ bond can only be cleaved by a few specialized organisms and only by using large amounts of metabolic energy.

FURTHER READINGS

Birge, E. A., and C. Juday. 1911. "The Inland Lakes of Wisconsin: The Dissolved Gases and Their Biological Significance." *Bull. Wis. Geol. Nat. Hist. Surv.,* **22.** 259 pp.

Ohle, W. 1952. "Die hypolimnische Kohlendioxyd-Akkumulation als produktionsbiologischer Indikator." *Arch. Hydrobiol.,* **46:**153–285.

Pamatmat, M. M., and K. Banse. 1969. "Oxygen Measurements by the Seabed. II. *In Situ* Measurements to a Depth of 180 m." *Limnol. Oceanogr.,* **14:**250–259.

Shapiro, J. 1960. "The Cause of a Metalimnetic Minimum of Dissolved Oxygen." *Limnol. Oceanogr.,* **5:**216–227.

Sugawara, K. 1939. "Chemical Studies in Lake Metabolism." *Bull. Chem. Soc. Japan,* **14:**375–451.

Talling, J. F. 1976. "The Depletion of Carbon Dioxide from Lake Water by Phytoplankton." *J. Ecol.,* **64:**79–121.

Walker, W. W. 1979. "Use of Hypolimnetic Oxygen Depletion Rates as a Trophic State Indicator for Lakes." *Water Resour. Res.,* **15:**1463–1470.

Wetzel, R. G. 1960. "Marl Encrustation on Hydrophytes in Several Michigan Lakes." *Oikos,* **11:**223–236.

Chapter 8

Nitrogen

OVERVIEW

Nitrogen is always present in aquatic ecosystems and most abundantly as a gas. Relatively small quantities exist in the combined forms of ammonia (NH_4^+), nitrate (NO_3^-), nitrite (NO_2^-), urea ($CO[NH_2]_2$), and dissolved organic compounds. Of these, nitrate is usually most important. Living cells contain about 5 percent total nitrogen by dry weight. The availability of various nitrogen compounds influences the variety, abundance, and nutritional value of aquatic animals and plants. Nitrogen is often present in quantities which can limit plant growth. This condition is common in warm climates and where phosphorus and silicon are present in relatively large quantities due to natural erosion or pollution. The concentration of most nitrogen compounds in lakes and streams tends to follow regular seasonal patterns. Biological uptake lowers concentrations in spring and summer in the photic zone. During the fall and winter, releases from sediments, tributary inflows, precipitation, and replenishment from the hypolimnion increase the nitrate and sometimes the ammonia concentrations.

Nitrate is normally the most common form of combined inorganic nitrogen in lakes and streams. The concentration and rate of supply of nitrate is intimately connected with the land-use practices of the surrounding watershed. Nitrate ions move easily through soils and are quite rapidly lost from the land even in natural drainage systems. This contrasts with phosphate or ammonium ions which are retained by soil-particle charges. Natural changes in the vegetation of the drainage basin caused by fires, floods, or artificial clearing usually result in increased nitrate in streams. Even moderate environmental disturbances, such as sensible farming or logging without severe erosion, release much more nitrate than ammonia or phosphate. More severe land disturbance, which accelerates erosion, releases nitrate in solution and large quantities of phosphate which are bound to sediment particles. When present in sufficient quantities, ammonia is the preferred form for plant growth since the utilitization of nitrate requires additional energy as well as the presence of the enzyme *nitrate reductase*. Ammonia may become toxic to animals and plants, especially at elevated pH levels when ammonium hydroxide forms. Ammonia as a metabolic waste product of animals provides a source of recycled nitrogen to plants. Its speed of turnover has intrigued limnologists and is an explanation for the growth of phytoplankton and stream algae when nitrate is exhausted. The success of plants in competitive natural situations may depend on their abilities to take up nitrate, ammonia, or both at low concentrations.

Nitrogen gas, although abundant in water, is almost inert and is normally utilized for growth in lakes and streams by some blue-green algae and bacteria through the process of *nitrogen fixation*. In some lakes this is a very important source of plant nitrogen. *Denitrification,* the bacterial reduction of nitrate to nitrogen gas, occurs at low oxygen levels in the sediments and hypolimnia of some lakes. Denitrification may be important in the nitrogen budget of lakes and wetlands. All waters contain both dissolved and particulate organic nitrogen which are generally not available to higher organisms until modified by bacteria and fungi. The organic products range from readily available urea and proteins to complex humic acids which are biologically nearly inert. However, some of these large inert molecules do play a significant role in the chelation of dissolved metals. In lakes particulate organic nitrogen falls to the sediments, which are the major nitrogen sink.

INTRODUCTION

After carbon, hydrogen, and oxygen, the most abundant element in living cells is nitrogen, which is essential for most biochemical reactions. The quantity of nitrogen accumulated by each animal or plant varies from 1 to 10 percent of dry weight and to some extent reflects the availability of nitrogen in the adjacent environment (Gerloff and Skoog, 1954; Fitzgerald, 1969). However, most of the earth's nitrogen is present in the chemically inert gaseous nitrogen form. Nitrogen is not used directly by most organisms since it requires considerable energy to split the $N \equiv N$ triple bond.

It is a biochemical paradox that most organisms which use O_2 rather than SO_4 as an oxygen source do not use the omnipresent N_2 rather than NO_3 as a nitrogen source. Growth which is limited by nitrogen sometimes is supplemented by nitrogen fixation. This gives only minimal evolutionary advantage because limitations due to phosphorus, iron, and other nutrients then develop. It is interesting that nitrogen fixation most frequently occurs in situations where there is at least a small amount of combined nitrogen, which appears to stimulate the initial growth of the nitrogen-fixing plants.

MEASUREMENT

The concentrations of nitrate, nitrite, ammonia, and organic nitrogen are estimated by using traditional chemical analyses (Strickland and Parsons, 1972; Solórzano, 1969; Jenkins, 1975). At moderate to high levels inorganic nitrogen compounds are measured with specific ion electrodes. Particulate organic nitrogen can be measured with an automatic analyzer which allows simultaneous analysis of carbon and hydrogen. Many plants can take up nitrate or ammonia at very low levels (Table 8-3) which are at the limits of detection of most analytical methods. Concentrations of nitrate, nitrite, or ammonia should always be expressed as elemental nitrogen. Thus nitrate is expressed as NO_3-N in micrograms or milligrams per liter or in microgram atoms (that is, 14 g of nitrogen per mole of NO_3), and never as NO_3 (that is, 62 g per mole NO_3). In the older literature and in some engineering reports, the reader may not be certain how the nitrogen is reported. Dissolved nitrogen gas can be measured by complicated extraction methods (Benson and Parker, 1961), using mass spectrometry or gas chromatography for final assay. A less accurate method uses volumetric analysis (Sugawara, 1939). The major problem in lakes and rivers is collecting and keeping water at its original temperature and pressure until gases are analyzed. *Nitrogen fixation* can be measured in lakes by using the heavy nonradioactive isotope of nitrogen, $^{15}N_2$, which is incorporated into blue-green algae or bacteria (Neess et al., 1962; Fogg and Horne, 1967). This procedure requires several chemical transformations and the use of a mass spectrometer for detection of the ^{15}N atoms among the more numerous ^{14}N atoms. Most limnologists now use the acetylene-reduction method for measuring nitrogen fixation. The technique depends on the ability of the nitrogen-fixing enzyme to reduce acetylene which has a molecular shape ($H-C\equiv C-H$) similar to that of nitrogen ($N\equiv N$). Acetylene is reduced to ethylene and measured on a gas chromatograph (Stewart, Fitzgerald, and Burris, 1967; Horne and Goldman, 1972). Three molecules of ethylene are produced for every nitrogen molecule which would have been fixed.

Denitrification can be measured in a few aquatic ecosystems by using the isotopes ^{13}N (Gersberg et al., 1976) or ^{15}N (Keeney et al., 1971). Bacterial denitrification occurs in anoxic sediments where nitrate is present. When $^{15}NO_3$ or $^{13}NO_3$ is added to lake muds, the gas evolved is a measure of the amount of denitrification. Unfortunately, there are problems with the sediment disturbance, the short half-

life of ^{13}N, the large amounts of ^{15}N used, and the difficulty of balancing the gains and losses of nitrogen. An easier method using gas chromatography has been developed which shows some promise for aquatic systems. Quantitative estimates of denitrification in lakes are made by acetylene blockage techniques (Sørensen, 1978). The most obvious method, the collection of N_2 gas as it naturally bubbles out of the lake muds, is usually unreliable. Any gas leaving the mud comes into equilibrium with the other gases dissolved in the water. Thus bubbles of methane (CH_4), a common gas produced by anoxic lake muds, acquire N_2 gas from the water as the bubbles rise. The origin of gases collected in containers on the lake bed is uncertain, but in some shallow lakes denitrification is so vigorous that direct gas collection can be used. With all environmental measurements, care should be taken to obtain representative samples to adequately describe horizontal, vertical, daily, seasonal, and year-to-year variations.

THE NITROGEN CYCLE

In aquatic ecosystems the major forms of nitrogen which are available to bacteria, fungi, and plants are nitrate and ammonia, just as they are for terrestrial systems. Since nitrogen is often in short supply for plant growth on land, it is not surprising that it may be a growth-limiting factor in water. In practice, some aquatic ecosystems do show nitrogen as the nutrient element which most limits plant growth. This tends to occur most frequently in lakes at the eutrophic or oligotrophic ends of the trophic spectrum. Before considering the spatial and temporal variations of nitrogen in various aquatic ecosystems, the interactions between each form should be understood. This is best accomplished by examining the nitrogen cycle illustrated in Fig. 8-1. The cycle applies to all

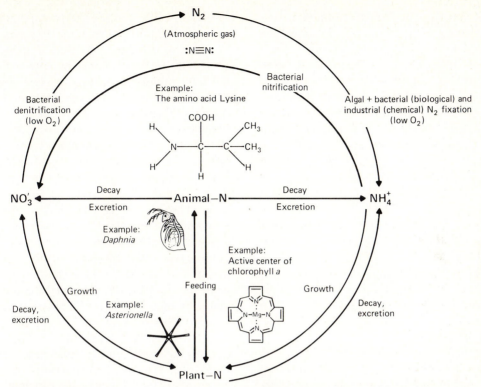

Figure 8-1 Diagrammatic representation of the nitrogen cycle. Note the two anoxic one-way sections of the cycle (nitrogen fixation and denitrification) which involve procaryotic cells, in contrast with the reversible flows which occur in oxygenated conditions in the rest of the cycle. Denitrification ($NO_3 \rightarrow NO_2 \rightarrow N_2$) and nitrification ($NH_4 \rightarrow NO_2 \rightarrow NO_3$) which involve separate bacteria (see Table 11-1) have been simplified in this cycle. The roles of sediments are not shown in this figure. Most nitrogen in aquatic ecosystems is present as plant and animal nitrogen, e.g., as part of chlorophyll in *Asterionella,* or amino acids in *Daphnia.*

ecosystems, but the presence of oxygen may be most important in aquatic systems. As previously discussed (Chap. 6), an oxygen shortage is more likely in water (~ 0.01 percent O_2) than in air (~ 20 percent O_2). In aquatic systems nitrogen fixation and denitrification are the ultimate sources and sinks of combined nitrogen available to algae. Both transformations involve enzyme systems requiring anoxic conditions, although specialized cells (heterocysts) or anoxic microzones may permit the processes to occur in the presence of oxygen.

FORMS OF NITROGEN IN LAKES

Of the commonly found nitrogen forms in lakes, nitrogen gas is the most abundant because it is little used and because the lake surface waters are in continual contact with the inexhaustible nitrogen reservoir of the atmosphere (80 percent N_2). Nitrogen is present in volumes comparable with those of oxygen despite its lower solubility because nitrogen has a greater partial pressure in the atmosphere. Nitrogen generally shows no seasonal or depth variations except those resulting from temperature changes. Removal of nitrogen is often facilitated in eutrophic lakes by the stripping action of methane bubbles released by anoxic decay in the sediments. Nitrogen gas enters the methane bubbles as they rise from the sediments through the water column. In some systems, well-illustrated by the case of Lake Mendota, anoxic sediments are supplied by nitrate-rich groundwater. Denitrification occurs in the sediments, producing considerable quantities of nitrogen gas (Brezonik and Lee, 1968). Nitrogen is a conservative gas, and its variations are biologically insignificant. An exception is the supersaturation with nitrogen gas occurring below some large dams where N_2 bubbles form in fish blood, resulting in high mortality. This is similar to "the bends" in human beings which occurs after breathing compressed air during diving and returning too rapidly to surface pressures.

NITROGEN FIXATION

This process is restricted to a few genera of bacteria and blue-green algae and is best defined as the transformation of nitrogen gas to ammonia by an enzyme. Nitrogen fixation occurs on land in symbiotic bacteria-plant systems, legumes such as alfalfa and peas, nonlegumes such as alder trees or *Ceanothus* bushes, as well as by free-living bacteria. Although not a universal phenomenon in lakes, nitrogen fixation is important because it is a major source of new, usable nitrogen. This process can accelerate lake eutrophication. A few genera of photosynthetic blue-green algae— *Aphanizomenon, Anabaena, Gleotrichia, Nodularia,* and *Nostoc*—dominate nitrogen fixation in lakes (Plates 6*a*, 7*b*) and streams (Horne, 1977). Under anoxic conditions found in the hypolimnion of some lakes, photosynthetic nitrogen-fixing bacteria may be important.

The enzyme responsible for nitrogen fixation is called *nitrogenase* and has two components. One has a molecular weight (m.w.) of $\sim 300,000$ and contains iron and molybdenum

in a ratio of 20:1; the other component has a m.w. of $\sim 35,000$ and contains only iron. It may seem paradoxical that nitrogenase cannot function in the presence of oxygen, but in blue-green algae it requires the energy generated by photosynthesis which produces oxygen. Further, nitrogen fixation occurs in well-oxygenated waters where oxygen can easily diffuse into the plant cells. This dilemma is resolved in special thick-walled cells called *heterocysts* which consume oxygen through very high rates of respiration (Figs. 11-1*c*, 12-1*d*). They also lack photosystem II, the part of the photosynthetic apparatus that produces oxygen. An anoxic microzone is established within the heterocyst which allows the nitrogenase to function despite the oxygen in the surrounding cells and medium. Heterocyst formation is induced by intracellular nitrogen shortages which generally occur when the C/N ratio exceeds 8:1. Nitrogenase is apparently synthesized in all cells but is most active in the heterocysts. In lakes and streams heterocystous blue-green algae, rather than bacteria, tend to dominate nitrogen fixation (Horne and Goldman, 1972; Horne and Carmiggelt, 1975).

Nitrogen compounds in lake water affect the nitrogenase enzyme in algae, and its activity is often repressed by the presence of ammonia. Nitrate acts mainly by repressing new enzyme synthesis. However, nitrogen fixation can occur despite the fact that nitrate levels are high. This occurs after the fall overturn when nitrate increases. Because nitrogenase contains considerable quantities of iron and molybdenum, nitrogen fixation places a demand on these two metals which must be extracted from the environment. These and other micronutrients, called *transition metals,* are also needed for the increased ferridoxin and cytochromes involved in the electron-transport pathway of respiration. The micronutrients needed may be in very short supply in lakes, and their availability can influence both algal species composition and productivity (Chap. 10). For example, an iron limitation might cause a switch away from nitrogen-fixing *Aphanizomenon* to a flagellated alga.

The significance of nitrogen fixation in lakes of various types is shown in Table 8-1. In very eutrophic lakes nitrogen fixation may be the major source of nitrogen, but this process plays a minor role in the nitrogen budget of oligotrophic lakes. A classic case illustrating the importance of nitrogen fixation is Clear Lake, California, where nitrogen fixation provides about half the lake's annual nitrogen budget (Table 8-1). The nitrogen fixation en-

ables the spring bloom of *Aphanizomenon* (Fig. 8-2) to persist until midsummer (Horne and Goldman, 1972; Horne, 1979*b*). Without nitrogen fixation the bloom would collapse in late spring after the winter accumulation of nitrate became depleted. Similarly, in autumn *Anabaena* dominates the plankton with its ability to grow in a nitrogen-depleted situation (Fig. 8-2). The ability to change from the use of NH_4 to NO_3 and then to N_2 as each nitrogen source is depleted has been demonstrated in an elegant experiment on an *Anabaena* bloom in Smith Lake, Alaska, using ^{15}N labeling (Billaud, 1968;

Table 8-1   Percent Contributions of Nitrogen Fixation to Nine Lakes

Note that the eutrophic polymictic lakes are normally the ones with the highest percent contributions. The absolute maximum rate of nitrogen fixation is often similar in lakes but the percent contribution varies with the amount of water inflow each year. For example, very low, nitrogen-deficient inflows to terminal, saline Pyramid Lake produce high percent figures even though the algal bloom is short-lived.

Lake	Date	% contribution	Present tropic and mixing status	Method used	References*
Windermere, South Basin, U.K.	1965 1966	< 0.5 < 0.2	Mesotrophic, monomictic	$^{15}N_2$	1
Smith, Alaska, U.S.A.	1963	5–10	Eutrophic, dimictic	$^{15}N_2$	2
Mendota, Wisconsin, U.S.A.	1970–1971	5–10	Eutrophic, dimictic	Acetylene	3
Tschornoje, Russia	1937	8	Eutrophic, dimictic (?)	Cell counts	4
Mize, Florida, U.S.A.	1969 1970	14 56	Dystrophic	Acetylene	5
George, Uganda, Africa	1969	33	Eutrophic, polymictic	$^{15}N_2$	6
Clear, California, U.S.A.	1970 1971 1972	43 30–50 30–50	Eutrophic, polymictic	Acetylene reduction	7
Erken, Sweden	1970	≈80†	Eutrophic, dimictic	Acetylene reduction	8
Pyramid, Nevada, U.S.A.	1979	~90	Mesotrophic monomictic	Acetylene reduction	9

*References: (1) Horne and Fogg, 1970; (2) Alexander, personal communication; (3) Torrey and Lee, 1976; (4) Kusnezow, 1959; (5) Keirn and Brezonik, 1971; (6) Horne and Viner, 1971; (7) Horne and Goldman, 1972; Horne, 1978; (8) Granhall and Lundgren, 1971; (9) Horne and Galat, unpublished.
†Values for Lake Erken have been recalculated.

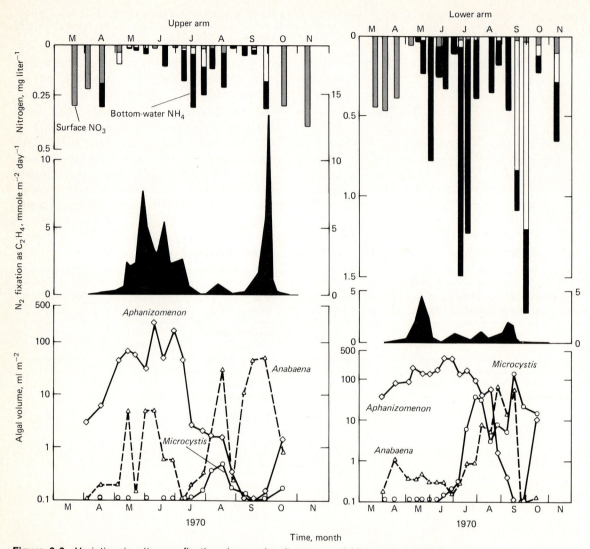

Figure 8-2 Variation in nitrogen fixation, inorganic nitrogen, and blue-green algae in two basins of eutrophic and polymictic Clear Lake in 1970. Nitrogen fixation is shown in solid black in the center of the figure. Surface nitrate (stippled), bottom-water ammonia (black), and surface ammonia (open) concentrations are shown by the histograms. Total algal cell volumes are shown for *Aphanizomenon flos-aquae* (◇—◇), *Anabaena solitaria* and *Anabaena* sp. (△—△), and *Microcystis aeruginosa* (○—○). Note the low summer nitrate levels and the large ammonia flux from the sediments (black bars) in the intermittently stratified Lower Arm relative to the well-stirred Upper Arm. Note also how this ammonia depresses nitrogen fixation. (*Modified from Horne and Goldman, 1972.*)

Fig. 8-3). In mesotrophic Windermere, nitrate and ammonia become depleted in late summer when the nitrogen-fixing alga *Anabaena*

reaches its maximum. Although amounting to only 1 percent of the lake's annual nitrogen budget, nitrogen fixation supplies between 10 and 70 percent of the nitrogen used by the alga, depending on the year (Horne and Fogg, 1970; Table 8-1).

Some bacteria and attached blue-green algae fix small amounts of nitrogen in lakes. Also important are the various nitrogen-fixing organisms living in close proximity to higher plants. The most common are the symbiotic algae and bacteria in the root zone or rhizosphere of *Lemna*, the duckweed, and the blue-green alga *Anabaena* associated with *Azolla*, the water fern (Chap. 11). Nitrogen-fixing organisms secrete some of the nitrogen which they fix (Jones and Stewart, 1969; Walsby, 1974). This may provide a significant source of nitrogen to the associated plants. For centuries nitrogen fixation associated with *Azolla* has provided a nitrogen source for rice culture in Asia. Nitrogen fixation also occurs in flowing waters where planktonic algae are absent. *Nostoc*, growing attached to the rocks, is common even in fast-flowing streams throughout the world up to altitudes of 2000 m and fixes considerable quantities of nitrogen in California streams (Horne and Carmiggelt, 1975).

DENITRIFICATION

Many facultatively anoxic bacteria can denitrify nitrate to nitrogen gas. The reaction enables the bacteria to donate electrons to nitrate during respiration at the low oxygen levels which occur in lake muds or anoxic hypolimnia. The optimum temperature for most denitrifying organisms is about 17°C; greatly reduced activity occurs in colder water. No known organisms denitrify ammonia. In lakes, reducing conditions and nitrate occur

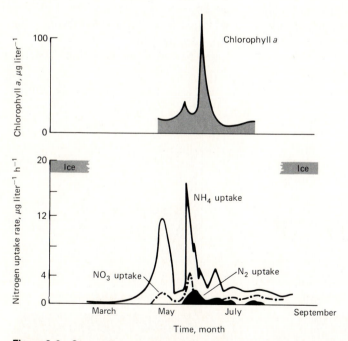

Figure 8-3 Sequential use of ammonia, nitrate, and nitrogen gas (N_2) by a bloom of nitrogen-fixing algae in a shallow eutrophic Alaskan lake. As one nitrogen source becomes depleted (not shown but see Fig. 8-5 for typical seasonal changes), the next source available with least energy expenditure is used. The uptake rates were determined by using a ^{15}N tracer. (*Redrawn from Billaud, 1968.*)

together only at the mud-water interface. Nitrate-rich groundwaters permeate anoxic soil pockets to create ideal microzones for denitrification. When nitrate-rich groundwater upwells through anoxic lake or marsh sediments, it may be largely denitrified before it reaches open water.

Losses of nitrate in oxygen-depleted waters can occur by denitrification or by some other biological transformations. In shallow estuaries where bacterial nitrification often results in abundant nitrate, denitrification may be more important than in lakes. Denitrification in the open water of lake hypolimnia may occur for short periods until nitrate is depleted (Brezonik and Lee, 1968). The conversion of nitrate to ammonia involves uptake and then decomposition and usually accounts for all observed losses of nitrate. Thus open-water denitrification is insignificant. Nitrate may increase in lakes when oxygen is abundant and ammonia is available. In this process of *nitrification,* bacteria convert ammonia to nitrate and obtain energy as a result.

NITRATE AND NITRITE

Nitrate, in contrast to ammonia, phosphate, or metal ions, moves freely through soils along with subsurface waters. For example, if water rich in both nitrate and phosphate passes through soil, the outflowing water will become relatively richer in nitrate than in phosphate. Nitrate is the most highly oxidized form of nitrogen and is usually the most abundant form of combined inorganic nitrogen in lakes and streams (Table 8-2). Nitrite, the partially reduced form of nitrate, is usually present in insignificant quantities (Fig. 8-4a). The distribution of nitrate and nitrite with depth is shown for eutrophic and oligotrophic lakes in Fig. 8-4, and the changes in overall concentrations with season in Fig. 8-5.

Plant cells use reduced nitrogen which is often transferred intracellularly as the amino group $-NH_2$. The electron transfer involved which reduces the oxidation-reduction state of the nitrogen atom from $+5(NO_3)$ to $-3(NH_4)$ takes place in only two steps. The transfer in-

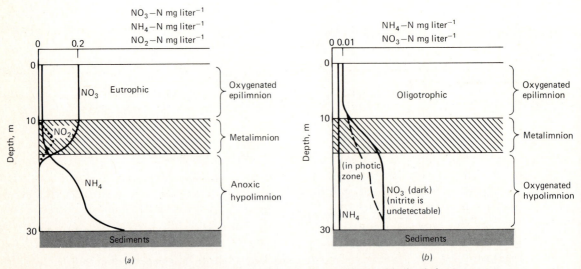

Figure 8-4 Idealized distribution diagram of nitrite, nitrate, and ammonia with depth for two stratified lakes (oligotrophic and eutrophic) in midsummer. Note that the oligotrophic example shown here has low nitrate levels (e.g. Lake Tahoe) but could in some cases have high nitrate levels (e.g. Lake Superior) since lack of other nutrients may cause the low productivity.

Table 8-2 Concentration of Inorganic Nitrogen (μg liter^{-1}) in the Surface Waters of Various Lakes and Rivers

Note the very wide range of total inorganic nitrogen ($NO_3 + NH_4$) available for plant growth. Values less than 100 μg liter^{-1} may limit growth, while levels above 400 would not. Both eutrophic and oligotrophic lakes may have very low or very high levels of total inorganic nitrogen.

Lake or river	Relative trophic state and mixing type	NO$_3$-N*		NH$_4$-N*		References†
		Summer	Winter	Summer	Winter	
Tahoe, Calif.	Oligotrophic, monomictic	4	μ-25	<2		1
Castle, Calif.	Mesotrophic, dimictic	<5	10-50	<5	10-50	2
Clear, Calif.	Eutrophic polymictic	μ-100	400-600	μ-300	μ-20	3
Superior	Oligotrophic monomictic	≈230	≈280	<10		4
Windermere	Mesotrophic monomictic	100-200	300-400	≈10		5
Esthwaite Water	Eutrophic monomictic	μ-100	400-500	≈30		6
George, Uganda	Eutrophic polymictic	μ		<10		7
Baikal, U.S.S.R.	Oligotrophic dimictic	0-20	45-80	μ		8
Titicaca, Andes	Mesotrophic monomictic	40-110	100-200	μ		9
Cayuga, N.Y.	Eutrophic monomictic	50-800	≈800	100-300	≈80	10
Uganda Rivers (annual mean)		530		24		11
Truckee River at km 3	···	20	···	<10		12
Hubbard Brook	···	440	2500	40		13

*μ = undetectable, generally < 10 μg liter^{-1}.

†References: (1, 2) Goldman, various sources; (3) Horne and Goldman, 1972; (4) Dobson et al., 1974; (5, 6) Heron, 1961; Horne and Fogg, 1970; (7) Viner, 1969; (8) Kozhov, 1963; (9) Richerson et al., 1977; (10) Oglesby, 1978; (11) Viner and Smith, 1973; (12) McLaren, 1977; (13) Likens et al., 1977.

volves two enzymes, *nitrate reductase* and *nitrite reductase*. Nitrate reductase catalyzes the reaction $NO_3 \rightarrow NO_2$, has a molecular weight of about half a million, and contains molybdenum at its reactive center. The nitrite reductase enzyme which catalyzes the reaction $NO_2^- \rightarrow NH_4^+$, is a relatively small molecule (m.w. = 63,000). This enzyme contains iron at its reactive center and requires copper as a cofactor.

The major summer inputs of nitrate to lakes and streams are rainfall and runoff. Nitrate can only be metabolized after transformation by nitrate reductase. The induction period for this enzyme is quite long, taking several hours to a few days. Thus an increased nitrate uptake is slow relative to uptake of ammonia. Nitrate reductase is low in algal populations growing where nitrate levels are low, but is present even if the plants are utilizing ammonia. The pres-

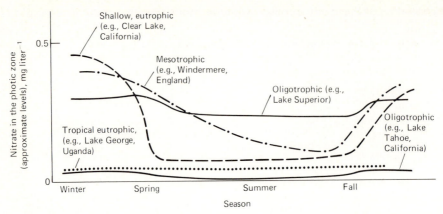

Figure 8-5 Idealized representation of seasonal changes in nitrate available for plant growth in temperate-zone eutrophic, mesotrophic, and oligotrophic lakes and a tropical eutrophic lake. Actual data are shown for some lakes in Figs. 8-2, 8-7, and Table 8-2.

ence of high levels of ammonia causes direct feedback inhibition and repression of nitrate reductase synthesis. Fortunately for the plants, the levels of ammonia in most aquatic systems are too low to cause this repression.

Competition for nitrate often proceeds by the type of enzyme kinetics described in Chap. 15. The rate of nitrate uptake v and the lower limits of uptake (indicated by K_s) vary with species and cell size. They also vary with the nitrate preconditioning which the plants have experienced (Table 8-3). For example, some algal species can use low levels of nitrate if grown in a low-nitrate environment, but may have lost the ability if found in an area of high-nitrate concentrations. These biochemical adaptations may be short-lived and reversible. Table 8-3 shows some of the contrasts for nitrate uptake in algae. In general, the smaller species have a competitive advantage in oligotrophic nitrogen-limited systems. In addition, some algae simply have more enzymes than others. For example, the green alga *Chlorella* may have 3 times more nitrate reductase per unit biomass than the blue-green alga *Microcystis* (Pomiluyko and Ochkivskaya, 1970).

Nitrate is usually not toxic in the quantities found in lakes and rivers (up to 1 mg liter^{-1}). Drinking water standards for human beings are set at about 10 mg liter^{-1} NO_3-N. Even at the higher levels the health hazard of nitrate itself is apparently small. Limits are placed on nitrate to ensure that water containing high levels of nitrate are not fed to babies because on rare occasions bacteria in a baby's gut reduce nitrate to nitrite. Nitrite can cause a respiratory problem by forming methemoglobin, which reduces the oxygen-carrying capacity of the blood. This oxygen starvation in the tissues can produce a fatal condition known as *blue baby* (Lee, 1970). In most human beings methemoglobin is broken down quickly, but very young babies lack the necessary enzyme.

Nitrite is generally present only in trace quantities in water exposed to oxygen, where it is transformed to nitrate. Nitrite is converted to ammonia in anoxic waters. Well waters in areas of manure accumulation may occasionally have a serious nitrite pollution problem. Polluted streams can contain up to 2 mg liter^{-1} NO_2-N. A small area near the thermocline of lakes may contain relatively large quantities of nitrite (Fig. 8-4a).

Table 8-3   Enzyme Uptake or Growth Kinetics of Nitrate and Ammonia for Some Natural Populations of Plankton and Attached Diatoms and Some Cultures of a Flagellate, a Dinoflagellate, Some Diatoms, and Blue-Green Algae from Waters of Differing Fertilities

Note that the half-saturation constant K_s generally increases with cell size and with increases in the average ambient nutrient levels (Table 8-2). Thus small algae in oligotrophic waters usually have low K_s values. The change in K_s in response to the environment is even shown by different races (ocean, estuaries) of the same species *Cyclotella nana*.

Dominant algal type/or habitat	Cell size, μm	K_s, μg liter^{-1} NO$_3$	NH$_4$	Max. growth rate, doublings day^{-1}	References[*]
Oligotrophic					
General marine	. . .	3	5	. . .	1
Chaetoceros gracilis	5	3	6	3.2	2,3
Cyclotella nana (from ocean)	5	7	5.6	. . .	3,4
Coccolithus huxleyi	5	4	1.4	1.7	3
Mesotrophic					
General marine	. . .	14	18	. . .	1
Asterionella japonica	10	14	15	2.0	2,3
Cyclotella nana (from estuary)	5	21	. . .	. . .	4
Skeletonema costatum	8	6	28	2.4	3
Melosira in aufwuchs	10–20 × 1000	. . .	70	. . .	5
Eutrophic					
Gonyaulax polyedra	45	130	80	0.5	4
Pseudoanabaena caterata	. . .	. . .	170	. . .	6
Anabaena sp.	100	70	. . .	. . .	7
Oscillatoria aghardhii	. . .	420	180	0.4	8
Castle Lake					
3 m, 50% blue-greens, 50% dinoflagellates	. . .	9–14	2–7	. . .	9
20 m, 75% dinoflagellates	. . .	10–16	3–14	. . .	9
25 m, 75% dinoflagellates		35–80	6–9		
Periphyton		200–500	200–1000		

[*]References: (1) MacIsaac and Dugdale, 1969; (2) Eppley and Thomas, 1969; (3) Carpenter and Guillard, 1971; (4) Eppley et al., 1969; (5) Horne and Kaufman, 1974; (6) Healy, 1977; (7) Mitchell, personal communication; (8) Zevenboom and Mur, 1981; Zevenboom et al., 1980; Zevenboom, 1978; (9) Priscu, 1982; Axler et al., 1982; Reuter, 1982.

Seasonal Cycles of Nitrate

The major changes occurring between autumn and winter and between winter and spring are considered in three categories (Figs. 8-5, 8-7):

1 Oligotrophic lakes in which nitrate levels remain almost constant (Lake Superior)

2 Eutrophic lakes in which nitrate levels fall to almost zero (Clear Lake)

3 Mesotrophic lakes in which nitrate levels fall but may not become limiting (Windermere, Lake Mendota)

The seasonal cycle of nitrate tends to be similar in most lakes. In winter, inflow exceeds uptake by algae and is supplemented by nitrate released from the sediments. In summer, nitrate uptake by plants is faster than inflow, and recycling from the hypolimnion is physically limited by the thermocline. In the hypolimnia of deep oligotrophic lakes nitrate persists unchanged or may even increase near the lake bed as organic nitrogen is mineralized to ammonia which is then nitrified to nitrate and diffuses from the sediments. In eutrophic lakes demand denitrification will often completely remove nitrate. A similar condition exists in eutrophic lakes under winter ice. In most oligotrophic and mesotrophic lakes the photic zone extends below the thermocline, and algal growth occurs in the hypolimnion. Algal depletion of nitrate will then extend into the hypolimnion (Fig. 8-4b).

The major source of nitrate for lakes is river inflow, although in the case of large lakes, such as the Laurentian Great Lakes, direct precipitation is also important. In addition, some lakes, especially those on alluvial soils, have considerable nitrate input from groundwater. Groundwater below the root zone is usually a more concentrated solution of nitrate than stream water since there is little opportunity for its removal by plants. Rainwater contains ammonia as well as nitrate and other nitrogen compounds. The other compounds are usually converted to nitrate in passing from the watershed to the lake. Uptake for plant growth enroute, together with any input from geological weathering, sewage, or other wastes, will further modify the water's nitrogen content. A major source of recycled nitrate is the fall overturn of the hypolimnion.

AMMONIA AND THE AMMONIUM ION

Ammonia, present in aquatic systems mainly as the dissociated ion NH_4^+ (ammonium), is a much more reactive compound than nitrate due to its higher chemical energy. Its positive charge enables it to form bonds with negatively charged clays which are common in many inflowing streams. Ammonia differs from nitrate in toxicity and mobility. It has a higher toxicity and is retained by most soils.

The ammonium ion is rapidly taken up by phytoplankton and other aquatic plants (Toetz, 1971). It persists in small quantities because it is the major excretory product of aquatic animals. The actual amount of ammonia present at any time will depend on the balance between animal excretory rates, plant uptake, and bacterial oxidation. Ammonia in the epilimnion may vary considerably with daily vertical migrations of zooplankton and fish through the thermocline. For example, fish are known to excrete sufficient ammonia to stimulate phytoplankton metabolism. The variations in ammonia with depth for typical oligotrophic and eutrophic lakes are shown in Fig. 8-4.

Cellular enzymes are needed to metabolize ammonia once it is taken into the cell. Three major pathways are reductive amination of some keto acids into amino acids, further amination to form amides, and the production of carbamoyl phosphate. The first alternative is probably the most important (Morris, 1974). Regardless of the relative importance of each pathway, the uptake of the ammonium ion, like nitrate, often follows classic enzyme kinetics (Chap. 15). Some available values of kinetic constants representative of major algal groups are given in Table 8-3.

The toxicity of ammonia to aquatic animals and plants is of great practical importance. The gas ammonia (NH_3) dissolves very readily in water and forms ammonium hydroxide (NH_4OH), which dissociates to give ammonium (NH_4^+) and hydroxyl ions (OH^-) as shown below:

$$NH_3 + H_2O \rightleftharpoons NH_4OH \rightleftharpoons NH_4^+ + OH^- \quad (1)$$

The reaction equilibrium lies far to the right at neutral pH 7 and 25°C, where only 0.55 percent of total ammonia is present as NH_4OH, almost none as NH_3, and the rest as NH_4^+. Changes with pH and temperature are shown in detail in Table 8-4. Under acid conditions the percentage of NH_4OH decreases; in alkaline conditions it increases. Undissociated NH_4OH is toxic, but the ion NH_4^+ is almost harmless. Ammonium hydroxide toxicity to aquatic organisms varies not only with pH but with temperature, dissolved-oxygen levels, the hardness or salt content of the water, and animal species and age. For example, even under normal

Table 8-4 Relationship of pH and Temperature to the Percentage of Un-ionized Ammonia [$NH_4OH + NH_3$ (dissolved)] in Freshwater.*

Note the very large increase in this toxic fraction as pH and temperature rise to levels often found in productive lakes and streams.

| pH | Temperature, °C | | | | |
	5	10	15	20	25
6.5	0.04	0.06	0.09	0.13	0.18
7.0	0.12	0.19	0.27	0.40	0.55
7.5	0.39	0.59	0.85	1.24	1.73
8.0	1.22	1.83	2.65	3.83	5.28
8.5	3.77	5.55	7.98	11.2	15.0
9.0	11.0	15.7	21.4	28.5	35.8

*Trussell, 1972.

physiochemical conditions (pH = 6–7, temperature = 5 − 10°C), young rainbow trout fry are rapidly killed by the toxic effect of ammonium hydroxide if the total ammonium concentration is 0.3 mg liter^{-1} NH_4-N. In contrast, minnows and other nonsalmonid fish are generally more tolerant, surviving to 10 times the quantity toxic to trout. Aquatic invertebrate zooplankton such as *Daphnia* apparently tolerate levels as high as 8 mg liter^{-1} NH_4-N.

Ammonia in most lakes and streams is generally well below 0.1 mg liter^{-1} (Fig. 8-6), and detrimental effects of naturally occurring ammonium are uncommon. Most toxic effects of ammonium on animals are due to pollution, generally sewage outfalls that contain 10 to 30 mg liter^{-1} NH_4-N before dilution by the receiving water. In lakes and rivers, at least a few days are required before this ammonia is removed by plant growth or transformed to nitrate. Plant photosynthesis due to the increased nutrient supply removes carbon dioxide and increases the pH of the water which greatly increases the toxicity of ammonia (Table 8-4). During the night when oxygen levels are low or when the oxygen demand of wastes exceeds photosynthetic production of oxygen, the susceptibility of animals to ammonia poisoning is increased.

The ammonium ion NH_4^+ is not toxic to most plants except at very high concentrations or elevated pH values. However, values of as little as 0.5 mg liter^{-1} are known to slightly inhibit photosynthesis in a species of blue-green algae and some estuarine diatoms at pH 7.5 to 9 (Horne and Kaufman, 1974). For most waters, values in the hundreds of milligrams per liter are needed for toxic effects since some algae can live in ponds where concentrations exceed 500 mg liter^{-1} NH_4-N.

Seasonal Cycles of Ammonia

In general, two types of patterns are followed depending on the trophic state of the lake (Figs. 8-6, 8-7). In oligotrophic and mesotrophic

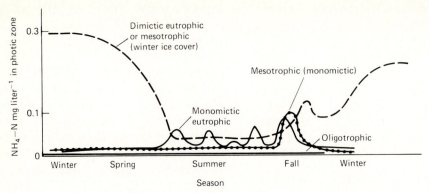

Figure 8-6 Idealized diagram of seasonal changes in photic-zone ammonia for various lake types. Some actual data are given in Figs. 8-2, 8-7, and Table 8-2. Random summer increases in ammonia occur in eutrophic lakes due to decay of algal blooms or irregular inputs from the hypolimnion. Most productive lakes show an increase in surface ammonia at the fall overturn.

lakes, ammonia in the epilimnion varies around a low value of about 5 μg liter^{-1} throughout spring and summer and may show an increase during the autumn overturn. In winter, ammonia decreases, but if the lake is ice-covered, moderate levels may persist until spring. For eutrophic lakes, summer values of ammonia in the epilimnion may fluctuate considerably over periods of a few days. At autumn overturn, ammonia levels rise considerably but then fall. In winter, ammonia may increase to very high levels (> 1 mg liter^{-1}), particularly under ice.

The major source of ammonia is from inflowing rivers, precipitation, atmospheric dust, or indirectly from nitrogen fixation. Most of the ammonia in rain is probably derived from aerosols originating in ocean, animal, or bacteria excretions rather than in volatilization of ammonia gas from the lake surface. Direct volatilization does not seem to occur even in warm lake water at high-pH values. Losses of ammonia do occur from stockyards or intensively used cattle feedlots. The excreted urea is converted to ammonia by soil bacteria. Hutchinson and Viets (1969) have demonstrated that ammonia from such feedlots can be detected in large quantities several kilometers downwind. Some soils and natural concentrations of animals such as seal wallows produce ammonia-rich aerosols. Flooded streams and rivers also often contain relatively large quantities of ammonia because there is insufficient time for plant uptake or microbial transformation of ammonia to nitrate.

DISSOLVED ORGANIC NITROGEN (DON)

All natural waters contain some DON, which is more abundant in eutrophic than oligotrophic systems (Manny, 1971). DON ranges from simple nutrients like urea to large complex molecules whose roles are little known. Urea is a common animal excretory product which serves as an excellent nitrogen source for phytoplanktonic growth (McCarthy, 1972; Carpenter et al., 1972). Urea is rapidly broken down to ammonia by bacteria or by the extracellular enzyme *urease* which is present in most waters.

Simple DON compounds like amino sugars are excreted by some plants (Jones and Stewart, 1969; Walsby, 1974) and can be utilized as both energy and nitrogen sources. The uptake of organic compounds for energy, termed *heterotrophy,* is discussed in Chap. 11. However, most algae and other aquatic plants are not heterotrophic, and thus DON plays a minor role in the energetics of lakes.

The other major role of DON is as a chemical

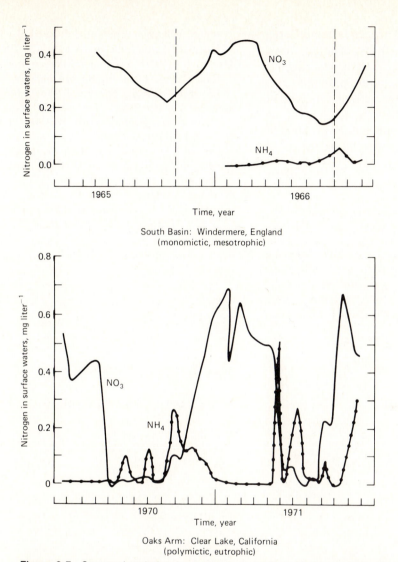

Figure 8-7 Seasonal variation in ammonia and nitrate in the euphotic zone of various lakes (see also Table 8-2). Vertical dashed line represents the fall overturn for the stratified lake. (*Windermere data by courtesy of J. Heron, Freshwater Biological Association, United Kingdom. Clear Lake data from California Department of Water Resources, and Horne, 1975.*)

modifier which alters the ionic state of many metals. The ability of nitrogen (as $-\ddot{N}=$) to form complexes with metal ions is enhanced by DON where N is usually present as the amine group $-NH_2$. The presence of large quantities of DON is apparently a prerequisite for the growth of many algae, particularly nitrogen-fixing blue-greens (Pearsall, 1932; Horne and

Fogg, 1970; Fogg, 1975). This may be due to the toxic and stimulatory roles played by several metals whose state of chelation depends on organic compounds such as DON (Chap. 10).

Nitrogen and phosphorus are nutrients required in moderate quantities in aquatic and terrestrial ecosystems, but both may become limiting for plant growth in lakes and streams.

The water chemistry of nitrogen is complicated by its positive and negatively charged forms (NH_4^+, NO_3^-, NO_2^-) as well as its gaseous phase. In contrast, phosphorus in aquatic systems has no gaseous form and only one type of ion (PO_4^{3-}) but, as will be seen in Chap. 9, the limnological chemistry of phosphorus is complicated by its adsorption onto solid particles.

FURTHER READINGS

Brezonik, P. L. 1973. *Nitrogen Sources and Cycling in Natural Waters*. Ecological Research Series. EPA-660/3-73-002. 167 pp.

Corner, E. D. S., and A. G. Davis. 1971. "Plankton as a Factor in the Nitrogen and Phosphorus Cycles in the Sea." *Adv. mar. Biol.*, **9**:101–204.

Ganf, G. G., and P. Blažka. 1974. "Oxygen Uptake, Ammonia and Phosphate Excretion by Zooplankton of a Shallow Equatorial Lake (Lake George, Uganda). *Limnol. Oceanogr.*, **19**:313–325.

Horne, A. J. 1978. "Nitrogen Fixation in Eutrophic Lakes," pp. 1–30. In R. Mitchell (ed.), *Water Pollution Microbiology*, vol. 2. Wiley, New York.

Keeney, D. R. 1972. "The Fate of Nitrogen in Aquatic Ecosystems." *Univ. Wis. Water Resour. Cent. Lit. Rev.* 3. 59 pp.

Kusnezow, S. I. 1959. *Die Rolle der Mikroorganismen im Stoffkreislauf der Seen*. Deutcher Verlag der Wissenschaften, Berlin.

National Academy of Sciences. *Nitrates: An Environmental Assessment*. Environmental Studies Board. National Research Council. Washington. 750 pp.

Chapter 9

Phosphorus

OVERVIEW

Although needed in small amounts, phosphorus is one of the more common phytoplankton growth-limiting elements because of the geochemical shortage of phosphorus in many drainage basins together with the lack of a phosphorus equivalent to nitrogen fixation. The phosphorus cycle in lakes involves only phosphates and organic phosphorus, but it is complicated by reactions in which phosphate is precipitated or sorbed by lake sediments and many common minerals such as clays. Phytoplankton are only able to use phosphorus in the phosphate form (PO_4) for growth. Intensive algal growth in spring usually depletes lake phosphate to low levels. In deep stratified lakes there is limited replenishment, and the quantity of "available" phosphorus in late winter may determine the maximum phytoplankton standing crop that can develop in summer. Growth during summer usually occurs using phosphate excreted by animals feeding on phytoplankton. In eutrophic lakes, animal recycling, especially that due to zooplankton, may supply all the daily phosphate needs for phytoplankton growth. Direct sediment resupply is important in summer in shallow areas. After migration through the thermocline to the epilimnion, zooplankton and fish return phosphorus to the euphotic zone. When surface phosphate levels are low, phytoplankton excrete extracellular enzymes called *alkaline phosphatases*. These enzymes have the ability to free phosphate bound to organic molecules. *Luxury consumption* of phosphate by algae results in the storage of polyphosphate granules in the cell. These may contain sufficient phosphorus for many cell divisions and help to carry phytoplankton through short periods of phosphorus depletion.

Phosphate, in contrast to nitrate, is readily adsorbed to soil particles and does not move easily with groundwater. High inflows of total phosphorus are due to erosion of particles from steep slopes with easily erodible soils. Agricultural, domestic, and industrial wastes are major sources of soluble phosphate and frequently contribute to lake eutrophication. Phosphate-containing detergents, for example, commonly contribute about half the phosphorus contained in domestic sewage.

INTRODUCTION

Phosphorus is not needed for growth in large quantities like carbon, oxygen, hydrogen, and nitrogen, but it is one of the more common limiting elements on land and in freshwater. The main reasons for this are threefold. (1) Phosphorus-containing minerals are sometimes geochemically scarce, and thus the normal nutrient supply derived from rock breakdown will be phosphorus-poor. (2) There is no gaseous phase in the phosphorus cycle so there is no equivalent to nitrogen fixation, and (3) phosphorus is sufficiently reactive to be tightly bound to a variety of soils.

Average concentrations for phosphorus in some of the world's rivers and lakes are given in Table 9-1. Comparison with similar figures for nitrogen (Table 8-2) shows that the N/P ratio varies widely. Since a plant typically requires a ratio of 7:1 by weight or 16:1 by element, it can be seen that phosphorus depletion is likely in many freshwaters. However, limitation of plant growth by deficiencies in nitrogen, silica, iron, or some trace elements is more important than limitation by phosphorus in a variety of different lakes, reservoirs, estuaries, and coastal waters.

MEASUREMENT

Three types of phosphorus are normally measured in aquatic ecosystems: dissolved phosphate (PO_4), dissolved total phosphorus, and particulate phosphorus. Full details are given in Strickland and Parsons (1972), Golterman and Clymo (1967), and the current edition of APHA. Only *dissolved phosphate* can be used directly for algal growth, and in this form it is called *reactive phosphorus*. *Total phosphorus,* whether filterable (particulate) or soluble (filtrate), must be first mineralized to phosphate by using perchloric acid. Other analytical techniques include the use of ion-exchange resins and the radioactive isotope ^{32}P.

THE PHOSPHORUS CYCLE

Phosphorus in lakes occurs in both organic and inorganic forms. The majority of inorganic phosphate present is in the form of orthophosphate (PO_4^{3-}), with lesser amounts in the form of monophosphate (HPO_4^{2-}) and dihydrogen ($H_2PO_4^-$) phosphate. Linear chains of condensed soluble polyphosphates are common algal storage products but are scarce in lake waters. Dissolved cyclic metaphosphorus compounds are also uncommon in lakes. Dissolved organic phosphorus (DOP) usually represents the bulk of the total soluble phosphorus. This organic fraction is made up of many different classes of compounds, but there is no general agreement as to which are the most important. The best estimates are that DOP is dominated by various types of nucleic acids (for example, RNA, DNA). A small fraction of the total phosphorus is usually present in colloidal form with very high molecular weight running to the millions. In most aquatic environments, total particulate phosphorus is present in much larger quantities than soluble phosphorus. Particulate phosphorus encompasses bacterial, plant, and animal phosphorus as well as that attributable to suspended inorganic particles such as clays or other minerals.

Most lakes are characterized by low levels of available phosphate; when added to lake waters, it soon disappears from solution. Most of the phosphate goes into bacteria, algae, or other plants, and some is precipitated or sorbed

Table 9-1   Concentrations of Inorganic Phosphorus (μg liter^{-1}) in the Surface Waters of Various Lakes and Rivers.

Note the wide range of PO_4-P available for plant growth from limiting levels (< 5 to abundance, > 10). Also, note that eutrophic and oligotrophic lakes may have similar levels.

Lake or river	Relative trophic state and mixing type	Soluble PO_4-P Summer	Winter	References*
Tahoe, Calif.	Oligotrophic monomictic	≈ 2		1
Castle, Calif.	Mesotrophic dimictic	≈ 2	. . .	2
Clear, Calif.	Eutrophic polymictic	≈ 20	≈ 10	3
Superior	Oligotrophic monomictic	5		4
Windermere	Mesotrophic monomictic	5	30	5
Esthwaite	Eutrophic monomictic	5	≈ 30	5
George, Uganda	Eutrophic polymictic	< 2		6
Baikal, U.S.S.R	Oligotrophic dimictic	≈ 2	6	7
Titicaca, Andes	Mesotrophic monomictic	≈ 15	≈ 15	8
Cayuga, N.Y.	Eutrophic monomictic	> 5	≈ 12	9
Uganda rivers	. . .	80–230		10
Truckee River at km 3	. . .	≈ 10	. . .	11
Hubbard Brook	. . .	3	2	12

*References: (1, 2) Goldman, various sources; (3) Horne, 1975; (4) Schelske and Roth, 1973; Dobson et al., 1974; Ragotskie, 1974; Bennett, 1978; (5) Heron, 1961; (6) Ganf and Viner, 1973; (7) Kozhov, 1963; (8) Richerson et al; 1977; (9) Oglesby, 1978; (10) Viner, 1973; Golterman, 1975; (11) McLaren, 1977; (12) Likens et al., 1977.

by physicochemical processes. Excellent proof of this has been given by using radioactive ^{32}P (Hayes and Phillips, 1958). Any phosphate added by animal excretion or stream and sewage inflow will also be taken up.

Figure 9-1 shows major pathways of the phosphorus cycle in lakes and the approximate quantity of phosphorus to be found in the various fractions.

In Pelagic Waters

Almost all phosphorus in lake water is present as organic phosphorus in living or dead biomass. A small fraction is excreted as various soluble organic phosphorus compounds. Figure 9-1 shows a possible mechanism in which the initial excreted product is of phosphate of a low molecular weight designated as X-P. This X-P may be rapidly transformed into a high-molecular-weight colloidal phosphorus compound which in turn can be transformed into phosphate for direct use by plants. Most plants can only utilize phosphate but are able, by releasing enzymes called *alkaline phosphatases,* to break down many forms of DOP to phosphate. Finally, decomposition of dead plants and animals releases DOP.

The major losses of phosphorus from open

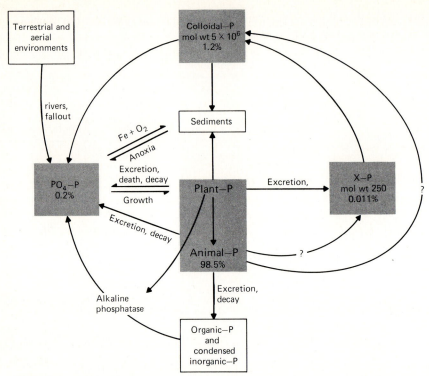

Figure 9-1 The phosphorus cycle. Plant phosphorus is mainly organic compounds and inorganic polyphosphate. Animal phosphorus is mostly organic. X phosphorus (X-P) is an unidentified soluble compound. Sediment phosphorus is largely inorganic phosphate, for example, $Ca_3(PO_4)_2$, $FePO_4$, or phosphate sorbed onto clays and various types of gels. River inflows are especially high in particulate phosphorus and contain many other types of phosphorus. For simplicity only PO_4 is shown in the cycle. Components inside shaded areas are present in the cycle in approximately the percent amounts indicated (sediments, terrestrial, and aerial fractions are excluded from the percent values). (*Extensively modified from Lean, 1973.*)

water are by sedimentation of the biota and by chemical precipitation. The net result is an accumulation of dead and dying phosphorus-containing material on the lake bed. There it provides a food source for various *detritivores* such as insect larvae, worms, and crustaceans, which digest and recycle a portion back into the overlying waters. The loss of soluble phosphate-phosphorus from the open water by chemical processes is both simple and complex. A pure sodium phosphate solution will precipitate iron phosphate ($FePO_4$) upon the introduction of iron. This process can be shown to vary

with the oxygen level of the water and the pH. In lake systems sedimentation of phosphate occurs but is dependent on a host of other substances which may also vary in their ability to precipitate phosphorus with oxygen and pH changes. Inorganic solid phosphate phases are formed by direct precipitation with calcium, aluminum, and iron, while clay particles "scavenge" phosphate by sorption. It should be noted that many of these solid-liquid phase equilibria can reach completion in the time required for the particle to sink. However, as the particle sinks the equilibrium conditions change as it passes through various physically and chemically distinct water strata before arriving at the lake bed.

In the Sediment

It is more important to know how phosphate is released in the sediments than exactly which of the many potential mechanisms removes phosphate from solution. Lake sediments are better understood than the water with regard to the phosphorus cycle because the sediment has only one active layer, the sediment-water interface. Sediments consist of solid particles separated by liquid-filled *interstitial spaces*. Although sediments are mixed by the larger benthic organisms, they are more physically stable than turbulent lake waters. Thus chemical processes in the sediments can be investigated in laboratory experiments with isolated sediment cores. In contrast, it is difficult to create an artificial water column in the laboratory that approximates lake conditions. The most important reactions in the sediments are those which change phosphorus from a solid phase into soluble phosphate in the interstitial waters, which may then in turn be released to overlying waters.

Sediment samples reflect the relative fertility of the lake and contain from 0.06 to 10 mg liter^{-1} of soluble interstitial phosphate. These levels are many times greater than those of the overlying waters. The amount of clay in a lake sediment is perhaps the most important factor in determining its phosphorus-holding capacity. Unfortunately, clay is a rather imprecise chemical compound. Clays consist of complex silicates of aluminum and iron together with their oxides. The purest form of clay, china clay or kaolin, has the formula $Al_2O_3 \cdot 2SiO_2 \cdot 2H_2O$. Clays sorb phosphorus by the specific chemical interaction of PO_4^{3-} with Al^{3+} on the edges of the clay plates (Stumm and Leckie, 1971). Phosphate is also sorbed directly onto hydrous iron or aluminum oxides, particles of calcite (impure calcium carbonate), and apatite (im-

pure $Ca_3[PO_4]_2$). Phosphate may also be occluded within iron oxides and some sediments.

Precipitated phosphate, as distinct from sorbed phosphate, is controlled primarily by pH and the redox potential E_h. The phase equilibria for several common precipitates is given in Fig. 9-2. This diagram shows how changes in the concentration of phosphate and pH precipitate different metallic phosphorus compounds. The quantity of phosphate released into sediment interstitial waters is in a heterogeneous equilibrium, with various minerals buffering releases from other minerals. In addition, organic phosphorus released from decaying organic matter further modifies the actual amount of phosphate present.

UPTAKE OF PHOSPHORUS AND INDUCTION OF PHOSPHATASES

Devices for overcoming phosphorus deficits have been evolved by algae. These are (1) luxury consumption, (2) an ability to use phosphate at low levels (low K_s), and (3) alkaline phosphatase production.

Luxury consumption of phosphate is probably found in all phytoplankton. The process entails the uptake of more phosphate than is required for growth and its storage within the cell. Microscopic examination of algal cells grown under nonlimiting conditions reveals highly refractive cellular inclusions called *polyphosphate granules*. They tend to form very rapidly when phosphate is added to a phosphorus-deficient culture (Stewart and Alexander, 1971). Experimental algal physiologists have used the amount of phosphate released from phytoplankton on boiling as a measure of actual cellular phosphorus depletion (Fitzgerald and Nelson, 1966). In the same fashion, decreases in the pigment phycocyanin-c show nitrogen depletion in blue-green algae. Polyphosphate granules disappear rapidly when the surrounding medium becomes depleted in phosphorus, although it appears that there may

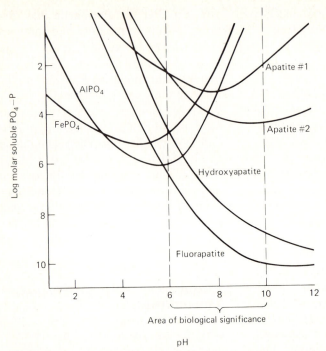

Figure 9-2 Solid-phase solubility diagram for various metal phosphate compounds in lake waters. The solid phase or precipitate occurs in conditions in the areas above the line, while its soluble form occurs below the line. The chemistry of metallic phosphate is complicated since reactions shown are influenced by losses of phosphate through sorption onto clays and onto iron hydroxide gels. Apatite no. 1 = $CaHPO_4$; no. 2 = $Ca_4H(PO_4)_3$; hydroxyapatite = $Ca_{10}(PO_4)_6(OH)_2$; fluorapatite = $Ca_{10}(PO_4)_6F_2$. (*Modified from Stumm and Zollinger, 1972.*)

sometimes be sufficient phosphorus in the granules to support up to 20 cell divisions. An obvious advantage to the cell-containing poly-phosphate granules is their lack of osmotic or toxic effects. This is in contrast with most ni-trogenous products, which do not readily con-dense to a suitable storage form and must be kept as protein or pigments.

In most lakes the phosphate growth constant K_s is very low for natural phytoplank-ton (see Chap. 15). It is generally in the region of 1 to 3 μg liter^{-1} phosphate-P, which means that the enzyme system in algae is not saturated for much of the time under natural lake condi-tions (see Tables 9-1 and 9-2). Particularly in summer, phosphate levels in unpolluted lakes often fall to extremely low levels. In contrast,

lakes polluted with sewage or those with high natural phosphorus inputs usually have phos-phate levels which will saturate the phosphorus uptake system of the phytoplankton (Fig. 9-3). There may be considerable variation in K_s be-tween species, and as available phosphate decreases, this might play a role in species suc-cession (Figs. 15-7, 15-8). Because phosphate is recycled rapidly, the rate of phosphate uptake is also important. A higher uptake rate can thus compensate, to some degree, for the lack of a mechanism to remove phosphate at very low levels.

The enzyme *alkaline phosphatase* cleaves the bond between phosphate and the organic molecule to which it is attached. The result is free phosphate available for plant growth. The

Table 9-2   Enzyme Growth and Uptake Kinetics of Phosphate for Two Diatoms and a Green Alga.

Note the superiority of *Asterionella* and *Selenastrum* at low levels of lake phosphate. They can take up PO_4-P at lower levels and use it faster.

Species	K_s, μg liter^{-1}	Max. growth rate μ, doublings day^{-1}	Uptake rate V_{max}	Luxury uptake coeff. R*	References†
Asterionella *formosa*	0.6	0.9	13×10^{-9}	82	1
Cyclotella *meneghiniana*	8	0.8	5×10^{-9}	6.6	1
Selenastrum *capricornutum*	1.3	1.2	. . .	57	2

*The luxury uptake coefficient R is the ratio of the cell quota of a nutrient when it is limiting to when it is abundant (Droop, 1974). Here *Asterionella* and *Selenastrum* are superior to *Cyclotella*.

†References: (1) Tilman and Kilham, 1976; (2) Brown and Button, 1979.

production of alkaline phosphatases by planktonic algae is a remarkable adaptation to an environment low in phosphate but comparatively rich in larger phosphorylated compounds. These may be either organic or condensed inorganic polyphosphates. The enzymes are originally attached near the cell surface and are produced in response to phosphorus deficiency. Synthesis of new enzyme is repressed by the addition of phosphate to the medium. Algae deficient in phosphorus may contain up to 25 times more alkaline phosphatase than algae with surplus phosphorus (Fitzgerald and Nelson, 1966; Bone, 1971). Even more remarkable is the release of phosphatases in free dissolved form into the environment, a process which quickly hydrolyzes much of the previously unavailable total phosphorus (Berman, 1970). The releases are roughly proportional to algal biomass in the lake and thus to its trophic state (Jones, 1972). The use of extracellular enzymes by phytoplankton is almost unique to phosphorus metabolism since the only similar mechanism is for iron (see Chap. 10). Both these elements are among the few most likely to be limiting under natural situations. It is of interest that insoluble inorganic phosphates such as $FePO_4$ and $Al_2(PO_4)_2$ can be solubilized by sediment bacteria. This is accomplished by the production of organic acids rather than phosphatases (Harrison et al., 1972). Such free-solution reactions are a major reason for the short turnover time of most phosphorus compounds in lakes, estuaries, and seas.

RECYCLING OF PHOSPHORUS

Zooplankton and fish excrete much of their waste directly among the phytoplankton. Zooplankton excrete approximately 10 percent of their body phosphorus daily, but there is considerable variation depending upon feeding rates, temperature, type of food, time of day, and larval stage (see review by Corner and Davis, 1971). Of the phosphorus excreted, roughly half is phosphate and the remainder is organic. In most cases zooplankton phosphorus excretion varies widely but may supply most of the daily demand of the phytoplankton. The effect of a drastically lower zooplankton excretion rate has been discussed for Lake George, Uganda, a nonseasonal equatorial lake. Ganf and Blažka (1974) predict that decreased zooplankton grazing would result in a "bloom and crash" phytoplankton population in place of the present constant algal densities (Fig. 19-5). The idea is that with few grazing zooplankton, the phytoplankton will grow until all phosphorus

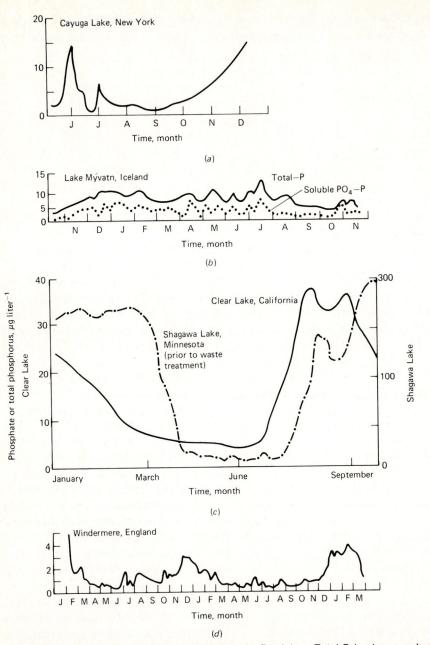

Figure 9-3 The seasonal cycles of phosphate in five lakes. Total-P is shown only for Lake Mývatn. In most lakes soluble phosphate concentrations fall from relatively high winter levels at the start of the spring bloom and continue low until early winter. Luxury uptake by phytoplankton (see text) may use up many times the requirement for one cell division (Table 9-1) and may account for the rapid decline in spring levels. Clear Lake, California, has naturally high phosphate levels; Shagawa Lake, Minnesota, was polluted by sewage; the levels in the other lakes are similar to the low levels found in many lakes. (*Modified from (a) Oglesby, 1978; (b) Olafsson, 1979; (c) Horne, 1975; Lallatin, 1972; Malueg et al. 1973; (d) Heron, 1961.*)

supplies are exhausted. The bloom will then decay rapidly, and phosphorus will be released in huge quantities by cell lysis and bacterial decay to start the cycle again. With more zooplankton there is always a supply of phosphate from excretion, and phytoplankton growth continues at a high but controlled rate.

While zooplankton feed on phytoplankton they excrete large amounts of plant nutrients. Perhaps to avoid consumption by predators which rely on sight, most herbivorous zooplankton move into the surface waters at night. In the day they remain in the dark, cool deeper water where digestion and excretion are slowed by the lower temperatures. Thus return of phosphorus to the euphotic zone is favored by their nightly vertical migrations to the warmer surface zone.

Oxygen plays an important role in controlling the rate of phosphorus release from the sediment to the photic zone. When the sediment-water interface becomes anoxic, phosphate passes rapidly into the water above. Anoxic sediments release phosphate at rates as much as 1000 times faster than releases from oxygenated sediments. This is attributable to both classical chemical bonding and physicochemical sorptive mechanisms. The reactions

$$Fe^{3+}PO_4 \text{ (insoluble)} \rightleftharpoons Fe_3^{2+}(PO_4)_2 \text{(soluble)} \rightleftharpoons 3Fe^{2+} + 2PO_4^{3-} \text{ (free)} \qquad (1)$$

may be less important than reactions of the type

$$Fe^{3+}(OH)_3 + PO_4^{3-} \text{ (sorbed)} \rightleftharpoons Fe^{2+}(OH)_2 + PO_4^{3-} \text{ (free)} \qquad (2)$$

where the ferric hydroxide (or oxyhydroxide) may be replaced by numerous other sorption sites, including clays. For this reason Eq. (2) is not balanced stoichiometrically. The equilibria in Eq. (2) are sensitive to E_h, pH, and dissolved oxygen. Unfortunately, the heterogenous nature (solid-liquid phases) of the reactions prevents easy extrapolation of laboratory results to real-lake sediment systems.

Some phosphorus may be made available from hypolimnetic entrainment, thermocline descent in the autumn, and leaking upward around the edges of the thermocline (Chaps. 4 and 5). Phosphate released from the hypolimnion sediments is largely unavailable for growth in most lakes. Even in the deeper lakes which possess a large illuminated hypolimnion, the algae are too far above the bottom to benefit directly from phosphate release from the sediment except during spring and fall overturn.

Role of Rooted Macrophytes and Algal Decomposition

Many lakes are edged with reeds and submerged macrophytes (Fig. 11-5). In small lakes these plants may be the dominant primary producers and constitute the largest biomass. It is well-established that uptake through the leaves of macrophytes occurs when the water is rich in phosphate (McRoy et al., 1972). However, as we see in Table 9-1, lake waters are usually low in available phosphate. The main method by which rooted macrophytes obtain phosphorus is absorption of phosphate directly from the interstitial soil water. There are two possibilities for loss of this plant phosphorus: direct excretion by the plants or eventual loss by death and decomposition. In estuarine eelgrass, McRoy and Barsdate (1970) found that some 30 percent of the phosphorus taken up daily was released to the surrounding water. Once the system is in steady state, as in mature *Phragmites* beds, there will also be a continuous release of phosphate from decay.

When algae are placed in darkness, as when they settle to the deeper waters of the lake, there are three stages in their decompostion: (1) within 24 h both releases and uptake of phosphorus occur depending on the original conditions of the cells, (2) over the next several days a stage of little net activity occurs, (3) over a few hundred days active nutrient regeneration takes place with releases to solution (Foree et al., 1971). Much of the cellular phosphorus, however, is not normally released. This is called *refractory phosphorus* and constitutes roughly one-third of the initial particulate phosphorus. It is the refractory phosphorus which constitutes the bulk of the sedimented organic phosphorus. This fraction requires further processing by bacteria or benthic organisms in the sediment before it becomes available for new plant growth.

SOURCES OF PHOSPHORUS

The main inputs of phosphorus into lakes are from inflowing rivers and precipitation. Rainfall is a much less important source of phosphorus than in the case of nitrogen. As already noted, most of the phosphorus is carried into unpolluted lakes as organic and inorganic particulates. This transport contrasts with nitrogen, much of which arrives as soluble nitrate. Even soluble phosphate does not move well through most soils so leachates will be relatively lower in phosphate than in nitrate. It has often been demonstrated that when water polluted with nitrogen and phosphorus from domestic sewage is passed through field soil, as in a septic tank leach field, the percolate is much lower in phosphate than nitrate. We saw earlier that phosphate becomes strongly sorbed onto clay particles and other minerals in soil and lake sediments.

Where the soil is geologically unstable, or ground cover absent, erosion occurs (Figs. 5-23, 6-2). Since erosion is the removal of solids, much of which are clay, the phosphate sorbed to the clay is carried off as silt. Such erosion is common in mountainous areas such as the southwest of North America, Alaska, the Rift Valley of Africa, or the Himalayas. In these areas we would expect nitrogen to be more limiting to algal growth than phosphorus. In contrast, in the older mountain areas of middle and eastern North America or much of northern Europe, particulate phosphorus inflows would be relatively low and phosphorus limitation would be expected. In semiarid zones,

Table 9-3 Increase in Transport of Nitrogen and Phosphorus with Increase in the Slope of Arable Land in Wisconsin Drainage Basins*

Note that as slope increases relatively more phosphorus is lost than nitrogen. This is because phosphorus is transported mostly as particulates (Table 16-1), and soil erosion also increases with slope. Nitrogen, in contrast, is usually lost as soluble nitrate.

Slope	N Kg ha^{-1}	% increase	P Kg ha^{-1}	% increase
8°	16	...	0.45	...
20°	34	210	1.6	360

*Modified from MacKenthun et al., 1964.

where rain, and thus nitrogen carried in it, is scarce, the trend is also toward high-silt (i.e., high-phosphorus) inflows and a relative deficiency in nitrogen. The flow of phosphorus into rivers is highly correlated with the average slope of the drainage basin (Table 9-3) and also in at least one case to the length of stream relative to the drainage area (Kirchner, 1975).

We have now discussed in detail the *major nutrients,* C, O, H, N, and P. The next chapter deals with the equally essential *minor* and *trace elements.* This diverse group has historically received less attention but may provide the key to some of the yet-unexplained fluctuations in plant and animal productivity.

FURTHER READINGS

Berman, T. 1970. "Alkaline Phosphatase and Phosphorus Availability in Lake Kinneret." *Limnol. Oceanogr.,* **15:**663–674.
Confer, J. L. 1972. "Interrelations among Plankton, Attached Algae, and the Phosphorus Cycle in Artificial Open Systems." *Ecol. Monogr.,* **42:**1–23.
Ganf, G. G., and P. Blažka. 1974. "Oxygen Uptake, Ammonia and Phosphate Excretion by Zooplankton of a Shallow Equatorial Lake (Lake George, Uganda)." *Limnol. Oceanogr.,* **19:**313–325.
Griffith, E. J., A. Beeton, J. M. Spencer, and D. T. Mitchell (eds). 1973. *Handbook of Environmental*

Phosphorus. Wiley, New York. (See especially chapters by H. L. Golterman, "Vertical Movement of Phosphate in Freshwater," and F. H. Rigler, "A Dynamic View of the Phosphorus Cycle in Lakes.")

Lean, D. R. S., and M. N. Charlton. 1976. "A Study of Phosphorus Kinetics in a Lake Ecosystem," pp. 283–294. In J. O. Nriagu (ed.), *Environmental Biochemistry*. Ann Arbor Science, Ann Arbor, Mich.

Pomeroy, L. R., H. M. Matthews, and H. S. Min. 1963. "Excretion of Soluble Organic Compounds by Zooplankton." *Limnol. Oceanogr.*, **8:**50–55.

Porter, R., and D. W. Fitzimons (eds.) 1978. *Phosphorus in the Environment*. Ciba Foundation, no. 57 (new series). Elsevier, Amsterdam. 330 pp. (See especially chapter by C. S. Reynolds, "Phosphorus and the Eutrophication of Lakes," pp. 201–228.)

Rigler, F. H. 1956. "A Tracer Study of the Phosphorus Cycle in Lake Water." *Ecology*, **37:**550–562.

Chapter 10

Other Nutrients

OVERVIEW

Silicon, calcium, magnesium, sodium, potassium, sulfur, chlorine, iron, and the minor metals constitute most of the plant and animal nutrients not covered in Chaps. 7 to 9. Diatoms require large quantities of silica for their cell walls. Because of this demand, silica can be a limiting element for phytoplankton growth where diatoms are the predominant algae. This element is derived from the weathering of soils, especially those containing the class of minerals called *feldspars*. Silica incorporated into many species of diatoms is insoluble and therefore less easily recycled than most other elements. Diatom shells, called *frustules,* accumulate in sediments and the changes in different species with time are used by the paleolimnologist to reconstruct the trophic history of lakes.

Calcium is the main skeletal component of many animals and some plants, but it is also important in buffering lake waters. Precipitation of $CaCO_3$ during intensive photosynthesis and its solution by rainwater in the watershed control the supply of this element. Magnesium, which has a similar water chemistry to calcium, is vital for energy transfer in any cell since it catalyzes the change from ATP to ADP. Plants also require magnesium to form the active center of their major pigment, chlorophyll *a*. Unlike terrestrial plants, algae do not need potassium in large quantities, and both potassium and sodium are usually present in excess of biological requirements.

Sulfur is an essential structural component of proteins and has an important gaseous phase when hydrogen sulfide is released from anoxic decaying sediments. Hydrogen sulfide is usually oxidized to sulfate at the aerated micro-zone of the sediments of some lakes and estuaries. Chlorine, as the chloride ion, is present in small quantities in freshwater but becomes a dominant ion in estuaries and the sea. Disinfection during municipal and industrial water treatment produces chlorinated compounds which can be very toxic to animals and plants. Iron is needed in relatively large quantities by plant and animal cells although it is often considered a minor element. The two or three outer-shell electrons of the iron atom dominate the electron-transport pathway of respiration. Iron is essential to hemoglobin in blood and for enzymes involved in nitrogen metabolism. Phosphate in lakes is often lost from the photic zone by precipitation or by sorption onto the outside of complex iron-containing particles. Iron availability may limit growth of algae in lakes and streams, especially when nitrogen fixation is important. Molecules called *siderochromes* are produced by some algae when dissolved iron is scarce. These make organically bound iron available for algal growth.

Natural waters are extremely dilute chemical solutions with very small quantities of several essential metals including Mn, Cu, Zn, Co, and Mo. These elements are required by animals and plants in minute quantities and for this reason are often called *trace elements* or *micronutrients*. Their main role in the cells is at the active center of enzymes or as cofactors in enzyme reactions. Algae in some oligotrophic lakes show growth limitation from shortages of one or more trace metals. Toxicity to plants and animals can result from high concentrations of trace metals. Toxicity is most likely to occur if the metal is in an ionic *unchelated* state. Animals and plants tend to accumulate trace metals far in excess of their immediate needs. Where trace-element limitation occurs, the basic cause is a geochemical shortage of the element in the drainage basin.

ELEMENTS SOMETIMES REQUIRED IN LARGE QUANTITIES: Si, Ca, Mg, Na, K, S, Cl

Silica (SiO_2) and Silicon (Si)

In lakes and seas silica plays an intriguing role since it apparently accounts for the success of diatoms which dominate most aquatic systems (Fig. 12-3). Most algae and animals have at best only a minor need for silicon, but in the diatoms, silica (SiO_2) forms the rigid algal cell wall or *frustule* which may account for half the cell's dry weight. Some chrysophycean flagellates and a few other algae also possess silicon cell walls. The form of silicon when used as a structural component in algae is hydrated to form amorphous silica ($SiO_2 \cdot nH_2O$), or opal. This rigid material is highly perforated and surrounded on both sides by a thin cell membrane (Figs. 10-2, 11-1). Freshwater sponges contain needlelike inclusions called *spicules* which are also almost pure silicon and, like diatoms, leave a permanent record of their presence. Lake Baikal has large numbers of sponges, many of which are endemic to this ancient lake, and masses of spicules are to be found on its beaches.

Silica is normally measured as "reactive" silicate, i.e., the molecule H_2SiO_4 and its short-chain polymers. Longer polymers of three or four units are not measured although they are present in natural waters (Strickland and Parsons, 1972). Reactive silica is probably the only form available for diatom growth. The most common test used is spectrophotometric measurement of the blue color produced when a silicomolybdate complex is reduced.

The Silica Cycle The *silica cycle* illustrates the nearly one-way flow of this element from rocks in the watershed to the lake sediments (Fig. 10-1). Such a pattern is very different from the cycles of nitrogen, phosphorus, iron, or other nutrients where plant and animal cells

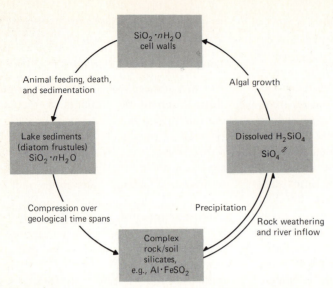

Figure 10-1 The silica cycle in lakes, emphasizing the role of diatoms. Although SiO_2 is not digested by any predator, it does not always protect against crushing or digestion of the cell contents by many invertebrates or fish that excrete SiO_2 fragments unchanged chemically. This is in contrast to nitrogen, iron, phosphorus, and most other elements.

take up and excrete large amounts in various forms. There are only two major sources of silica in lakes, from inflows and from below the photic zone. Animal recycling of silica is generally believed to be unimportant in lake waters, but some release of silica occurs from anoxic sediments. Certain diatoms release up to 15 percent of the silica they take up (Nelson et al., 1976).

The world average concentration for dissolved silica in large rivers is about 13 mg liter⁻¹ while lakes show values from less than 0.5 to 60 mg liter⁻¹ (Table 10-1). Silica-containing rocks make up 70 percent of the earth's crust and therefore provide the major minerals in contact with streams and groundwater. Rainwater, springs, and the leachate from soils are all high in carbon dioxide, which weathers rocks to release soluble silica (SiO_2). Feldspar is an important class of minerals which contain large amounts of silicon. White or pinkish feldspar crystals are a main component of granite. The reaction can be illustrated as follows,

where feldspar is weathered to china clay, or kaolinite:

$$2NaAlSi_3O_8 + 2CO_2 + 3H_2O \rightleftharpoons 4 \ SiO_2$$

Sodium   Rainwater,   Dissolved
feldspar   streams, springs   silica

$$+ Al_2Si_2O_5(OH)_4 + 2Na^+ + 2HCO_3^-$$

Kaolinite   (1)

This weathering is also an important source of sodium ions, and in the buffer system of natural waters, bicarbonate produced in silicon release is another important by-product (see Chap. 7 and Bricker and Garrels, 1967).

Silica exists in lakes in many forms, but only silicic acid (H_2SiO_4), which is partially dissociated at normal lake pH values, can be used by algae (Lewin, 1962; Darley, 1974). Other forms such as colloidal silica and clays play a physicochemical role by providing sorption sites for phosphate and ammonia.

Table 10-1 Concentrations (mg liter⁻¹) of Soluble Major Nutrients (Other than Nitrogen and Phosphorus) in the Surface Waters of Various Lakes and Rivers.

The very dilute lakes, such as Tahoe or the Experimental Lakes Area (ELA) lakes in Ontario, contrast with terminal lakes such as Mono, where the conservative elements accumulate. Note how conservative elements increase downstream in rivers, indicating that dilute lakes need small watersheds to remain low in conservative elements.

Lake or river	Conductivity, μmho cm^{-1}	SiO_2	Ca	Mg	Na	K	SO_4	Cl	HCO_3	References*
Tahoe	92	…	9.4	2.5	6.1	1.7	2.5	1.9	40	1
Castle, Calif.	30	1.3	1.6	2.6	1.1	0.2	0.2	0.1	20	2
Clear, Calif.	250	14	23	15	10	2.0	9	6	145	3
English Lake District, ave.	…	…	4.5	1.0	3.9	0.5	6.3	7.1	7.8	4
South Basin, Windemere	…	…	6.2	0.7	3.8	0.6	7.6	6.7	11	4
Erie	6	0.3	38	8.5	7.2		22	15	118	5
Superior	79	2	12.4	2.8	1.1	0.6	3.2	1.9	28.1	5
ELA lakes, ave.	19	≈1	1.6	0.9	0.9	0.4	3.0	1.4	3.8	6
Cayuga, N.Y.	56	…	44	10	51	2.6	36	81	122	7
Mono, Calif.	…	…	…	60	28,000	1400	9000	17,500	18,800	8
George, Uganda	223	20	17.2	7.4	20	4.2	14.6	8.4	99	9
Biwa, Japan	…	≈1	≈10	≈2	≈5	≈2	≈6.8	≈7	…	10
Titicaca, Andes	…	0.07–1.1	66	34	176	14	282	260	…	11
Tanganyika	…	0.3	11	39	63	33	6.3	26	…	12
Baikal, Siberia	…	≈3	15	4.2	6.1		4.9	1.8	≈60	13
World ave. lakes and rivers	…	12	15	4.1	6.3	2.3	11.2	7.8	58.6	14
North American rivers, ave.	…	9	21	5	9	1.4	20	8	68	14
European rivers, ave.	…	7.5	31	5.6	5.4	1.7	24	7	95	14
Nile, Khartoum	…	26	17.4	5.2	30.7	11.8	0.44	8	149	14
Rhine, Netherlands	…	5.7	42	6.1	10.1	6.4	19.5	11.3	113	14
Amazon (Narrow Santarem)	…	11.1	12.5	1.5	1.1	1.4	4.3	2.3	41	14
Truckee at km 10	…	…	8.5	4.3	6.1	1.7	2.9	≈3	…	15
Truckee at km 15	…	16	9.5	4.8	12.4	1.8	3.5	5.9	48	14,15
Lower Congo River, Kinshasa	105	7.5	10.8	3.9	14.2	…	7.8	6.1	…	16
Hubbard Brook, N.H.	≈25	4.5	1.7	0.38	9.1	0.21	6.3	5.4	<1	17

*References: (1,2) Goldman, various sources; (3) Horne, 1975; Lallatin, 1972; (4) Macan, 1970; (5) Schelske and Roth, 1973; Dobson et al., 1974; Ragotzkie, 1974; Bennett, 1978; (6) Armstrong and Schindler, 1971; (7) Oglesby, 1978; (8) Mason, 1967; Livingstone, 1963; (9) Viner, 1973; Ganf and Viner, 1973; (10) Itasaka and Koyama, 1980; (11) Richerson et al., 1977; (12) Hecky et al., 1978; (13) Livingstone, 1963; Kozhov, 1963; (14) Livingstone, 1963; (15) McLaren, 1977; (16) Visser and Villeneuve, 1975; (17) Likens et al., 1970.

Silica and Diatoms   Diatom productivity can be very great. For example, the production of 15 million tons of opaline diatom frustules each year in the inner Amazon estuary removes 25 percent of the river's dissolved silica (Milliman and Boyle, 1975). Since diatoms contain 25 to 60 percent of their dry weight as silica, silicon can become a limiting nutrient for diatom growth, primarily because of the absence of significant recycling. Green algae contain about 50 percent of their dry weight as carbon in the cellulose cell wall. There is no mechanism for silica storage when it is abundant although some diatoms may be able to make a few further divisions by thinning their walls. The interactions between silica and diatom blooms are discussed in more detail for *Melosira* and *Asterionella* in Chap. 12. Like phosphorus, the silica growth constant K_s indicates that the enzyme system often operates below saturation in some lakes (Tables 10-1, 10-2).

Compared to the levels of other pollutants such as nitrogen or phosphorus, waste discharges of silica are minor. A shortage of silica

relative to the other nutrients has major consequences in cultural lake eutrophication (Chap. 18). Lakes may change from aesthetically pleasing diatom waters to less desirable ones dominated by blue-green algae. The high loading of nitrogen and phosphate favors algae which do not require silica. The nearshore waters of Lake Michigan show this shift in algal species composition (Schelske and Stoermer, 1971). A few wastewaters contain elevated levels of silica. Swamp-cooler types of air conditioning systems that work by evaporating water concentrate silica to near the precipitation point. When the systems are cleaned the wastewater, containing relatively high levels of silica, is discharged to sewers or rivers.

After death, bacterial and fungal decay removes the organic material, leaving only the stable diatom frustule. These may be heated to redness or boiled in acid to remove any remaining organic material. Diatom frustules have beautiful symmetry. They are laced with minute holes and often crowned with sharp spines (Fig. 10-2). Since diatom frustules remain unchanged

Table 10-2   Enzyme Uptake Kinetics of Silica for Three Diatom Species and Three Clones of the Same Species from Different Environments.

Note the similarity in K_s but difference in V_{max} for the clone from the relatively eutrophic estuary compared with those from the oligotrophic Sargasso Sea. The freshwater diatom *Asterionella* has a higher K_s than *Cyclotella* and will thus outcompete it when nutrients are high, for example, in spring (Chap. 12).

Species	K_s, μg liter^{-1}	Max. growth rate $\hat{\mu}$, doublings day^{-1}	Uptake rate V_{max}, μg SiO$_2$ cell^{-1}h^{-1}	References*
Asterionella formosa	230	1.1	36×10^{-3}	1
Cyclotella meneghiniana	84	1.3	15×10^{-3}	1
Thalassiosira pseudonana	102	...	...	2
T. pseudonana (estuarine clone)	90	3.2	16×10^{-3}	3
T. pseudonana (ocean clone)	90	1.0	60×10^{-3}	3

*References: (1) Tilman and Kilham, 1976; (2) Paasche, 1973; (3) Nelson et al., 1976.

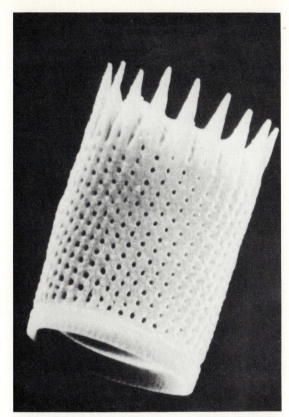

Figure 10-2 Scanning electron micrograph of the silica cell wall of half a small diatom, *Melosira crenulata,* from oligotrophic Lake Tahoe, California (magnification X4000). The cell has been treated with acid to clear the cell membrane (see text). Note the elaborate structure where the spines of the "crown" connect with the adjacent cell (not shown) and the numerous wall pores which allow nutrients to flow freely into the cell. (*Photo by H. W. Paerl.*)

over millenia in the sediments, their types and numbers can be used to help the paleolimnologist interpret the trophic status of a lake through time (Chap. 18). In parts of the ocean bed, frustules constitute most of the sediment, and diatomaceous earth is produced from previous deposits (Chap. 11).

Calcium

This element is essential for metabolic processes in all living organisms and as a structural or skeletal material in many. An example of its

biochemical role is the relaxation of animal muscle after contraction, which depends on depolarization of the cell membranes by an inflow of Ca^{2+} ions in the same manner that nerves require an inflow of Na^+ ions. Virtually all vertebrates, mollusks, and certain other invertebrates require large quantities of $CaCO_3$ as a major skeletal-strengthening material. In addition, there are some marine and marsh algae which use $CaCO_3$ in the cell wall. Coral reefs represent the ultimate expression of $CaCO_3$ skeletons. Since carbonate is more soluble in cold than warm waters, arctic mollusks have greater difficulty precipitating it and may substitute strontium in their shells. The fragile, paper-thin shells of dead snails in some Alaskan lakes contrasts with the rocklike masses of shells found in desert Pyramid Lake. Since the need for calcium in animals is large, growth limitation might be anticipated for some invertebrates. In contrast, the nonskeletal demands by aquatic plants are met by normal lake calcium levels.

An important role of calcium is its effect on pH and the CO_2-HCO_3^- system (Chap. 7). Calcium is present in ionic form and as suspended particulates, mainly $CaCO_3$. Calcium salts are the main cause of hard water, which can be softened by substituting sodium for calcium and magnesium, or the hardness can be circumvented by the addition of detergent phosphates. About half the phosphate in sewage is derived from the detergents used, which contributes to lake eutrophication (Chap. 18).

Rain and soil water contain weak carbonic acid, which is a major carbon source for receiving waters [see Eq. (1), Chap. 7]. Basic carbonate rocks, such as limestone in the lake watershed and sediments, neutralize this acid by the reaction

$$CaCO_3 + H_2CO_3 \rightleftharpoons CaCO_3 + CO_2 + H_2O \rightleftharpoons$$
$$Ca(HCO_3)_2 \rightleftharpoons Ca^{2+} + 2HCO_3^- \qquad (2)$$

Although this is the main buffering system in natural waters, it is overcome when the bicarbonate or CO_2 is removed by photosynthesis, and $CaCO_3$ is precipitated. This is a common phenomenon in hard-water lakes, especially the previously mentioned marl lakes (Chap. 7). In these lakes the water precipitates so much $CaCO_3$ at the height of algal blooms that the sediments show alternate layers of $CaCO_3$ and darker organic material deposited in periods of lower productivity. The precipitation of $CaCO_3$ has dramatic effects on both the organic and inorganic nutrient levels in marl lakes.

Calcium, bicarbonate, pH, and specific conductivity are only approximately correlated in lake waters. Conductivity measures the number of charged ions in the water (for example, Ca^{2+}, Na^+, SO_4^{2-}). Calcium in most temperate waters is often correlated with most other minor and major ions (Table 10-1). This gives rise to a general lake classification by ionic strength. Soft-water lakes can be distinguished from hard-water ones by their lower conductivity (Fig. 10-3). Since calcium is one of the most

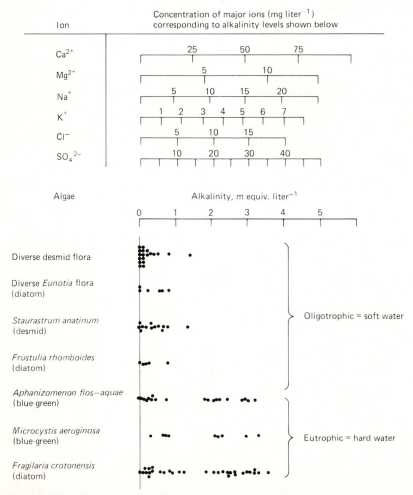

Figure 10-3   The distribution of species of algae with respect to alkalinity and major ions in freshwater. Four examples are indicators of oligotrophy; three are typical of eutrophic waters. Desmids are a type of green algae, Chlorophyceae (Table 11-3). (*Modified from Moss, 1972.*)

abundant and easily measured ions, it is often used as an indication of lake hardness.

One might expect an analogy between terrestrial and aquatic ecology through the presence of aquatic *calciphobes* (calcium-hating organisms) and *calciphiles* (calcium-loving organisms). There is little evidence to show that planktonic algae show such traits (Fig. 10-3). The apparent calciphobic distribution of oligotrophic desmids appears to be a chance association having no metabolic cause. Lund (1965) suggests calcium's main effect on phytoplankton to be the buffering of lake pH. With a few exceptions, invertebrates with high-calcium demands, such as mollusks, are apparently favored in high-calcium waters.

Magnesium

This element is often associated with calcium primarily due to its similar chemistry (Table 10-1). Dolomite, the major rock of the Dolomite Mountains in the European Alps, is a magnesium calcium carbonate. Dolomite is a fairly common rock in many parts of the world although it is rare in central and southern Africa.

Magnesium is needed by all cells for phosphate transfer involving adenosine triphosphate and diphosphate (ATP and ADP) as shown in the following equation:

$$ATP \overset{Mg^{2+}}{\rightleftharpoons} ADP + P + energy \qquad (3)$$

This step is the major short-term energy-transfering reaction in living cells. The exact role of magnesium is unknown for many other reactions, but the magnesium requirement has been well-established. Magnesium in plants serves as the transition metal at the heart of the reactive center in the chlorophyll molecule (Fig. 10-4). Despite its major role in algal photosynthesis, there are only a few instances of either magnesium deficiency or toxicity in lakes (Goldman, 1960).

The element is usually present in aquatic systems in large amounts relative to plant needs (Table 10-1). From this we must conclude that

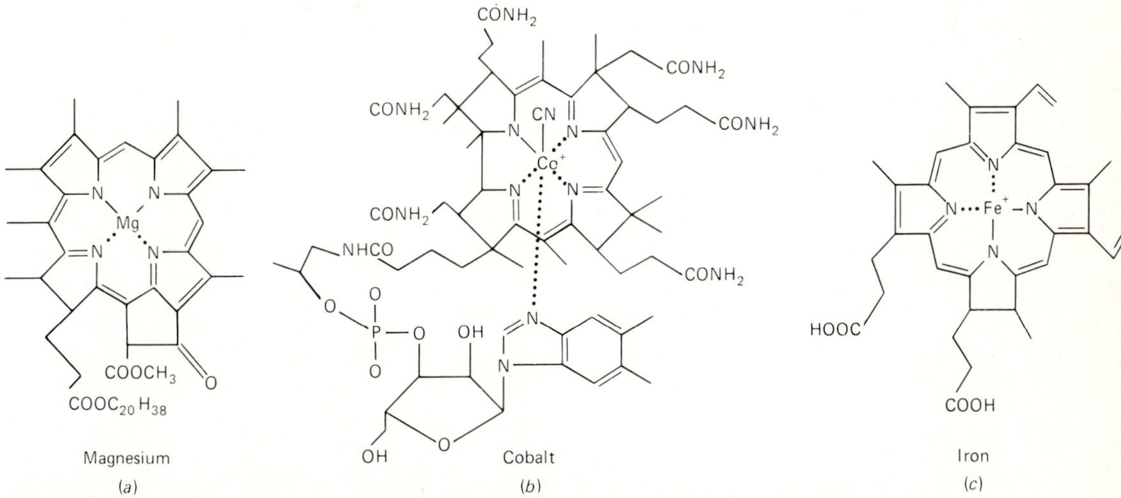

Figure 10-4 Examples of how iron and trace metals play a major role in aquatic biota. Note the metal ion in the heart of the molecule, surrounded by a porphyrin ring made of four pyrrole rings in a proline chain of alternating double and single carbon bonds which hold the resonating electrons. (*a*) Magnesium in the photosynthetic pigment chlorophyll. (*b*) Cobalt in vitamin B_{12}. (*c*) Iron in the blood pigment hemoglobin. Copper in the molluscan blood pigment hemocyanin does not contain a porphyrin ring, and its detailed structure is uncertain.

magnesium does not play a major role in limiting the growth or distribution of animals or plants in most waters. In Brooks Lake, Alaska, magnesium limitation of algal growth was attributed to the higher-than-normal sodium levels. Plants may show magnesium deficiency when sodium levels are high because sodium competes for binding sites in the cell.

Sodium and Potassium

One of the greatest chemical differences between terrestrial plant growth and that of phytoplankton is the minor role played by potassium in waters. In general, agricultural land is fertilized with large amounts of nitrogen, phosphorus, and potassium, but cultivated lakes such as fishponds need only nitrogen and phosphorus. Potassium is required for all cells principally as an enzyme activator and is present in larger quantities inside the cells of aquatic biota than in the surrounding medium. Since lakes are much less concentrated solutions of potassium, cell membranes must continually pump in potassium and pump out sodium, a process requiring the expenditure of large amounts of energy. In saline lakes (Chap. 19) and estuaries (Chap. 17) too much sodium, along with other ions, restricts the biota to a few species which can tolerate the osmotic stress.

Most water appears to possess adequate supplies of sodium for plant growth (Table 10-1), but nitrogen metabolism in blue-green algae may benefit from additional quantities. It is difficult to see any biochemical reason for this since the experimental evidence (Ward and Wetzel, 1975) does not provide a clear picture of the role of sodium in blue-green algal ecology.

Sulfur and Chlorine

Sulfur is important in protein structure but rarely limits the growth or distribution of the aquatic biota. This is due to the abundance of the element, primarily in its most energetically stable form, sulfate (SO_4^{2-}) (Table 10-1). The complex three-dimensional structure of enzymes and other proteins is partially due to bridges between two sulfur atoms which stabilize the geometry of the enzyme. This is the main use of sulfur in all cells, but the element plays a role in other metabolic processes such as cell division.

Anoxic sediments rich in organic matter release gaseous hydrogen sulfide (H_2S) to produce the familiar rotten-egg smell of decaying vegetation. In lakes and estuaries this H_2S is oxidized to SO_4^{2-} at the mud-water interface so long as there is an oxygenated microzone. Once this layer is reduced by bacterial respiration and chemical oxygen demand, H_2S is released into the air or overlying water (Fig. 10-5). Hydrogen sulfide may form a highly insoluble black ferrous sulfide precipitate, effectively removing available iron from solution (Fig. 10-5).

Industrialization has greatly affected aquatic sulfur cycles. In recent times much sulfur in some rivers and lakes has originated from the burning of fossil fuels (Kellogg et al., 1972). The major source in most rivers remains that derived from the breakdown of parent material in the drainage basin. Atmospheric pollution caused by burning fossil fuels is largely in the form of SO_2 which forms sulfuric acid (Fig. 10-6) and is often the major contributor to acid rain. Tide flats and salt marshes release enormous quantities of sulfur to the atmosphere from bacterial reduction of sulfate to hydrogen sulfides in the mud.

Among the halogens (Cl, Br, I, F), chlorine is the most abundant. The chloride ion is required by photosynthesizing cells for the photolysis of water to release oxygen, for ATP formation, and for certain phosphorylation reactions. In contrast, free chlorine is a very toxic substance, even at low concentrations. Because chloride is easily measured, it is frequently used to identify water masses in pollution studies of chlorinated waste. Toxic chlorinated wastes are discharged to lakes and seas in ever-increasing quantities. Free chlorine is used as a disinfectant for waste

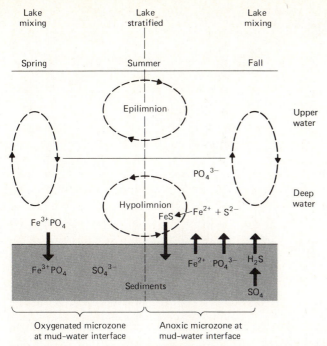

Figure 10-5 The iron, sulfur, and phosphorus interaction in eutrophic lakes. During overturn and whole-lake mixing, oxygenated water precipitates ferric (Fe^{3+}) phosphate. During the summer anoxic period, eutrophic lake sediments release ferrous (Fe^{2+}) ions, phosphate ions (PO_4^{3-}) and hydrogen sulfide (H_2S). The dissociated sulfide ion (S^{2-}) from H_2S can precipitate ferrous sulfide, a very insoluble substance, and remove, virtually permanently, iron and sulfur from the system.

discharges primarily to kill harmful bacteria and in power plants to curb slimes in the heat-exchange pipes. Chlorine kills by oxidation of the cell membranes but lasts only a few minutes as free chlorine in most systems. It is then either converted to the harmless chloride ion by sunlight or combined with organic compounds. Of the resulting compounds, chloramines are the most toxic in freshwater and last for several days before conversion to less harmful states. In estuaries, toxic bromamines are produced since seawater contains abundant bromine. Even at great dilution, chlorinated and brominated organics may harm or even kill algae, zooplankton, and the larvae of commercially important fish and shellfish. To prevent toxicity, sewage and power plants should de-

chlorinate with bisulfite, which nullifies most of the chloramine or bromamine toxicity (Stone, Kaufman, and Horne, 1973).

Iron, the "Trace" Element Needed in Moderate Quantities

Iron is needed by most living organisms in larger quantities than any other element apart from the structural atoms C, O, H, N, and P. Clifford Mortimer's classic studies are good examples of the major sediment-water interactions of dissolved iron (Mortimer, 1941–1942, 1971). As noted, iron is an essential constituent of many enzymatic and other cellular processes. Oxidative metabolism in all organisms and photosynthesis in plants involves cytochromes which contain iron. Iron is needed in the hemo-

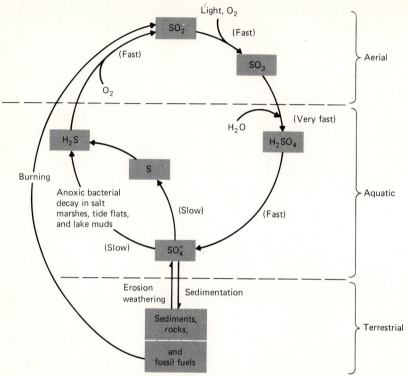

Figure 10-6   Sulfur cycle in water, on land, and in the air. Sometimes all the aerial stages may take place in oxygenated mud or water. In the figure, slow means days and fast means hours. Sulfur occasionally produces a white or yellowish precipitate on the surface of exposed mud at lake edges or in salt marshes. Hydrogen sulfide (H_2S), a poisonous gas, is a major nuisance for human beings near rotting vegetation and is extremely toxic to fish and other animals. Lower plants, however, may use H_2S if a little light is available. Blue-green algae grow in the upper layers of anoxic hypolimnia or as felts in salt marshes. (*Greatly modified from Kellogg et al., 1972.*)

globin of blood and has a vital role in the nitrogenase enzyme of some lower plants and bacteria. Plants and animals have evolved to take advantage of the oxidation-reduction reaction of iron. The unique feature of iron is the little energy required to change from ferrous (Fe^{2+}) to ferric (Fe^{3+}), and vice versa. It is not surprising that iron is also easily transformed by changes in lake oxygen concentration.

Because iron is so important to all organisms, there is considerable demand for it in the environment. Lakes in semiarid climates may show an iron shortage, and iron limitation in temper-

ate lakes was first described long ago (Rodhe, 1948; Gerloff and Skoog, 1957; Schelske, 1960; Goldman and Mason, 1962). Iron is one of the most common metals in the earth's crust and occurs in small amounts in almost all clays, soils, and granite rocks. Organic and inorganic forms of iron exist in particulate and dissolved phases in most water. For convenience, we will consider all iron attached to organic material as "organic" even though the attachment may be sorption rather than chemical bonding. Lakes generally have particulate iron on the order of a milligram per liter, most of which is contained in the lake biota. Oligotrophic lakes have less, since they have sparse populations of algae and zooplankton.

Measurement Dissolved iron (inorganic plus organic) is generally measured by using acid and Ferrozine or bathophenanthroline (Strickland and Parsons, 1972). Most natural chelating or sorption reactions which bind soluble organic iron are broken down by the acid reaction. Sodium EDTA, a common metal chelating agent used in laboratory cultures, is so strong that it prevents breakdown to free iron. Soluble iron ranges from 100 to 3,000 μg liter^{-1} in most waters (Table 10-3) but may drop below 2 μg liter^{-1} in some lakes, particularly in semiarid regions (Fig. 10-9). Under normal oxidizing conditions (Fig. 10-7) ferric iron (Fe^{3+}) predominates over ferrous iron (Fe^{2+}). Because ferric iron forms insoluble compounds, it rapidly disappears from the lake's mixed layer and is

deposited on the sediments as a rust-colored layer called *ocher* ($Fe(OH)_3$).

Iron Cycles The loss of ferric iron from oxygenated water occurs as precipitation of ferric hydroxide (Fig. 10-7). Phosphate ions are adsorbed onto this hydroxide. Approximately equal quantities of Fe^{3+} and PO_4^{3-} in the lake epilimnion could result in precipitation of both elements. In many temperate and polar lakes where iron predominates, phosphorus may limit phytoplankton growth. This principle is occasionally used to clean up eutrophic lakes where the phosphate can be removed by precipitation with added metals. Iron salts can be used, but it is cheaper to use the mixture of metals in fly ash or salts of aluminum or zirconium. In semiarid zone lakes there is often sufficient phosphate to precipitate most soluble iron, and soluble iron may then limit plant growth. The next stage in the iron cycle occurs

Table 10-3 Concentration of Soluble Minor Nutrients (μg liter^{-1}) in the Surface Waters of Various Lakes and Rivers.

Although present in relatively low concentrations, most are not limiting or toxic to the biota. Both ionic and chelated metals are present in unknown proportions in the soluble, or filterable, fraction.

Lake or river	Fe	Mn	Cu	Zn	Co	Mo	References*
Tahoe	< 10	2.6	Trace	<14	<0.6	0.5	1
Castle, Calif.	< 10	< 1	< 0.5	< 2	<1	< 0.5	2
Clear, Calif.	5–20	4.6	2–30	<14	<1.4	< 0.3	3
Windermere	...	...	...	8.0	>0.1	> 0.1	4
Cayuga, N.Y.	3–80	1–30	0.6	2.7	0.005	...	5
Biwa, Japan	40	5–17	> 2.5	5.30	0.03	...	6
Titicaca, Andes	...	...	2.5	28	...	...	7
Schöhsee, Germany	15	4.5	1.0	1.8	0.03	0.2	8
World ave.	≈ 40	35	10	10	0.9	0.8	9
Sacramento, Calif.	...	6.3	2.9	...	<1	0.4	9
Truckee, at km 10	110	9	4	5	2	>10	10

* References: (1,2) Goldman, various sources; (3) Goldman and Wetzel, 1963; Horne, 1975; (4) Macan, 1970; (5) Oglesby, 1978; (6) Itasaka and Koyama, 1980; (7) Richerson, et al., 1977; (8) Groth, 1971; (9) Livingstone, 1963; (10) McLaren, 1977.

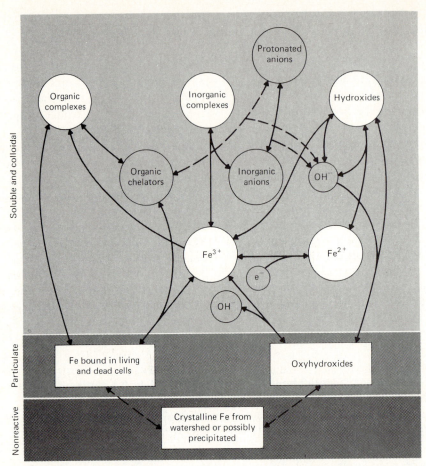

Figure 10-7 Iron cycle in natural aquatic systems. Solid enclosures and arrows indicate iron-containing components and pathways; dashed enclosures and arrows are indirectly involved but do not contain iron. The exchanges between the nonreactive and particulate components are unknown. The vertical arrangement of the three phases has no significance other than simplicity. The particulate phase could be suspended throughout the soluble phase.(*From Elder and Horne, 1978a; for a more detailed description see Elder, 1974.*)

when ferric precipitate reaches the sediments. In oligotrophic lakes where the benthic region is always highly oxygenated, little change occurs. It is generally believed that iron ore beds were deposited in ancient seas by this process. In many more productive lakes the benthic region can become anoxic in summer. Once the lake sediments become anoxic, the $Fe^{3+} \rightleftharpoons Fe^{2+}$ equilibrium shifts to the right, releasing soluble ferrous iron. Unfortunately

the exact level of dissolved oxygen or redox potential E_h at which this equilibrium shifts is difficult to measure in lake sediments.

A diagrammatic representation of the iron cycle is given in Fig. 10-7. Much of the crystalline form of iron flowing into lakes from the watersheds is unavailable. Lake Tahoe, which has very low levels of dissolved iron, has large particles of insoluble iron pyrite, or fool's gold, along its beaches. Crystalline iron, alone or mixed with clays and soils, may pass directly to the sediments without dissolution. The iron cycle consists of two major compartments, the particulate fraction and the soluble fraction, which includes most colloidal iron. The particulates are represented by iron contained in organic matter and hydroxides. As mentioned previously, phosphate is commonly considered an important precipitating agent of iron, and the two nutrients may scavenge each other from solution. Solubility considerations suggest, however, that if such scavenging occurs, it is not by the mechanism of direct precipitation but rather by an indirect process such as adsorption of phosphate on ferric hydroxide as suggested by Mortimer (1971) or uptake of iron and phosphate by microorganisms which Shapiro et al. (1971) observed at Lake Washington.

The soluble or colloidal compartment contains all the iron normally measured in soluble form (Fig. 10-7). The hydrated ferric iron (symbolized Fe^{3+} for simplicity) occupies a central position in the system. It is in equilibrium with nearly every other iron species. In oxygenated waters, the oxidation-reduction reaction favors the ferric form (Elder and Horne, 1978a). At pH levels encountered in most natural waters, the ratio of ferric to dissolved iron is near unity. This, coupled with the fact that ferrous iron does not easily form complexes, renders the ferrous contribution to the iron cycle small in oxygenated waters. Anoxic conditions and low redox potentials reduce ferric iron to the more soluble ferrous iron and alter the balance of the various fractions. In the hypolimnia of eutrophic lakes such conditions result in release of large quantities of ferrous iron.

Most animals acquire iron directly from their food, while phytoplankton must remove iron from solution or suspension. Perhaps the most intriguing problem concerning iron uptake is its possible role in algal competition. Little ionic iron is available in the oxygenated mixed layer of lakes, so most phytoplankton must ei-

ther take up chelated iron directly or break the bond between iron and organic matter. Some algal species use only inorganic iron, others use chelated iron, and some both. Nevertheless, all algae do not have an equal ability to take up chelated iron. Murphy, Lean, and Nalewajko (1976) describe how blue-green algae can gain a competitive advantage over other plankton by an extracellular secretion of powerful iron chelators called *siderochromes*. Iron deprivation in these blue-green algae induces hydroxamate chelators which suppress the growth of other algae by depriving them of iron (Fig. 10-8). Blue-green algae have this advantage because their chelators are stronger than those produced by competing algae (Groth, 1971). Bacteria also produce siderochromes and, when growing on particles in lakes or on dead leaves in streams, may compete with algae for available iron. Extracts from blue-green algal blooms are sometimes toxic to other phytoplankton (e.g., Williams, 1971; Lange, 1971; Lefevre et al., 1952). The inhibition of competing algae by blue-greens may be the result of direct toxicity or iron limitation induced by the production of siderochromes.

Seasonal Cycles Seasonal changes in the concentration of particulate iron are much smaller than those of soluble iron and often follow the rise and fall of the plankton. Figure 10-9 shows seasonal iron cycles in several different types of lakes. These lakes range in mean depth from 2.4 to 15 m and range in area from a few hectares to over 17,000 ha. The lakes show large differences in absolute levels of soluble iron but have rather similar levels of particulate iron. At present one can only speculate that iron limitations and severe interspecific algal competition for iron may occur in those lakes with low levels of soluble iron. Where iron levels are high it seems unlikely that blue-green algae could chelate and thus make unavailable all soluble iron (Fig. 10-9). As mentioned previously, stimulation of natural or pure algal cultures by iron has often been shown in the

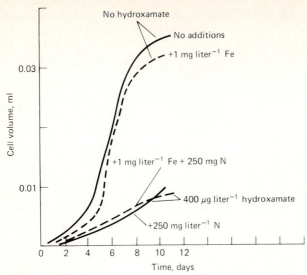

Figure 10-8 Suppression of growth of the alga *Scenedesmus basiliensis* by hydroxamate chelate (iron-removing organic) secreted by the blue-green bloom-forming alga *Anabaena flos-aquae*. The very large addition of nitrogen (e.g., versus Fig. 10-2), may also have been the cause of inhibition of growth in the lower two curves since nitrogen always accompanied hydroxamate additions. (*Modified from Murphy, Lean, and Nalewajko, 1976.*)

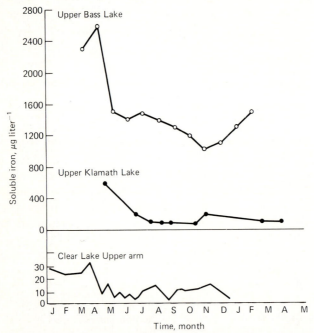

Figure 10-9 Seasonal cycles of dissolved iron in the upper mixed layers of three lakes. Upper Bass Lake (a small dimictic lake in Ontario, Canada), Upper Klamath Lake (midsized, very shallow, and polymictic, Oregon, United States), Clear Lake (a midsized polymictic lake in California, United States). Notice the great differences in absolute quantities of soluble iron in each system but the generally similar pattern of lower soluble iron in summer. [*From McMahon, 1969; Phinney and Peek, 1960 (cf Horne, 1975); W. A. Wurtsbaugh and A. J. Horne, unpublished.*]

laboratory. Dramatic increases in nitrogen fixation can be produced by addition of iron to samples taken during blooms of the blue-green algae *Aphanizomenon* or *Anabaena* (Fig. 10-10). This is due to the need for iron in the nitrogenase enzyme and in the respiratory system of the heterocyst which protects nitrogenase from being denatured by oxygen (Chap. 8). However, studies in enclosed vessels may overestimate iron limitation. Like phosphate limitation, iron deficiency may be satisfied

under natural conditions by materials excreted by zooplankton.

Diel Variations Since iron may play a role in interalgal competition, daily changes in its concentration are important. Iron levels are usually highest between noon and midafternoon (McMahon, 1969; Elder and Horne, 1978*b*). In stratified lakes and in rivers, animal excretion and lake mixing constitute the main sources of daily variations in iron levels. Inflowing rivers and atmospheric fallout of dust and rain may

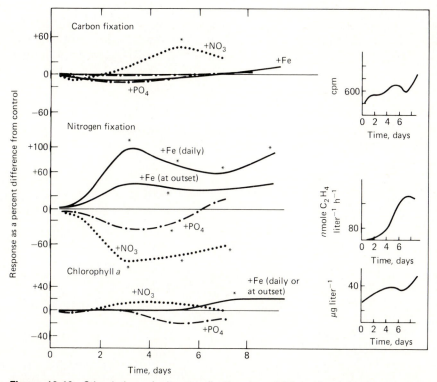

Figure 10-10 Stimulation of nitrogen fixation by addition of soluble iron to natural phytoplankton assemblages. The experiment was carried out at the end of the spring *Aphanizomenon* bloom in Clear Lake, California. The initial soluble iron concentration was 4 μg liter^{-1} and total filterable iron was 890 μg liter^{-1}. At this time the lake contained 80 μg liter^{-1} $NO_{3-} - N$, 225 μg liter^{-1} $NH_4 - N$ and 37 μg liter^{-1} $PO_4 - P$. 20 μg liter^{-1} Fe, 200 μg liter^{-1} $NO_3 - N$ and 100 μg liter^{-1} $PO_4 - P$ were added only at day 0. In the case of daily addition of iron, 20 μg liter^{-1} was added on each day. Small graphs show absolute levels of the control, large graphs percent difference from control, * = significance P > 0.01. Note that only iron, not nitrogen or phosphorus, stimulated nitrogen fixation and that photosynthesis and chlorophyll a were almost unaffected. c.p.m. = ^{14}C-uptake in counts per minute. (*From W. A. Wurtsbaugh and A. J. Horne, Can. J. Fish. Aquatic Sci., in press.*)

also contribute considerable quantities, but these do not fluctuate so rapidly. Diel vertical migrations of zooplankton, fish, and some algae change iron levels by uptake or excretion. Afternoon convection and wind mixing of the epilimnion may disrupt the thermocline, bringing hypolimnetic iron up from below (Chap. 5 and Fig. 10-11).

Finally, mention should be made of one of the most familiar iron-related sights in nature—that of the rusty-brown iron deposits in small streams and marshes. These deposits are composed of iron oxide or hydroxide which forms when seepage from acid and anoxic areas such as peat bogs encounters the well-aerated water. These deposits are often made slimy by a covering of iron bacteria (Chap. 11).

TRACE ELEMENTS: Mn, Zn, Cu, Mo, Co

Trace metals are important in aquatic ecosystems. However, technical difficulties involving the chemical form and biotic availability of the minute quantities present have hindered ecological interpretation (Table 10-3). Annual, seasonal, and daily variations in temperature, pH, dissolved oxygen, and the chelating capacity of the water may rapidly change both the absolute and biologically available quantities of most metals.

Nevertheless, some generalizations can be made regarding how these metals influence aquatic plant growth. First, trace-metal limitations are most likely to occur in oligotrophic lakes, particularly those with small drainage

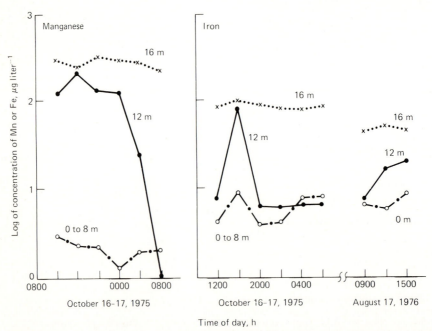

Figure 10-11 Daily changes in concentrations of two soluble trace metals with depth. Note the low constant surface (0–8m) values in the oxygenated epilimnion and the constant higher values in the anoxic hypolimnion. The variable levels at 12m in the thermocline may be due to the breaking of internal waves or leakage from the shoreline area where the stratified layer meets the lake bed. The lake is Perris Reservoir in southern California. (*From Elder et al. 1979.*)

basins, on basic or ultrabasic rocks. Micronutrient limitation in lakes was first discovered in Castle Lake, California, utilizing a then-new radiocarbon bioassay technique (Goldman, 1963, 1972). Second, the actual concentration of a trace metal often gives little information on its role as a limiting factor as a result of algal luxury consumption, recycling, and the variation in the ability of organisms to extract chelated and unchelated forms. Third, effects of additions of one trace metal which stimulates growth are likely to reveal a spectrum of other elements which are very close to limiting plant growth at the time. When one element is limiting it is likely that others are also approaching limiting levels. Physical and chemical availability of metals is often controlled by the same mechanism, such as the redox potential E_h at the sediment-water interface (Chap. 7). Living organisms are chiefly responsible for the cycling of copper, zinc, cobalt, and molybdenum, whereas redox potential in the sediments is more significant in the cycles of manganese and iron.

At the higher trophic levels there is no aquatic analog of the well-known specific nutrient deficiencies found in terrestrial systems. For example, the growth of cattle or sheep is commonly reduced by lack of cobalt, molybdenum, or magnesium even when grass is plentiful. In contrast, it is usually a simple lack of food which limits fish productivity in trace-element–deficient oligotrophic mountain lakes. Trace-element fertilization of lakes simply provides more plant food without relieving any obvious animal metal deficiency.

Toxic effects of trace metals on some animal populations in lakes are well known, particularly that of copper on young salmonids in unbuffered mountain streams (Sellers et al., 1975). Dramatic poisoning of streams by zinc, copper, and other metals in runoff from mining areas, or chromium from industrial use, are also well known (Weatherley and Dawson, 1973; Whitton, 1970; Bryan and Hummerstone, 1971).

In oxygenated waters the soluble forms of trace metals are present in microgram quantities and rarely exert toxic or stimulatory effects. One reason for this is the evolution of biotic mechanisms for uptake or chelation of metals. The presence of organic matter in all lakes, together with inorganic particulate debris, produces stable chelates of trace metals and prevents their precipitation as insoluble inorganic salts (Groth, 1971). Most organisms can accumulate metals far in excess of their immediate needs. The actual concentrations of dissolved metals found in the photic zone depend on the interplay between uptake by the phytoplankton, release by decay from sediments, and input from rivers. In stratified eutrophic lakes, decay and subsequent release of metals exceeds uptake, increasing the soluble-metal content in the hypolimnia. Manganese and iron (Fig. 10-11) are easily released from anoxic sediment, while zinc, cobalt, molybdenum, and copper are less mobile.

Manganese

This element, whose cycling closely parallels iron (Fig. 10-11), is needed by plants for photosystem II, the oxygen-evolving process in photosynthesis. Animals and plants need manganese as a cofactor in several enzyme systems, including those involved in respiration and nitrogen metabolism. Manganese toxicity has been reported only in streams polluted by mine wastes.

Considerable effort has gone into determining the effects of manganese on aquatic algal communities since the early work of Guseva (1939). She suggested that blue-green algal blooms in a Russian reservoir were sometimes inhibited by naturally occurring levels of manganese. Later work by Patrick et al. (1969) suggested that this conclusion was also applicable to stream algae. Although there has been no successful repetition of these studies, it now appears that calcium normally prevents any measurable toxic effect. The levels of manganese (> 2 mg liter^{-1}) which were toxic in these laboratory experiments are larger than those usually encountered. In contrast to its toxic effect, Goldman (1966) and Lange (1971) have shown that some algal populations increase with small manganese additions. In general, it is apparent that manganese plays a minor role in regulating the growth of algal populations in most lakes.

Zinc

Zinc serves as an activator in some enzymatic reactions and is a cofactor for the enzyme carbonic anhydrase. This enzyme catalyzes a critical rate-limiting step for carbon use in photosynthesis (Chap. 7). Apart from the normal flow of trace-element–loaded sediment caused by land disturbance, the zinc plating of pipes, gutters, and culverts can add greatly to zinc levels in streams. As with most trace elements, serious limitation of phytoplankton growth by zinc deficiency, although known to occur (Goldman, 1965), has not been widely studied. In Lake Tahoe, only unchelated zinc is inhibitory to algal growth, while chelated zinc can be stimulatory (Elder, 1974). The quantities of soluble zinc produced by mining wastes are often large enough to decimate the flora and fauna of the receiving water in the vicinity of the discharge.

Copper

In phytoplankton, copper functions mainly as a metalloprotein component and is probably a catalyst. Enzymes concerned with nitrate transformations require the presence of copper. Copper forms the active center of the molluscan blood pigment hemocyanin. Despite the frequently low availability of lake copper,

deficiencies of this metal have rarely been demonstrated (Goldman, 1965; Horne, 1975). Even in the case of eutrophic Clear Lake, California, stimulation was small and the onset of toxic effects rapid (Fig. 10-12).

Copper toxicity is well known and appears to exert its major effect on algae by interfering with the activity of enzymes situated on cell membranes (Steeman Nielsen and Wium-Anderson, 1971). This interference prevents cell division and leads to an eventual cessation of photosynthesis by product inhibition. Toxic effects are much more pronounced at lower trophic levels, especially in phytoplankton and zooplankton, than at higher trophic levels. As little as 0.1 μg liter^{-1} ionic copper can kill some algae in water with low-chelation potential, and even in normal lakes, 5 to 10 μg liter^{-1} affects blue-green algae. In contrast, most fish are almost unaffected at 100 to 500 μg liter^{-1} copper. Even trout, which are very sensitive to metal poisoning, especially in soft waters, often survive levels of 30 μg liter^{-1}. The relative tolerance of fish to copper makes it possible to control nuisance algae with copper sulfate applications. The method has been used for over 60 years on some lakes, with no significant long-term changes in benthic animal populations. Algae may develop a resistance to copper, which often necessitates the use of other methods of algal control (see Chap. 20). The major problem in the use of copper sulfate is its rapid loss from solution. At pH levels of 6 to 8.5, most ionic copper is lost by precipitation as the mineral malachite, which is hydrated copper carbonate ($CuCO_3 \cdot Cu[OH]_2$). Some loss of ionic copper by sorption or chelation by organic materials is probable. Low levels of copper (5–20 μg liter^{-1}) can persist in soluble form for several days, if not weeks (Elder and Horne, 1978b). If low levels of soluble copper are ionic, they will depress blue-green algal growth while not affecting other phytoplankton. However, in nature most copper

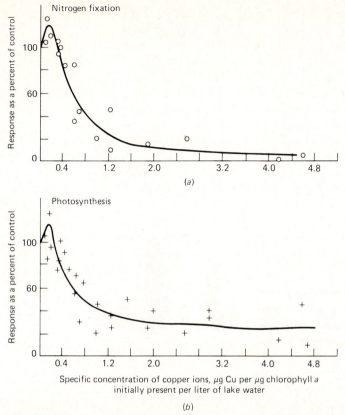

Figure 10-12 Effect of added ionic unchelated copper on photosynthesis and nitrogen fixation by the Clear Lake, California, phytoplankton population. Note the apparent slight stimulation response at very low additions and the rapid toxic effect soon after. (*From Horne, 1975; W. A. Wurtsbaugh and A. J. Horne, 1982. Can. J. Fish. Aquatic Sci.*)

is chelated, nontoxic, and bears no apparent relation to algal growth patterns.

Molybdenum

This metal is needed for several enzymatic processes, particularly those involved in the nitrogen cycle, since molybdenum is part of the active center of nitrate reductase and nitrogenase enzymes (Chap. 8). Molybdenum deficiency, but not toxicity, for phytoplankton growth is well known. This may reflect the emphasis placed on this element, since it was the first micronutrient found to limit algal

growth in lakes (Goldman, 1960). In Castle Lake, California, a clear picture of its role has emerged. Bottle bioassays indicated that natural phytoplankton populations were limited by molybdenum at certain times of year (Fig. 10-13d). The main inflow of molybdenum is via springs which flow through an alder stand prior to entering the lake (Fig. 6-1). Since these trees also require molybdenum for nitrogen fixation as well as for nitrate reductase, virtually all the molybdenum is stripped from the groundwater inflows. Paleolimnological studies of the lake sediments (Chap. 18) indicated that in the past a

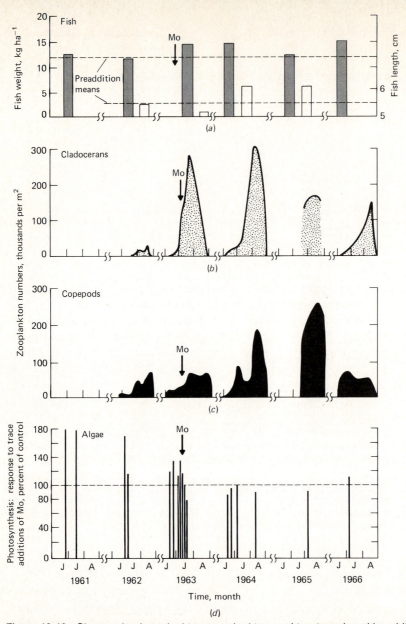

Figure 10-13 Changes in phytoplankton, zooplankton, and trout produced by addition of small quantities of a limiting trace metal, molybdenum, to Castle Lake, California. (*d*) Note immediate cessation of stimulation of photosynthesis by further additions of molybdenum in bottle assays (*c*) delayed increase in populations of the longer-lived copepod zooplankton, but (*b*) almost immediate response for the rapidly reproducing cladoceran zooplankton which were mostly *Daphnia rosea*. Despite natural year-to-year variations in cladoceran numbers, the prolonged increase over 4 years after molybdenum addition was probably due to the addition. (*a*) Note rapid response in fish catches or yield and increase in average fish length in years following molybdenum addition. This response by fish is similar to that found in fish ponds fertilized with nitrogen and phosphorus where effects usually occur in the first year after addition and last for 3 to 4 years. Arrow indicates date of molybdenum addition to the epilimnion on July 1963. Lake epilimnion concentrations rose from less than 0.2 to 7 μg liter^{-1}, while hypolimnion levels remained at less than 0.2 μg liter^{-1}. Phytoplankton numbers and primary production (not shown) rose less dramatically in the years after molybdenum addition, presumably due to rapid grazing of the algae by the increased zooplankton crop.

Table 10-4   Numbers of Cells per Liter of the Major Castle Lake Phytoplankton from June 29, 1959, Cultures Incubated for Five Days with and without Molybdenum*

Species	Control	+100 mg liter^{-1}Mo
Dinobryon sertularia	99	255
Cyclotella meneghiniana	401	301
Synedra radians	268	345
Total algae counted	1984	2376

*Modified from Goldman et al., 1975.

larger molybdenum supply, presumably now eroded away, had provided adequate amounts of molybdenum.

In one of the first whole-lake experiments, molybdenum was added to increase the lake concentration to 7.7 μg liter^{-1} from less than 0.2 μg liter^{-1}. Most interesting were the responses of the higher trophic levels which could not be assessed adequately by bottle tests. Cladocerans increased 10-fold, and fish yields also exceeded the average of the preceding quarter century (Fig. 10-13a to c). A species change was observed after molybdenum addition to Castle Lake, when the chrysophyte *Dinobryon sertularia* increased rapidly relative to other algae (Table 10-4).

Cobalt

Cobalt is needed in vitamin B12 or as a metal with which to synthesize that vitamin. Cobalt in vitamin B12 lies at the active center of four pyrrole rings in a fashion almost identical to magnesium in chlorophyll (Fig. 10-4) or iron in the blood pigment hemoglobin. Like molybdenum, cobalt has been shown to stimulate production in some oligotrophic lakes, and it is certainly present in very small quantities in almost all lakes. The actual amounts in lakes are largely controlled by biological cycling.

Other Trace Metals

As bioassay experiments are more widely applied, it is likely that specific trace-element responses will appear from a greater diversity of aquatic habitats. A recent example is the discovery of a specific requirement for selenium in the growth of the dinoflagellate *Ceratium* in Lake Kinneret (Lindström and Rodhe, 1977). Selenium, like arsenic, can be very toxic to cattle or people. It is an element deserving further study in aquatic systems.

At this point the reader has been exposed to the physical and chemical structure of aquatic ecosystems. This is the environment which forms the habitat for the host of organisms living in lakes, streams, and estuaries. These organisms will be discussed next in general terms in Chap. 11 and in more detail in subsequent chapters.

FURTHER READINGS

Egeratt, A. W. S. M., and J. L. M. Huntjens (eds.). 1975. "The Sulphur Cycle." *Plant Soil*, **43**:228 pp.

Goldman, C. R. 1960. "Primary Productivity and Limiting Factors in Three Lakes of the Alaska Peninsula." *Ecol. Monogr.*, **30**:207–230.

Goldman, C. R. 1972. "The Role of Minor Nutrients in Limiting the Productivity of Aquatic Ecosystems." *Symposium on Nutrients and Eutrophication*, pp. 21–33. American Society of Limnology and Oceanography, Special Symposia vol. 1.

Groth, P. 1971. "Untersuchungen über einige Spurenelemente in Seen" ("Investigations of Some Trace Elements in Lakes"). *Arch. Hydrobiol.*, **68**:305–375. (Summary in English.)

Kellogg, W. W., R. D. Cadle, E. R. Allen, A. L.

Lazrus, and E. A. Martell. 1972. "The Sulfur Cycle." *Science,* **175:**587–596.

Stewart, W. D. P. (ed.) 1974. *Algal Physiology and Biochemistry*. Botanical Monographs, vol. 10. University of California Press, Berkeley. (See especially J. C. O'Kelley, "Inorganic Nutrients," chap. 22, pp. 610–635; W. M. Darley, "Silicification and Calcification," chap. 24, pp. 655–675; and E. A. C. MacRobbie, "Ion Uptake," chap. 25, pp. 676–713.)

Whitton, B. A. 1970. "Toxicity of Heavy Metals to Freshwater Algae: A Review." *Phykos,* **9:**116–125.

Chapter 11

Organisms in Lakes, Streams, and Estuaries

OVERVIEW

The reader has been exposed to the physical and chemical nature of inland waters, and it is now appropriate to consider the plants, animals, fungi, and bacteria that give life and a great deal of interest to limnology. Here, limnology diverges sharply from classical zoology and botany and assumes the more general view of the ecologist. The relationships of the components of the aquatic ecosystem may, in this approach, be considered of greater significance than their taxonomic status. Their trophic relationships, in terms of which are the producers, which are the consumers, and which are the recyclers, achieve the greatest importance. Recognition of the individual species, however, is likely to prove essential once the more general relationships are understood. This chapter, especially Tables 11-1 to 11-9, is intended as a reference for organisms mentioned in other chapters, and only those taxonomic groups most familiar to limnologists have been included.

INTRODUCTION

The compartmentalization of plants and animals into trophic levels in the aquatic ecosystem has become very useful in understanding their general relationships (Chap. 15). Although the levels themselves are often not discrete entities, they do provide a certain order which the limnologist has found convenient. If we consider the pyramid of biomass and energy in the aquatic system (Chap. 15), we see that most of the organic material resides at the level of the primary producers and that there is a progressive decrease at each successive level.

FUNCTIONAL CLASSIFICATION

Organisms can be classified at the functional level by their sources of energy, sources of carbon, and in the case of some lower organisms, such as sulfur bacteria, by the molecule which serves as the electron donor. Classified on the basis of their energy sources are the *phototrophs,* which derive energy directly from sunlight in photosynthesis, and the *chemotrophs,* which utilize a chemical energy source. If the cell receives electrons from an organic compound it is an *organotroph;* if it derives them from inorganic matter it is a *lithotroph.* The prefix *litho* originates from the Greek word for "stone" and also provides the root of the word *lithosphere* which is the nonaquatic portion of the earth. We also make a general division of organisms on the basis of the type of carbon used for food. *Autotrophic* organisms utilize inorganic carbon dioxide to produce organic matter, and *heterotrophic* organisms depend on preformed organic carbon such as glucose or pyruvate.

The six terms—phototroph or chemotroph, organotroph or lithotroph, and autotroph or heterotroph—can be used in various combinations. For example, the base of most aquatic food chains is formed by algae and higher aquatic plants which use light energy, carbon dioxide as a carbon source, and the oxygen in water as the electron donor. They could be called photolithoautotrophs but are referred to by the shortened term *photoautotrophs.* This is because the electron donor for almost all photosynthetic organisms is an inorganic compound.

Most heterotrophs use organic carbon for both energy and as an electron donor, while most autotrophs do not. For convenience the food chain is simply divided into autotrophs (plants and chemolithotrophic bacteria) and heterotrophs (some bacteria, fungi, and animals). As we go higher in the food chain, we encounter heterotrophic animals called *herbivores,* which utilize plant material for their growth, and

higher level heterotrophs, the *carnivores,* which feed on the herbivores and, sometimes, on each other. *Omnivores* eat both plants and animals and thus function at more than one trophic level (Chap. 15).

We will begin with a discussion of the simplest communities of lakes and streams. With the exception of higher aquatic plants that can be cut or dragged from the depths by divers the general methods for their collection, identification, and preservation are included here or in the three chapters that follow.

THE MAJOR GROUPS OF ORGANISMS

Viruses

The role of viruses in the aquatic environment has largely been associated with problems of public health. Such well-known diseases as hepatitis can be transmitted by ingestion of natural waters contaminated with the feces of infected people. Viruses are obligate intracellular parasites. They are very small, usually about 0.02 μm across, although some forms are 10 times that large. They can only be seen individually by electron microscopy. Viruses include phages which attack only bacteria and blue-green algae. Viruses are usually identified by their protein shell or the disease they cause. Little is known about their role in limnology other than that cyanophages are associated with the decline of blue-green algae growing as periphyton or phytoplankton (Chap. 12). The use of cyanophages as an algal nuisance-control technique has been attempted, but blue-green algae may build up a resistance to the virus just as human beings develop immunity to influenza virus. Presumably, virus attacks affect many aquatic organisms and may play an important role in their population dynamics. The role of viruses remains an area needing further study.

Bacteria

Although there may be a million bacterial cells per cubic centimeter in the open waters of lakes and rivers, these constitute relatively low numbers when compared with bacteria in soil. Bacteria are unicellular organisms which can multiply very quickly by simple division but which occasionally reproduce sexually. Most do not contain chlorophyll or carry on photosynthesis, although a special group of anoxic photosynthetic bacteria are found in some lakes and estuaries.

The cell wall of bacteria is made up of a mixture of acetylated sugars and amino acids called a *peptidoglycan.* When present in abundance in the cell wall, it is stained by a crystal violet dye–iodine complex which is not removable by alcohol. These bacteria are called *gram-positive*. If little peptidoglycan is present, the dye complex is easily removed by alcohol and the bacteria are referred to as being *gram-negative*. This method is extensively used in identifying various groups of bacteria (Table 11-1).

Bacteria have a great variety of shapes (Table 11-1), but small coccoid, rod-shaped, single, and chain-forming bacteria are the most common. Most taxonomic groups contain many examples of cocci, rods, and spiral forms. The limnologist is also likely to encounter bacteria embedded in thick mucus sheaths. A common example is *Sphaerotilus,* the "sewage fungus," which is quite common in streams where organic matter is present and may become superabundant in polluted waters. Like some of the unicellular algae, many bacteria can swim by using hairlike flagella which may be attached at either or both ends of the cell or all around the cell. Unfortunately, the shape of bacteria provides little insight into their role in aquatic ecosystems.

In addition to morphology, bacteria are classified by mode of nutrition, or by serology which detects characteristic chemicals such as peptidoglycan. Most bacteria are very small, between 0.2 and 5 μm in length, and often appear only as dots when viewed under the highest power of the light microscope. Consequently, their ability to perform chemical transformations has become the standard method for their identification and classifica-

tion. Although this functional classification requires time-consuming procedures, it should be familiar to limnologists who are already acquainted with the functional role of higher organisms (Chaps. 15, 16). The nutritional requirements are found by giving the bacteria a range of substrates and determining which promote growth and what products are released. Bacteria have very specific requirements for organic or inorganic nutrients and release characteristic products. For example, some can only convert acetate to carbon dioxide, some convert nitrite to nitrate, some require anoxic conditions, and still others require light. Collectively, bacteria can mediate more chemical transformations than any other group of organisms. These include the nitrogen transformations such as nitrogen fixation, denitrification and nitrification, breakdown of cellulose, mineralization of carbon and sulfur, spoilage of food, as well as disease production in plants and animals. Selective breeding and genetic engineering concentrate on enhancing the transformations most useful to human beings. The oil-degrading bacteria helpful in cleaning up oil spills provide a good example. A general bacterial classification, based on nutritional requirements, is Bergey's manual (Buchanan and Gibbons, 1974), and a summary of the parts most useful to limnologists is given in Table 11-1. Oxygen plays a major role in this classification, because some bacteria require oxygen, others cannot tolerate it, while yet others called *facultative* forms can grow with or without it.

The main role of bacteria in nature is in the recycling of organic and inorganic materials. Heterotrophic bacteria cause decay and provide a nutritious layer for detritivorous animals feeding on decaying organic particles in rivers and lakes (Chaps. 15, 16). Bacteria sometimes produce outbreaks of disease which decimate the host populations. For example, the "salmon disease" in New England has been attributed to *Cutophage* spp., members of the gliding bacterial group (Table 11-1).

Table 11-1 Partial Classification of the Bacteria Common in Aquatic Habitats

Taxonomic division	Aquatic habitat	Substrate for energy, electrons, or carbon	Example and some products
Requiring light			
1. Photolithotrophic or photoautotrophic	Dim, anoxic layer below thermocline in deeper lakes or mud surfaces in marshes and estuaries	H_2S, CO_2	Green and purple sulfur bacteria, e.g., *Chromatium* ($H_2S \rightarrow S$)
2. Photoorganotrophic or photoheterotrophic		Organics, CO_2	Purple, nonsulfur bacteria, e.g., *Rhodospirillum*
Not requiring light			
3. Gliding	Detritus, sediments, parasitic on other organisms, sulfur springs	Cellulose, chitin, H_2S	*Cytophaga* (cellulose, chitin) *Beggiatoa* ($H_2S \rightarrow S$)
4. Sheathed	Attached to rocks, logs, in flowing water	Organics	*Sphaerotilus,* the sewage fungus
5. Budding or appendaged	Streams, water pipes	H_2S, predatory on other bacteria	*Thiodendron* ($H_2S \rightarrow S$) *Caulobacter* (predatory)
6. Spiral or curved	Eutrophic fresh or saline water	Organics, N_2, predatory	*Spirillum* *Bdellovibrio* (predatory)
7. Gram-negative, oxic, rod and cocci-shaped	Fresh, saline water, and living organisms	Organics, N_2, CH_4	*Azotobacter* ($N_2 \rightarrow$ org-N) *Methylomonas* ($CH_4 \rightarrow$ org-C)
8. Gram-negative, facultatively anoxic, rod-shaped	In plants, animal intestinal tracts, fresh or salt water	Parasitic, disease-causing in animals and plants	*Escherichia coli*
9. Gram-negative anoxic	Anoxic sediments	SO_4, S, N_2, parasitic	*Desulfovibrio* (SO_4, S $\rightarrow H_2S$)
10. Gram-negative, chemolithotrophic	Oxic lake sediments, rivers, sulfur springs	Reduced N or S compounds	*Nitrosomonas* ($NH_4^+ \rightarrow NO_2$) *Nitrobacter* ($NO_2 \rightarrow NO_3$) *Thiobacillus* (H_2S, S $\rightarrow SO_4$)
11. Methane-producing	Anoxic hypolimnia, muds, marshes	CO_2	*Methanobacterium* ($CO_2 \rightarrow CH_4$)

Chemolithotrophic bacteria found in the sediments or on particles suspended in natural waters are responsible for the oxidations of ammonia first to nitrite and then to nitrate, hydrogen sulfide oxidation to sulfate, and oxidation of ferrous (Fe^{2+}) iron to ferric (Fe^{3+}) iron (Table 11-1). They produce the familiar whitish-yellow sulfur patches on rich organic mud flats of estuaries or saline lakes and form the rust-colored gelatinous masses often found in streams.

Layers of chemolithotrophic purple photosynthetic bacteria are often found in anoxic conditions in lakes or on the mud surface of tide flats. Unlike algae, photosynthetic bacteria cannot tolerate oxygen. These bacteria are known for their ability to exist in platelike layers at very low light levels in or just below the metalimnion. The plate is usually found below the zone where algae are likely to grow. Photosynthetic bacteria occur in many productive lakes, particularly saline ones, and, because of their narrow vertical distribution, may go unnoticed during routine sampling.

Little is known about the role and numbers of bacteria in aquatic ecosystems. The *coliform bacteria* that are particularly well known to public health officers and sanitary engineers are the facultative, gram-negative, rod-shaped, non-spore-forming ones which can ferment lactose. They are associated with sewage since many, but not all, are derived from the digestive tract of mammals. Although there will always be coliform bacteria present in natural waters, high levels provide evidence of water pollution from sewage.

Limnologists are becoming increasingly familiar with the morphology of bacteria due to the use of the scanning electron microscope (SEM), but only dead bacteria can be observed. SEM observation has shown epiphytic bacteria apparently in the process of digesting moribund diatoms (Fig. 11-1a,b) or growing on living blue-green algae (Fig. 11-1c). Unlike other plankters, most planktonic bacteria are in a rest-

ing stage. Of the millions of bacteria present in a cubic centimeter of water, only a few thousand, at most, are likely to be metabolically active. The remainder wait for a solid surface or suitable nutrient conditions to initiate enzyme activity. The physiological response can be rapid, and bacteria may grow unchecked until their food supply is exhausted. An example of the division between total and viable bacteria is given in Fig. 11-2 for relatively unproductive Windermere and more eutrophic Esthwaite Water in the English Lake District. Very few bacteria were viable in the oligotrophic lake relative to the more eutrophic one, but the total number of bacteria was similar in both lakes (Fig. 11-2). The presence of more viable bacteria in the epilimnion of Esthwaite Water than in that of Windermere correlates well with the number of phytoplankters. Similar results have been found in German lakes (Overbeck and Babenzien, 1964) and a tropical lake in the Amazon region (Schmidt, 1969). In general, exponentially growing phytoplankton are relatively bacteria-free, whereas dead or dying algae are rapidly attacked (Fig. 11-1). It has even been suggested that diatoms growing far from shore and organic substrates in the oligotrophic, tropical oceans are virtually bacteria-free.

Fungi and Fungi-like Organisms

Like viruses and bacteria, the aquatic fungi participate in the decomposition and recycling of vegetable and animal matter. An obvious example, known to fisheries biologists, is *Saprolegnia*, a frequent parasite of dying or injured fish. Spawned-out salmon may be completely encased in a white cocoon of *Saprolegnia* before they die. Primitive parasitic fungi such as chytrids are common in many lakes and may alter the species composition of spring algal blooms (Chap. 12).

All fungi have rigid cell walls and are either *saprophytic,* using organic substances for growth, or *parasitic* (Table 11-2). The thin film

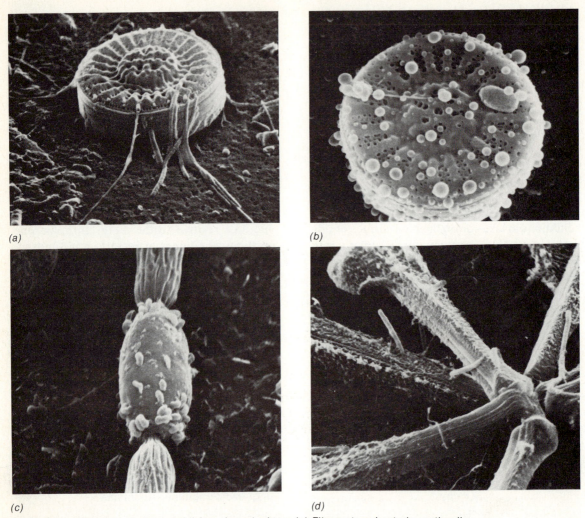

(a)

(b)

(c)

(d)

Figure 11-1 Microbes on dead and living phytoplankton. (a) Filamentous bacteria on the diatom *Cyclotella* from Lake Michigan. The algal cell is about 15 μm in diameter. (b) Coccoid bacteria on *Cyclotella* from Lake Tahoe. (c) Bacteria on the heterocyst of the blue-green alga *Aphanizomenon flos-aquae* from Clear Lake, California. Magnification ×9000. The bacteria are clustered around the junctions with adjacent vegetative cells, possibly leaking organic material. (d) Bacterial infection on the diatom *Asterionella* late in the spring bloom in Lake Tahoe. (*SEM pictures by H. W. Paerl.*)

of bacteria and fungi present on virtually all submerged organic detritus is a major food source for river and lake invertebrates. Although most bacteria can use detrital cellulose, only fungi and a few bacteria possess the special enzymes that can break down lignin, which is the skeletal material of leaves. Fungi

may digest the majority of plant cellulose and the chitinous skeletons of insects. Fungi grow and follow the bacteria in a second phase in the decomposition of detritus. True fungi and the Actinomycetales are classified by their sexual structures and shapes. In water the lower fungal orders, phycomycetes, some yeasts, and the

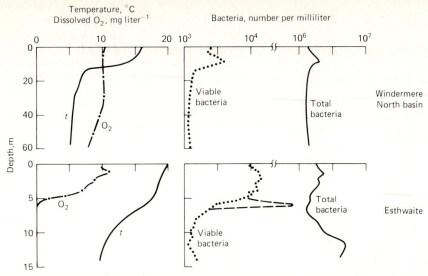

Figure 11-2 Total and viable bacteria in a deep mesotrophic lake, (Windermere, North Basin) and a shallow more eutrophic lake (Esthwaite Water) during the summer stratification. Dashed line indicates short-term maximum in the epilimnion. Although the total bacterial numbers are similar in both lakes, note that numbers of viable bacteria are much greater in the eutrophic lake. (*Redrawn from Jones, 1977.*)

Fungi Imperfecti are the most common (Table 11-2).

Green Plants: Algae and Macrophytes

Algae Shallow habitats may have extensive growth of higher aquatic plants, but algae dominate primary production in most aquatic ecosystems. They are present as attached algae or as free-floating phytoplankton (Fig. 11-3a,b). The former dominate the shallow areas of clear-water lakes and streams, and the latter are most important in larger lakes and in the slowest reaches of rivers. The algae, an extremely diversified group, are separated from higher plants more upon what they lack than what they have. Photosynthetic algae lack roots, stems, and leaves. Algae may be unicellular, colonial, or filamentous in form. They can also develop special reproductive cells, but rapid growth occurs by vegetative reproduction.

The primary classification of algae into different divisions is accomplished on the basis of pigment composition, how they maintain their energy reserve, cell wall composition, locomo-tory organs, and their general structure (Table 11-3). The algae contain two main groups of pigments, chlorophylls and carotenoids. Chlorophyll is a complex molecule composed of four carbon-nitrogen rings surrounding a magnesium atom (Fig. 10-4). The loss of the metal atom produces the common degradation product phaeophytin. Carotenoids are composed of linear unsaturated hydrocarbons called *carotenes*. The orange color of carrots is caused by this pigment and is the basis of the name. The red color of some zooplankton, salmon, and trout is derived from a high carotenoid diet. Chlorophyll *a* is the only form which can pass electrons, excited by light energy, to produce chemical energy in photosynthesis. All other pigments including the widespread *b* and *c* forms of chlorophyll, carotenes, xanthophylls, and the less common phycocyanins, phy-coerythrins, and phycopyrrins, are accessory pigments. These gather light in regions of the spectrum where chlorophyll *a* does not absorb light. The accessory pigments then pass elec-trons to chlorophyll *a*. This allows growth in the

Table 11-2 Classification of Fungi and Fungilike Organisms Common in Aquatic Habitats

Taxonomic division	Aquatic habitat	Example
Organism with true branching mycelium but a prokaryotic cell organization		
Actinomycetes (Actinomycetales)	Damp leaves, streams, benthic sediments of lakes	*Actinoplanes* *Streptomyces* (odor-producing form)
Organisms with eukaryotic cells and often a true branching mycelium		
1. Phycomycetes: a. Chytridiomycetes (water molds)		*Rhizophydium* (parasitic, single cell) *Allomyces* (saprophytic, filamentous)
b. Oomycetes (water molds and mildews)	Most fresh or saline waters, on living or dead animals or plants	*Saprolegnia* (parasitic on fish or their eggs) *Aqualinderella* (on fruits)
2. Ascomycetes and Fungi Imperfecti		*Cryptococcus* (aquatic yeast) *Dactylella* (captures nematode worms)
3. Basidiomycetes	Uncommon in free water, common on damp wood	*Aureobasidium* (from streams)

dim light deep in lakes where most red and blue light is absent (Fig. 3-6*b*).

The cell walls of algae are composed of cellulose and other polysaccharides, silica, proteins, and lipids. All these combine in various proportions to produce distinctive cell walls which may form the basis for taxonomic classification (Table 11-3). Silicon compounds form the cell walls of diatoms. These frequently have beautiful and intricate ornamentation, which

has added to the popularity of diatom taxonomy. The high-resolution light microscope and the scanning electron microscope (Fig. 10-2) reveal details of the three-dimensional structure of diatom shells or *frustules*. Taxonomists have classified over 10,000 species of diatoms on the basis of their frustules alone without the need to see the cells alive or preserved. The science of paleolimnology (Chap. 18) uses frustule classification because diatom shells may be the only recognizable algal cells which survive in the often acidic, bacteria-rich lake sediments. The concentrated deposits of marine diatoms now uplifted above sea level are called *diatomaceous earth* and are of commercial importance as an inert filtering substance for a variety of liquids. Swimming pool filters are traditionally packed with diatomaceous earth. Other algae have hardened but less long-lasting shells. The dinoflagellates, for example, include armored forms which have several distinctively ornamented plates (Fig. 11-6b).

Many planktonic algae have the ability to move. The most common swimming device uses one or more flagella which are whiplike structures containing contractile fibrils. These serve as oars or like a ship's propeller depend-

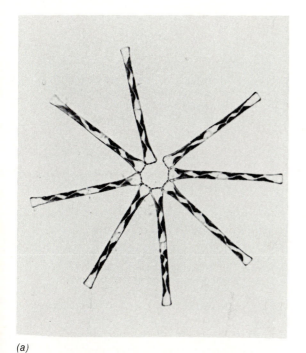

(a)

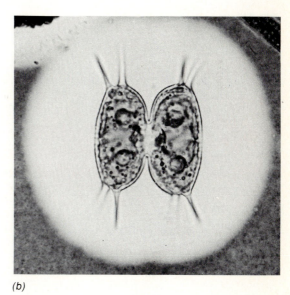

(b)

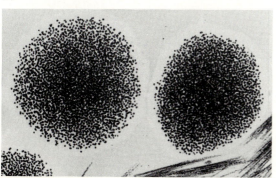

(c)

Figure 11-3 Common phytoplankton. (a) The colonial pennate diatom *Asterionella formosa* which occurs in the spring and fall plankton of most moderately productive lakes (see Chap. 12). Magnification ×520. (b) A representative of the summer plankton, the green alga *Xanthidium*. A ball of gelatinous, very hydrated mucus surrounds the cell but can only be seen if the background water is stained with india ink. These mucus balls are common in planktonic algae and may serve as protection against digestion when swallowed by animals. Magnification ×610. (c) *Microcystis aeruginosa* ×80. This colonial coccoid blue-green phytoplankton occurs in the summer and fall in more productive lakes. It is held together with mucus. Blue-green algae may form spectacular water blooms (Plates 6b, 7b; Chap. 12). (*All photographs by H. Canter-Lund.*)

Table 11-3   Classification of Algae Common in Aquatic Habitats

Name and typical color	Most common aquatic habitat	Common morphology and loco-motion	Cell wall* composition and major pigment	Examples
1. Diatoms (golden-brown) Bacillariophyta	Oceans, lakes, rivers, estuaries; attached or planktonic.	Usually microscopic, filamentous, or unicellular	Opaline silica	*Asterionella Melosira Nitzschia Navicula*
2. Green algae (grass-green) Chlorophyta	Lakes, rivers, estuaries; planktonic or attached	Microscopic or visible; filamentous, colonial, or unicellular; some are flagellated	Cellulose	*Cladophora Oocystis Ulva*
3. Dinoflagellates (red-brown) Pyrrhophyta	Oceans, lakes and estuaries; planktonic	Microscopic; unicellular or small chains; all are flagel-lated	Cellulose (when present), some silicated forms	*Peridinium Ceratium*
4. Blue-greens (blue-green) Cyanophyta	Lakes and oceans; planktonic or attached	Microscopic or visible; usually filamentous; some can float and glide	Mucopeptide (amino sugar + amino acids)	*Anabaena Nostoc Phormidium Oscillatoria*
5. Chrysophytes (yellow- or brown-green) Chrysophyta	Lakes, streams, oceans; planktonic	Microscopic; unicellular, or colonial; some are flagellated	Pectin, sometimes silicified or cellulosic	*Mallomonas Dinobryon Tribonema*
6. Cryptomonads (various colors) Cryptophyta	Lakes; planktonic	Microscopic, unicellular, flagellated	Cellulose	*Rhodomonas*
7. Euglenoids (various colors) Euglenophyta	Ponds, lakes, oceans; planktonic	Microscopic, unicellular, flagellated	Proteinaceous pellicle	*Euglena*
8. Red algae Rhodophyta	Oceans, estuaries; lakes, streams; attached	Visible or micro-scopic; colonial or unicellular	Cellulose + gels	*Gigartina Batrachospermum*
9. Brown algae Phaeophyta	Oceans, estuaries; attached	Visible, often long fronds	Cellulose	*Gracilaria Sargassum*

*Not all listed are true cell walls, but have a similar function.

movement

ing on the type of organism. Dinoflagellates such as *Peridinium* or *Ceratium* (Fig. 11-6*b*) are able to migrate very rapidly in response to changes in light, while most flagellates, for example, the chrysophyte *Mallomonas,* or *Euglena,* appear to swim randomly. Many otherwise nonmotile forms such as diatoms and attached filamentous green algae may have flagellated sexual gametes. Some pennate diatoms and filamentous blue-green algae have the ability to move by gliding when next to a solid surface. All planktonic blue-green algae contain vacuoles, minute air-filled sacs which can be created or collapsed (Fig. 12-1). This alters their buoyancy and allows them vertical movement in the water column (Fig. 12-14). Motile forms have the obvious advantage of being able to adjust their position to optimize the availability of light and nutrients. It should be recog-

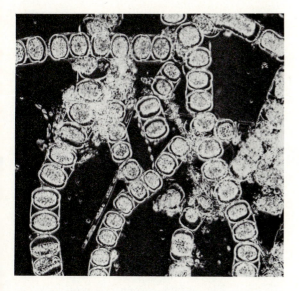

Figure 11-4 Attached algae. Long chains of the filamentous centric diatom *Melosira* scraped from rocks in San Francisco Bay. The chains are 20 to 40 μm in diameter. A long solitary pennate diatom is visible at bottom center, and bacteria, detritus, and broken cells compose the whitish amorphous areas on the right. This mixed assemblage is often called aufwuchs. See also Plate 4*b*. (*Photograph by H. M. Anderson.*)

nized, however, that this has the disadvantage of using up a portion of their available energy.

The green algae contain both chlorophyll *a* and *b,* carotenes, and some xanthophylls, but lack the phycobilins. Their photosynthetic storage is in the form of microscopically visible structures called *pyrenoids,* which consist of starch plates surrounding a protein core. The presence of starch is the basis for the commonly used starch-iodine test. The cell walls consist of an inner cellulose and outer pectinaceous layer. Green algae frequently have one phase of their life cycle which is flagellated. This enables attached filamentous forms such as *Ulothrix* to colonize rocks and logs in streams and lakes.

Blue-green algae possess chlorophyll *a,* β-carotene, and a number of unique xanthophylls and phycobilins, particularly phycocyanin-c, which gives them a distinctive blue-green color. Glycogen granules serve as their main energy store, together with globules of lipid. Granules containing cyanophycin, a polypeptide consisting of argenine and aspartic acid, provide a unique storage of nitrogen. The cell wall of blue-green algae is made of mucopeptides similar to those contained in bacteria, and many species are surrounded by a gelatinous sheath or matrix (Fig. 11-3*c*). Sexual reproduction is very rare in blue-green algae, and vegetative reproduction can be more rapid than in most other phytoplankton.

The diatoms contain both chlorophyll *a* and *c,* carotene, and a few xanthophylls, but lack phycobilins. They store their energy as fat and oil in large globules and sometimes as chrysolaminarin, a polysaccharide. The cell wall is made of silica embedded in a pectinaceous matrix and is constructed of overlapping halves called *valves.* This gives the diatoms a pillboxlike structure (Fig. 11-1*a, b*). They have no flagella except during sexual reproduction when uniflagellated spermatozoids may be produced.

Other algal classes also contribute to the diversity of the freshwater plankton flora. Examples include the dinoflagellates *Peridinium*

Table 11-4   Classification of Aquatic Macrophytes Common in Aquatic Habitats

Aquatic habitat	Taxonomic group*	Examples
Free-floating		
1. Subtropical, tropical lakes, slow streams	MA MA F	*Pistia* (water lettuce) *Eichornia* (water cabbage) *Salvinia* (water fern)
2. Temperate ponds and backwaters	F MA	*Azolla* (water fern) *Lemna* (duckweed)
Rooted		
1. Temperate marshes, lake and stream edges	MA DA MA	*Phragmites* (reed) *Rorippa* (watercress) *Scirpus* (bulrush)
2. Tropical marshes and rivers, lakes/edges	MA DA	*Papyrus* (reed) *Victoria* (water lily)
3. Ponds, slow streams	DA MA DA DA	*Nuphar* (water lily) *Potamogeton* (pond weed) *Ceratophyllum* (coontail) *Myriophyllum* (milfoil)
4. Estuaries	MA MA	*Zostera* (eel grass) *Ruppia* (widgeon grass)
5. Deep in unproductive lakes; also near lake edges	L A A B	*Isoetes* (quillwort) *Chara* (stonewort) *Nitella* (stonewort) *Fontinalis* (willow moss)

*MA = monocotyledenous angiosperm (angiosperms are higher, or flowering, plants); DA = dicotyledenous angiosperm; F = fern; L = Lycosida; B = moss; A = large algae.

and *Ceratium* which become very abundant in some temperate lakes (Figs. 11-6*b*, 12-15). Most freshwater planktonic algae are photoautotrophic, and a good deal of effort in limnology has been directed toward quantitative estimation of the energy flows associated with their photosynthesis (Chap. 15).

Algae which grow on submerged rocks, plants, or debris in lakes and streams are often called *periphyton* (meaning "on plants"), but are best referred to as *attached algae* (i.e., on any substrate) (Fig. 11-4, Plate 4*b*). These algae are most commonly green, blue-green, or diatoms. Among the attached diatoms, unicellular pennate forms dominate over filamentous centric ones. The golden-brown color of rocks in streams and lakes is often due to a film of diatoms, while green or blue-green algae make up the long, often brilliant green, filamentous streamers. Anglers wading in streams know that algae-covered rocks are slippery! Attached algae occur in all streams, but growth is restricted in shaded areas. In sunny areas productive streams may have luxuriant growth of the blanket weed *Cladophora*. This filamentous green alga is a *mesophyte*, intermediate between true higher plants or *macrophytes* and diatoms or other small algae referred to as *microphytes*. Submerged higher plants and mesophytes provide a substrate for dense *epiphytic* diatom growth. Historically, less attention has been paid to the ecological role of attached algae relative to the more easily studied phytoplankton, but in situ methods for measuring the productivity of periphyton are now available (Marker, 1976a, b).

Macrophytes Large plants, the aquatic macrophytes, may dominate in shallow lakes and streams. Most aquatic macrophytes are flowering plants (angiosperms), but aquatic ferns, mosses, liverworts, and even the large algae of the Charophyceae group may be abundant in particular habitats.

Macrophytes are classified according to their habitat since they comprise such a diverse taxonomic group. The major division is based on their attachment by roots to a solid substrate. The two divisions, free-floating or rooted, are common to lakes, rivers, marshes, and estuaries (Table 11-4). Rooted macrophytes may have all

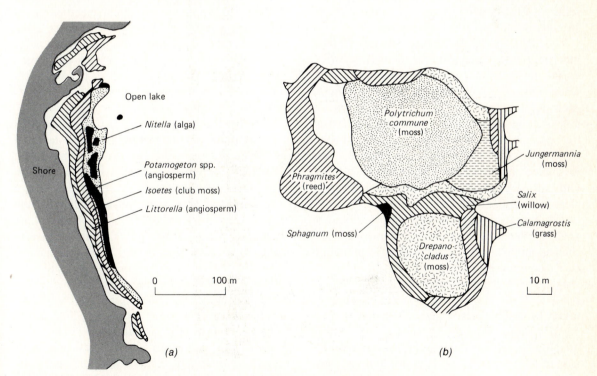

Figure 11-5 (a) Zonation of rooted aquatic macrophytes in unproductive Windermere. The more primitive Charophyte alga *Nitella* and the club moss *Isoetes* (quillwort), extend into deeper water than the angiosperm *Littorella* (shoreweed). The long stems of *Potamogeton* (pondweed) permit growth at depth. There is no emergent, rooted vegetation on this windswept, rocky shore. (*Redrawn and modified from Pearsall and Pennington, 1973.*) (b) Zonation of rooted macrophytes in a shallow naturally acid lake (pH = 4.2), Lake Ao-numa, Honshu, Japan. As in Windermere the primitive mosses occupy the deepest zones and submerged and emergent angiosperm reeds occupy the lake margins. (*Redrawn from data provided by Yamamoto, Kata, and Kashimura, Takai University, Japan.*)

or part of their vegetative and sexually reproductive parts above the water or may be completely submerged. Many have some emergent leaves or flowers for only a short time. Rooted plants have the advantage of being able to mine the littoral sediments for nutrients as well as removing them directly from the water. They remove nutrients, particularly phosphorus, from the sediment and return them to the surrounding water through excretion or decomposition (see Chap. 9). The epiphytic algae are likely to be the most immediate beneficiaries of this process since they are in direct contact with plant surfaces. Rooted macrophytes do not always use the roots to supply nutrients. Their role as attachment organs is important in streams or along waveswept shores.

The most striking feature of an aquatic macrophyte assemblage is the distinct zonation as one moves from land to progressively deeper water. One or two species will occupy a band at each depth, although variations in the type of substrate and exposure to waves may interrupt the regularity of distribution (Fig. 11-5). Macrophytes can cover the entire lake bed if the water is shallow and transparent. Even deep lakes with sparse phytoplankton may be sufficiently clear so that light reaches the bottom.

All freshwater angiosperms are restricted to shallow water, and growth at depths greater than about 10 m may be prevented by an inability to compensate for the increased pressure. Angiosperms are also not found much below 2 percent of surface illumination. Their place is taken by totally submerged lower plants (Fig. 11-5). These species appear to be shade-adapted to the dim blue-green light characteristic of the deeper waters in unproductive transparent lakes (Fig. 3-6b).

A typical member of this deep community is the primitive and cosmopolitan club moss, the quillwort *Isoetes,* which grows up from the bottom sediments of clear mountain lakes in spiky clumps. Also common are mosses, the charophyte algal genera *Chara* and *Nitella,* and

mesophytic algae. *Chara* is known as *skunk weed* in the United States because of the strong smell of some species. In both deep and shallow lakes it often becomes enveloped by a calcium carbonate crust which may protect it from excessive solar radiation and animal grazing. Species of *Nitella* together with aquatic mosses such as *Sphagnum* and *Fontinalis* and occasionally the liverwort *Marchantia aquatica* are also shade-adapted plants. The aquatic mosses in particular survive only where there is a lack of grazing and an absence of competition from angiosperms. Most mosses are unable to use bicarbonate and prefer fast-flowing streams where atmospheric carbon dioxide is abundant (Chap. 16). Some mesophytic algae grow at extraordinary depths. In very transparent Lake Tahoe, one can observe the green alga *Gongrosira* covering rock surfaces like green paint even at 160 m. Growth in the very low light at this depth may be subsidized by facultative heterotrophy.

Emergent monocotyledonous angiosperms are the dominant macrophytes in marshes and along many stream courses. Wetlands represent the transitional zone between land and water, and plants growing in coastal zones may have a cyclic exposure to fresh and salt water through tidal intrusion (see also Chap. 17). Where wave action does not uproot plants, many temperate lakes have extensive littoral stands of macrophytes. There are frequently growths of *Phragmites australis,* reeds which make excellent roof thatch. Other common freshwater-rooted plants include cattail or reed mace *Typha,* the American bulrush *Scirpus,* and *Carex,* the sedge. In the tropics enormous stands of emergent plants, especially papyrus, noted in biblical times along the Nile Delta and once used to make paper, serve to buffer lakes and river systems from the inflow of nutrients and sediments. In saline waters the widgeon grass *Ruppia,* the eel grass *Zostera,* and the alkali bulrush *Scirpus robustus* are food for ducks, geese, and other animals.

One of the most important of the emergent macrophytes is cultivated rice, which contributes most of the caloric intake of the world's population. Wild rice, the aquatic grass *Zizania aquatica,* is native to shallow northern lakes of the United States and Canada. It was an excellent and important food for the original native tribes and is now cultivated. Another delicious plant is the dicotyledonous angiosperm watercress (*Rorippa nasturtium-aquaticum*) which is found in permanent and temporary streams and springs where cool well-oxygenated waters flow. It is cultured extensively for salad.

If algal concentrations or sediment turbidity are too high, higher plants may be shaded out. Fertilizing to promote an algal bloom can be prescribed for macrophyte weed control in farm ponds. Among the rooted macrophytes which have much of their foliage at or near the surface, are the highly successful monocotyledonous pond weeds, *Potamogeton.* These are found in temperate waters of both high and low fertility. Rooted floating plants are typically the more-advanced dicotyledonous angiosperms with well-developed leaves. Water lilies of the genera *Nymphaea* or *Nuphar* are well-known examples. The giant water lily, *Victoria,* of the Amazon and Paraná river systems of South America is a spectacular representative of this growth form. The underside of the leaf is protected from grazing animals by sharp spines, and its shape and flotation assured by an intricate series of rigid gas-trapping cells. Most lakes, ponds, and streams have some representatives of dicotyledonous angiosperms and most have common names. Dense beds of water milfoil (*Myriophyllum*) and coontail (*Ceratophyllum*) are food for muskrat and their seeds are eaten by birds. An interesting turnabout in the food chain is provided by the carnivorous bladderwort, *Utricularia,* which, like Venus's flytrap, catches small animals in pouches to supplement its nitrogen supply.

There are relatively few pelagic or free-floating macrophytes. This may be because they require rather fertile water and are subject to stranding in large lakes when exposed to wind or water-level fluctuation. The free-floating water fern *Salvinia* spp. and the monocotyledonous angiosperms, water cabbage *Eichornia crassipes* and water lettuce *Pistia stratiotes* (Plate 4a), may completely cover the surface of even very large water bodies in the tropics and subtropics and produce a variety of water management and public health problems (Chap. 20). In Lake Kariba, Africa, a third of the reservoir may be covered by over 10 billion (10^{10}) *Salvinia* plants. In temperate and subtropical regions much smaller free-floating plants such as the duckweed *Lemna,* a monocotyledonous angiosperm, and the fern *Azolla* completely cover the shallow wind-protected areas of ponds, irrigation ditches, and backwaters. *Azolla* contains the symbiotic nitrogen-fixing blue-green alga *Anabaena azollae* that helps give the combined plant system value as cattle feed and as fertilizer for rice fields. The suspended roots of *Lemna* may be covered with nitrogen-fixing bacteria. *Lemna* and *Azolla* appear yellow-green, but late in the growing season *Azolla* may have a red coloration.

Most aquatic macrophytes are not used for food, as their common name waterweed implies. However, the seeds of the angiosperms are a favored food for many birds, and the shelter provided by their dense foliage is often vital for small invertebrates and the eggs and young of fish. Some macrophytes break down rapidly and contribute to the detritus pool. For example, 80 percent of the nitrogen in leaves and stems of watercress is returned to the stream within 3 weeks of the plants' deaths in autumn. Throughout the world, rooted macrophytes are managed for fisheries enhancement or flood control. Dense growths of macrophytes effectively dam swamps and rivers, sometimes increasing the depth of a river fourfold. In this way, macrophytes assist in spreading flood waters onto the surrounding land to increase its fertility and provide additional areas for fish and amphibians to feed and spawn.

Protozoans

These unicellular organisms range in size from a few microns in length up to 5 mm. Protozoans are found in almost all aquatic habitats (Table 11-5) and tend to be most abundant in waters where organic matter, bacteria, or algae are abundant. Farm ponds which receive drainage from manure heaps and even antarctic pools near seal wallows teem with rapidly moving protozoans. Benthic algal felts in lakes or vegetation and rocks in slow-moving streams are also good sites for protozoans. Due to the wide dispersal of resting stages, most genera and even some species are found worldwide. Protozoans move slowly, using amoeboid motion, or rapidly, using flagella or cilia. Their clas-

sification is largely based on locomotion (Table 11-5).

Protozoan ecology has been rather neglected, perhaps because many are soft-bodied and difficult to preserve. An exception occurs in some saline lakes and the oceans where sediments consist of the remains of the calcified or silicified skeletons of radiolarians and foraminiferans. The well-known freshwater ciliate *Paramecium* is a particularly common form in temporary pools. Some species of this genus may contain live cells of the green alga *Chlorella* which are an integral part of the animal's metabolism. *Vorticella,* an attached filter-feeding ciliate, is probably the most familiar protozoan encountered when examining detritus from streams or ponds. Protozoans feed on detritus and also consume free-living bacteria, fungi, yeasts, algae, and other protozoans (Fig. 11-6a).

Table 11-5 Classification of Protozoans Common in Aquatic Habitats

Name	Aquatic habitat	Examples
1. Flagellates	Lakes, oceans; some parasitic forms	*Synura** (Chrysomonad) *Ceratium** (Dinoflagellate) *Euglena** *Oikomonas*
2. Ciliates	Ponds, streams, detritus; also parasitic forms	*Paramecium* *Vorticella* *Ichthyophthirius* (parasitic on fish)
3. Amoeboid	Testate and naked forms in most waters; also parasitic	*Difflugia* *Globigerina* (Foraminifera) *Actinosphaerium* (Radiolaria) *Vampyrella* (algal parasite)
4. Sporozoans (all endoparasitic)	Aquatic organisms	*Henneguya* (on fish) *Plasmodium* (human malaria)

*Indicates that photosynthetic forms are also classed as algae (see Table 11-3).

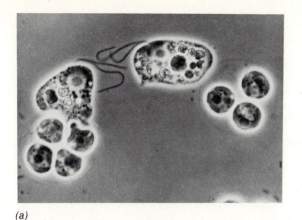

(a)

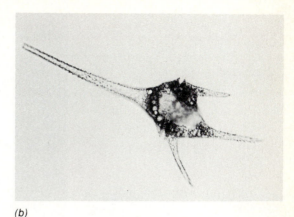

(b)

Figure 11-6 (a) Flagellated protozoans of the *Pseudospora* type which are invading the colonial green algal *Paulschultia*. Magnification ×685. (b) Photograph of the large dinoflagellate protozoan *Ceratium,* a common genus in lakes, and the open ocean (magnification ×265). The central groove which holds the flagellum, four long spines, and the small armored plates are visible. This genus can be photosynthetic. (*Photographs by H. Canter-Lund.*)

Some protozoans are strictly parasitic. Examples are the ciliate *Ichthyophthirius,* which causes white pustules on freshwater fish, the orange-colored *Vampyrella,* which feeds on algae, and *Plasmodium,* which causes human malaria. Some flagellated protozoans such as *Ceratium* or *Peridinium* (Fig. 11-6b) can photosynthesize and are sometimes classified with the algae (Chap. 12). Protozoans which parasitize freshwater phytoplankton may modify algal species composition (Fig. 12-7).

Rotifers

Rotifers are often as small as the larger ciliated protozoans and also inhabit a wide range of aquatic habitats. Much is known about the biology of laboratory cultures but perhaps because of their small size these metazoans, like the protozoans, are still somewhat neglected in limnology. Reviews of the taxonomic and life histories of rotifers is given in Hutchinson (1967), Ruttner-Kolisko (1972), and Pennak (1978).

There are over 1800 species of rotifers; many are cosmopolitan and most occur only in freshwater. The majority are sessile, but rotifers also form an important part of the zooplankton community in lakes (Fig 13-4), although many pass through coarse plankton nets (Chap. 13). They usually dominate the zooplankton in rivers. Most rotifers have a crown of cilia which is used both for movement and for drawing in suspended particles (Fig. 11-7). Food is captured and macerated by hard structures called *trophi.* A few predatory species can extend this structure outside the body to penetrate a prey and suck out its contents. Some rotifers, such as *Asplanchna,* are permanently planktonic but most attach to solid substrates with their foot and creep in leechlike fashion. The rotifer body is protected by a cuticle which may be thickened to produce a distinctly ornamented *lorica.* Rotifers are usually classified by the distinctive shape of their trophi and lorica (Table 11-6).

Crustaceans

Most larger zooplankton belong to the class Crustacea which also includes primarily benthic species, such as shrimps, crabs, and crayfish of lake edges, streams, and estuaries (Table 11-7). Because of their abundance, size,

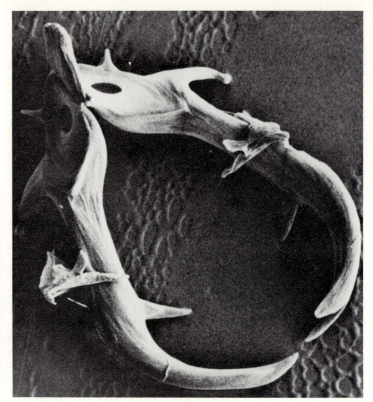

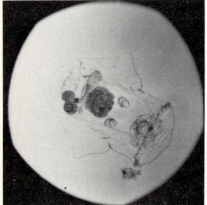

(a)   (b)

Figure 11-7 (a) A relatively large, predatory rotifer, *Asplanchna,* which is exclusively planktonic. (Another planktonic rotifer, *Kellicottia,* is seen in scale with other zooplankton in Fig.
13-1b.) The cilia, at the right in this picture, are not very obvious in most rotifers. Only in a few attached forms does it somewhat resemble the rotating wheels after which the group is named.
(*Photograph by G. W. Salt.*) (b) The antlerlike structures, called trophi, are hard and their pointed
ends assist rotifers in the capture of prey. They are visible in (a) just below the crown of cilia.
(*Photograph by M. L. Commins.*)

and ease of preservation, two crustacean
groups within the plankton, the cladocerans and
the copepods, have received particular attention from limnologists (Chap. 13). Crustaceans
are a major group within the largest of all the
animal phyla, the Arthropoda, and are organisms with jointed appendages. All crustaceans
are enclosed in a protective exoskeleton made
of chitin, a polysaccharide of acetylglucosamine
residues. The exoskeleton may be hardened
with calcium carbonate or made strong but flexible by a process similar to the tanning of

leather. Their classification is based on shape of
the exoskeleton and the number of segments
and appendages (Table 11-7). One major aquatic group consists of filter-feeding cladocerans, of which the common "water flea"
Daphnia is the best known representative
(Figs. 11-8, 13-9e). The other major group, the
copepods, is divided into the mostly filter-feeding calanoids and the cyclopoids which feed
raptorially or occasionally by filtration (Table
11-7). The cladocerans and copepods, together
with the rotifers and protozoans, dominate the

Table 11-6 Classification of Rotifers Common in Aquatic Habitats

Taxonomic group	Common aquatic habitat	Food/ feeding type	Examples
1. Seisonidea	Marine epizoic	Sessile	*Seison*
2. Bdelloidea	Freshwater planktonic, sessile, creeping	Suspension feeders	*Philodina* *Rotaria*
3. Monogonata	Freshwater planktonic	Suspension feeders	*Brachionus* *Keratella* *Polyarthra* *Floscularia*
		Raptorial predator	*Asplanchna*

Table 11-7 Classification of Crustaceans Common in Aquatic Habitats

Taxonomic group	Most common aquatic habitat*	Feeding method	Examples
1. Cladocera (water fleas)	fw,p	Predatory	*Leptodora*
	fw,p,b	Filters	*Sida*
	fw,p	Filters	*Daphnia*
	fw,p,b	Filters	*Chydorus*
	fw,p	Filters	*Bosmina*
	fw,s,p	Predatory	*Polyphemus*
2. Copepoda: Harpacticoida	fw,s,b	Filters and parasitic	*Nitocra*
Cyclopoida	fw,s,b,p	Raptorial on animals	*Cyclops* *Diaptomus* *Megacyclops*
		Raptorial on algae	*Thermocyclops* *Eucyclops*
Calanoida	fw,s,p	Filters	*Limnocalanus* *Boeckella* *Diaptomus*
3. Mysidacea (opossum shrimps)	fw,p	Predatory	*Mysis*
4. Amphipoda	s,fw,b	Raptorial	*Gammarus*
5. Decapoda (freshwater crayfish)	s,fw,b	Raptorial	*Astacus* *Pacifastacus* *Cancer*

*p = planktonic; b = benthic or attached; fw = freshwater; s = saline water.

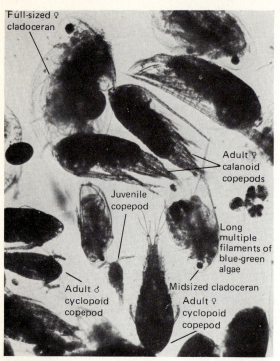

Full-sized ♀
cladoceran

Adult ♀
calanoid
copepods

Juvenile
copepod

Long
multiple
filaments of
blue-green
algae

Adult ♂
cyclopoid
copepod

Midsized cladoceran

Adult ♀
cyclopoid
copepod

Figure 11-8 Common crustacean zooplankton as encountered after collection in a net and preservation with formalin. Indicated are zooplankton of various ages (sizes). Cladocerans, cyclopoid copepods, calanoid copepods, and algae are visible in this tow from the summer plankton of a productive lake. (*Photograph by M. L. Commins and G. W. Salt.*)

zooplankton of freshwaters. In estuarine and marine waters a greater diversity of copepods together with other groups such as the jellyfish and chaetognaths constitute the zooplankton.

Aquatic Insects

Another arthropod group, the insects, have been enormously successful in invading almost all regions of the earth's surface with the exception of the marine environment. In rivers and streams, aquatic insects and their larvae dominate the trophic level between primary producers and fish (Chaps. 15, 16). Stream sampling may yield several hundred individuals of a few dozen species of aquatic insect larvae in 1000

cm² of river bed. The adults are usually short-lived aerial forms. Although some aquatic insect larvae are predaceous, many filter or graze detritus and attached algae. Most possess hooks and streamlined contours to withstand even torrential stream currents (Fig. 11-9). Commonly encountered forms in streams are the mayflies *Ephemeroptera*, the stoneflies *Plecoptera*, caddis flies *Trichoptera*, and the dragonflies *Odonata* (Table 11-8). Insects are an important component of the lake benthos, and emergence of the adults often triggers a frenzy of surface feeding by fish.

The hatch is an important event on a trout stream; it is a time when stoneflies or hellgrammites (dobsonflies) are emerging as adults. The avid dry-fly fisherman attempts to mimic the emergent insect and becomes a rather competent amateur entomologist in the process.

Giant water bugs, *Belostomatidae*, are large enough to capture small fish. Among the benthic insects in lakes, the chironomid larvae are probably the most widespread and, since they tolerate low oxygen concentrations, may even be found in the anoxic hypolimnion of some eutrophic lakes (Figs. 13-14 to 13-16). In the Arctic enormous numbers of blackfly larvae *Simuliidae* coat the surface of rocks in swift streams, and swarms of adults together with mosquitoes can at times make Arctic limnology uncomfortable. In fast-flowing tropical water, blackflies are the carriers of river blindness, onchocerciasis. Control of blackfly larvae remains one of the most important public health problems in many areas of Africa. The predatory *Chaoborus* or phantom midge with its ghostly transparent larvae is the only major insect member of the plankton.

Certain aquatic beetles may actually be used for higher plant control, and we can expect further advances in this area of applied aquatic entomology. Because of the proximity of small streams to land, the falling terrestrial insects can serve as an important food resource for fish (Figs. 14-4 and 14-5, 15-2 to 15-4). These or-

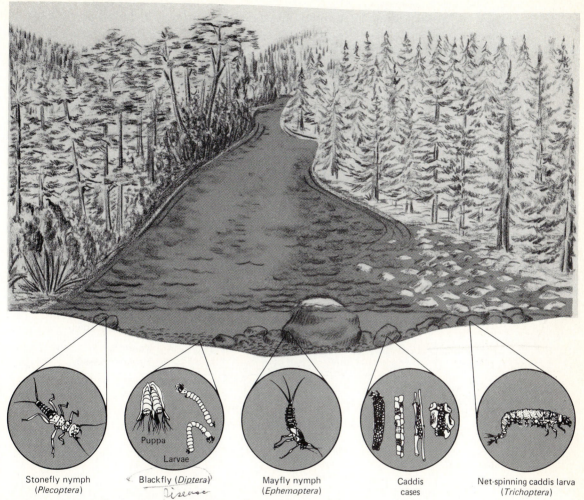

Stonefly nymph
(*Plecoptera*)

Blackfly (*Diptera*)

Puppa

Larvae

Mayfly nymph
(*Ephemoptera*)

Caddis
cases

Net-spinning caddis larva
(*Trichoptera*)

Figure 11-9 The major groups of stream insects. Most of the benthic insects in streams are larval forms called *nymphs*, which do not resemble their adult flying forms. Most stream invertebrates are only 1 or 2 cm long. The mobile ones are flattened to resist the water current and hide from predatory fish during the day. At night they emerge to graze on algae or on each other (see Fig. 16-16). The fixed forms like the blackfly, Simuliidae, or the net-spinning caddis fly live in protected cracks or other places where predators are few.

ganisms, whether they be of aquatic or terrestrial origin, are collectively called *drift* (Fig. 16-1).

Worms and Mollusks

No review of the common animals which inhabit natural waters would be complete without mention of worms and mollusks. These include nematode and oligochaete worms, leeches, flatworms, lamellibranch (bivalve) mollusks, and the gastropod snails (Table 11-8). Most of these organisms, like aquatic insects, convert detritus and living plant material to food for fish. Some, such as leeches, may be active predators on large animals or fish, and nematode worms are common internal parasites of

Table 11-8 Classification of Benthic Invertebrates Common in Lakes and Streams

Taxonomic group	Common aquatic habitat	Food/feeding method	Examples
1. TURBELLARIA (flatworms)	General	Carnivores	*Dugesia*
2. NEMATODA (round worms)	General	Carnivores, herbivores parasites	*Dolichodorus*
3. ANNELIDA Oligochaeta (segmented worms)	General	Sediment grazers	*Tubifex* (red worms)
Hirudinea (leeches)	General	Carnivores, detrivores	*Haemopis* (horse leech)
4. MOLLUSCA Gastropoda (snails)	General	Grazer	*Limnaea* (pond snail) *Planorbis* (ramshorn snail)
Pelecypoda (bivalves)	Streams	Filterers	*Anodonta* (swan mussel)
5. CRUSTACEA Malacostraca (crayfish, amphipods)	General	Detrivores	*Gammarus* *Astacus*
6. INSECTA Plecoptera (stoneflies)	Well-oxygenated waters	Omnivores	*Nemoura* *Isoperla*
Odonata (dragonflies)	Ponds, streams	Raptorial carnivores	*Libellula*
Ephemeroptera (mayflies)	General	Mostly scrapers, grazers	*Baetis* *Ephemerella*
Hemiptera (water bugs)	General	Beaked carnivores, herbivores	*Notonecta* (backswimmer) *Gerris* (water strider)
Megaloptera (hellgrammites, alderflies)	General	Carnivores	*Sialis*
Trichoptera (caddis flies)	General	Mostly filterers, scrapers	*Limnephilus* *Hydropsyche*

Table 11-8 Classification of Benthic Invertebrates Common in Lakes and Streams (*Continued*)

Taxonomic group	Common aquatic habitat	Food/ feeding method	Examples
Coleoptera (beetles)	Pools	Raptorial carnivores	*Dytiscus* (diving beetle)
Diptera (two-winged flies)	Pools	Filterers	*Culex* (mosquito)
	Lakes	Raptorial carnivores	*Chaoborus* (phantom midge)
	Fast-flowing water	Filterers	*Simulium* (blackfly)
	General	Filterers, scrapers, carnivores	*Chironomus* (bloodworm, true midge)

many aquatic animals. A scoop of mud from lake, estuarine, or stream sediment may contain dozens of worms, most of which are small and difficult to identify. Their ecological role has been poorly quantified.

Large mollusks such as the swan mussel are susceptible to pollution, and bivalves in general are becoming rare in many U.S. rivers. The huge masses of snails found along every pond edge can decimate attached algal crops in the same manner as stream insect larvae do. Snails serve as hosts for parasitic schistosomes responsible for the disabling tropical disease bilharzia as well as the annoying swimmers' itch of temperate climates. Macrophyte beds and irrigation canals promote snail growth, which is a major management problem in the tropics.

Fish

Fish are divided taxonomically into the Cyclostomes, Chondrichthyes, and the Osteichthyes, and most of those mentioned in this text are listed in Table 11-9. The primitive, jawless, cyclostome vertebrates Cephalaspidomorphi

include the lampreys Petromyzontidae and the hagfishes Myxini. There are a number of freshwater lampreys, the most primitive of freshwater fish, and like the marine hagfish they lack a biting jaw and are adapted primarily to life as a parasite or scavenger. The sea lamprey normally lives in the sea as an adult but eventually returns to freshwater to spawn. In the Great Lakes this lamprey has substituted lakes for the ocean during its adult life (Chap. 14) and, like the landlocked red salmon called kokanee (*Oncorhynchus nerka*), ascends tributary streams of lakes to spawn in the spring. The sharks and their relatives are the Chondrichthyes of which there are only a few freshwater examples, such as the shark in Lake Nicaragua, Central America.

The bony fish Osteichthyes are the advanced fish with which we are largely concerned (Table 11-9). Most species have gas bladders which enable them to be slightly less dense than water and not expend as much energy in swimming as heavy fish such as sharks. Sturgeon are primitive fish, greatly valued for the caviar they

Table 11-9 Taxonomic Classification of Fish Referred to in This Text*

Only some of the more than 300 families of freshwater and marine fish are listed.

Taxonomic classification	Selected examples	Particular references in this text
Agnatha (jawless fish)		
ORDER: PETROMYZONIFORMES Family: Lampreys (Petromyzonidae)	Sea lamprey, *Petromyzon fluviatilis* Brook lamprey, *P. branchialis*	Changes in Great Lakes fisheries, Chap. 14
Gnathostoma (jawed fish)		
ORDER: ACIPENSERIDAE Family: Sturgeons (Acipenseridae)	Lake sturgeon, *Acipenser fluvesens*	Changes in Great Lakes fisheries, Chap. 14
Paddlefish (Polydontidae)	American paddlefish or spoonbill cat, *(Polydon spathula)*	Plankton feeding, Chap. 16.
ORDER: ANGUILLIFORMES Family: Eels (Anguillidae)	Freshwater eel, *Anguilla anguilla*	A catadromous fish, Chap. 16.
ORDER: CLUPEIFORMES Family: Herrings (Clupeidae)	Alewife, *Alosa pseudoharengus*	Introduced to Great Lakes fishery, Chap. 14

ORDER: SALMONIFORMES
Family: Salmon, trout,
     whitefish, chub,
     char and greyling
     (Salmonidae)
	Kokanee salmon,
	Oncorhynchus nerka
	Chinook or king salmon,
	O. tshawytscha
	Trout, *Salmo salar*
		Top predators in many cold lakes and rivers, Chap. 14.

	Lake whitefish, *Coregonus clupeaformis* Blackfin cisco or chub, *C. nigripinnis* Bloater, *C. hoyi* Lake herring, *C. artedii* Lake trout, *Salvelinus namaycush.*	Changes in Great Lakes fisheries, Chap. 14
Family: Smelts (Osmeridae)	Rainbow smelt *Osmerus eperlanus (mordax)*	
Family: Pike (Esocidae)	Pike, *Esox lucius.*	Population dynamics in Windermere, Chap. 14
ORDER: CYPRINIFORMES Family: Minnows, carp (Cyprinidae)	European carp, *Cyprinus carpio*	Introduced to North America and Great Lakes fisheries, Chap. 14.

Table 11-9   Taxonomic Classification of Fish Referred to in This Text* (*Continued*)

Only some of the more than 300 families of freshwater and marine fish are listed.

Taxonomic classification	Selected examples	Particular references in this text
Gnathostoma (jawed fish)		
	European roach, *Rutilus rutilus*	Food-chain dynamics in River Thames, Chap. 15
	Bleak, *Alburnus alburnus*	
	California roach, *Hesperoleucus symmetricus*	Feeding in rivers, Chaps. 14, 16
	European minnow, *Phoxinus phoxinus*	
	European chub, *Squalius cephalus*	
Family: Suckers (Catostomidae)	Tahoe sucker, *Catastomus tahoensis*	Feeding and distribution in lakes, Chap. 14, and river ecology, Chap. 16.
	Mountain sucker, *C. platyrhynchus*	
Family: North American catfish (Ictaluridae)	Channel catfish, *Ictalurus punctatus*	
ORDER: CYPRINODONTIFORMES Family: Killifishes, pupfish (Cyprinodontidae)	Desert pupfish, *Cyprinodon nevadensis*	Saline habitats, Chap. 19
ORDER: GASTEROSTEIFORMES Family: Sticklebacks (Gasterosteidae)	Three-spined stickleback, *Gasterosteus aculeatus*	
ORDER: MUGILIFORMES Family: Mullet (Mugilidae)	Grey mullet, *Mugil cephalus*	Breeding migrations to estuaries, Chap. 17
ORDER: PERCIFORMES Family: Temperate basses (Percichthyidae)	Striped bass, *Morone saxatilis*	Introduction to San Francisco Bay, Chap. 17.
Family: Sea basses (Serranidae)	White bass, *Roccus chrysops*	Feeding methods in lakes, Chap. 14
Family: Sunfish (Centrachidae)	Sunfish, *Lepomis* spp. Largemouth black bass, *Micropterus salmoides*	Food webs, Chap. 15, introduction to Lake Gatun, Chap. 14
Family: Perch (Percidae)	Perch, *Perca fluviatilis*	Year-class survival in Windermere, Chap. 14
Family: Archerfish (Toxotidae)	Archerfish, *Toxotes jaculator*	Special feeding methods, Chap. 14

Table 11-9   Taxonomic Classification of Fish Referred to in This Text* (*Continued*)

Only some of the more than 300 families of freshwater and marine fish are listed.

Taxonomic classification	Selected examples	Particular references in this text
Gnathostoma (jawed fish)		
Family: Cichlids (Cichlidae)	Tilapia, *Tilapia* spp.	Phytoplankton feeding, Chap. 14, and tropical lakes, Chap. 19
Family: Icefish (Channichthyidae)		"Bloodless" fish, Chap. 19
ORDER: SCORPAENIFORMES Family: Sculpins (Cottidae)	Miller's thumb, *Cottus gobio* Piute sculpin, *C. beldingi*	Resource partitioning in streams, Chap. 14
ORDER: GADIFORMES Family: Cod (Gadidae)	Burbot, *Lota lota*	Great Lakes fisheries, Chap. 14

*Modified from Lagler et al., 1977.

produce. They are threatened on a worldwide scale by pollution of their marine and freshwater habitat and by blockage of migration routes by dams. Overfishing has exterminated the Great Lakes sturgeon, *Acipenser fulvescens* (Harkness and Dymond, 1961), in Lake Erie and threatens the famous source of beluga caviar in eastern Europe. Sturgeon were often considered pests because they interfered with fishing gear during capture of other species. In San Francisco Bay similar problems caused migrating sturgeon to be caught in large numbers, stacked like cordwood on the beaches, and left to rot. Sturgeon are now protected in San Francisco Bay, and populations are slowly increasing there.

The closely related paddlefish *Polyodon*, once abundant in the Mississippi River drainage of North America, comprised a significant portion of the commercial river fishery of the central United States. In the same general habitat are to be found a single species of bowfin,

Amidae, and numerous representatives of gars, *Lepisosteidae*, which are covered with an armor of primitive scales.

Among the progressively more advanced bony fish, the great stocks of shad and herring, Clupeidae, are best known from marine areas but have several freshwater representatives such as the gizzard shad and goldeye herring. The salmon family, Salmonidae, includes the lake-dwelling trout and whitefish which are of great commercial value in northern lakes. The pike, Esocidae, and perch, Percidae, discussed in detail later in Chapter 14, are circompolar in distribution (Fig. 11-10). They are valued as food, and are important in lakes as predators. The minnow family, Cyprinidae, has many representatives in Europe, North America, and Asia (Fig. 11-11) and dominates China's freshwaters. The Cyprinidae include the carp, originally native to Asia, which have long been utilized for aquaculture, and are greatly valued as food in many countries. The carp's ability to endure poor water quality makes it a natural survivor in polluted situations and introductions to Europe and North America have made it widespread. The freshwater suckers, Catostomidae, are close relatives of the carp and are widespread in North America (Fig. 11-10). Although very bony they are excellent food if properly prepared. They, like members of the salmon and minnow families, have round, cycloid scales. The family Centrarchidae (Fig. 11-11) is among the most successful of modern North American fishes. It contains the black basses, sunfish, and crappies which, like the perch family, have ctenoid scales. Both families are prized as sports fish, but some species overpopulate artificial lakes. A similar family, Cichlidae, which includes the numerous species of *Tilapia,* is extremely abundant and important as food in Africa, South and Central America, and India (Figs. 11-11, 14-6).

Pacific and Atlantic salmon, shad, and smelt are *anadromous* fish. Feeding in fertile areas of the ocean, salmon grow to large adults which *Frsh +* must return to freshwater streams to spawn. In *salt* so doing, the Pacific salmon which die after spawning carry in their bodies nutrients from the sea to unproductive lakes and streams where their progeny will eventually hatch. These species may be introduced to lakes where they are able to complete their life cycle by spawning in the lakes' tributaries and spending their adult life in the lake in the same manner that their marine relatives spend theirs in the sea. Fish which live in freshwaters but must go to the oceans to spawn are called *catadromous;* the eel, Anguillidae, which returns from both Europe and North America to spawn in the waters near Bermuda is the best-known example. After hatching, their larvae are carried by the Gulf Stream to the shores of North

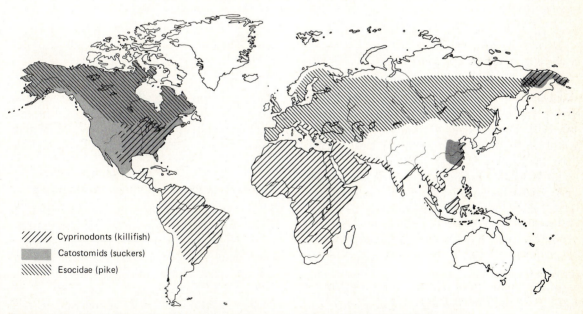

////// Cyprinodonts (killifish)

▨ Catostomids (suckers)

\\\\\\ Esocidae (pike)

Figure 11-10 World distribution of three common groups of fish. The catostomids, or sucker family, are mainly confined to North America. In Africa and the tropical coast of Asia, suckers are replaced by cyprinodonts (killifish). The two families overlap in the southwestern United States. In contrast the pike are a cold-water circumpolar group. (*Redrawn and modified from Lagler et al., 1977.*)

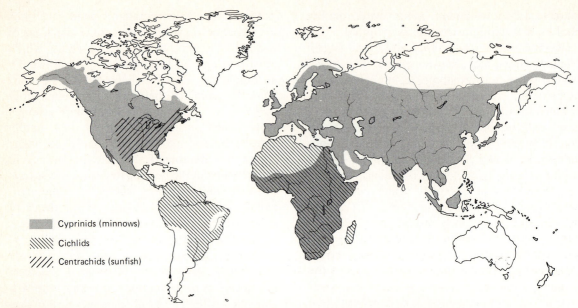

Figure 11-11 World distribution of three more common groups of fish. Minnows (cyprinids) are widespread throughout most of the world. The centracids or sunfish such as the crappie or black bass are confined to the central and southeastern United States, although they have been widely introduced to warm lakes elsewhere. The cichlid family is common over most of Africa and South America and is famous for rapid speciation after an initial species invades a new lake (see Fig. 14-6). (*Redrawn from Lagler et al., 1977.*)

America and Europe where they ascend the rivers as elvers to grow to maturity in freshwater. Both the elvers and the adult eels are highly prized as food and are extensively cultured for food in Japan.

Since anadromous or catadromous fish must move between fresh water and salt water, they are specially equipped metabolically to adjust their salt balance, or *osmoregulate*. Estuaries usually provide the transitional zone for a more gradual adjustment to salinity changes (more details are given in Chap. 17). Freshwater fish entering salt water must be able to eliminate salts effectively to remain *hypotonic* to their environment, while the saltwater fish entering fresh water must conserve the salts in their body fluids to remain *hypertonic* to their new, less-saline habitat. Except for sharks and their

allies, which maintain a high urea concentration in the blood, the bony fish balance salt uptake and loss with their gills and kidneys. Some additional salt loss occurs in marine fish through the feces.

Amphibians, Reptiles, Birds, and Mammals

The amphibians require water in their early life stages and are therefore inseparably tied to the aquatic habitat. Even the terrestrial salamanders found in moist habitats beneath logs and leaves must return to water to reproduce. Newts, salamanders, and frogs inhabit the water's edge, and their young spend their preadult period beneath the surface. The mud puppy, *Necturus*, with external gills at the neck, spends its entire life in midwestern streams of North America in fairly active competition with benthic fish. The stream angler often captures them when fishing with worms.

Reptiles are able to leave water since their amniotic eggs do not need external water and their impervious skin retains body fluids in terrestrial and even desert environments. Many reptiles, however, remain closely associated with the water. In Central America, a light-footed lizard, the Jesus lizard, can actually run across the surface of the water. Small snakes are common predators of fish and amphibians in weedy streams and ponds. *Natrix,* known as the water snake in America and the grass snake in Europe often delight the limnologist with their graceful swimming motion. Among the more dramatic examples of aquatic snakes is the Amazonian giant anaconda, whose size and strength is legendary. Major carnivorous reptiles in tropical rivers and lakes include the caimans of South America, a number of species of the more widely distributed crocodiles, and the American alligators. Although feeding mainly on fish, larger individuals also take small to

moderate-sized mammals, including an occasional human being. In a sense, crocodiles are to a limnologist what man-eating sharks are to the oceanographer—they provide a certain element of danger and excitement during the study of tropical habitats. In fact, because of the value of their hides, they are an endangered species almost everywhere.

Perhaps the best-documented attack of a crocodile on limnologists occurred during one of the expeditions of Daniel Livingstone on Lake Chishi, Zambia. He had set out in a rubber boat powered by a small outboard motor. In the middle of this large, shallow lake, an enormous Nile crocodile, *Crocodilus niloticus* (Fig. 11-12), suddenly attacked the bow of the boat from beneath. The bow chamber collapsed, and despite blows from the aluminum oars wielded by Livingstone and a kick from his companion Richardson, the giant reptile maintained the attack. The creature finally climbed on top of the disabled craft and drove the limnologists into the water. From the size of

Figure 11-12 Several very large Nile crocodiles, *Crocodilus niloticus,* at rest near the plunge pool below Murchison Falls on the River Nile, Uganda.

the crocodile, estimated to be a meter across the back, both men might well have been recycled into the aquatic food chain had it not been for the crocodile's apparent fixation on sinking the rubber boat. As they swam toward shore, the crocodile was last seen still atop the boat (Richardson and Livingstone, 1962). An explanation for this aggressive behavior was provided from observations in a marsh near Lake Chishi where a wooden scow of 4 m length, also powered by an outboard motor, was repeatedly attacked by Nile crocodiles. They appeared to be defending their territory from the noisy craft in an area where they had not been hunted (Jackson, 1962).

Wetlands are the traditional nesting and feeding grounds for many species of water fowl. They exhibit a great variety of feeding activities and adaptations, since they feed at every level of the aquatic food chain. Specific examples are referred to throughout this text. The various herons and egrets stalk their prey on long legs, while the tropical fish eagle, fish hawk, and osprey pluck living fish when any venture near the surface. Among the most interesting avian fish predators are the kingfishers and Arctic terns that hover over the water and dive to capture minnows and young salmon during their migration to the sea. Whether straining algae out in flamingo fashion (Fig. 11-13) or pecking along the shore in search of snails and small crustaceans, the diversity of bird life associated with the aquatic environment provides another realm of interest for limnologists. Perhaps the most truly aquatic birds are grebes and cormorants which, like the penguins, obtain food by directly pursuing and capturing fish under the water. The cormorant literally flies through the water. In rivers and streams the water ouzel also uses its wings to swim as it moves from rock to rock in the fast-flowing rapids.

Mammals have many representatives associated with inland waters. Much of the early exploration of North America was carried out

Figure 11-13 A small part of the huge flock of lesser flamingoes, *Phoenicopterus* (= *Phoeniconais*) *minor*, feeding on the blue-green alga *Spirulina* and benthic diatoms (Tuite, 1981) in alkaline Lake Nakuru, Kenya. The bird's bill is specially adapted to remove algae while the head is inverted (Jenkin, 1936).

Figure 11-14 African hippopotami in a *Pistia*-covered pool near Lake George, Uganda. These animals are normally out of the water only during darkness, but this shallow, muddy pool allowed them to show more of themselves than the limnologist usually sees.

in search of new beaver-trapping grounds, since this animal had been hunted to extinction in much of Europe. Their luxuriant pelts once formed an important part of the New World's currency. Closely associated with beavers are muskrat and mink and their South American relatives, the coypu, which are also highly valued for their fur. Muskrats abound in marshy areas, along small streams, and particularly in irrigation ditches. Like the beaver, they operate through the winter, under the ice, and live in bank burrows. Beavers are limnologically important as they create their own aquatic environment throughout the northlands of the New World. Their small ponds often provide the only relatively still water in upland areas.

Although seals are usually considered to be marine, endemic populations of the freshwater seal, *Phoca (Pusa) sibirica,* exist in Lake Baikal, Siberia, and another species is found in Lake Iliamna at the base of the Alaskan peninsula. Otter have representatives in both the Old and New Worlds. They seem to play as much as they hunt for fish, and make excellent pets. Among the larger mammals, the hippopo-

tamus has a role in the freshwater ecology of many African lakes and marshes (Fig. 11-14, Table 19-5). They consume enormous quantities of terrestrial grasses around the lake, fertilize the nearshore waters with their excrement (Horne and Viner, 1971), and keep waterways open in the shallower portions of their habitat. During periods of drought or overpopulation they may denude the shore zones and accelerate erosion. They are considered by some limnologists to be even more dangerous than crocodiles, and both authors have had their boat in Lake George, Uganda, attacked by these impressive herbivores. Elephants, cape buffalo, and bison make a great deal of use of water, and in North America and Africa some ponds owe their origins to their wallowing.

Like the seal, the manatee is a truly aquatic mammal inhabiting areas of Central and South America as well as Africa. They are frequently found in clear streams or estuaries and consume higher aquatic plants. A number of attempts have been made to utilize their grazing abilities to control aquatic plant infestations, but their slow reproduction and their susceptibility to injury by boats have greatly limited their success. Their numbers have declined drastically since they were extensively harvested for food, and they are now considered endangered animals.

Having considered the broad range of bacteria, fungi, plants, and animals, the next three chapters provide more detail on phytoplankton, zooplankton and zoobenthos, and fish and fisheries. In these chapters we have selected a few examples from among the enormous diversity which characterizes aquatic environments.

FURTHER READINGS

Brinkhurst, R. O., and D. G. Cook (eds.) 1980. *Aquatic Oligochaetes*. Plenum, New York. 529 pp.

Holm, L. G., L. W. Weldon, and R. D. Blackburn. 1969. "Aquatic Weeds." *Science,* **166**:699–709.

Hutchingson, G. E. 1975. *A Treatise on Limnology,* vol. III: *Limnological Botany,* 660 pp. Wiley, New York.

Jones, J. G. 1971. "Studies on Freshwater Bacteria: Factors Which Influence the Population and Its Activity." *J. Ecol.,* **59**:593–613.

Mitchell, D. S. 1969. "The Ecology of Vascular Hydrophytes on Lake Kariba." *Hydrobiologia,* **34**:448–464.

Pennak, R. W. 1978. *Freshwater Invertebrates of the United States.* 2d ed. Wiley, New York. 803 pp.

Russell-Hunter, W. D. 1968, 1969. *A Biology of Lower Invertebrates* (1968), 181 pp.; *A Biology of Higher Invertebrates,* (1969), 224 pp. Macmillan, New York.

Ward, H. B., and G. C. Whipple. 1959. In W. T. Edmondson (ed.), *Freshwater Biology.* 2d ed: Wiley, New York. 1248 pp.

Willoughby, L. G. 1969. "A Study of the Aquatic Actinomycetes of Bleham Tarn." *Acta Hydrobiol. Hydrographia Protistol.,* **34**:465–483.

Chapter 12

Phytoplankton

OVERVIEW

Phytoplankton in temperate regions usually grow in a series of pulses or blooms. The first blooms are initiated in spring by the increase in sunlight, and autumn growth is terminated as light decreases in winter. In tropical regions growth may be nearly continuous when sufficient nutrients are available. In polar regions where sunlight and ice-free periods are brief, there is only a single short period of growth. The life cycle of each species is further modified by the availability of nutrients, the degree of thermal stratification, algal movements relative to the water, zooplankton grazing, interalgal competition, and parasitism by protozoans, fungi, bacteria, or viruses. Algae have evolved various strategies to overcome nutrient depletion and grazing. These include production of special enzymes which release chemi-

cally bound nutrients or allow the alga to take up nutrients at low concentrations. Movement by swimming or a change in cell density may allow them to reach new sources of nutrients. Some algae form resting stages to overcome unfavorable conditions. Other species produce protective spines, gelatinous coats, or grow large or fast to survive zooplankton grazing. Grazing delays the time of maximum algal density and also may change species composition to favor some unpalatable species such as blue-green algae.

Holoplankton, which are always in the plankton, include algae such as the diatoms *Asterionella, Fragilaria,* or *Tabellaria.* These dominate the spring bloom in many productive lakes because they grow faster than competing algae. In some lakes blooms may be slowed by predation from chytrid fungi, zooplankton grazing, or protozoan infestations. Growth eventually stops after thermal stratification because diatoms deplete the nutrients in the epilimnion. The actual limiting nutrients may be silica (SiO_2), which is needed for diatom cell walls, phosphorus, nitrogen, or even trace metals depending on the geographical and geological location of the lake.

The meroplankton, which are only sometimes in the plankton, include the common diatom *Melosira*. In stratified holomictic lakes this alga succeeds because it is present in large numbers in the winter plankton. Resuspension of live cells from the sediments during fall overturn is responsible for this winter population. *Melosira* has a low growth rate but is able to take advantage of the high nutrient levels in early spring and benefits from low levels of competition and grazing. Suspension of heavy filaments of this centric diatom depends on vigorous mixing. A few days after thermal stratification, and well before nutrients are depleted, healthy *Melosira* cells sink to the bottom and form resting stages.

The meroplanktonic blue-green algae such as *Aphanizomenon, Anabaena,* and *Microcystis* usually do best in warm lakes in summer and fall. They overwinter as "spores" or other resting stages. Although they grow much slower than diatoms at low temperatures, they may divide faster at higher temperatures. Planktonic blue-green algae succeed in the summer-fall plankton because they regulate the depth at which they grow by production and collapse of minute gas-filled vacuoles. The ability to float near the surface shades out competing algae. Large colonies rise or sink much faster than single filaments. Large size and unpalatability prevent serious loss to zooplankton grazing. Many blue-green algae have an ability to take up nutrients like phosphate and ammonia at low levels. A few genera of blue-green algae, such as *Aphanizomenon,* can fix dissolved atmospheric nitrogen gas. All other algal groups as well as many blue-greens lack this ability. Blue-green blooms decline if their flotation fails, if they are stranded on the shore, destroyed by summer sunlight, or deplete some essential resource. Rapid destruction of large blooms may also occur from cyanophage virus or cytrid fungus attack.

Dinoflagellates like *Peridinium* and *Ceratium* generally grow best in summer and fall and succeed because they can actively swim to positions of favorable light and nutrients. The algae are positively phototactic and may form reddish-brown surface patches, called *red tides*. Dinoflagellate nutrient requirements are complex and may include organic substrates. Their populations may decline due to heavy zooplankton grazing, competition from other algae, and possibly nutrient depletion. The cycles of other algae in the plankton are not well-understood. These chlorophyte, chrysophyte, cryptophyte, and euglenophyte algae often dominate the reduced summer plankton due to an ability to take up nutrients at low levels and to maintain their position by swimming. In some lakes they also exist below the thermocline and at great depths.

The spatial distribution of phytoplankton is uneven for almost all species. Surface patches or streaks of blue-green algae or dinoflagellates are often visible in oceans and lakes. Variations in phytoplankton populations also occur with depth. Rapid growth produces a clump of algae if growth rate exceeds the dispersion of the patches by water turbulence. Uneven grazing by zooplankton can also produce patches. Horizontal patches and vertical layers are frequently produced by active motion of the algae attempting to optimize their light and nutrient conditions.

INTRODUCTION

Phytoplankton, like terrestrial plants, are seldom distributed completely at random due to variations in reproductive patterns, microhabitat preferences, or grazing. Most have an uneven distribution despite the fact that they are constantly mixed by water movement. The seasonal variations that are so obvious in the growth of terrestrial plant communities are just as apparent in the phytoplankton. However, the continuous annual accumulation of biomass in the trunks and branches of forest trees has no counterpart in the plankton. Almost all phytoplankters are dependent upon a single season's growth or a physical redistribution of resting stages to reach their annual maximum. Following a bloom, phytoplankton biomass is reduced to almost nothing. Virtually all the dynamic features of lakes such as color, clarity, trophic state, water taste, or animal plankton and fish production depend to a large degree on the phytoplankton. This chapter focuses on some well-known examples of phytoplankton ecology in temperate and Mediterranean climates. Tropical, polar, and alpine phytoplankton are covered in Chap. 19. Physiological adaptations to overcome low nutrient levels are discussed in Chaps. 8, 9, and 15.

MEASUREMENT

Phytoplankton samples in relatively shallow water are best collected with a long tube, but most limnologists prefer a small sampling bottle which encloses water at a given depth. Either method collects all phytoplankton, regardless of size. Where only large algae are of interest, plankton nets are often adequate. Certain delicate algae may be retained in the mesh or damaged. In some lakes more than 90 percent of the phytoplankton will pass through even the finest mesh nets. Some details are given in the *IBP Handbook No. 12* (Vollenweider, 1969).

An integrated water sample which contains a representative sample of phytoplankton may be collected with a flexible tube. It is lowered through the water column, closed at the top, and hauled to the surface by using a rope tied to

the lower end. Alternatively the tube can be attached to a pump which continually draws up water from selected depths. A fluorometer may be connected to the flow of water from the tube. This instrument measures the photons emitted by even low levels of chlorophyll *a* excited by light of a particular wavelength. Fluorescence can be calibrated to give a continuous record of living phytoplankton, but its use has limitations in eutrophic systems where chlorophyll *a* exceeds about $10 \mu g$ liter^{-1}. Underestimates result from light absorbed by overlapping cells. The light emitted per unit of chlorophyll *a*, called the *fluorescent yield*, varies with the physiological state of the phytoplankton. This variation requires frequent recalibration of the fluorometer against known chlorophyll *a* standards. Algae can also be detected with the submarine turbidimeter which measures the intensity of a horizontal beam of light as it is lowered through the water. High light extinctions are often due to dense algal concentration (Baker and Brook, 1971), but interference by organic debris or muddy water is a major drawback. Another method of measuring the surface layer of algae over large areas is by remote sensing. This technique usually involves airplanes or satellites equipped with specialized cameras. Limnologists can obtain useful data with a camera in a small airplane or even from nearby mountains (Wrigley and Horne, 1974). Algae and other vegetation reflect certain specific wavelengths (Fig. 3-7). Best results so far have been achieved from eutrophic lakes using the near-infrared band (700–800 nm). The striking details of horizontal distribution of algae revealed by this method provide an important new dimension in synoptic sampling (Plate 6a, Fig. 5-3).

Microscopic identification and counting of natural populations of phytoplankton is difficult to perform accurately. Although algae are best identified alive, this is rarely possible. Consequently, methods have been developed for preservation. Iodine, formalin, gluteraldehyde, mercuric chloride, or chloroform are most frequently used. Iodine, usually prepared as Lugol's solution, offers the advantage of making cells heavy enough to settle out for counting. Settling chambers are an excellent method of concentrating the normally dilute algal populations of lakes for microscopic examination and counting (Lund, Kipling, and LeCren, 1958). Small cells may settle so slowly that alternative concentration techniques are required. Sufficient samples should be taken to determine daily and seasonal vertical and horizontal variations in numbers, particularly for motile dino-flagellates, blue-green algae, and ciliates.

ALGAL MOVEMENTS IN WATER

Two main factors control the rate of growth of phytoplankton populations: (1) a genetically determined maximum growth rate at a particular temperature and (2) the ability to reach optimum light and nutrients. Some species adapt their enzymatic systems for low nutrient levels (Chap. 15), while low light often produces more chlorophyll per cell. Where nutrient limitation or shading is severe, physical movement toward more suitable conditions is another strategy of motile forms.

Many phytoplankters exhibit purposeful daily and seasonal cycles as well as movements which appear random. Although they are largely at the mercy of the currents in the surrounding water mass, most algae are slightly heavier or lighter than water and therefore show some vertical movement relative to the surrounding media (Walsby and Reynolds, 1979). Algae move by swimming or changing their density. Sinking is slowed down by the viscous drag on their cell walls.

Algae deplete the water immediately around them of nutrients. This forms a layer or diffusion shell, through which fresh nutrients must pass to reach the cell. The rate of supply or transport of nutrients through the diffusion shell is affected by movement. When the cell moves relative to the water, the diffusion shell becomes stretched and thinner, allowing more rapid contact with a fresh nutrient supply. This is similar to the way the windshield of a moving car collects more insects than a stationary one. For much the same reason, attached algal growth may be enhanced in the strong currents of streams or littoral zones of lakes even when nutrients are scarce (Chap. 16). Studies by Pasciak and Gavis (1974) indicate that transport limitation across a diffusion shell may be unimportant for some algae. In these cases, the inherent half-saturation constant K_s for nutrients (Chaps. 8, 9, and 15) determines uptake, and vertical movements are made to reach optimal light levels rather than fresh nutrient sources.

Effect of Cell Shape

The diatoms, which include at least 10,000 valid species (Guillard and Kilham, 1977), are the most abundant algae of the plankton and are important in almost all food chains. The common phytoplankton diatoms have no active form of movement except during sexual reproduction in some genera.

Because most algae are slightly heavier than water, they sink at an angle to the mean water current. Many phytoplankton are not simple spheres or cylinders, and their shapes, together with spines in some forms, slow sinking and even cause rotation. Very large spines, such as those in some diatoms, may actually halve the sinking rate (Walsby and Xypolyta, 1977). The absolute rate of sinking is crucial for some diatom life cycles. For example, the meroplankton *Melosira* is much heavier than the holoplankton *Asterionella,* which allows *Melosira* to sink out of the epilimnia of stratified lakes before nutrient depletion occurs. The spines and protrusions so characteristic of diatoms are sometimes covered by mucus, which is clearly visible when the samples are stained with india ink (Figs. 11-3*b*, 12-2*b*). This mucus apparently negates any major hydrodynamic function of the spines. Some green algae, such as *Botryococcus braunii,* may accumulate oil in their cells and float on the lake surface (Fogg, 1975).

Change in Density

The sudden appearance of blue-green algae as thick films or scums on the surface of lakes (Plate 7*b*) was recorded as long ago as 1188 in an English lake (Griffiths, 1939). It has often been erroneously thought that these surface blooms of algae grew overnight. We now know that these algae developed over a few weeks and then floated up to concentrate on the surface in calm periods. Blue-green algae float because they contain gas vacuoles. All common planktonic blue-green algae possess gas vacuoles which are small cylinders with cone-shaped ends (see review by Walsby, 1972). Unlike many cellular inclusions, they are hollow and the gas inside is in equilibrium with the surroundings. Gas vacuoles can be collapsed by the application of relatively small external pressure (Fig. 12-1*a* to *d*). The gas vacuoles do not all collapse at one pressure but do so progressively. As flotation gives an ability to intercept incoming light, blue-green algae can remain near optimum light levels and also may shade out competing algae. This ability to maintain optimum position is probably the most important reason why they dominate the phytoplankton of many eutrophic lakes.

The intense sunlight at the lake surface can damage and even kill algae in a few hours. The gas vacuoles of blue-green algae collapse in high light, and new ones are produced in the darkness of the deep waters. At the lake surface algae photosynthesize and produce sugars which increase the cell pressure. This pressure collapses the weakest gas vacuoles and the cell sinks. Eventually it reaches a level where light and perhaps nutrients are optimal. On summer mornings shoreline hotel owners at Clear Lake, California, observe the surface film of blue-green algae. If it does not sink by 11 A.M. they must stir the water mechanically to dissipate the algal nuisance. The speeds with which the algae can rise or sink are quite fast. Reynolds (1973), in his studies on the Cheshire meres in England, has measured a rate of rise in excess of 10 cm h^{-1} for *Microcystis.*

Effect of Size

The rate of rising or sinking of any object in water is given by a modification of Stoke's law:

$$V = \frac{2}{9} \frac{(p - p')r^2 g}{\eta \phi} \qquad (1)$$

where V = speed of rising or sinking
p = density of water
p' = density of the object
r = radius of a sphere of equivalent volume (effective radius)
η = viscosity of water
g = acceleration due to gravity
ϕ = the coefficient of form resistance which depends on the shape of the object

Equation (1) demonstrates that only altered density or size will change buoyancy of a fixed shape. Density changes have been discussed, but the size of the algae, or at least the algal colonies, is also important. The radius of the object is important because it is a squared term (r^2). Algae do not usually change their cell sizes appreciably, but many diatoms, green algae, and blue-green algae form chains and clumps. Figure 12-2*b* shows large colonial forms of some planktonic blue-green algae. The consequent high r^2 values produce high rates of sinking or rising depending on the number of gas vacuoles. Fig. 12-1*e* shows an SEM of large flakelike clusters of *Aphanizomenon.* Obviously, the effective radii for these flakes are much larger than that of single filaments. Only algae with large colony size have sufficient buoyancy to rise to the surface and "bloom" overnight.

Colony size and gas vacuoles also control the

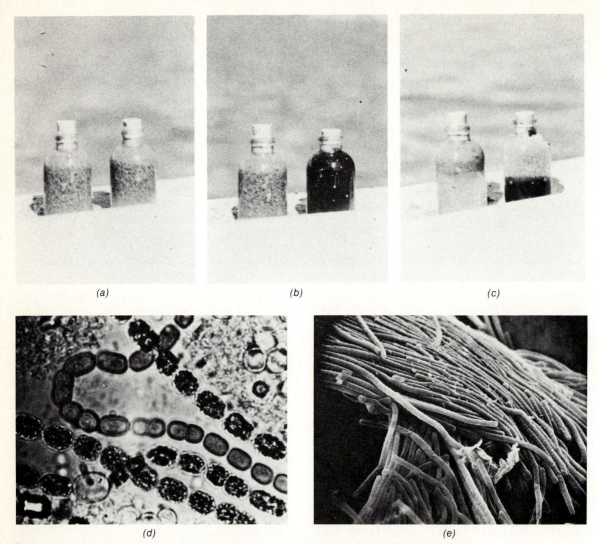

(a)   *(b)*   *(c)*

(d)   *(e)*

Figure 12-1   Sinking and rising of gas vacuolate blue-green algae. (*a*) Before, both 60-cm³ bottles contain equally distributed algae. (*b*) Immediately after collapse of vacuoles in the right-hand bottle by pressure applied by a hammer on the cork. The dark (green) color of the bottle is due to the lack of reflection of white light when vacuoles are collapsed. (*c*) Thirty seconds after: the control algae have floated to the surface and those with collapsed vacuoles have sunk. This experiment mimics natural diel changes (see text). The algal population composed mostly of *Microcystis, Aphanizomenon* flakes, and *Anabaena* from Clear Lake, California. Large colonies are present [see (*e*)] which increase the speed of rising or sinking. (*d*) Photomicrograph originally made using color infrared film showing filaments of *Anabaena* with intact gas vacuoles (filaments above and below) and with vacuoles collapsed (central clear filament). Also visible is a hetrocyst (large clear cell at lower right). The infrared light is reflected more from gas vacuoles than from other cellular material (Fig. 3-7) and enhances the contrast between the filaments which have gas vacuoles and those which do not. (*Photograph by H. M. Anderson.*) (*e*) SEM photograph of a bundle of *Aphanizomenon* filaments in typical flake form (×370). (*Photograph by H. W. Paerl.*)

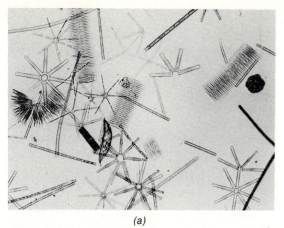 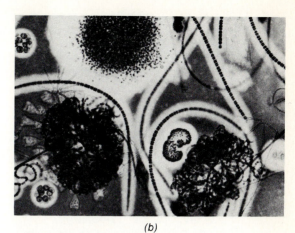

(a)   (b)

Fig. 12-2 Phytoplankton from (a) the early spring bloom and (b) fall blooms in a moderately productive lake, the South Basin of Windermere, England. See also Fig. 11-3. Part (a) shows the dominance of the colonial pennate diatoms, *Asterionella* (slender stars), *Fragilaria* (comb-like), *Tabellaria* (stout stars), and filamentous diatoms (*Melosira*) typical of spring blooms (see Fig. 12-3). Part (b) shows the condition in October just before overturn. The plankton is dominated by large colonies of blue-green algae, but diatoms and green algae are also present. (*Photographs by H. Canter-Lund.*)

position of some blue-green algae which live permanently in the weak light near the bottom of the photic zone (Baker, Brook, and Klemer, 1969). These algae usually occur only in single filaments, have few gas vacuoles, and rise or sink very slowly. They may be abundant near the thermocline of lakes where nutrients are plentiful. Although there is little chance of their rising to the surface by their own buoyancy, they may reach the surface during autumn overturn. This occurred suddenly with *Oscillatoria rubescens* in Lake Zurich, producing a reddish-colored mass. Photosynthetic bacteria, growing in single filaments near the bottom of the photic zone when oxygen is in short supply, also use a gas-vacuolate mechanism to maintain their position. Buoyancy regulation for these bacteria is vital for their survival, as they require low-oxygen conditions and light (Chap. 11).

An important property in the ecology of planktonic blue-green algal cells is another form of motion—the *gliding* of single filaments

against each other on a firm substratum. Gliding can easily be seen in a colony of planktonic *Aphanizomenon,* where filaments slide back and forth at random (Fig. 12-1e). Gliding occurs using contractile filaments in the walls of the algal cells (see review by Castenholz, 1973). Gliding allows a filament in the center of a colony to gain access to improved light and nutrients. Gliding also occurs in benthic blue-green algae and pennate diatoms.

Flagella and Cilia

Random and directional movement using flagella occurs in dinoflagellates, ciliates, flagellates, and many green algae. Random motion thins diffusion shells and increases the availability of nutrients. The phototactic directional movement of some dinoflagellates and ciliates gives rise to spectacular red tides in oceans and lakes. Their movement allows access to optimum light and temperature and may shade out competition.

THE SEASONAL VARIATION OF PHYTOPLANKTON

The phytoplankton of most of the world's lakes are subject to strong seasonal influences. In the temperate and polar zones there is great contrast between summer and winter, and in the tropical climes between rainy and dry seasons. The phytoplankton respond to this constant rearrangement of the physical and chemical structure of their environment with characteristic population fluctuations. A typical composite picture of algal seasonal cycles in the temperate zone is given in Fig. 12-3.

The most obvious features of the seasonal cycle are the large spring diatom blooms, the smaller irregular summer peaks of various flagellates, and the large autumnal bloom of diatoms, blue-green algae, and dinoflagellates. A bloom is not a precise term and has different definitions. Blooms occur when cell numbers exceed their annual average or background concentrations manyfold, or when a certain high cell number is reached, for example, 5×10^6 cells liter^{-1}. A bloom is also evident when algae color the water. Figure 12-3 describes a generalized seasonal bloom pattern, but there are lakes where other algal types replace those shown and other seasonal cycles where peaks of algal abundance are less pronounced. We will describe well-known examples of spring, summer, and fall blooms and trace some of the causes. Reviews on seasonal variation of phytoplankton are given in Fogg (1975), Lund (1965), and vol. 19 of *Mitt. Int. Ver. Theor. Angew. Limnol.* (1971).

Changes in phytoplankton numbers (dC/dt) can be expressed by the following conceptual equation:

$$\frac{dC}{dt} = \frac{dP}{dt} C - (S + G + Pa + D) \qquad (2)$$

where C = algal cell concentration
$\frac{dP}{dt}$ = rate of photosynthesis

S = sinking out of algae to below the photic zone
G = zooplankton grazing
Pa = parasitism or disease
D = natural death or senescence

Each of these factors will be considered as the text progresses. However, it is important to realize that it is not yet possible to solve the above equation since S, G, Pa, and D have rarely been measured in natural plankton.

The Spring Bloom of a Holoplanktonic Diatom: *Asterionella formosa*

Holoplankton (holo = totally) are always present in the plankton, as distinct from *meroplankton* which spend some of their life cycle in a dormant form, usually in the sediments. Most open ocean algae are holoplanktonic, while lake and coastal forms may be either holoplanktonic or meroplanktonic.

Asterionella is a common diatom in the spring bloom of temperate lakes throughout the world (Figs. 12-4, 12-5). The spring bloom provides grazing for zooplankton that, in turn, feed young fish during the crucial posthatching period (Chap. 14). The conditions necessary for the initiation of blooms of a holoplanktonic diatom like *Asterionella* may be summarized as follows. Enough light and nutrients are needed plus freedom from severe parasitism or predation. The increasing sunlight in spring, not temperature rise, remains the key force, just as it is for the spring growth of land plants.

A problem in describing algal blooms arises from the use of the term *growth rate,* which is often inferred from the rate of population increase in the lake. Although very rapid cell division rates often occur, the lake population may actually rise quite slowly. During periods of rapid growth, losses from an increase in sedimentation, grazing, and parasitism tend to create the fairly uniform increase in population numbers observed.

Asterionella will continue to increase and dominate the spring population as long as it has the necessary light, nutrients, and freedom from

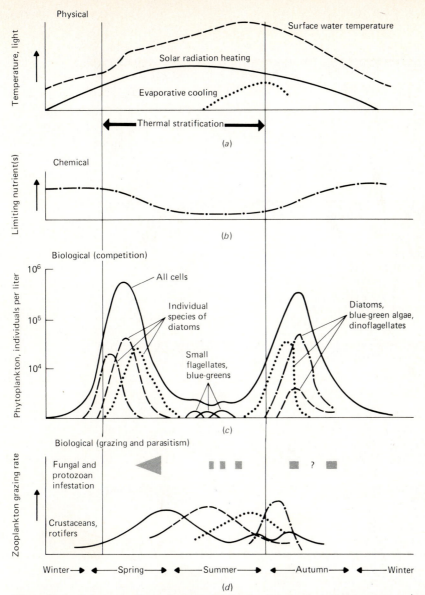

Figure 12-3 Schematic diagram of the interacting (a) physical, (b) chemical, and (c) and (d) biological forces which control phytoplankton seasonal cycles. The major physical effect is the increase of sunlight which is sufficient for high rates of photosynthesis even in late winter. Not until the sunlight heats the lake to allow thermal stratification can high growth rates occur since until then plankton are often mixed below the photic zone. Temperature alone has little effect on phytoplankton cycles. Once stratification has occurred the supply of nutrients to the epilimnion is cut off, and one or more may successively or alternately limit daily growth. Temperature is important in controlling zooplankton grazing and parasitic attacks which occur mostly during the stratified period and may sometimes deplete algal populations. The diatoms in spring enjoy high nutrients, adequate light, and, at least initially, low grazing or predation. Zooplankton do not usually feed on large algae. Small summer blooms of large blue-green algae or rapidly reproducing small flagellates avoid or outgrow predation, and nutrients may recycle rapidly at low concentrations. The loss of thermal stratification in the fall due to evaporative cooling increases nutrients, light is still adequate, but the higher temperatures allow zooplankton grazing and perhaps parasitism to play a larger role. Thus large algae such as blue-greens and dinoflagellates are present as well as diatoms.

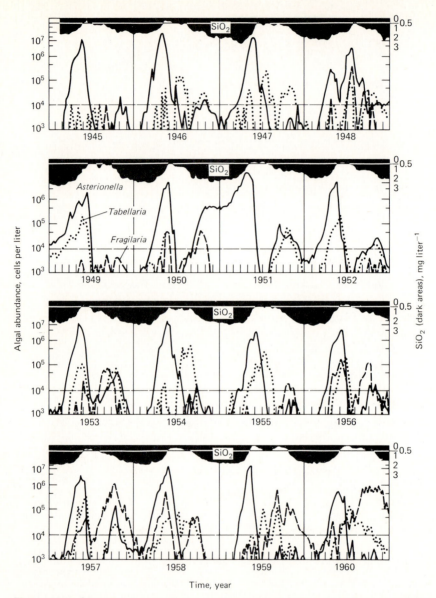

Figure 12-4 Seasonal changes in the limiting nutrient, silica and in the major diatom genera in the North Basin of Windermere, 1945–1960. Solid line = *Asterionella formosa*, dashed line = *Fragilaria crotonensis*, dotted line = *Tabellaria flocculosa var. asterionelloides*, black = dissolved silica in surface waters. (*Modified from Lund, 1964.*)

predation. This dominance over potential competition from other diatoms, such as *Fragilaria* and *Tabellaria*, or blue-green algae, such as *Oscillatoria*, is maintained because *Asterionella* is able to grow faster than the competing algae. In addition, *Asterionella*, like many algae, has the ability to store one of the major growth-limiting nutrients, phosphorus, in a nontoxic form. As much as 100 times the amount needed for cell structure and function can be stored as poly-

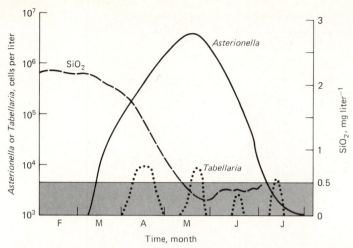

Figure 12-5 The spring bloom of the holoplanktonic diatom *Asterionella* in the North Basin of Windermere in relation to the critical limiting nutrient silica. The similar diatom *Tabellaria* is uncommon and sporadic, and *Fragilaria* is usually found later in the year—but also see Fig. 12-6 for some exceptions to this typical seasonal distribution. Note that the maximum bloom population in the water increases to 500–2,000 times that of the winter population (exceptionally, 50,000 times). This is typical for holoplankton which must have fast growth rates (high μ). (*Data redrawn from Lund, 1964; Heron, 1961.*)

phosphate granules by the process of "luxury consumption" (Chap. 9). *Asterionella* can reduce phosphate in lakes to very low levels, and the population can be sustained on stored phosphorus for a considerable time even when external phosphate is apparently limiting. The role of nitrogen as a limiting element in some blooms of *Asterionella* appears small. As in most temperate climates, rain falls in the English Lake District all year, and nitrate is continually fed to the lake via the rivers. As discussed later, silicon, which falls from 3.0 to 0.5 mg liter^{-1} (as SiO_2), is believed to play the major role in limiting the *Asterionella* bloom in Windermere (Fig. 12-5).

Without parasitism, *Asterionella* will grow faster than competing phytoplankton such as *Fragilaria*. However, when parasitism does occur (Fig. 12-6), the result may favor the competing algae. Important phytoplankton parasites in Windermere are primitive aquatic fungi called *chytrids*, which are host-specific (Chap. 11). Chytrids are single cells that attach to one member of an *Asterionella* colony (Fig. 12-7). The fungi penetrate algal cells using their pseudopodia, eventually killing the algae. The parasites then form spores that are released to the water to infect other cells. Chytrid attacks, like many other infectious epidemics, occur when a large number of individuals live close together as occurs during the spring bloom. While *Asterionella* is infected, its population growth rate is reduced to below that of some other algae like *Tabellaria* and *Fragilaria*, which then overtake the *Asterionella* population (Fig. 12-6). Nevertheless, when *Asterionella* eventually recovers from the chytrid infection, it may once again outgrow other plankton algae so long as nutrients remain plentiful.

Diatoms also compete with other algae. In Windermere there is competition for light between diatoms and the blue-green alga *Oscillatoria*. In its colonial filamentous form *Oscillatoria* is holoplanktonic in Windermere and Esthwaite Water. Dense *Oscillatoria* blooms can

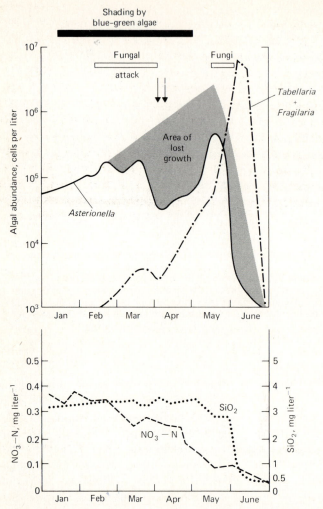

Figure 12-6 Interactions of physical, chemical, and biological factors to shape the spring diatom bloom. See Fig. 12-3 for simplified version. (↓ ↓ = floods). The fungal attack by chytrid fungi is species-specific and is confined here to *Asterionella*. Thus the slower-growing competing diatoms *Tabellaria* and *Fragilaria* become more abundant than normal. Floods wash out and thus reduce all algae. The nutrients NO_3 and SiO_2 have no effect until SiO_2 reaches low levels (0.5 mg liter^{-1}). Blue-green algal (*Oscillatoria*) shading hinders a rapid start to the diatom bloom since *Oscillatoria* floats near the lake surface. (*Based on Lund, 1949–1950.*)

shade *Asterionella,* reducing their growth (Fig. 12-6). When the *Oscillatoria* populations decrease, *Asterionella* can resume vigorous growth.

An important consideration in phytoplankton ecology is the dilute nature of the cell suspension in lakes compared to those grown in culture flasks. Optimal light and high nutrients enable laboratory cultures to reach 10^9 or even 10^{10} cells per liter. Under natural conditions in lakes, 10^4 cells per liter may be the maximum for any one species and 10^7 cells per liter an

Plate 1 *(a)* The deep blue-violet color of a very pure lake—Crater Lake, Oregon. This picture was taken by G. Malyj using a polarizing filter to enhance color and reduce rim and cloud reflections. *(b)* Another very pure blue-water lake—Lake Tahoe, California–Nevada, showing the more productive Emerald Bay in the foreground (see Chap. 13). No filters were used in this picture so the lake looks much less blue than Crater Lake due to sun and cloud reflections. Oligotrophic lakes and oceans appear blue due to the backscattering of short wavelengths of light by water molecules. In most lakes, particles absorb blue light and give a greenish cast to the water (see Chap. 3, Fig. 3-6, and Plate 2*b*).

Plate 2 The colors of a chain of glacial lakes, Grinnell Lakes, Glacier National Park, Montana. *(a)* The upper lake's grey-brown color is due to a heavy suspension of ground-up rock called glacial milk or glacial flour. *(b)* The next two lakes in the series are colored blue-green and blue, respectively. The blue-green color is not due to algae but the suspension of smaller amounts of glacial flour. This series of grey, green, and blue is common in glacial areas. The trend from blue to green or brown is shown by all lakes as particulates increase (Figs. 3-6, 3-7).

Plate 3 Unusual lake colors. *(a)* Deep green color probably due to iron salts in very acidic Okama Lake (pH 3), a large crater lake in Japan. Fifty years ago, before the last volcanic eruption, this lake had a pH of 2 and, although green, was much more transparent than at present. *(b)* Red color produced mostly by bacteria and blue-green algae in a shallow, very alkaline (pH~11) crater lake, Mahiga Lake, Uganda.

Plate 4 Larger aquatic vegetation. *(a)* Floating macrophytes typical of tropical lakes. Here is a dense stand of *Pistia*, the water cabbage, that covers large areas and attracts herons and storks to feed. *(b)* Attached algae, or mesophytes, dominating the autochthonous primary production of a small stream in California. This algae, *Ulothrix zonata*, is common in unshaded streams in spring but is less common in shaded streams (see Plate 5*b*).

Plate 5 The colors of streams in semiarid and well-wooded temperate climates. *(a)* The brown color of a small stream draining an undisturbed chaparral area in California during its flood stage. The brown color is due to silt particles which are only carried during floods (Figs. 16-5, 16-6). Phosphorus is also transported as particulates mostly during this time of flood (Fig. 16-9 and Table 16-1). *(b)* The clear-water color characteristic of most streams in well-wooded areas. Much less particulate material and, thus, less phosphorus is carried relative to soluble nitrate (Table 16-1). The differences in stream color are a clue as to whether phosphorus or nitrogen will limit growth in the lake into which the streams flow (Chap. 6).

(b)

(a)

Plate 6 Remote sensing of lakes. *(a)* The Oaks Arm of Clear Lake taken using infrared color film during a blue-green algal bloom. About 80 percent of the lake's algae are present in the red mass which is primarily reflecting near-infrared light at 715 nm (Fig. 3-7). Boat tracks are visible through the algae; the white area is dead algae. *(b)* The eutrophic Clear Lake, California, and adjacent lakes from about 2000 m above the lake using a normal lens and film modified only by a polarizing filter to enhance the lake color. The grey-green color of parts of the lake is due to blue-green algae floating near the lake surface (Fig. 12-14). The brown color of Thurston Lake at the extreme left is due to erosion of brown silt caused by thoughtless land drainage. The blue color of small Borax Lake in the central peninsula is due to a lack of planktonic algae in this lake, which is very high in borax. The Oaks Arm of Clear Lake with its small island shows bluish on the far right of the picture.

(a)

(b)

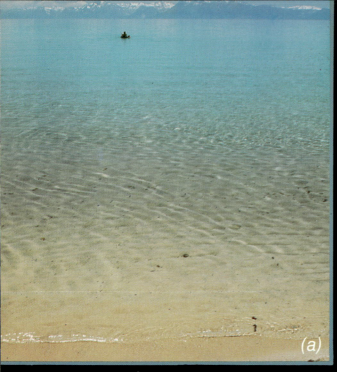

(a)

Plate 7 The colors of the edges of oligotrophic and eutrophic lakes. *(a)* The extreme transparency of Lake Tahoe where Secchi disk depths reach over 30 m. *(b)* A windblown bloom of the blue-green alga *Aphanizomenon flos-aquae* in Clear Lake, California, where Secchi disk depths in this area may only reach 10 cm. Both photographs were taken with polarizing filters to minimize any reflection. Without the filter, Tahoe would appear blue and Clear Lake a yellow-green.

(b)

(a)

(b)

Plate 8 Special features of glacial lakes. *(a)* The cirque face (see Fig. 18-1*d*) of Castle Lake, California. The steep, rocky area is present only at one end of the lake where a glacier has ground down the mountainside. *(b)* The central massif of the English Lake District from which more than 17 lakes radiate like the "spokes of a wheel" (Fig. 19-1). The relatively oligotrophic North Basin of Windermere is in the foreground.

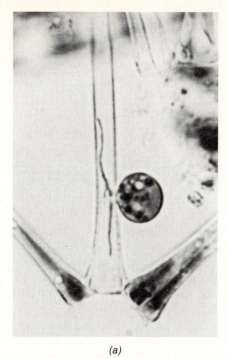 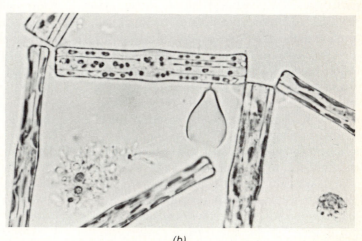

(a)   (b)

Figure 12-7 Predation of the spring diatom bloom by chytrid fungi. (a) *Zygorhizidium affluens* infecting *Asterionella*. The young sporangium of this primitive fungus has two internal rhizoidal threads which have digested most of the cell, leaving the two adjacent ones alive and still full of cell material ($\times$1900). (b) The dehisced sporangium of *Chytridium versatile* on a *Tabellaria* colony ($\times$800). The rhizoidal threads have destroyed one individual, but again adjacent cells are still alive as shown by their large chloroplasts. (*Photographs by H. Canter-Lund.*)

upper limit for the total population. At typical lake phytoplankton densities, there is sufficient light and nutrients for *Asterionella* and *Oscillatoria* to coexist. Coexistence of this sort occurs in many algal populations.

Algal blooms generally end quite suddenly (Figs. 12-3, 12-4). The possible reasons for bloom decline are physical (e.g., no light), chemical (lack of nutrients or the accumulation of toxic waste products), or biological (grazing). There is no shortage of light when *Asterionella* declines in summer, and this species can grow at temperatures from 0.5 to 24°C. However, higher temperatures promote higher growth rates which increases nutrient consumption. The effect of grazing by zooplankton or parasitism is apparently negligible for *Asterionella* in

Windermere, although this may not be true for some other lakes. Nitrogen and phosphorus are not below critical levels inside the cell, and in the English Lake District, Lund found that silicon is the critical factor which eventually limits the bloom of *Asterionella*. The role of silicon as a limiting nutrient is confined to diatoms because the cell wall is made from silica (SiO_2) (Chap. 10). Silica is not readily recycled in lakes and comes mostly from winter river inflows. Therefore the amount of silica present in midwinter determines the maximum spring diatom population if another nutrient does not become limiting first. When silica drops below 0.5 mg liter^{-1} in Windermere (Fig. 12-5), growth of *Asterionella* stops. Below this level the enzyme uptake systems in *Asterionella* are unable to extract silica from the water. *Asterionella* blooms decline rapidly because each cell requires a minimum level of about 140 pg silica per cell. Cells continue to divide even without sufficient silica, but the resultant daughter cells have thin walls and die. In some English lakes, this chemical factor, rather than physical conditions, grazing, or competition, terminates the *Asterionella* bloom in late spring.

Asterionella in Windermere and Esthwaite Water provides a good example of silicon as a growth-limiting element. Other diatoms, for example, the common marine diatom *Skeletonema,* have more flexible requirements for silica and a thinner cell wall can be tolerated. *Asterionella* populations also may be controlled by other factors in some lakes. For example, lake residence time (Chap. 2), nitrate, and phosphate levels can be important, and the rule that *Asterionella* growth stops in all lakes at 0.5 mg liter^{-1} SiO$_2$ is not true. In other lakes in England, nitrogen, competition with *Stephanodiscus,* parasitism, or physical factors may limit growth before lack of silica at the 0.5 mg liter^{-1} level occurs (Hutchinson, 1957). In the Laurentian Great Lakes phosphate may limit phytoplankton growth in unpolluted diatom-dominated areas (Fig. 15-8), while silica or nitrogen may play a similar role in the more eutrophic zones. Regardless of which factor limits growth at any one time, the interaction of the three structural elements—physical (light), chemical, and biological (chytrid predation and competition from other diatoms and blue-green algae) factors—makes the spring bloom of *Asterionella* in the English lakes a valuable case study in limnology.

The Spring Bloom of a Meroplanktonic Diatom: *Melosira italica*

Melosira is a large chain-forming diatom abundant in many lakes and estuaries. The seasonal cycle we describe here is derived from studies of monomictic English lakes; it has many features in common with cycles in dimictic or polymictic lakes elsewhere in the world. Various species of *Melosira* are common in lakes as varied as Siberian Lake Baikal, tropical Lake Tanganyika, and temperate Lake Biwa, Japan (Kozov, 1963; Skabitchewsky, 1929; Mori and Miura, 1980). This cycle is chosen, once again, because of the detailed studies carried out over the last 30 years by Lund (1954, 1971).

The bloom cycle of *Melosira* in the English lakes (Fig. 12-8) is quite different from that of *Asterionella*. For example, *Melosira* is most abundant in the plankton during the period from October to May, and numbers peak well before the annual maximum of *Asterionella*. Despite a low growth rate, the increase of *Melosira* numbers in the fall is more rapid than for *Asterionella*. Because of a massive resuspension of cells from the sediments, *Melosira* numbers increase up to 100,000 times that of the non-bloom population, and during winter there is little loss to grazing or parasitism. *Melosira* is meroplanktonic, which means that cells are not always found in the plankton. At these times living cells accumulate on the bottom of the lake. Unlike *Asterionella*, *Melosira* has cells in a physiologically resting state which are found on top of the sediments. These cells are morphologically distinct from the bloom population.

Comparing the cycle of *Asterionella* (Fig. 12-5) with Fig. 12-8, one can see that the sudden large increase in numbers of *Melosira* in autumn coincides with the breakdown of the summer thermal stratification. Simultaneously, the number of *Melosira* cells in the lake sediment decreases by some 200 times. In fact, the number of cells in the sediment is inversely related to that in the plankton (Fig. 12-8). The cells first found after the fall overturn are visibly distinct from the later winter-spring population and are accompanied by fungal threads characteristic of bottom deposits. *Melosira* is a heavy diatom with a thick silica cell wall and

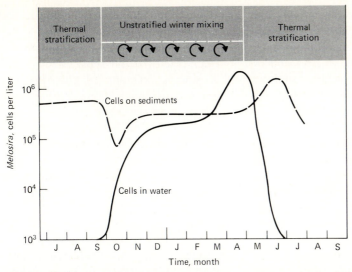

Figure 12-8 Typical seasonal cycle of the meroplanktonic diatom, *Melosira italica*, in Blelham Tarn, English Lake District. Note inverse correlation between suspended cells and those on the sediment for this particularly heavy diatom. Note also that the bloom occurs during the winter and early spring unstratified conditions. (*Modified from Lund, 1954.*)

cannot remain in the plankton without vigorous mixing. Holomictic stirring (Chap. 4) occurs in monomictic lakes throughout the winter, which keeps *Melosira* suspended. Despite the high nutrient levels, lack of light slows growth rates. In spring, light provides the necessary energy for increased photosynthesis and growth. In dimictic lakes with winter ice cover, *Melosira* must wait until after ice-out before holomictic stirring and growth can begin. Resuspension of *Melosira* in very deep monomictic lakes, such as Loch Ness, Scotland, is also delayed until early spring when the lakes may finally mix top to bottom.

Growth of *Melosira* in spring is slow relative to *Asterionella* or other holoplankton and increases about 200 times in comparison to increases of 500 to 2000 times for holoplankton. Even before nutrients become limiting, bloom decline occurs when thermal stratification cuts off deep mixing and *Melosira* cells fall through the thermocline during the first calm periods (Fig. 12-8).

Melosira survives anoxic conditions in the lake sediments for up to 2 years, while other algae usually die. Because *Melosira* cells drop to the bottom for physical reasons, they are usually healthy at the time and not nutrient-starved. These healthy cells have the ability to form a physiological resting stage with a low respiratory rate which is biochemically akin to a spore. The low oxygen demand enables *Melosira* to survive virtually anoxic conditions in the lake mud surface.

Although *Melosira* is unlikely to experience a nutrient shortage before stratification occurs, if a lake is destratified by artificial mixing, nutrient depletion could occur. This was demonstrated by Lund (1971), who induced nutrient depletion by mixing and caused mass deaths of *Melosira* cells. He concluded that continual year-round mixing prevented many healthy resting cells from reaching the sediments to supply the innoculum for the next season's bloom.

In the English Lake District chytrid fungal

infections and zooplankton grazing do not influ-ence the seasonal cycle of *Melosira*. *Melosira* is larger than most chain-forming diatoms and is not ingested by most zooplankton. In some es-tuaries, however, waves produce shearing forces which break up long diatom chains that are more easily ingested by zooplankton.

Because the sediment supplies all the cells of meroplanktonic algae for initiating the next year's spring bloom, the size of one year's max-imum can influence the initial population avail-able the following year. Late winter storms, if severe, may mix deeper into the sediment and increase the percentage of cells resuspended re-gardless of the last season's maximum. In the deepest lakes, sediment may only be disturbed in years of the strongest winter storms. Large winter floods may cause a burying under the sediment or washout of much of the winter pop-ulation. Since cells are not growing at this time, they do not replace themselves. The resulting lower spring innoculum in turn produces a smaller spring crop. The mixing, burying, and washout effects reduce the value of conclusions about plankton ecology based on studies of a single year.

Eutrophic polymictic lakes with continual vigorous wind-mixing and ample nutrients should provide ideal conditions for one or more of the many species of the genus *Melosira*. However, there are few naturally eutrophic lakes with continuous large supplies of all nu-trients, and in Clear Lake, California, nitrogen, rather than turbulence, probably limits the ul-timate growth of *Melosira*.

Seasonal Cycles of Blue-Green Algae

Unlike most other phytoplankton, blue-green algae are unimportant in planktonic food chains although there are exceptions, particularly in some tropical waters. Like thistles left alone by cows in a grassy field, blue-green algae are weeds among the more nutritious populations of diatoms or green algae. Further, blue-green algae cause unsightly algal scums on eutrophic lakes while shading out other algae that con-tribute more to the productivity of the fi-sheries. A blue-green algal bloom in a lake is often the first obvious sign of cultural eu-trophication (Chaps. 18, 20). For this reason, it appears paradoxical that blue-green algae are often abundant when nutrient concentrations are low. Cultural eutrophication (Chap. 18) and the inflow of sewage distort nutrient cycles; it is better to consider the interaction of blue-green algae with pollution using applied limnology and lake management techniques (Chap. 20).

Most blue-green algal blooms occur in the late summer or autumn (Fig. 12-3), although spring blooms and year-round growths are known. Blue-greens in lakes may be either holoplanktonic or meroplanktonic, but as many species readily form spores or resting stages, a meroplanktonic mode is probably more typical.

We will now discuss spring and fall blooms of two types of blue-green algae: those such as *Aphanizomenon* and *Anabaena* that often fix ni-trogen and those like *Microcystis* and *Oscilla-toria* that cannot fix nitrogen.

The Spring Bloom   A bloom of blue-green algae occurs in polymictic Clear Lake, Califor-nia. This lake has a small overwintering pop-ulation and a large, almost uni-algal, spring bloom of *Aphanizomenon*, a sum-mer bloom of *Anabaena* followed by *Microcystis*, ending with an *Aphanizomenon* bloom in early winter (Fig. 12-9). Similar pat-terns are shown for Crose Mere, which is a monomictic, stratified lake in England (Fig. 12-10).

The spring *Aphanizomenon* population in Clear Lake initially dominates because of buoy-ancy due to gas vacuoles. Nutrients are plen-tiful at this time. Rain and muddy inflows make the lake turbid in winter, and only floating algae can get enough light to grow in the early spring. Their growth continues into early summer even

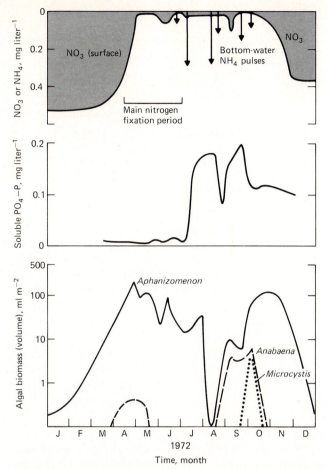

Figure 12-9 Spring and fall blooms of the blue-green algae *Aphanizomenon, Anabaena,* and *Microcystis* in warm unstratified Clear Lake, California, in relation to the cycles of nitrogen and phosphorus. Note that fixation allows continued spring growth after nitrate is used up (see also Fig. 8-2). Ammonia is low at this time. Phosphorus is always naturally in excess in this lake, which is surrounded by easily eroded soils (see Chap. 2). Sediment releases of ammonia in summer are caused by the decomposition of the spring bloom and allow growth of the non-nitrogen-fixing *Microcystis*. (*Redrawn from Horne, 1975.*)

when nitrate and ammonia are depleted, due to nitrogen fixation and an ability to shade out competing algae. The necessary phosphorus is supplied from animal recycling and from the sediments. In Clear Lake and most other lakes, parasitism or animal grazing of blue-green algae is unimportant in regulating bloom size or timing. Growth of *Aphanizomenon* ceases in summer, probably due to lack of soluble iron

needed for nitrogen fixation (Figs. 10-10 and 12-11) and the intense sunlight that kills this alga when near the surface (Plate 6b). *Aphanizomenon* forms large-radius flakelike colonies that can float rapidly to the surface (Fig. 12-1). The overwinter population of *Aphanizomenon* is large compared to that of any other alga and can grow rapidly when light increases in the spring. The only other common

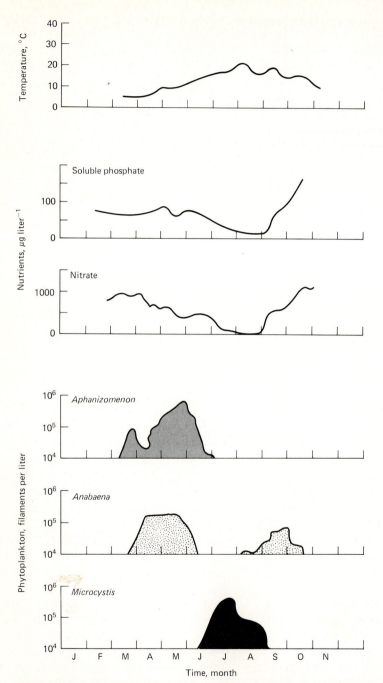

Figure 12-10 Seasonal cycles of three blue-green algae, nutrients, and temperature in the mixed layer of Crose Mere, a small, temperate stratified lake. As in Clear Lake, nitrogen may be limiting for blue-green growth as phosphate is plentiful. Ammonia (not measured in this study) may account for the bloom of *Microcystis* as it did in Clear Lake (Fig. 12-9). (*Modified from Reynolds, 1971.*)

Algal genus	Suggested controlling factors					
Aphanizomenon	1 Algal buoyancy 2 Low sunlight 3 Water turbidity	1 Buoyancy	1 Nitrogen fixation 2 Lack of grazing	1 Lack of Fe 2 Light induced death	1 Buoyancy	1 Algal buoyancy 2 Low light
Anabaena	1 Low sunlight 2 Water turbidity	Poor buoyancy relative to *Aphanizomenon*		? | Lack of Fe	Nitrogen fixation | ?	1 Low light 2 Water turbidity
Microcystis	1 Low sunlight 2 Water turbidity	Lack of NH$_4$			Rate of supply of NH$_4$	1 Low light 2 Water turbidity

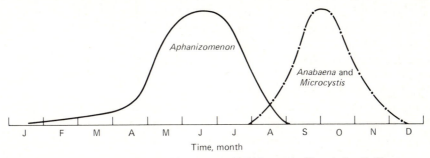

Figure 12-11 Suggested factors controlling the blue-green algal blooms in Clear Lake, California. (*Modified from Horne, 1975.*)

alga present is the dinoflagellate *Peridinium,* which is able to swim to the surface to obtain optimum light. Eutrophic lakes such as Upper Klamath Lake, Oregon, and Smith Lake, Alaska, have a winter ice cover but quickly become warm in spring. In these lakes, spring blue-green blooms after ice-out are often preceded by a short period of diatom growth. Periods of low surface temperatures (4–8°C) and high spring light conditions favor diatom growth over that of the blue-green algae.

The buoyancy of flakes of Aphanizomenon, large coils of *Anabaena,* or clumps of *Microcystis* provide another advantage for blue-greens in eutrophic lakes. The top few centimeters of some patches of water containing algae often move with the wind at a much higher speed than the rest of the water column. After a few minutes, these algae are transported over a different water mass (Figs. 5-3, 5-4). If a colony then sinks, it reaches a new water mass which may provide a better source of

nutrients. Horizontal movements of surface algae, together with the ability to rise and sink, provide a unique advantage.

In late spring or early summer, nutrients in eutrophic lakes become limiting for blue-green algae despite the advantage given by gas vacuoles and luxury phosphorus consumption. The nutrients most likely to become depleted are nitrogen, phosphorus, and iron. Which of these becomes depleted first differs from lake to lake, but nitrogen is often the first nutrient to be depleted in natural or culturally eutrophic lakes. Under such circumstances, blue-green algae often fix nitrogen gas. The role of nitrogen fixation in lakes, including Clear Lake, was described in Chap. 8. We need only mention here that nitrogen fixation allows *Aphanizomenon* to continue growth when it has depleted the winter accumulation of nitrate. Figure 12-9 shows the close correlation of *Aphanizomenon* growth, nitrogen fixation, and the decline of nitrate. Also correlated is an increase in heterocysts, which are the actual site of nitrogen fixation. The ability to fix nitrogen allows *Aphanizomenon* to extend its period of dominance and increase its maximum standing crop above that which could be sustained by the nitrate in the lake.

The very dense spring-summer bloom of *Aphanizomenon* in Clear Lake declines rapidly in midsummer, probably from a lack of iron. Nitrogen fixation is heavily dependent on iron, and in this lake, concentrations of soluble iron are particularly low (Fig. 10-9), while phosphate supplies from animal excretion and from the sediments are high (Fig. 9-3c). Cessation of growth does not explain the decline of a population that is not grazed or parasitized. One possible reason is death by overexposure to sunlight. Damage to the photosynthetic apparatus by sunlight in oligotrophic high-altitude or high-latitude phytoplankton is well known (Figs. 19-8 to 19-10) and a similar effect may occur in Clear Lake (Plate 6b). The initial effect is a bleaching of the chlorophyll and then destruction of the DNA in chloroplasts by ultraviolet light. This short-wavelength light penetrates only a short distance in lakes so surface blooms are most affected. Death is probably due to the photooxidative disruption of the cells followed by lysis and disintegration. As explained previously, blue-green algae can modify their buoyancy to maintain themselves close to optimal light intensities. Buoyancy is maintained by active photosynthesis and thus buoyancy regulation is poor when nutrient depletion occurs. Without the ability to sink, the algae remain on the lake surface. This occurs in midsummer in Clear Lake when the sun is at its maximum and a rapid destruction of surface algae by sunlight can occur. Two hours of sunlight can destroy as much as 80 percent of the *Aphanizomenon* population (Horne and Wrigley, 1975). The *Aphanizomenon* population of Upper Klamath Lake, Oregon, has an adequate nutrient supply and survives summer conditions without the spectacular mortality found in Clear Lake. This presumably results from a continuous ability to regulate buoyancy.

The Autumn Bloom In Clear Lake, the spring bloom is followed by summer-autumn blooms of *Anabaena* and *Microcystis*. Fall blooms of blue-green algae are common in many temperate lakes (Figs. 12-3, 12-10). The unicellular *Microcystis* aggregates into large clumps containing as many as 10^6 cells. It is a major nuisance species and is of particular interest because it cannot fix nitrogen. The two factors explaining its abundance are increased animal and bacterial recycling of ammonia in summer and, once again, shading out of other algae.

Blue-green algae usually possess very efficient mechanisms for the uptake of nutrients at low concentrations. In warm lakes, almost all planktonic blue-green algae can compete successfully with other species for nutrients. Where nitrogen limits summer growth, *Microcystis* is dependent on ammonia-nitrogen supplied by animal and bacterial excretion in the water or sediments. Large populations cannot be created by recycling the small pool of euphotic-zone nutrients, so a bloom of *Microcystis* must be supplied partially by ammonia from the sediments. In stratified lakes nitrogen is supplied to the epilimnion from the lowering of the thermocline in autumn (Fig. 12-10, Chap. 4). The amount of ammonia released in two contrasting basins of Clear Lake was shown in Fig. 8-2. The ammonia level in the bottom waters which soon mixes into the euphotic zone is much greater in one basin than the other. The size of the *Microcystis* bloom is well-correlated with the larger benthic releases of ammonia.

SPATIAL VARIATIONS OF PHYTOPLANKTON

Almost all phytoplankton show vertical variations in distribution throughout the euphotic

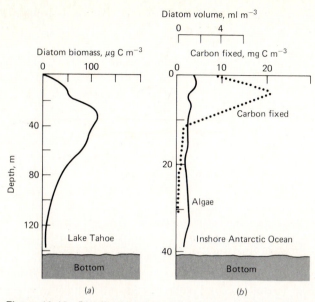

Figure 12-12 Distribution of two diatom species with depth in the photic zone of unstratified systems. (*a*) Uneven distribution of *Cyclotella ocellata* in Lake Tahoe in April 1976. In this transparent lake 20–60 m may be the optimal depth for growth at this time. Primary production often shows a similar distribution (see Fig. 18-7). (*Redrawn from Lopez and Richerson, unpublished.*) (*b*) Uniform distribution of the giant chain-forming diatom, *Biddulphia striata*, in very well-mixed Antarctic waters. Even though primary production peaks near the surface, the constant storms (winds up to 160 km h^{-1}) and the very cold water (0°C) mix the water column continuously and maintain a rare uniform diatom distribution. (*Redrawn from Horne et al., 1969.*)

zone. For diatoms, green algae, or small flagellates, variations are usually small except at the density gradient of the thermocline where sinking cells may accumulate. This is a concentrated source of food for zooplankton, bacteria, or fungi. A typical vertical distribution of a diatom species in unstratified waters is shown in Fig. 12-12*a*, and a rare example of almost uniform distribution is shown for isothermal waters during the summer season in Fig. 12-12*b*.

A very different picture is presented by motile algae which may form concentrated thin layers sometimes associated with the thermocline. One good example of such microstratification is shown in Fig. 12-13 for blue-green algae and cryptomonads in Josephine Lake in Minnesota. Blue-green algae grow best in low-oxygen, high–carbon dioxide concentrations and where nutrients are high near the thermocline in lakes with anoxic hypolimnia. Algae in these dense microlayers must often grow at low-light levels and compensate by producing more pigments which cover a wider range of wavelengths than near-surface forms. Some microlayers are very stable and are not disturbed by vigorous surface mixing. As mixing normally is an advantage for phytoplankton, the stable microlayers must provide other benefits. The infamous reddish-colored *Oscillatoria rubescens* grows at thermocline depth but becomes visible on the lake surface at overturn. The sudden unsightly appearance of this alga drew attention to the serious effects of waste disposal in Lake Zurich.

Other, more buoyant, blue-green algae do not

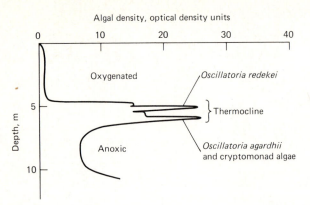

Figure 12-13 Vertical distribution of motile phytoplankton—cryptomonads and blue-green algae in a very strongly stratified kettle lake. (*Redrawn from Baker and Brook, 1971.*)

need overturn to form surface blooms. Figure 12-14 shows the changes in the vertical distribution of the alga *Anabaena circinalis* in Crose Mere, England which occurred over only a few days (compare with Fig.12-10). In small lakes like Crose Mere or even larger ones like Windermere, surface scums frequently wash up on shore, dry out, and die.

DINOFLAGELLATES AND RED TIDES

Unlike the blue-green algae, dinoflagellates can quickly regulate their position by swimming.

They are often large, can move at speeds of meters per hour, and are phototactic. The normal pattern is for these algae to swim up to the surface in the mornings for photosynthesis and then swim down again in late afternoon. Occasionally these movements, when combined with large populations, form surface masses in lakes, estuaries, and the ocean which are visible to the naked eye. These blooms are usually called *red tides*, but are actually more brown or vermillion than red and are occasionally spectacular in lakes (see color picture of the dinoflagellate *Peridinium* on an alpine lake, Gessner, 1955).

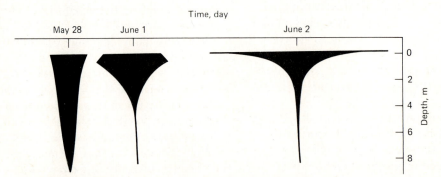

Figure 12-14 Change in vertical distribution of *Anabaena circinalis* in Crose Mere. Diagrams represent "cylindrical curves." The horizontal axis at each depth corresponds to the cube root of the number of colonies recovered at that depth. Note that the volume of algae per unit area of surface is about the same for each day; i.e., this surface accumulation cannot be ascribed to growth. (*Redrawn from Reynolds, 1971.*)

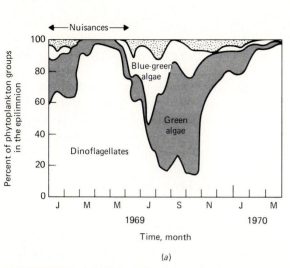

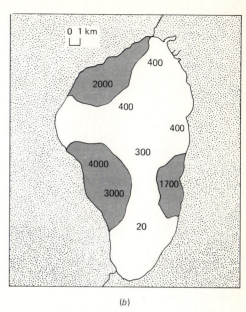

(a)   (b)

Figure 12-15 (a) Seasonal distribution of phytoplankton in Lake Kinneret. Note the dominance of dinoflagellates (mostly *Peridinium*) and the virtual absence of diatoms (dotted area). Lake users notice the color of dense surface patches in spring which may become nuisances. (b) Horizontal distribution of *Peridinium* cells (shaded areas) in Lake Kinneret. Numbers are in cells per milliliter, March 5, 1970. Note the extreme horizontal patchiness of these algae. (*Both figures after Berman and Rodhe, 1971.*)

The movements, physiology, and physical processes that concentrate and dissipate red tides are described for estuarine waters by Seliger et al. (1970) and Fig. 17-14. The salt wedge in estuaries (see Chap. 17) produces some special conditions that do not apply in lakes. In Clear Lake a red tide of *Peridinium pernardii* may form a dense band some 20 m wide and about 40 km long near the shore in spring (Horne et al., 1971). The bloom forms each morning and breaks up each afternoon and may persist for a few weeks. This bloom was probably due to a combination of late floods that retarded the growth of its spring competitor *Aphanizomenon* and a period of calm sunny weather that warmed the nearshore lake sediments. Heating the mud increases the bacterial release of nutrients and perhaps such growth factors as vitamins essential for dino-

flagellate growth but which are not necessary for most other phytoplankton.

Dinoflagellate blooms are not well-understood in lakes or oceans, but considerable knowledge has been gained from some rather unusual lakes in the Austrian Alps (Fig. 19-10) and the desert Lake Kinneret (also known as Lake Tiberias or Sea of Galilee) which is fed by the Jordan River. *Peridinium* blooms are common and quite regular in Lake Kinneret (Berman and Rodhe, 1971) and may reach nuisance levels in spring (Fig. 12-15a) at the nearshore surface (Fig. 12-15b). Lake Kinneret, like Clear Lake, California, shows interalgal competition in spring between *Peridinium* and the blue-green alga *Microcystis*. Even though the mechanisms may not be the same in all lakes, the species that blooms first prevents the bloom of the other. Unlike marine red tides, whose initiation is still largely a mystery, some freshwater red tides can now be predicted with near certainty (Serruya and Pollingher, 1971).

PATCHINESS

Using remote sensing, the detailed patterning of the horizontal patchiness is revealed by algal reflectance in the invisible near-infrared band. The intricate swirls, plumes, and simple blocks of reflected light are shown in Figs. 5-3, 5-4, and Plate 6b. Surface algal concentrations shown in these aerial photographs of Clear Lake, California, vary by factors of 10^2 within a meter and 10^4 over less than 1 km. It is probable that similar complex patchiness occurs for other types of algae in near-surface phytoplankton in many water bodies.

Although surface patches may be advantageous for blue-green algae, this is not the case for most other phytoplankton. High concentrations of algae attract herbivorous zooplankton in the same way that windrows of seaweed and debris in the oceans attract fish, birds, and even sea snakes. If growth of the algae is faster than zooplankton grazing or if turbulent diffusion does not erode it away, the patch will expand. If, as is usually the case, water turbulence is much greater than can be overcome by growth, the patch is dissipated. Only large patches that are more than a kilometer long are likely to persist for several days in lakes. Light and strong winds affect the patchiness of blue-green algae in Clear Lake. When the night was calm very distinct patterns were observed on the surface, but the following day nocturnal winds destroyed the surface patterns (Horne and Wrigley, 1975).

The distinction between some phyto- and zooplankton is confused taxonomically (Table 11-5) but is ecologically distinct. Algae are the producers, and zooplankton the consumers. In the next chapter we discuss how the animal plankton contends with the spatial and temporal variations in phytoplankton just considered.

FURTHER READINGS

Boney, A. D. 1975. *Phytoplankton.* Institute of Biology Study 52. Crane, Russak Co., New York. 116 pp.

Fogg, G. E. 1975. *Algal Cultures and Phytoplankton Ecology.* 2nd ed. The University of Wisconsin Press, Madison. 175 pp.

Hutchinson, G. E. 1957. *A Treatise on Limnology,* vol. II: *Introduction to Lake Biology and the Limnoplankton.* Wiley, New York. 1115 pp.

Lund, J. W. G. 1965. "The Ecology of the Freshwater Phytoplankton." *Biol. Rev.,* **40**:231–293.

S.I.L. (Societas Internationalis Limnologiae) Symposium No. 19. 1971. "Factors That Regulate the Wax and Wane of Algal Populations." *Mitt. Int. Ver. Theor. Angew. Limnol.* 318 pp.

Zooplankton and Zoobenthos

OVERVIEW

Lake zooplankton is ubiquitous. Its main components are protozoans, rotifers, and crustaceans including cladocerans, cyclopoid, and calanoid copepods. In certain lakes larvae of the insect *Chaoborus* and fairy and opossum shrimps are common. Most zooplankton are about 0.5 to 1 mm in length; few are smaller than 0.1 mm or larger than 3 mm. They may be herbivorous on bacterial clumps and algae about 5 to 50 μm in diameter or predaceous on other zooplankton. Zooplankton abundances range from up to 500 individuals per liter in eutrophic lakes to less than one per liter in the most oligotrophic waters. Adaptations of zooplankters to the aquatic habitat include rapid reproduction, small size, and spine formation. Vertical migration to deep water during daylight and a tendency to be transparent reduce

predation on zooplankton by fish and other sight-feeding predators. Zooplankters may have spines which reduce their sinking rate and provide protection from predation. Rapid parthenogenetic reproduction can reduce the impact of predation and permits some zooplankton to exploit short-lived algal blooms. Cladocerans and rotifers tend to be more abundant in summer, probably due to the greater availability of food. Copepods and opossum shrimps are generally perennial, with active overwintering populations.

Zoobenthos are those animals which spend all or most of their existence in or near the sediments. They are a common and widespread group of small animals that includes insect larvae, crustaceans, and mollusks. Benthic invertebrates which dominate the biomass of streams and rivers are discussed in the chapter on flowing waters (Chap. 16). Quantitative sampling of patchily distributed benthos is difficult, but the knowledge of qualitative distribution of species is easier to obtain and has been historically valuable in the classification of lake types.

Benthic animals transform the fine particulate detritus they utilize for food into an animal protein supply for larger carnivores such as the bottom-feeding fish. Zoobenthos may recycle nutrients to the open water by their excretion. Feeding mechanisms range from those of clams which filter out algae, bacteria, and detritus to those of worms and insect larvae which prey on other animals or feed directly on the bottom sediments.

The distribution and abundance of zoobenthos is primarily determined by environmental oxygen concentrations and the type of bottom sediment. The most dense and diverse zoobenthos populations are found in the sediments above the thermocline and along the lakeshore where turbulence provides oxygen and an abundance of food. Below the thermocline, they tend to be less abundant because temperatures are low. Only a few specialized organisms exist in the profundal zone of eutrophic lakes where oxygen levels are also low.

INTRODUCTION

Crustacean zooplankton are small, often transparent organisms which, along with the pelagic protozoans and rotifers, comprise most of the freshwater animal plankton (Figs. 11-6 to 11-8, 13-1). The larvae of marine benthic crustaceans, mollusks, and worms are conspicuous components of zooplankton in estuaries and the sea. In fresh waters this is not the case. True zooplankters are independent of the lakeshore or bottom and inhabit all depths if oxygen conditions are suitable. Some species occupy the nearshore zone, often in association with higher aquatic plants and the littoral substrate. Some species possess resting stages which enable them to reappear suddenly after extended periods of absence. *Daphnia,* the common

"water flea," has an overwintering resting egg stage, the *ephippium,* which is produced in summer or fall as the adult population dies off.

In contrast to the diverse assemblages of the pelagic marine ecosystem, freshwater zooplankton populations in the pelagic zone are usually characterized by only a few dominant species. Typically in a single lake there are a few species of cladocerans, one to three copepod species, and three to seven species of rotifers at any one time. Protozoans are little studied by ecologists. Larger lakes may have as many as 25 crustacean species, and tropical oceans have more than 50 species of copepods as well as many species of planktonic mollusks, annelids, radiolarian protozoans, tunicates, and medusae.

MEASUREMENT

Early biologists like Ernst Haeckel took qualitative zooplankton samples by dragging through the water a net of the same silk bolting cloth used for sieving flour. The more difficult quantitative sampling of zooplankton is a problem that has not yet been completely solved. The objective is to obtain a representative animal sample from a known volume of water. Most nets, traps, or pumps either fail to collect all the organisms or to measure the actual volume of water filtered, or both.

Nets of various kinds are employed for both horizontal and vertical tows (Fig. 13-2). They strain the larger zooplankters and phytoplankters from the water and are normally hauled at a speed of about 1 m s^{-1}. The smallest mesh nets, 60 to 80 μm, are required to retain the minute rotifers or juvenile stages of copepods and cladocerans, but they may rapidly clog with algae. A larger mesh of 200 μm or greater minimizes clogging but may not collect smaller forms, so, in practice, mesh-size selection involves a compromise. The entrance to the net may be restricted to provide a larger filtering area of the net surface relative to the water entering the net. This reduces the back

(a)

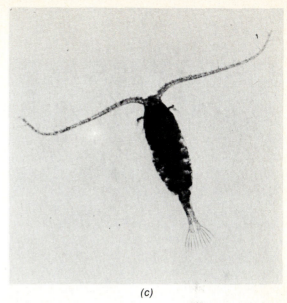

(c)

Rotifers

(b)

Figure 13-1 Zooplankton from (a) a productive lake, Clear Lake, California, and (b) an unproductive one, Lake Tahoe. The pictures represent organisms as normally seen after collection by net tows which may damage some individuals, and preservation with formalin, which causes the antennae to fold down alongside the body. The dominance of cladocerans over copepods typical of productive lakes in summer is shown. Part (b) shows some of the dozen or so moult stages of copepods and a few rotifers; copepods usually outnumber cladocerans in unproductive lakes (Fig. 13-4a). Cladocerans disappeared from Lake Tahoe after the introduction of mysid shrimps (Fig. 13-8). In both pictures, large colonial algae are just visible, indicating the approximate upper limit of size of herbivorous zooplankton food. Most algae eaten will be smaller than those shown here. (c) A copepod with antennae in a typical position when seen alive. (*Photographs by M. L. Commins and G. W. Salt.*)

223

(a)

(b)

Figure 13-2 Photographs of a large zooplankton net (a) in the lake and (b) just after a haul. Note the flowmeter in the net opening in part (a) and the weighted cod end in (b) where the sample accumulates. Such large nets are used in oligotrophic waters, but much smaller ones are adequate for more productive lakes. [*Photograph (a) by R. Richards.*]

pressure that builds up within the net. However, the pressure head formed by the net can produce a shock wave ahead that may warn the more mobile zooplankton. Since all nets are subject to clogging as organisms or detritus accumulate in the mesh, sampling efficiency is highest at the start of the tow. For quantitative sampling, nets should have a flowmeter at the throat which measures the actual volume of water passing through (Fig. 13-2a). Water-sample bottles and plankton traps are useful, but avoidance by the stronger-swimming zooplankton is a potential problem. Their small volume is often statistically inadequate. Trap

avoidance is difficult to evaluate but can be reduced by employing a large-volume clear plastic plankton trap which is lowered at a high speed and closed quickly. Integrated water samples taken with vertical tubes of plastic, or even garden hose, are particularly useful for slow-moving zooplankton in shallow environments. Pumping from specific depths may also be an effective method since it has the advantage of sampling continuously with sufficient intake velocity to prevent escape. Nets mounted on benthic sleds, mechanical dredges, or direct collection by divers may be required to sample amphipods and other zoobenthos which are associated with the lake sediments. Sleds and dredges are difficult to operate quantitatively but may be the only way to collect large numbers of organisms near the bottom.

Variable-frequency echo sounding is used for detecting dense concentrations of zooplankters which, like fish, form *deep scattering layers* (Fig. 14-1*a*). High-speed plankton samplers such as the Hardy plankton recorder have been developed for marine work and are towed routinely along sea lanes by commercial ships.

Field sampling errors invariably exceed subsampling errors. Therefore it is advisable to sample a large volume of water, then subsample before counting. Subsampling may be done with a pipette or with a mechanical plankton splitter. Zooplankton are usually easier to identify soon after collection than after extended storage in neutralized formalin where they lose color and shape. A binocular dissecting microscope is convenient for identifying and counting most individuals, although rotifers and small or juvenile stages of copepods and cladocerans may require higher magnification.

ZOOPLANKTON POPULATION STRUCTURE

Much of the fascination in the study of lakes lies in the structure and the dynamics of zooplankton populations in both space and time. The zooplankton community in most lakes is composed of five to eight dominant species and several rarer forms. Factors such as oxygen, light, temperature, food, and water movements influence the number of species found. Low pH, for example, may reduce species diversity or abundance. The more species of fish that are present in a lake, the greater the number of different zooplankton that are likely to occur. Why this relationship holds is not clear, but a reduction in interspecific competition among the grazing zooplankton might be involved. Another explanation is merely that a habitat suitable for a variety of fish is also suitable for a variety of zooplankton. Large oligotrophic lakes have more species than smaller lakes; for example, the Great Lakes have about 25 species of crustaceans (Patalas, 1969).

Predation by fish and invertebrates is a primary mechanism behind the seasonal changes in zooplankton shape called *cyclomorphosis* (e.g. Grant and Bayly, 1981) and in the zooplankton size distribution. Diel vertical migrations are often a response to predation. This idea was initiated by the postulation of an experimentally testable size-efficiency hypothesis (Brooks and Dodson, 1965). It led to an exciting phase in experimental zooplankton ecology. At present, alternate hypotheses explain the relative abundance of large and small zooplankton as a function of selective predation by fish and predatory zooplankton (Dodson, 1974; Hall et al., 1976). In general fish select the more visible large zooplankton such as *Daphnia rosea* (Fig. 13-9*e*), allowing for a relative increase in the smaller species. Fish fry as well as invertebrate predators prey on smaller zooplankton such as *Bosmina, Ceriodaphnia,* and others less than 1 mm long, but the impact of small adult fish on these forms is minor. Above 1 mm length, the loss to fish predation increases greatly. An early demonstration of this effect was given for Czechoslovakian fishponds by Hrbáček (1962).

Large zooplankton dominate when zooplanktivorous fish are absent but may soon be eliminated if predators are introduced. Exceptions may occur if large size is due to spines or other features which are invisible to the predator. Fish can easily see the pigmented eye spot or the dark gut contents but may be unable to see the body and carapace which are virtually transparent (Figs. 13-1*a*, 13-9*e*). An important feature of this predation-based hypothesis is that it is experimentally confirmable. Experiments on zooplankton feeding and predation are a particularly exciting area of contemporary research.

Since fish and algae suffer from diseases and parasites (Chaps. 11, 12, and 14), it is also likely that zooplankton populations are similarly affected, but few studies have documented the quantitative significance.

Migrations

In deep lakes, the zooplankton may be segregated at different levels during the day. Most respond to changes in light intensity by migrating upward in the evening and downward around dawn. The downward movement may be passive sinking or active swimming away from light. During thermal stratification, strong swimmers pass through the thermocline to graze in the epilimnion during the darker hours, with reduced danger from predators. They then return to the cooler hypolimnetic waters by day. Figure 13-3 illustrates the vertical stratification of four different species in Lake Tahoe. Although this is the most frequent pattern, some zooplankters exhibit rather different diel vertical distribution.

The speed and causes of migratory responses are varied but presumably provide some advantage to the zooplankton. Large species such as the opossum shrimp, *Mysis relicta,* migrate at speeds up to 1.7 cm s^{-1} during the evening hours, preying on other zooplankters as they go. Most freshwater copepods and cladocerans swim at about half the speed of *Mysis.* Migration by planktonic crustaceans leads to a much more pronounced layering than occurs with the less-mobile phytoplankton. Optimal depth varies with different species and also with different life stage. This type of spatial separation provides an element of structure to aquatic ecosystems (Chap. 2). Zonation occurs in shallow lakes but is shown best in a long, transparent water column. Such a structure is shown in the upper 160 m of Lake Tahoe in Fig. 13-3. In its nocturnal migration the large predator *Mysis relicta* avoids warm near-surface water and remains below the thermocline in summer. The copepods *Diaptomus,* and especially

Epischura, migrate upward at night into the warmer surface waters. The *Diaptomus* nauplii remain deeper in the cold water, below the peak of *Mysis* abundance. Thus prey distributions minimize the predator's impact.

The nutrients in a particular water layer may be increased by zooplankton excretion and by lower rates of nutrient uptake due to reduced numbers of algae. Metalimnetic oxygen depletion (Fig. 7-11c) may result from zooplankton respiration and reduced phytoplanktonic oxygen production (Shapiro, 1960). The excretion of ammonia and orthophosphate in the euphotic zone is directly available for primary production and can constitute a substantial fertilization (Johannes, 1968; Hargrave and Green, 1968; Axler et al., 1981).

Some planktonic forms exhibit *littoral avoidance* in which they make horizontal migrations away from the lake edge. They tend to cluster as they actively swim away from the littoral shelf. This may result in patchiness of organisms along the drop-off.

Annual and Seasonal Variations

Seasonal zooplankton cycles can be described in a fashion similar to phytoplankton cycles (Fig. 12-3). Overwintering populations of zooplankton in temperate lakes consist of a small number of individuals which grow slowly since food is scarce. This is analogous to overwintering holoplankton algal populations which grow slowly in winter, but contrasts with the riverine invertebrates which often grow considerably in winter because fallen leaves are a major food (Chap. 16). Overwintering zooplankton are mostly adult and immature copepods (Fig. 13-4). Cladocerans are rare in winter although considerable numbers may be present in the sediments in a resting state or *diapause.* This phase is similar in principle to the resting cells of the meroplanktonic alga *Melosira* previously discussed. Rotifers are also less abundant in the

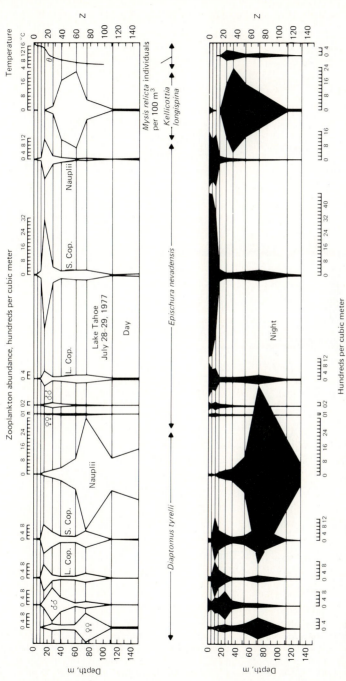

Figure 13-3 Day and night vertical distributions of several growth stages of four species of zooplankton in Lake Tahoe. Similar distributions of different species occur in most lakes. Temperature is indicated by t, and the thermocline is at about 20 m. Note the daily migration of *Epischura* copepodites from 20 m in the day to the surface at night and the very different positions of females (♀), males (♂), copepodites, and nauplii of the other calanoid copepod, *Diaptomus*, which may reflect temperature preferences. *Mysis*, the opossum shrimp, migrates to 300 m during the day so is not shown in the upper figure. Note that the scale for *Mysis* differs by an order of magnitude from the rest. The four species shown constitute the entire lake zooplankton in 1977. (*From Roth and Goldman, unpublished.*)

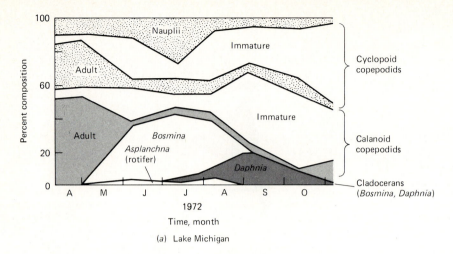

(a) Lake Michigan

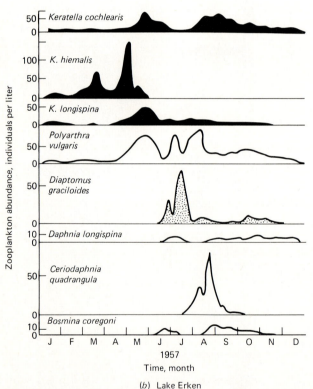

(b) Lake Erken

Figure 13-4 Seasonal cycles of major zooplankton (a) at an offshore Lake Michigan station and (b) in temperate Lake Erken, Sweden. Note in Lake Michigan the dominance of adult calanoid and cyclopoid copepods in winter and the rise in cladocerans and rotifers in summers. This is typical of many lakes (see text). Nauplii of calanoid and cyclopoid copepods were counted together in (a). In Lake Erken note how the different species of *Keratella*, the calanoid copepod *Diaptomus*, and the cladocerans *Ceriodaphnia* and *Bosmina* peak at different times of year (*Part (a) modified from Roth and Stewart, 1973; part (b) from Nauwerck, 1963.*)

winter plankton, remaining as resting eggs in the sediments.

In spring as the lake temperature increases, zooplankton become more active. There is a fairly linear relationship between the rate of feeding and temperature, so that growth increases rapidly if food is available. Once a favorable temperature has been reached in spring, the eggs of cladocerans and rotifers hatch, producing large numbers of young which soon mature. Copepod eggs also hatch in spring, but the maturation rate of the young is much slower. The result is a very rapid spring increase in the number of cladocerans and rotifers with a slower rise in copepods and mysids (Fig. 13-4). Cladocerans and rotifers in temperate lakes in late spring can increase their biomass by a few percent each day, partially because they reproduce asexually (see below). The average size of crustacean zooplankton increases at this time, making them more important as grazers than their numbers would suggest. Seasonal fluctuations in zooplankton are, like those of phytoplankton, less dramatic in the tropics than in temperate regions. However, individual species of zooplankton may vary seasonally over two orders of magnitude or more, especially in deeper tropical lakes.

The spring bloom of diatoms is grazed on by a large number of young and overwintered zooplankton. After the decline of the spring bloom, algae are quite scarce and often consist mostly of indigestible large green algae or small flagellates (Fig. 12-3). Spring zooplankton populations may then decline, to be replaced by different species adapted to feed on summer phytoplankton. This is illustrated for the rotifer *Keratella* and other genera of zooplankton in Lake Erken, Sweden (Fig. 13-4b).

In the English Lake District large annual changes in the populations of limnetic crustaceans have occurred over the last 20 years (Fig. 13-5). In mildly eutrophic Esthwaite Water (Tables 19-1, 19-2), two cyclopoid copepods, *Cyclops abyssorum* and *Mesocyclops*

leuckarti, coexist in widely varying numbers. The less common *Cyclops* exhibited several years of very small populations followed by dramatic increases in the 1960s when its numbers almost equaled those of *Mesocyclops*.

It is difficult to know if this variation is a typical part of a long-term population change. *Cyclops abyssorum* may be an indigenous ice-age species which is being replaced by the warm-water *M. leuckarti*. This zoogeographic explanation needs further support from investigations in other lakes, since it is unwise to generalize from one lake (see Chap. 19). These two species coexist in only 3 out of 18 lakes in the English Lake District. This tends to support a theory of gradual replacement, possibly by interspecies predation on the nauplier or copepodid instars. These two adaptable species have similar life cycles but are different in body size, periodicity, and vertical distribution. Both species can switch from an herbivorous diet to a carnivorous one in the copepodid stage. If copepodids of one grew faster they could switch to a carnivorous diet and prey on the slower-growing species changing the relative abundance of the two species. This explanation assumes that temperature is a major factor since cooler water favors *C. abyssorum* and warmer water, *M. leuckarti*. Thus, long-term trends would be partially obscured by short-term variations in climate.

Two of the most common zooplankton are cladocerans of the genus *Daphnia* and calanoid copepods of the genus *Diaptomus*. The seasonal cycles of several species of these genera are given in Fig. 13-6 for Lake Washington in the northwestern United States. More details of this well-studied lake are given in the chapter on applied limnology (Chap. 20). Five years of measurements selected from a much longer data run show significant changes in population structure. *Diaptomus ashlandi* was the dominant species from 1972 to 1975 when five species of *Daphnia* were present in low numbers. In 1976 a remarkable change oc-

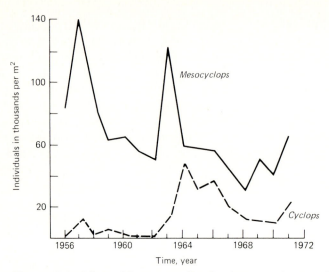

Figure 13-5 Effect of long-term climatic change and competition between species on the mean annual totals of the two cyclopoid copepods, *Cyclops abyssorum* and *Mesocyclops leuckarti,* in Esthwaite Water, English Lake District, from 1956 to 1971. Note the large variation from year to year. (*From Smyly, 1978.*)

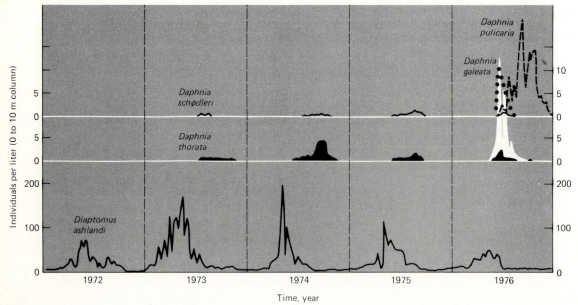

Figure 13-6 Effects of multiple changes in Lake Washington on the seasonal and annual variations of the zooplankton population dominated by a calanoid copepod, *Diaptomus,* and various species of the cladoceran *Daphnia.* Note how two *Daphnia* species suddenly invaded the plankton in 1976 and the differences in scales. *Daphnia ambigua* (not shown) was present only in 1972 and 1973. (*Redrawn from Edmondson and Litt, 1982.*)

curred. Two species of *Daphnia* suddenly increased and remained abundant, while *Diaptomus* was reduced to less than half its 1972–1975 levels. Even after these changes, *Diaptomus* was about as abundant as *Daphnia* in the upper 10 m of water. However, there was a definite reduction in the peak numbers of *Diaptomus* from 200 per liter in 1973–1974 down to 50 per liter by 1976. The changes in Lake Washington are an extreme example of the kind of variation discussed for *Cyclops* and *Mesocyclops* in the English Lake District. *Cyclops* in Esthwaite Water suddenly rose from very low numbers, as did *Daphnia* in Lake Washington (Figs. 13-5 and 13-6). In Lake Washington the dominant zooplankter *Diaptomus* decreased, but the dominant species in Esthwaite Water was unaffected by the rise in *Cyclops* numbers.

The reasons for the sudden increase in *Daphnia* in Lake Washington are not fully understood but may involve interactions at all trophic levels. Prior to 1965 *Daphnia* populations were low, presumably due to grazing by the opossum shrimp *Neomysis*. In some Californian mountain lakes, *Mysis relicta* has caused similar effects after its artificial introduction. In Lake Washington, *Neomysis* decreased after 1965, probably due to an increase in fish predation. Between 1965 and 1975, *Daphnia* was apparently not able to take advantage of reduced predator pressure due to mechanical interference in its feeding. This was caused by an abundance of the blue-green alga *Oscillatoria rubescens* whose long filaments are collected during feeding but then rejected. By 1975, according to Edmondson, the restoration of the lake (Chap. 20) had progressed so far that *Oscillatoria* no longer was present in sufficient amounts to prevent the increases in *Daphnia* (Fig. 13-6).

Changes in populations may occur for reasons other than temperature or competition. For example, following trace-element fertilization of Castle Lake, California, with molybde-

num, *Daphnia rosea* and *Diaptomus novamexicanus* increased dramatically (Fig. 10-13). The cladoceran response to an increase in algal growth was almost immediate since their generation time is about 14 days. The copepod *Diaptomus* responded by producing large numbers of eggs and showed an order-of-magnitude increase when these eggs hatched the following year. An increase in trout yield due to increased zoobenthic production occurred for several years after the trace-element fertilization (Fig. 10-13).

Rotifers constitute the third major group of freshwater zooplankters. The seasonal cycles for three herbivorous *Keratella* species and for *Polyarthra* in Lake Erken, Sweden, are shown in Fig 13-4*b*. The cycle of the relatively large predatory rotifer *Asplanchna priodonta* in small, productive Donk Lake in Belgium is shown in Fig. 13-7. Two main peaks in abundance usually occurred in early spring and in autumn. The maximum populations varied greatly from year to year, which is typical of zooplankton. In other lakes the cycles may differ considerably; for example, maximum abundance of *Asplanchna* may occur in summer. In dimictic lakes some rotifers reach their peak abundance in winter under ice cover. These seasonal patterns in Donk Lake may be due to interaction between *Asplanchna* and large filter-feeding zooplankton such as *Bosmina* and *Daphnia*. There is apparently little direct food competition between some rotifers and cladocerans; *Asplanchna* feeds mostly on other rotifers like *Keratella* as well as on phytoplankton. However, *Keratella* may be outcompeted for food by *Bosmina* since both feed on small algae and bacteria.

A fourth group of zooplankton are generally known as *inland water shrimps*. They include the fairy shrimp found in many temporary pools and one genus, the brine shrimp *Artemia*, which lives in saline lakes. Also common are the opossum shrimps. One of these, *Mysis relicta*, is a holarctic form common in the

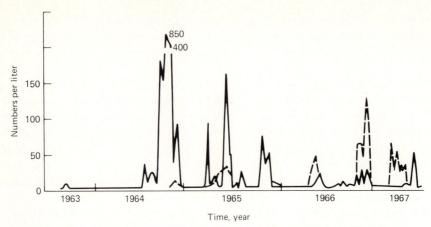

Figure 13-7 Seasonal cycles of the predatory rotifer *Asplanchna priodonta* in Donk Lake, Belgium, at two stations (dashed and continuous lines) in this small (43-ha) lake. Note that if only one station had been sampled, a decline could have been shown from 1964 to 1966. Actually no such trend occurred. (*Modified from Dumont, 1972.*)

Great Lakes. It was introduced to Lake Tahoe as a forage food for juvenile lake trout. By the early 1970s *Mysis* had increased dramatically but the cladocerans *Daphnia pulicaria, D. rosea,* and *Bosmina longirostris* disappeared from the pelagic zone (Fig. 13-8). Predation by *Mysis* and by introduced Kokanee salmon, *Oncorhynchus nerka,* is believed to be the cause of the cladoceran decline. *Bosmina* now coexists with *Mysis* only in relatively shallow Emerald Bay, where primary productivity may be high enough to allow *Bosmina* to offset predation by having higher birth rates than is possible in the oligotrophic pelagic zone (Morgan, 1980).

The transparent "phantom" midge larvae of *Chaoborus* are the only truly planktonic insects. These larvae, close relatives of the mosquitos, are voracious predators. They are ordinarily benthic by day and planktonic at night.

Feeding

Many zooplankton are filter feeders on suspensions of mixed bioseston composed of bacteria, detritus, and algae. Their filtering rate is greatly influenced by temperature and generally is maximal close to the thermal death point.

Grazing zooplankton can be discriminating in their food selection. Most diatoms and many small green algae are excellent food, but not all the available phytoplankton are eaten, and of those ingested some may not be digested. Arnold (1971) studied zooplankton grazing on cultures of small colonies of various blue-green algae and concluded that planktonic blue-greens are seldom utilized by zooplankton. Most planktonic blue-green algae form large colonies and may be mechanically less available to filter feeders. In shallow tropical lakes and oceans, however, fish, invertebrates, and flamingos are specially adapted to consume them. Blue-green algae attached to coral reefs may be an important fish food. They also form a thick algal felt in many shallow polar lakes. Other algal types not greatly influenced by the grazing activities of zooplankton are dinoflagellates and desmids. However, zooplankton may have indirect effects on algae via excretion (Roth and Horne, 1981) or through removal of the smallest colonies by grazing (de Bernardi et al., 1981).

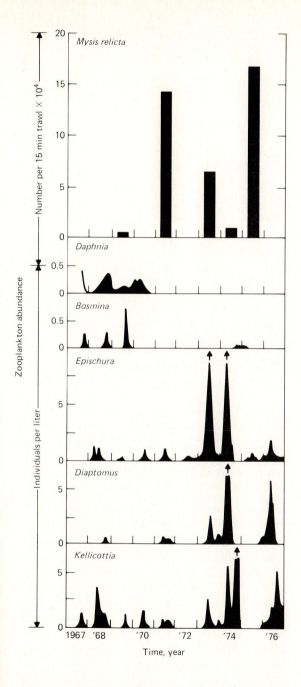

Figure 13-8 Effects of predator introduction on the seasonal cycles of zooplankton in Lake Tahoe. Note how the *Daphnia* (*D. rosea* and *D. pulicaria*) and *Bosmina longirostris* populations decreased drastically, while the copepods *Epischura nevadensis* and *Diaptomus tyrelli* and the rotifer *Kellicottia longispina* increased after the introduction of the opossum shrimp *Mysis relicta*. (*Redrawn and modified from Goldman et al., 1979.*)

for different species. Figure 13-9 illustrates the feeding method and filtering appendages of *Daphnia*.

Daphnia and other Cladocera such as *Ceriodaphnia* and *Bosmina* have legs bearing hairs and setae used to filter particles which are then collected in the ventral groove and moved to the mouth. *Polyphemus* and *Leptodora* are raptorial cladocerans predatory on other cladocerans, copepods, protozoa, and rotifers.

Calanoid copepods create a current by flapping four pairs of feeding appendages. Their second maxillae seize particles prior to filtering them (Koehl and Strickler, 1981). Calanoids filter out particles from about 5 to over 100 μm in size. Some, such as *Diaptomus shoshone*, are predaceous and are able to alternate between suspension feeding and predation.

The predatory species add greatly to the trophic complexity of the zooplankton community since many cyclopoid copepods are raptorial feeders on both animal and plant plankton. Their food includes the young stages of their own species, those of other copepods, and rotifers. Rotifers are extremely varied in both form and feeding. *Keratella, Filinia*, and *Brachionus* are omnivorous genera, and *Asplanchna* and *Synchaeta* are largely predatory. Rotifers use their anterior ring of cilia to direct particles to the mouth. The particles are sedimented out of the water rather than by true filtration. Because of their small size, they feed on smaller particles than most cladocerans and copepods. They utilize nannoplankton, bacteria, and detritus up to about 15 μm in diameter.

Zooplankton employ various feeding methods. The antennae and thoracic limbs of crustaceans differ considerably in their fine structure, providing a different filtering capacity

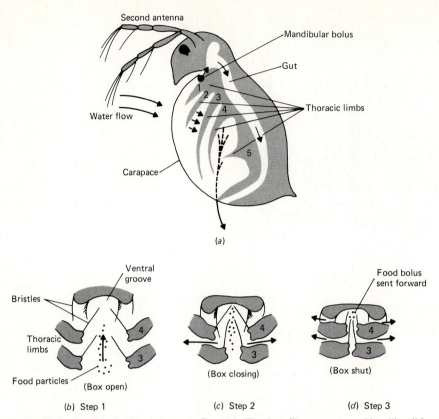

Figure 13-9 Feeding in the cladoceran *Daphnia*. The fast-filter pump used is a "box" formed by the body and carapace and the third, fourth, and fifth thoracic limbs. (*a*) When the limbs move out, food particles, carried in by a current of water, move into the box (*b*). As the limbs close in, particles are retained by the bristles as water is squeezed out (*c*) and particles are trapped, and swept up toward the mouth (*d*). Rejected particles are expelled through the lower carapace. The bristles overlap and are feathered for maximum food retention. (*Modified from Russell-Hunter, 1969.*) Laminar flow predominates at the small dimensions of the bristle apertures [see Chap. 5, Eq. (1)]. The limbs sweep the water like solid paddles rather than filters and in copepods may act as grabs which only separate particles from water during the final squeeze (Koehl and Strickler, 1981). (*e*) The "box" of the carapace (light stippled area) and the body and limbs. Several dark eggs are also visible inside the carapace.

Predatory rotifers such as *Asplanchna* draw small organisms into their mouths by using the water current from their cilia and then seize them with their trophi or mouth parts (Fig. 11-7). General food selection by zooplankton is discussed in detail in Kerfoot (1980).

For herbivorous zooplankton, the filtration rates determine their ability to affect phytoplankton and bacterial populations. Their ability to recycle nutrients into the euphotic zone has already been discussed. *Filtration rate* is defined as the volume of ambient medium per unit time that is cleared of particles. This is equivalent to the *grazing rate*. The *feeding rate*

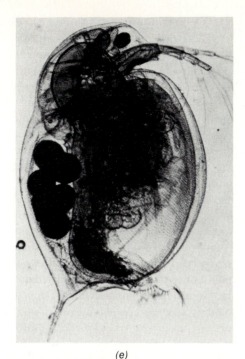

(e)

Figure 13-9 continued

is the quantity of food ingested per unit time, and both feeding rate and filtering rate are affected by cell or detritus concentration. The feeding rate is usually proportional to food concentration up to a limiting level where ingestion rate limits feeding. In general, the filtering rate of zooplankton increases with the square of the body length and also increases with temperature to an optimal level before declining. This varies with species and with acclimation temperature (Burns, 1969). In a classic study (Haney, 1973), ^{32}P-labeled yeast was used for measuring zooplankton feeding rates in situ within a special chamber in Heart Lake.

In nature, grazing by zooplankton may significantly reduce the size of phytoplankton populations. Because the major food source for herbivorous zooplankton is the ubiquitous diatom population, grazing pressure is likely to change

diatom species composition and numbers. There has been a good deal of speculation on the importance of grazing by zooplankters on phytoplankton density. Harvey (1937) proposed a grazing theory which is illustrated by the calculation that, beginning with 100 algal cells, after six divisions there are 6400 cells. If, however, 10 percent is grazed off between divisions, only 3410 cells would result, which amounts to a reduction in the potential population size by almost half. In Lake Washington in late spring each zooplankter can filter the algae from several ml of lake water per day. The number of algae removed is approximately the same as the rate of primary production. Thus the zooplankton may keep the algal population at a constant level.

Most filter-feeding zooplankton prefer smaller algal species, whose numbers tend to be reduced by grazing. Some algal species show little effect of grazing pressure, while others may even increase due to a reduction in the number of their competitors. Certain blue-green and green algae, particularly those with durable cell walls and gelatinous sheaths, may pass unharmed through the guts of zoo-plankters and even gain some nutrients during passage (Porter, 1976). Interaction between zooplankton and algae has long interested limnologists as well as oceanographers, and considerable theory has been developed. Grazing may influence the succession of phytoplankton species which gradually change from a largely edible group of cells to predominantly inedible ones, such as blue-green algae, during the course of a growing season. Differential nutrient excretion by zooplankton may further modify phytoplankton succession. Excellent examples can be found of severe zooplankton decline coincident with a shift from diatoms to flagellates or dinoflagellates (Russell et al., 1971; Reid, 1975). Studies with *Daphnia* fed on *Chlorella* show that the age of the phytoplankton crop can be important in determining its nutrient value and, on occasion, its toxicity. The tropical freshwater *Thermocyclops hyalinus* exists almost entirely by browsing on the colonial coccoid blue-green alga *Microcystis*. However, this is an apparent exception to the general rule that herbivorous zooplankton prefer a diet of diatoms or green algae.

As already noted, there is considerable difference in the digestibility and nutritional value of particles that zooplankton eat. In addition to the quantity of available food, variations in carbohydrate-to-protein ratios, and vitamin and trace-element content may also be reflected in the species composition and condition of the phytoplankton. This, in turn, may be important in determining the standing crop and recruitment of zooplankton.

Reproduction

That zooplankton are adapted to maximize utilization of the short-lived phytoplankton blooms is well-illustrated by their life cycles and reproductive strategies (Figs. 13-10, 13-12). Under favorable conditions rotifers and cladocerans have a life cycle of only a few days, so they are capable of producing many generations each year (Fig. 13-10). They are thus termed *multivoltine*. Some copepods are multivoltine but others are *univoltine* and produce only a single generation each year. Copepods and mysids grow relatively slowly because their metamorphosis requires several moults before sexually reproductive adults are produced (Fig. 13-12). Multivoltine zooplankton reach full size and begin to reproduce early in relation to their life expectancy. Most of the food they assimilate in their lifetime goes into egg production rather than into the growth of individuals. In contrast, univoltine lake copepods and benthic insects spend most of their lives and energy in growing to sexual maturity. Thus growth in more highly developed organisms takes a larger percentage of the total food than is utilized in egg production. Therefore, multivoltine rotifers and cladocerans multiply rapidly when food is available (Figs. 13-4 to 13-8), with their numbers rising almost as fast as those of phytoplankton (Fig. 12-3).

Rotifers and cladocerans can avoid the time-consuming process of finding a mate by reproducing rapidly through *parthenogenesis*. In this process the ovum develops without fertilization of the egg and only females are produced. Although the life cycles of rotifers and cladocerans are ecologically similar, they are quite different cytologically (Fig. 13-10).

After production the fate of the eggs varies. Most rotifers quickly release eggs but a few, like *Asplanchna,* brood them internally. Female cladocerans usually carry their eggs inside the carapace. Female cyclopoid copepods carry eggs in paired sacs attached to the abdomen (Fig. 13-11). Calanoid copepods typically have an unpaired egg sac, but a few genera, such as *Epischura* and *Limnocalanus,* shed their eggs one at a time into the water.

Unfavorable conditions, particularly falling temperatures, crowding, and changes in diet or

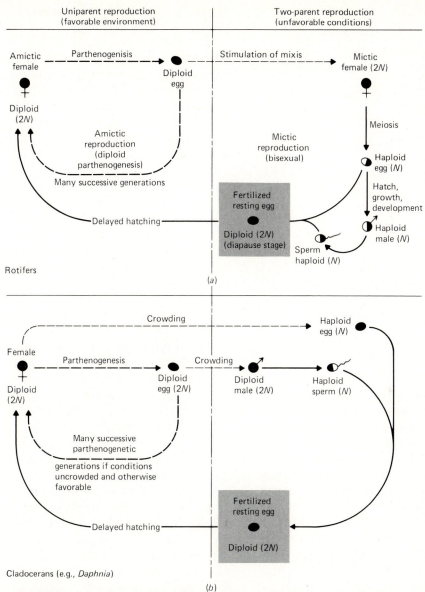

Figure 13-10 Life histories of zooplankton where uniparent, parthenogenetic reproduction is important. The cycles of (a) rotifers and (b) cladocerans such as *Daphnia* appear similar but are very different genetically. Both groups use parthenogenetic reproduction when conditions favor rapid growth, e.g., during phytoplankton blooms. In rotifers sex determination has a chromosomal basis; females are diploid (2N), and males are haploid (N). In cladocerans, as in most higher animals, sex is determined by the familiar XX chromosome (female) and XY (male). In both groups under favorable conditions diploid females produce diploid eggs which hatch into more diploid females without the ovum having to be fertilized by a male. Thus exact copies are produced. Under stress of crowding, temperature, photoperiod, or diet changes, males or haploid eggs are produced and sexual reproduction occurs. The fertilized resting eggs produced have a greater genetic variability to survive possible new conditions in the following years. (*For more details see Edmondson, 1955; Gilbert and Thompson, 1968; King, 1972.*)

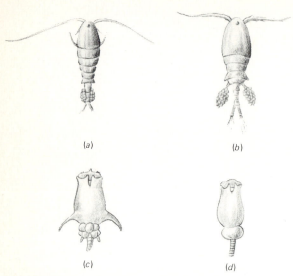

Figure 13-11   Zooplankton and their eggs. (a) calanoid copepod; (b) cyclopoid copepod; (c) rotifer, with summer eggs; (d) *Brachionus* with single autumn egg. (*Modified from drawings by H. Anton.*)

photoperiod, cause parthenogenesis to cease (Fig. 13-10). Males as well as females are now produced. Sexual reproduction can then occur, with the resulting fertilized egg usually thick-walled and resistant to desiccation and cold. This has an obvious advantage for the zooplankton of temporary ponds who must survive long periods without water. Copepods always reproduce sexually (Fig. 13-12). However, they can produce resting eggs, and under stress cyclopoids may enter a benthic resting stage while copepodids.

Population Dynamics

A study of population dynamics involves knowledge of intrinsic rates of growth, birth, and death. Rates of *natality* (births) and *mortality* (deaths) in natural populations reflect the seasonal dependence zooplankton have on food quantity and quality as well as on physical or chemical fluctuations. For example, food

supply greatly affects the number of eggs produced by zooplankton.

Birth (*b*) and death (*d*) rates together determine the instantaneous rate of growth in numbers (*r*) of a population, according to the general formula $r = b - d$. The exponential rate of growth per day over a short time interval is estimated from *r* and is easily calculated from the weekly population data by the equation

$$r = \frac{\ln N_{t_1} - \ln N_{t_0}}{t_1 - t_0}$$

where N_{t_1} = number of individuals at time t_1
$\quad\quad N_{t_0}$ = number at t_0
$\quad t_1 - t_0$ = length of time in days between the two sampling intervals

(For more details see Edmondson, 1974*a*, 1977.) The major assumption is that population changes are exponential between any two sampling dates. This is usually a realistic assumption for natural zooplankton populations. An example of the population dynamics of the common cladoceran *Daphnia* in a lake, showing *b*, *r*, and *d*, is given in Fig. 13-13.

While *r* is simply estimated from the weekly population estimates, the data required for the calculation of birth and death rates are more difficult to obtain. Birth rate *b* estimates the number of new individuals born to the population each day. Like *r*, it is an instantaneous, exponential rate. The calculation of birth rate requires weekly estimates of egg numbers and the density of females. In most zooplankton species, eggs are carried by the female and are easily identifiable if separated after collection. Eggs can also be counted within egg cases or brood pouches (Figs. 13-9*e*, 13-11).

If eggs and adults are sampled with different efficiencies, calculations of birth rates will be inaccurate. With the exception of planktonic protozoa, eggs, especially rotifer eggs, are the smallest zooplankton stage sampled. Most commonly used zooplankton sampling nets are too coarse for quantitative collection of zooplank-

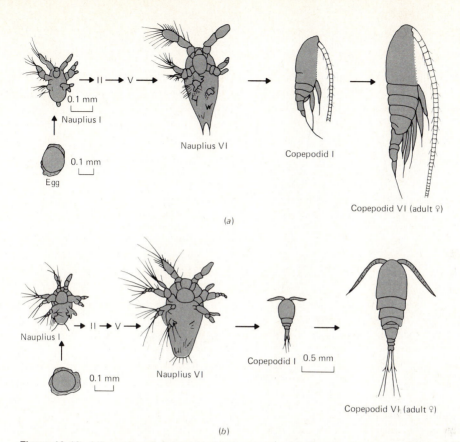

0.1 mm
Nauplius I

0.1 mm
Egg

Nauplius VI

Copepodid I

Copepodid VI (adult ♀)

(a)

Nauplius I

0.1 mm

Nauplius VI

Copepodid I 0.5 mm

Copepodid VI (adult ♀)

(b)

Figure 13-12 Life cycles of copepods. Freshwater copepods always produce sexually fertilized eggs. The free-swimming larva which hatches from the egg does not look like a small adult as do young rotifers and cladocerans. It is known as a stage 1 nauplius and has three anterior pairs of appendages. There are 11 subsequent molts in each of which there is an increase in size and development of appendages. The final stage is the adult; the first six are nauplii and the last six are copepodids since the change at nauplius VI is greater than other between-molt variations. (a) The calanoid copepod *Diaptomus vulgaris.* (b) The cyclopoid copepod *Cyclops strenuus.* Note scale changes. (*Redrawn from Ravera, 1953.*) There are considerable variations in the general pattern of metamorphosis among the groups of copepods.

ton eggs unless the eggs remain attached to females. Even rotifer adults are often underestimated when collected by nets. Sedimentation of unfiltered samples is an accurate way to estimate zooplankton, eggs, and rotifer densities.

Assuming that eggs and females are counted accurately, the instantaneous birth rate, in new individuals per day, is calculated as follows:

$$b = \ln\left(\frac{E}{D} + 1\right)$$

where E = egg ratio = number of eggs divided by number of females
D = egg development time, days.

The egg ratio is easily computed from samples, but D must be determined experimentally.

Conveniently, variation in egg development

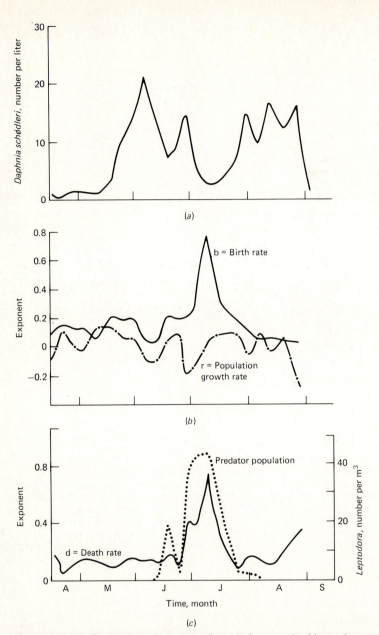

Figure 13-13 Population dynamics of a cladoceran herbivore in summer. (*a*) Measured changes in numbers of *Daphnia schødleri*. (*b*) Birth *b* and population growth rate *r*. (*c*) Death rate *d* and numbers of the large cladoceran predator *Leptodora*. Note the sudden rise in death rates and decrease in *Daphnia* populations caused by the increase in *Leptodora* in July. Some compensation for losses occurs from an increase in birth rate which eventually restores the *Daphnia* population to its spring level. (*Redrawn and modified from Edmondson, 1974; data from Wright, 1965.*)

time within a population is determined almost totally by temperature. D varies inversely with temperature, and it is preferable to determine D for individual populations. Egg development time can be estimated in the laboratory by isolating a group of 50 to 100 females with eggs at a temperature similar to the environment at the time and depth of capture. A graph of percent unhatched eggs versus time is extended to the point where no eggs remain unhatched. The resultant time is an estimate of D at that temperature (Edmondson, 1965). Alternatively, individual eggs can be followed from the time of their production to hatching. The major difficulty in calculating D for zooplankton is estimating the temperatures in the lake during hatching, since water temperature changes during the development of the eggs. Zooplankton are rarely evenly distributed with depth and usually migrate vertically through layers of various temperatures on a seasonal or daily basis (e.g., Fig. 13-3). Egg temperature may be estimated by using an average egg depth, calculated maximum and minimum egg development times (Goldman et al., 1979), or by a more exacting approach that takes into account zooplankton vertical distributions, migrations, and the volumes of water strata of different temperatures (Prepas and Rigler, 1978).

Once D is known, b can be calculated and r is determined from weekly population measurements. The instantaneous death rate d is then calculated by subtraction, since $d = b - r$. Negative death rates are, of course, impossible in nature but occur when b is underestimated or r overestimated. Additionally, d is subject to all the assumptions and problems inherent in the calculations of b and r.

Taken together, estimates of r, b, and d greatly expand an analysis of zooplankton population dynamics. Birth rates can be influenced by food supply as well as temperature and, as such, are useful indicators of relationships between zooplankton and their food supply (e.g., King, 1967; Frank et al., 1957). An analysis of death rates can establish relationships between zooplankton and higher trophic levels or the influence of major environmental fluctuations on zooplankton populations. Although the above discussion on population dynamics has been directed toward zooplankton, a similar approach may be applied to the zoobenthos of lakes, streams, estuaries, and even wetlands (Erman and Erman, 1975).

ZOOBENTHOS

The present classification of lakes by trophic status (Chap. 18) had its basis in the early investigations of Thienemann (1922), who classified lakes by the qualitative composition of their profundal benthic fauna. For example, his class of *Chironomus* lakes dominated by red midge larvae and tubifex worms resistant to low oxygen levels still remains the equivalent of the more eutrophic lakes. Similarly, lakes dominated by larvae of the chironomid midge *Tanytarsus*, which are intolerant of low oxygen levels, would now be regarded as moderately oligotrophic.

The sudden disappearance of the burrowing benthic mayfly population in western Lake Erie in 1953 was a dramatic example of a benthic response to changed lake conditions (Britt, 1955; see also Carr and Hiltunen, 1965). The accelerated eutrophication of the lake apparently increased the supply of algae settling to the lake bed. In turn, during infrequent episodes of thermal stratification the decay of these algae reduced oxygen levels in the temporary hypolimnion to below that tolerated by the mayfly *Hexagenia*. Such catastrophic conditions need occur only once to virtually wipe out an entire population. This process can be reversed in lake restoration by using artificial lake mixing or reaeration (Chap. 20). Benthic populations are usually the first and most prominent beneficiaries (Fast, 1973). The importance of benthic organisms in food-chain dynamics of whole-lake or river ecosystems is discussed in Chap. 15.

The Benthic Environment

The benthic environment can be divided into two quite distinct habitats, the profundal and the littoral. The profundal zone is below the thermocline and is physically and chemically uniform except for changes due to decreases in oxygen in mesotrophic and eutrophic lakes in summer. In contrast the littoral and sublittoral waters in the mixed layer show large seasonal and daily variations of physical and chemical factors.

In the profundal zone of lakes there are usually only four main groups of benthic organisms. These are chironomid midge larvae, oligochaete worms, the transparent "phantom" midge larvae *Chaoborus,* and the fingernail clams *Sphaeridae.* Some important exceptions occur. In particular, the amphipod crustacean *Pontoporeia affinis* and the mysid shrimp *Mysis relicta* dominate much of the deepwater benthos in the Great Lakes and some other northern cold-water lakes. Similarly, oligochaetes tend to be a cold-water group, and only a few species live in permanently warm lakes and rivers in the tropics (Timm, 1980). Occasionally, as in Lake Erie, the burrowing mayfly *Hexagenia* may dominate some areas of the profundal zone provided oxygen is high. Despite these exceptions, the structure of the profundal benthos is simple, with only predatory midges such as *Chaoborus* and *Procladius* differing from the detritus-feeding mode of the majority of forms.

In the littoral and sublittoral zones diversity and productivity are much higher. Most types of insects, snails, worms, crustaceans, and fish are found. All types of feeding are represented, from carnivorousness to grazing on attached algae and bacteria.

The patchy physical and chemical nature of the substrate strongly influences the structure of the benthic community. Factors directly related to lake bathimetry have given rise to the previously mentioned terminology of surf or littoral, the sublittoral, and profundal communities (Fig. 2-2). In shallow, transparent zones of lakes where aquatic macrophytes are abundant (Fig. 11-5), there will be a benthic community associated with the shelter and detritus produced on or by these larger plants. Most lakes have much more extensive profundal than littoral zones. For example, the profundal zone of Lake Baikal (> 250 m) constitutes three-quarters of the total bottom area. This means that even at low densities, profundal organisms can constitute the majority of the benthic biomass.

As in planktonic, stream, and estuarine communities, water motion plays a vital role in controlling the distribution of benthos. At the lake edge the continual turbulence produced by waves and currents imparts an energy subsidy for the biota by providing a stream of food particles, just as does tidal action in estuaries (Chap. 17) or the current in streams (Chap. 16). In the surf zone the wave motion also disturbs the tube building and food uptake of some benthos with the result that the deeper sublittoral zone has a more favorable habitat. The diversity, density, and productivity of most sublittoral organisms are much greater than in the nearly motionless profundal zone (Fig. 13-14, Tables 13-1 and 13-2). The productive littoral and sublittoral areas are most affected by diffuse chemical pollution from inflows which tend to follow the shoreline. The effect of reducing near-shore benthos by poisoning or smothering the animals with sediment often has a particularly damaging effect on fish stocks, since many open-water species feed and reproduce there.

In addition to efficiently supplying food, the turbulence of the mixed layer also provides oxygen. The high rates of feeding, respiration, and growth in the warm littoral zones are only possible if ample oxygen is available. For example, the surf-zone amphipod *Gammarus pulex* has a high oxygen requirement for its activity, which is linearly proportional to the amount of dissolved oxygen (Fig. 13-15). In

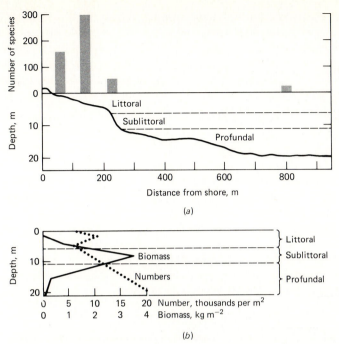

Figure 13-14 Distribution of zoobenthos in eutrophic Lake Esrom, Denmark. The near shore zone has (a) high species diversity and (b) low total numbers compared to the low diversity and high numbers of organisms in the almost stagnant low-oxygen environment of the profundal zone. Oligotrophic lakes with high oxygen levels in the sediments show higher diversity. (*Redrawn and modified from Jónasson, 1978.*)

contrast a typical member of the profundal benthic community of eutrophic lakes, the chironomid midge *Chironomus anthracinus,* shows a curvilinear relationship between respiration and dissolved oxygen. Its activity remains constant until low oxygen levels are reached; 75 percent of maximum respiration is possible even at very low oxygen levels (Fig. 13-15).

Feeding

Most of the zoobenthos are detritivores (Table 13-1), and in the food web their main role is to convert low-quality, low-energy detritus into better-quality food for higher trophic levels such as fish and crayfish (Chaps. 14, 15). For example, the tubifex worm ingests rich mud from eutrophic lakes (Ivlev, 1939). This in-

creases the energy content per unit weight from about 20 cal mg^{-1} of wet detritus to about 950 cal mg^{-1} of tubifex protoplasm. Selection of particular components of detritus, particularly sedimented algae, is probably common, and some zoobenthos, especially in littoral zones, are active carnivores. Phytoplankton and macrophytes in some lakes supply most of the detritus. The relationship between seasonal changes in phytoplankton primary productivity and that of the zooplankton was already discussed in this chapter, and a similar relationship occurs with zoobenthos (Fig. 13-16).

Life Cycles

Many benthic organisms live for more than a year and have several year classes in the population at any one time. For the *Chironomus anthracinus* larvae in eutrophic Lake Esrom,

Table 13-1   Type of Feeding of Zoobenthos in Relation to Depth in Lake Esrom, Denmark*

Approximate values in numbers of organisms per unit area or weight as a percent of total. Note the absence of herbivores and grazers in the deeper, darker waters.

Zone	Herbivores and grazers		Filterers		Detritivores		Carnivores	
	No.	Wt.	No.	Wt.	No.	Wt.	No.	Wt.
Surf and littoral	30	40	26	46	32	7	13	7
Sublittoral	0	0	63	99	30	1	7	1
Profundal	0	0	20	16	70	72	10	12

*Modified from Jónasson (1978).

Table 13-2   Distribution of Zoobenthos in Lake Baikal, Siberia*

Note that the lowest biomass occurs in deeper zones, as in shallow, eutrophic Lake Esrom (Fig. 13-14a, b) and that absolute quantities are about 100 times less in Baikal. Taken as a whole Baikal has only 5-6 gm m^{-2} of zoobenthos. Where there are extensive shallow areas zoobenthos thrive by using the littoral-zone algae and macrophytes to supplement their diet of phytoplankton.

Open lake regions		Area of extensive shallows	
Depth, m	Zoobenthos, g m^{-2}	Depth, m	Biomass, g m^{-2}
0–20	25–30	0–50	30
20–70	20–25	0–100	20
70–250	10–15	0–250	22
250–500	3–5		
>500	1–2		

*Modified from Kozhov (1963).

Figure 13-15 Adaptation to benthic environments. Zoobenthic activity, measured as oxygen consumption, can be related to dissolved oxygen (percent air) in the water for two animals from contrasting benthic environments. Note the linear relationship between dissolved oxygen and activity for the amphipod *Gammarus pulex* which lives in the well-oxygenated surf zone compared to the curvilinear relationship for *Chironomus anthracinus* from the stagnant low-oxygen profundal zone. The rapid rate of change in activity over small changes in dissolved oxygen allows *Chironomus* to utilize very low oxygen levels. A drop from 90 to 30 percent air saturation does not affect the activity of *Chironomus* but substantially reduces that of *Gammarus*. (*Redrawn and modified from Jónasson, 1978.*)

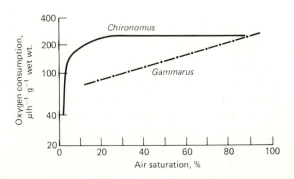

the spring bloom of diatoms is particularly important (Fig. 13-16). A similar relationship occurs between chironomids and other benthos and dinoflagellate plankton in a high alpine lake (Pechlaner et al., 1972). In Lake Esrom rapid growth of *C. anthracinus* in the first year of life occurs for limited periods just after the autumn overturn and during the spring algal maximum. In the second year only relatively small fluctuations in weight occur (Fig. 13-16). This species feeds on detritus at the mud surface. The adult midges emerge, mate, and lay their eggs during

the spring algal bloom of their second year. The numbers of emerging adults can be enormous. A particularly good example of mass emergence occurs in Lake Mývatn, Iceland. The word *mý* is the Icelandic name for "midges," and the adults form dense black columns at the lake edge which can be seen miles away (Lindegaard and Jónasson, 1979). The swarms are composed of *C. islandicus* and *Tanytarsus gracilentus* which feed among the abundant *Cladophora* growing on the bottom of this shallow, eutrophic lake. In shallow lakes

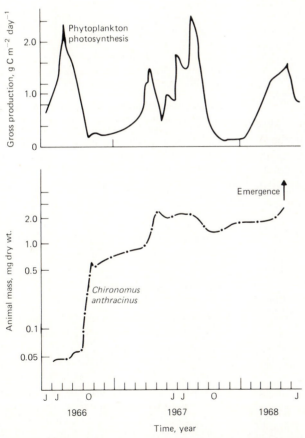

Figure 13-16 Growth of the profundal chironomid midge larvae *Chironomus anthracinus* in Lake Esrom in relation to the growth of its food—the phytoplankton in the epilimnion. Large increases in biomass occur following the spring diatom bloom dominated by *Asterionella*. Some midge adults emerge after only 1 year but most require 2 years. (*Redrawn from Jónasson, 1978.*)

246

CHAPTER 13 ZOOPLANKTON AND ZOOBENTHOS

the whole lake bed is equivalent to the littoral or sublittoral zone. The production of these midges has sustained a large duck population and even a significant wild duck egg industry for centuries (Gudmundsson, 1979).

Other benthic organisms also emerge in massive swarms. In Clear Lake, California, billions of adults of the small gnat *Chaoborus* may emerge on warm evenings and darken street lights. In central Africa near Lake Malawi, "Kungu cakes" made from compacted masses of adult *Chaoborus* are eaten by the local tribes.

Profundal species may experience low oxygen levels in the muds during summer and require special adaptations to survive. The ability of *C. anthracinus* to feed at low oxygen levels (Fig. 13-15) is partially due to hemoglobin (Fig. 10-4) present as its blood pigment. Few aquatic invertebrates possess this efficient oxygen-carrying system, but profundal organisms in eutrophic lakes often do.

Many zoobenthos do not spend their entire larval existence in the sediments. The predatory gnat *Chaoborus* usually is found in the sediment during the day and at night is planktonic. It feeds nocturnally on rotifers and small crustaceans in the surface waters. By day it returns to the low oxygen levels in the mud of eutrophic lakes to avoid fish predation (Fig. 13-17). *Chaoborus* larvae are uncommon in oligotrophic lakes because these unproductive waters provide no anoxic refuge from predation. *Chaoborus* is as well-adapted for respiration at low oxygen levels as is *Chironomus* (Fig. 13-15).

We have discussed the primary and secondary trophic levels of phytoplankton, zooplankton, and zoobenthos. The next chapter will consider the highest trophic level, the fish, which have a variety of roles in lake structure. Many rivers, lakes, and estuaries studied by limnologists have been preserved from total destruction through the efforts of anglers and their organizations, but unfortunately the average lim-

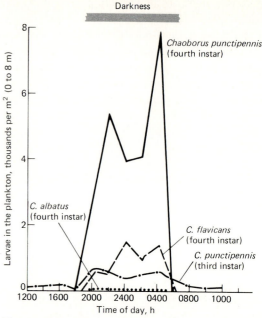

Figure 13-17 Diel migration of three species of the predatory gnat, *Chaoborus*, in a midwestern lake, Frains Lake, Michigan, in May. Fourth instar larvae remain in benthic darkness during the day and only swim up to feed at night. They accumulate at about 2–5 m in this small dimictic lake. (*Modified and redrawn from Roth, 1968.*)

nologist is often poorly informed about the fish and fisheries.

FURTHER READINGS

Brinkhurst, R. O., and B. G. M. Jamieson. 1971. *Aquatic Oligochaeta of the World*. Oliver & Boyd, Edinburgh. 860 pp.

Capblancq, J., and H. Laville. 1972. "Etude de la productivité du lac de Port-Bielh, Pyrénées centrales," pp. 73–88. In Z. Kajak and A. Hillbricht-Ilkowska (eds.), *Productivity Problems of Freshwaters*. PWN (Polish Scientific Publishers), Warsaw-Krakow.

Cook, D. G., and M. G. Johnson. 1974. "Benthic Macroinvertebrates of the St. Lawrence Great Lakes," *J. Fish. Res. Board Can.*, **31**:763–782.

Edmondson, W. T. 1974. "Secondary Production." *Mitt. Int. Ver. Theor. Angew. Limnol.,* **20:**229–272.

Jónasson, P. M. 1978. "Zoobenthos of Lakes." *Verh. Int. Ver. Limnol.,* **20:**13–37.

Kerfoot, W. C. (ed.). 1980. *Evolution and Ecology of Zooplankton Communities.* Special Symposium vol. 3, American Society of Limnology and Oceanography. University of New England Press, New Hampshire. 793 pp.

Lewis, W. M. 1979. *Zooplankton Community Analysis—Studies on a Tropical System.* Springer-Verlag, New York. 163 pp.

Porter, K. G. 1977. "The Plant-Animal Interface in Freshwater Ecosystems." *Am. Sci.,* **65:**159–170.

Chapter 14

Fish and Fisheries

OVERVIEW

Fish are often neglected by limnologists despite their commercial value and ecological importance at the apex of most aquatic food webs. One reason may be the difficulty in collecting truly representative samples of fish populations in lakes and large rivers. Fish in lakes, streams, rivers, and estuaries are dominated by the bony fish such as perch, trout, or minnows. Most possess gas bladders which allow them to maintain almost neutral buoyancy in the water column. Many fish live and breed in one area but a few migrate considerable distances. Anadromous fishes like salmon, shad, and smelt spawn in streams but mature in the ocean, while catadromous fishes such as eels mature in freshwater but return to the sea to reproduce.

Fish are usually the largest animals and influence the structure of aquatic ecosystems by their voracious predation; the available planktonic food resources are partitioned among various fish species and life-history stages which are adapted to specialized modes of feeding. They may also feed on small benthic animals or browse on aquatic vegetation. In their early life stages almost all lake fishes feed on small zooplankton and when inshore also eat planktonic protozoans and attached organisms such as rotifers. Juvenile stream fish have similar diets of attached biota. Most *planktivorous* fish eat only zooplankton. Phytoplankton form a major part of the diet of only a few species. Larger, carnivorous fish, called *piscivores,* eat smaller fish, sometimes even members of their own species. Fish may eat crayfish, frogs, and even small mammals. *Detritivorous* fish digest bacteria, fungi, and protozoans living on particles of detritus or mud. In flowing waters drifting invertebrates are eaten by some fish; others pluck insect larvae and crustaceans from the substrate, while still others feed on detritus in pools. Many fish are opportunistic feeders, and even large carnivores can sometimes be found with algae, detritus, and mud in their stomachs.

In any water body total biomass of fish may be fairly constant from year to year, but there is usually considerable variation in the size of the various year classes. The success or failure of a year class is dependent upon many environmental factors. For cold-water fish like trout summer temperatures are most important, but inter- and intra-species competition for food, predation, and cannibalism are also significant. In rivers, as well as lakes, floods and droughts can greatly alter the availability of sites for reproduction and feeding.

Most human modifications of streams, estuaries, and lakes have reduced native fish populations. These effects are often compounded by introductions of exotic fish which are esteemed by anglers or are considered commercially desirable. The carp has been introduced to almost all North American waters, and various sports fish have been moved far beyond their normal range. A sad example of cultural impact is found in the Laurentian Great Lakes, where two premium salmonid species, locally called lake trout and whitefish, were decimated by overfishing and predation by an invasion of sea lamprey. Introductions of the marine alewife and rainbow smelt may have reduced the stocks of other salmonid species, while the Great Lakes' sturgeon were overfished and their stream-breeding habitat destroyed. Only recently have long-term, expensive measures begun to restore some of these lost fisheries. In the Great Lakes most of the emphasis is now placed on sport fishing for introduced fish, particularly Pacific salmon.

Dams and artificially straightened channels on the watershed together with pollution in rivers and in the littoral zone of large lakes are major causes of fishery change. In reservoirs, careful management may be able to maintain or even increase fish yields over those of the original riverine system, but this usually involves considerable change from the endemic species composition.

INTRODUCTION

Historically, fish have been a neglected component of the majority of limnological investigations. Excluding the great commercial value of the water itself, fish are the main produce harvested from inland waters. Of lesser economic importance are rice, watercress, crayfish, turtles, frogs, and occasionally crocodiles and alligators.

Fish occupy several different levels of the aquatic food chain, make up over 40 percent of the earth's vertebrate species, and are preyed upon by a few predatory animals and one another.

The commercial and sports fishermen of the world represent a formidable body of public opinion for maintaining the quality of freshwaters. In the British Isles alone, 3 million people are involved in commercial or sports fishing and in the United States considerably more. Through increased cooperative efforts, fisheries biologists and limnologists alike should provide the information to more effectively conserve and manage our inland fisheries resources.

Limnologists have too frequently ignored fish and have left their life histories to the ichthyologists, their ecology to the fisheries biologists, and problems associated with their harvest and conservation to sport and commercial fishing interests. For obvious economic reasons the applied aspect of fishery biology has been emphasized relative to the theoretical trends emphasized in limnology. This difference in perspective may have contributed to the historic divergence between the two approaches.

We believe that the time has come for fish to have a more definite place in limnology, and we hope that fishery biologists will share this opinion. The gap between the two is narrowing as both recognize that the ecosystem approach provides a more balanced view of the theoretical as well as the applied aspects of aquatic biology. For some time fishery biologists have recognized that fish yield is a function of the whole-lake or stream production process. Limnologists in turn are increasingly aware that fish can be extremely effective in altering both the structure and function of primary and secondary components of the lake or stream system. We already discussed the importance of fish in altering the average size and species composition of zooplankton populations (see Chap. 13), and benthic invertebrates may be similarly influenced.

Basic and applied fishery biologists need to be more aware of the technical and statistical methods limnologists have applied to physical and chemical problems in lakes, streams, and estuaries. In particular the relationships between light, nutrients, and primary production are reasonably well understood but are seldom used as a basis for prediction of fish production (Melak, 1976). In part this is due to the complexity of the food web since fish often feed at several trophic levels (Chap. 15). The task of integrating fish yield, primary and secondary productivity, growth, mortality, competition, and population dynamics with basic limnological parameters challenges the modern fishery biologist.

One problem with discussions on fish and fisheries is the use of the same common name for entirely different fish. This is particularly true for North Americans who use the European-derived words *roach, chub,* and *herring* to describe similar looking fish as they were discovered in the New World. The taxonomy of fish is covered in general terms in Chap. 11. We attempted to avoid this name confusion by the use of both common and scientific names in Table 11-9. The complex taxonomy of fish has been simplified to cover only those referred to

in the text. A few general remarks and Figs. 11-10 and 11-11 will also assist the reader. The minnow family, Cyprinidae, is found in Europe, Asia, North America, and Africa, but not in South America, Australia, or New Zealand. The sucker family, Catostomidae, is particularly abundant in North American lakes and streams and is closely related to the minnow family, but is not found elsewhere except in a few places in eastern Asia. The cichlid family is abundant in Africa and South and Central America, but is replaced by its ecological equivalent, the sunfish (Centrarchidae) family, in North America. Neither suckers nor sunfish are present in Europe or Asia with the exception of southern India. The pike family, Esocidae, is widespread throughout the east and north of North America and in northern Europe and Asia but is absent elsewhere.

MEASUREMENT

Fishery studies involve collection and identification of the species present, determination of their age, growth rates, and population structure. Analysis of how fishes interact with other components of the ecosystem can be established by examining their stomach contents. Collecting procedures may involve netting, electrofishing, hook-and-line fishing, or even poisoning of lakes or streams with rotenone or other fish toxicants. The use of an echo sounder for fish location is important in commercial fisheries, and the scientist may use a specially equipped fathometer for population estimation, as well as for locating dense concentrations of zooplankton (Fig. 14-1a).

Rotenone, long used by native inhabitants of the Amazon basin, can be used for management or estimation of fish populations. It is prepared from the roots of *Derris* and other native plants. It blocks oxygen metabolism in fish. Two to five mg liter^{-1} of rotenone is more than enough to kill most fish and invertebrates, and is slightly selective against fishes with high oxy-

1 f = 1.83 m

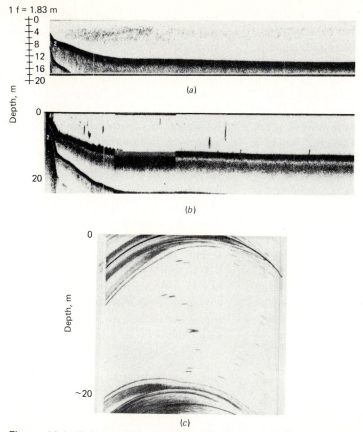

Depth, m

(a)

(b)

(c)

Figure 14-1 Echo-sounding charts which detect fish by reflection of sound waves from their gas bladders. (a) Schools of clupeid fish indicated by dark dots between 2 and 6 m in the Great Rift Lake Tanganyika, Africa, at night. Note the concentration of fish near the lake edges. (b) Similar chart for midday showing only a few fish remaining in the water column (thin, black blobs). (*Modified from Balon and Coche, 1974.*) (c) Small trout (thin marks) and large trout (larger dark marks) detected by echo sounding in small subalpine Castle Lake, California. (*Made by L. J. Paulson.*)

gen requirements. Most fishes arrive gasping at the surface where they can be collected with a dip net. In streams they are usually carried downstream by the current, and a seine or weir is placed below the poisoned area to collect the fish. In streams, fish kills may occur a considerable distance below the target areas, and oxidizing agents may be used to detoxify streams after rotenone has been applied. In large lakes "cove" rotenone application in bays

separated from the lake with nets gives the best estimate of fish composition and abundance. Rotenone is commonly used to remove nongame fish or "rough" fish before establishment of a sport fishery for trout or bass. This management technique should be used with caution since its indiscriminate application may further reduce endangered species. The importance of the *rough*, or *coarse*, fish, as they are called, should not be underestimated as they serve as food for predatory species as well as recycle nutrients. It should also be noted that large numbers of enthusiastic anglers pursue these "lesser species" with the same zeal and concentration that the fly angler practices with a dry fly or nymph for the most popular sport-fishes.

Nylon nets are commonly used in fish collection. The simplest approach is seining, where a long, fine-mesh net is worked toward shore and the fish are strained out as the net is removed from the water. Higher efficiency is achieved if the net contains a cod end, which is a finer-mesh sacklike extension from the midportion of the net to the end. Gill nets and trammel nets ensnare fish and can be placed at various depths or at an angle from the bottom to the surface. Fish are caught by their gills, teeth, or jaws when they attempt to pass through the mesh. In clear waters gill netting is most effective at night when the fish are less able to see and avoid the net. Experimental nylon gill nets are available with panels of different mesh sizes so that very small fish can be collected as well as the larger individuals. Fyke nets or hoop nets have long been used in fisheries (Fig. 14-2). They are fixed-trap nets with funnel-like throats through which the fish enter but have difficulty exiting. Although "shrimp dry nets" may be operated from small boats, trawl nets which are dragged at moderate speed have a very heavy drag and require more power than is usually available with small research vessels. As is the case with zooplankton caught by nets (Chap. 14), all nets used for fishing are selective not only for size but also for species. Estimation of selectivity is a major difficulty in estimating fish populations.

Electrofishing, a technique employed in nature by electric eels and a few tropical catfish, is particularly useful in streams and shallow regions of lakes. A portable generator or battery supplies current and develops an electric

Figure 14-2   Hoop nets used in the fishery on shallow dystrophic Tjeukemeer Lake in the Netherlands.

field between positive and negative electrodes held in the stream or suspended from a boat (Larimore et al., 1950). Alternating current between the electrodes electronarcotizes the fish passing between them and when direct current is used, attracts them toward the positive (anode) pole where they are easily netted. In ordinary practice, fish taken by electrofishing revive fairly quickly when returned to the water. Thus they may be identified, weighed, measured, tagged, and then returned without harm. By tagging or fin clipping, migrations and movements are studied through subsequent collections of marked and unmarked fish. The mark-and-recovery technique of population estimation may be applied to most kinds of animals as well as to sport and commercial fishes. Crayfish are marked by cauterizing their carapace with small dots or numbers which last through several molts.

Hook-and-line fishing, among the oldest of all fish-harvesting methods, is a slow but valuable means of sampling the population of a lake, estuary, or stream, especially during high waters. A diligent angler willing to use a variety of baits and hook sizes throughout the year can collect most of the species present.

In still waters reasonably devoid of snags, collections are easily made by a combination of beach seining and overnight gill netting. In small streams electrofishing is usually the most effective sampling method. However, since some species such as the sculpin have no air bladder and are less likely to surface when shocked, this method may not produce a truly representative sample of the stream population. In large rivers where gill nets are apt to be hopelessly snagged, trap or hoop nets are most effective, although beach seining is useful too.

As mentioned previously, there is no method which will collect all fish. K. F. Lagler examined fish populations in a small pond in Michigan by using various types of nets. Having estimated populations conventionally, he drained the pond and found that the methods produced gross underestimations of the total stock. Even then he believed that numerous fish remained uncounted in the mud.

Once fish have been caught they can be identified from taxonomic keys or, if they are unusual or are from a remote part of the world, sent to a specialist for identification. Fish are usually fixed with formalin and preserved in alcohol. Large specimens may require opening of the body cavity or hypodermic injection to prevent internal decomposition. Fish scales contain growth rings which, like tree rings, are used for determining the age and growth rate of many fishes. Species without scales can usually be aged by cross sections of vertebrae or spines. The limnology student should consult the comprehensive treatment of ichthyology by Lagler et al. (1977). Collections of type specimens are also available in some of the major museums, and the local curator of fish is usually both interested and helpful in identification.

In our industrialized society native fish stocks are displaced by exotic species and are destroyed or depleted by chemical and thermal pollution or habitat alteration. Fish are sensitive indicators of water temperature and chemical composition, and can be used to detect pollution. Bioassay procedures can determine the toxicity to fish of particular industrial chemicals or municipal effluents. Originally these tests involved only measuring acute toxicity where fish were killed by the pollutant. Modern tests measuring sublethal effects, including changes in metabolism or behavior, have given a much better estimate of the potentially harmful effects of low level pollutants.

FEEDING

Fish interact with various levels of the food chain (Figs. 15-3, 15-4) and influence the structure of lakes, streams, and estuaries since they are usually restricted to a particular mode of life related to their food source and reproductive requirements (Fig. 16-17a, b). Different fish occupy different levels of the food chain. Their importance in recycling nutrients to the primary producers is evident in the high fertility of most fishponds and the response of phytoplankton to the increased levels of phosphorus and nitrogen excreted by the fish (Paulson, 1980).

Fish are often the only significant large aquatic predators, and results of their feeding can be very dramatic. A small fish can eat hundreds more zooplankton or large algae than the largest zooplankter. There is usually a greater difference in community structure among waters with fish than those without fish. In nature, polar or alpine lakes may not contain fish. Some polar or alpine lakes are without fish because it is physically impossible for the fish to get to the lake. Temporary pools are also fishless.

Feeding may be conveniently divided among *pelagic* fish, which feed in open or surface water, and *littoral* and *benthic* fish, which feed at the lake edges or on the bottom of lakes and streams. Pelagic feeding is either *planktivorous* or *piscivorous*. Benthic and littoral feeding is often more generalized and can include grazing on aquatic plants as well as ingestion of bottom debris (*detritivorous* feeding) or ingestion of benthic invertebrates.

In general, pelagic fishes are associated with surface waters. Some feed very near the surface, and examples of these are the well-known mosquito fish, *Gambusia affinis,* or the archer fish, *Toxotes jaculator,* which is able to dislodge its prey from adjacent terrestrial vegetation with a jet of water from its mouth. The white bass, *Morone (Roccus) chrysops,* feeds on zooplankton. It has been seen to eat *Daphnia* from the surface of lakes when the zooplankters have become trapped in windrows produced by Langmuir spirals (Fig. 5-14). The surface-feeding habits of trout and various members of the sunfish family also make them excellent targets for surface fishing with fly and plug.

Planktivorous fish feed throughout the water column, and fish such as the shad, herring, or minnows feed on zooplankton for most of their lives. Almost all fish, even large predators like pike and lake trout, feed on tiny rotifers, protozoans, and other small plankton during their early development. The now endangered paddlefish, *Polydon spathula,* of the Mississippi River drainage occasionally reaches over 50 kg and must consume great quantities of plankton as well as benthic invertebrates.

A number of successful planktivores strain out the animal plankton upon which they feed. Many of the pelagic species like the whitefish, *Coregonus* spp., accomplish this with very fine gill rakers. Selective predation by fish may shift the zooplankton community to smaller species (Fig. 13-8), and the gill rakers of many predators may even allow the smaller zooplankton to escape.

Planktivorous fish usually feed on zooplankton, but some utilize phytoplankton. A large minnow, the Sacramento blackfish, *Orthodon microlepidotus,* once the dominant fish in lowland California, feeds partially on phytoplankton. *Tilapia nilotica* and *Haplochromis nigripinnis* in central Africa subsist predominantly on the planktonic blue-green alga *Microcystis aeruginosa.* They are able to do this by secreting extra acids into the stomach once algae have been ingested. Blue-green and gelatinous green algae may pass through most fish guts undigested. Stomach acidity in vertebrates drops to about pH 2 during digestion of proteins but is lowered to 1.4 in *Tilapia* in Lake George, Uganda (Fig. 14-3). Acid must be released only after food is in the stomach because undiluted acid would destroy the stomach walls. The low pH assists in breakdown of the blue-green algal cell wall which, unlike most algal walls, contains amino acids suitable for incorporation into fish protein. Nitrogen is typically in short supply in most tropical lakes, and this unusual feeding method may allow *Tilapia* to dominate the planktivorous fish populations in Lake George.

A variety of benthic fish feed principally on

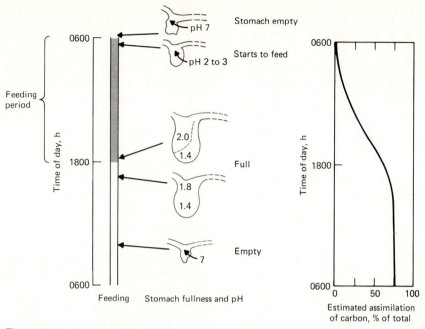

Figure 14-3 Feeding and digestion of the blue-green alga *Microcystis* by the planktivorous tropical fish, *Tilapia*. Note the extreme acidity of the lower stomach which enables breakdown of the normally resistant peptidoglycan cell wall. The stomach returns to a neutral pH when the fish is not feeding. (*Modified from Moriarty et al., 1973.*)

the organisms associated with the lake or stream bed. For these fish *detritus* is often the major source of food. Detritus, also discussed in Chap. 16, consists of dead leaves, twigs, and other organic debris as well as fine silt and mud. The debris is rapidly covered with a living film of bacteria, fungi, protozoans, small insect larvae, and worms. This covering provides the food for the detritivorous fish. Small stones ingested are often used to grind up detritus into an easily digestible paste. Almost all fish occasionally ingest detritus, either accidentally in the stomach of their prey or simply when there is nothing else to eat.

The sucker family, Catostomidae, and some of the freshwater catfish, Ictaluridae, have particularly well-adapted ventral mouths which are useful in ingesting detritus, mud, and insects from the bottom. The highly successful carp also has a mouth well-adapted for feeding on

the bottom. The Paraná River, Argentina, contains enormous numbers of sabalo, *Prochilodus platensis,* and gut-content analysis reveals only clay and sediment particles. Although the life history of this fish has not been studied in detail, it appears likely that it digests bacteria, fungi, and dissolved organic material attached to the clay particles. Other bottom feeders include sturgeon and the sculpins, the Cottidae, which ingest small insects attached to rocks and weeds. In African rivers the most important detritivores decline dramatically after dam construction since the supply of upstream river detritus is cut off.

A few species such as white amur, also known as the grass carp, concentrate on aquatic vegetation and others, on attached filamentous algae. The roach, *Rutilus rutilus,* a cyprinid, feeds almost exclusively on filamentous algae, but careful examination of the gut contents usually shows a significant proportion of invertebrates which are ingested along with the algae.

Although some fish are omnivorous, others are very specialized in their selection of food. Several kinds of fish graze the bottom of the lake or stream by selectively removing a single organism or vegetation type. In Castle Lake, California, two species of trout, the brook trout, *Salvelinus fontinalis,* and the rainbow trout, *Salmo gairdneri,* have partitioned the food supply. The brook trout feed mainly on chironomid and dragonfly larvae on the bottom, while the rainbow trout feed near the surface and depend principally on terrestrial organisms in summer (Fig. 14-4) (Swift, 1970; Wurtsbaugh et al., 1975). Another example of resource partitioning involving many fish species is shown for Lake Tahoe in Fig. 14-5a.

A similar division of food resources occurs in streams where the largest, most aggressive car-

nivores like trout pick off the drifting insects below riffles (see Chap. 16). Smaller fish such as sculpin and dace catch benthic invertebrates in the shallow water of the riffles, while suckers sort the detritus at the bottom of the pools (Fig. 14-5b). In some African lakes one species may rapidly evolve into a "flock" of different species, all utilizing a different food resource (Fig. 14-6).

Unusual feeding specialization is illustrated by an Indian catfish and an African cichlid which actually pluck scales from other fish for their food. Nipped fins of the sabalo in the Paraná River are purported to result from the attack of the piranha, *Serrasalmus.* The main enemy of the piranha in the Paraná River is the voracious dorado, *Salminus maxillosus,* which is the most important sports fish in this enormous river. In Brazilian reservoirs absence of the dorado has caused an unfortunate dominance of the piranha.

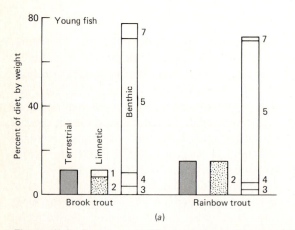 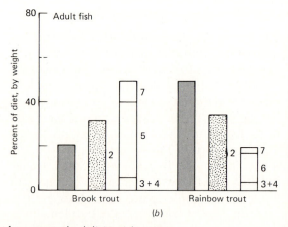

Figure 14-4 Resource partitioning between two species of young and adult trout in mesotrophic Castle Lake, California. (a) Young brook trout, *Salvelinus fontinalis,* lived near the lake bed and fed mostly on benthic animals. The young of the introduced rainbow trout, *Salmo gairdneri,* had a more pelagic habitat but still fed mostly on benthos. (b) The adults maintained similar habitats to their young stages but changed food preferences. Adult brook trout increased their consumption of cladocerans. Young, but adult, rainbow trout switched to mainly terrestrial food in summer and fed much less on benthos. Key: shaded = terrestrial origin; dotted = limnetic; white = benthic. 1 = copepods; 2 = cladocerans; 3 = chironomid larvae; 4 = chironomid pupae; 5 = mayfly larvae; 6 = dragonfly larvae; 7 = other benthos. (*Modified from Wurtsbaugh et al., 1975; Swift, 1970.*)

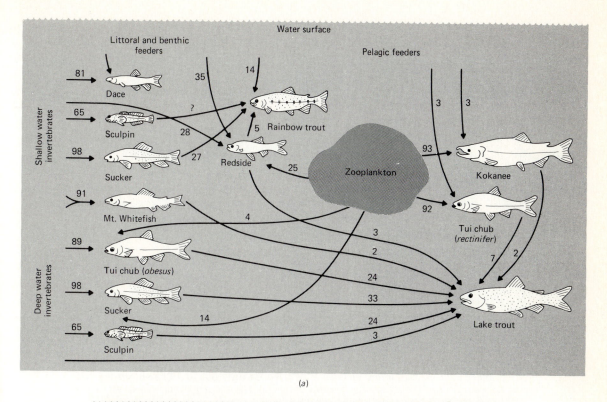

(a)

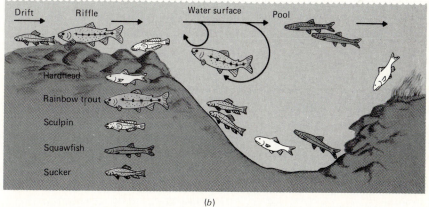

(b)

Figure 14-5 Resource partitioning by lake and stream fish. (a) Feeding relationship of fish in Lake Tahoe, California. Note how the whole-lake food supply is divided between the various types of fish. This division into pelagic and littoral, surface and deepwater, is typical of most lakes and oceans. Numbers next to arrows are percent diet by volume. (*Modified from Moyle, 1976; Miller, 1951; and Cordone et al., 1971.*) (b) Resource partitioning in flowing water: a mountain stream in the Sierra in Northwestern United States. Note trout feeding on drift above and below the riffle; sculpins and small suckers in the shallow riffle areas and amongst the rocks. Larger fish browse on detritus and submerged plants in pools, and some feed on drift invertebrates not taken by the more aggressive trout. (*Redrawn and modified from Moyle, 1976.*)

Figure 14-6 Adaptive radiation: species of the genus *Tilapia* and one species of another genus adapted for various forms of feeding in Lake George, Uganda. Shown are the phytoplanktivorous form *T. nilotica* which feeds on the blue-green alga *Microcystis,* as well as benthic, offshore and nearshore zooplanktivorous forms.

POPULATION CHANGES

In many lakes and estuaries annual phytoplankton blooms are similar from year to year. The remarkably constant annual maxima of the diatom *Asterionella* in Windermere (Fig. 12-4) or that of the blue-green alga *Aphanizomenon* in Clear Lake, California, demonstrate this phenomenon. Individual fish may live many years and the overall biomass can be fairly constant. However, large variations in annual numbers may occur in individual year classes. The year class is that *cohort* or group of fish which was hatched in a particular year. In the case of lake trout which may live as long as 20 years, there will be 15 to 20 year classes present at any one time, but there are progressively fewer individuals in the older age classes. Sometimes unu-

sually large numbers of a particular year class survive and continue to dominate the population for several years.

In Windermere, where some long-term records are available, the year classes for 1942–1972 have been recorded for two common fish, the large predatory pike, *Esox lucius,* and a smaller predator, the perch, *Perca fluviatilis.* The pike and near relatives of the perch are Holarctic in distribution and are abundant in lakes and streams in North America. These fishes can survive up to 20 years and specimens of all ages can be found in the lake. However, few survive beyond 8 years and most are sexually mature at 2 to 3 years. This means that the majority of the adult population belongs to only a few age groups. Most fish produce far more eggs than are required to replenish the population, and most young die or are eaten in the first year. Over the last 20 years the strongest pike year class was 7 times as large as the weakest, and the perch varied 400-fold between maximum and minimum year classes (Fig. 14-7).

Perch and pike spawn once per year. Perch attach strings of sticky eggs to submerged aquatic vegetation and stones, while pike spawn in weedy areas. The young fry feed on protozoans and rotifers and soon graduate to small insect larvae and crustaceans. Both grow quickly; perch reach about 8 cm in length after 1 year and pike may exceed 20 cm by fall. In 1963 the 4-year-old cohort dominated the perch population (Fig. 14-8). This cohort was the 1959 year class, and it continued to dominate all age classes, even in 1966 when it had reached an age of 7 years. There are two factors involved in the long-term dominance of abundant year classes: the initial survival of a large hatch and the factors which allow its continued success. Before pursuing this discussion of perch year-class success, the contrasting case of pike in Windermere should be considered. Pike year classes have more stability than those of perch, and no one year class dominated in the period from 1941 to 1964.

The abundance of perch and pike year classes in this cool, relatively oligotrophic lake has been attributed to temperature. A warm summer enables the cold-blooded fish to feed faster and grow more quickly. This same effect is shown for zooplankton. An index of the effect of temperature on growth conditions (Le-Cren, 1958) was devised which incorporates the increase in temperature above 14°C during the summer period May–September. For example, if the average epilimnetic temperature for 1 day is 17°C, then the index scores 3 (that is, $17 - 14 = 3$). The cumulative number for the whole summer is the temperature index. When this index is plotted against the relative abundances of the year classes for both fish, a significant relationship is shown for both pike and perch (Fig. 14-9).

For pike other factors, such as the number of adults at the time of egg hatching or the biomass of adult females which laid the eggs, showed no significant correlation with year-class abundance. It appears likely that temperature controls the major variation in year-class strengths for this fish. Older pike are cannibalistic on small pike if they are less than about 20 cm in length. In warm years, many more

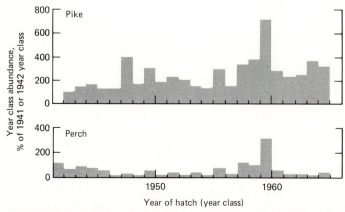

Figure 14-7 Reproductive success in fishes. The abundance of each year class for pike and perch in Windermere North Basin for the years 1941–1964. Note that only a few classes of perch are very successful. (*Redrawn mainly from Kipling and Frost, 1970.*)

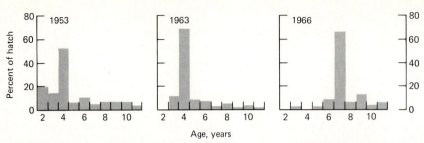

Figure 14-8 Cohort dominance. Age distribution of male perch in Windermere North Basin in 1953, 1963, 1966. Values are expressed as a percent of the estimated total male population. Note the persistence and dominance of the 1959 year class in 1963 (4 years old) and 1966 (7 years old) relative to 1953 when the 1949 year class dominated. In this lake cannibalism and other factors reduce the frequency of successful year classes. (*Modified from LeCren, 1958; and Kipling, 1976.*)

young are able to exceed this critical size by fall and thus avoid predation by larger fish during the winter.

In the case of perch, the relationship with temperature occurs only in some of the warmest years. The unusually warm 1959 season, as well as being a great wine year in Europe, produced a large class of perch, while the warm 1947 year was a complete failure for that year class. Other factors must modify the influence of temperature. These factors are the biomass of mature female perch, availability of suitable spawning areas, food supply for the fry, total biomass of perch more than 1 year old, cannibalism, predation by pike and fish-eating birds, and occasional winter ice. The most abundant perch year classes were produced when there were relatively few breeding perch, and the weakest year classes occurred when perch were most common. Thus there were always sufficient eggs but cannibalism may have effectively reduced the numbers of young fish. In Windermere, and perhaps many cool temperate lakes, the pike year-class abundance is primarily determined by temperature during the first few months of life, but perch appear to be controlled by more complex factors.

FISHERIES MANAGEMENT AND CONSERVATION

Water development throughout the world has had a profound effect on inland and estuarine fisheries. The construction of levees and artificial channels, removal of riparian vegetation, and draining of marshes for flood control is commonplace despite the resulting loss of fish habitat. The increased discharges from flood-control structures greatly increase the frequency and intensity of bed scour (Chap. 16) and can leave an impoverished benthic invertebrate food supply for the few fish which can survive in the less-favorable habitat.

Innumerable stream systems have been modified in their upper course by dams converting riverine areas to impoundments. These artificial structures are formidable blocks to either upstream or downstream migration. While effectively destroying one kind of habitat, another is created and, as discussed later, overall productivity may be changed (Fig. 14-12). Irrigation projects often divert great quantities of water, may dry up streams that once ran year-round, and can result in the extinction of lakes. Water diversion for irrigation and domestic water sup-

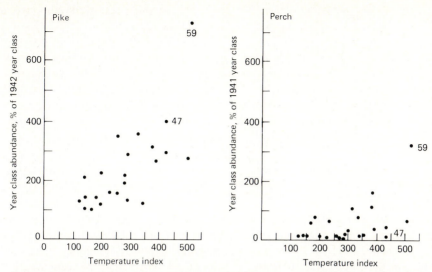

Figure 14-9 Year-class abundance of pike and perch in Windermere North Basin relative to temperature index (see text for explanation). Note the dominating effect of higher temperatures in summer, a high index, in the success of pike. For perch only the warmest years, such as 1959, produced high numbers for that year class since other factors usually control year-class abundance. (*Modified from LeCren, 1958; and Kipling, 1976.*)

ply may divert both eggs and young fish from the system.

In contrast, some properly regulated dams may provide excellent year-round flows and cool hypolimnetic waters immediately below the dam to improve the spawning opportunities for cold-water fish such as salmonids. The extensive Shasta Dam system ($\Sigma V = 8$ km³) on the Sacramento River in northern California, although cutting off extensive reaches of spawning gravels upstream, has improved downstream rearing areas with a continuous flow of cold water during the summer months. A nearby fish hatchery attempts to make up for the loss of the upstream spawning areas.

Breeding grounds, such as swamps flooded in spring, help to maintain the high fish production in the river system but are usually the first to be drained and lost forever. These swamps can be turned into valuable, flat agricultural land. Lumbering activities on many lake watersheds have

destroyed spawning grounds, and even in a lake as large and remote as Baikal in Siberia commercial fisheries are threatened. An extreme case is the Oconto River where sawdust from timber cutting extended 2 miles into Green Bay on Lake Michigan. By 1880 the spawning runs of whitefish had stopped (Beeton, 1969), and the Michigan grayling was exterminated by sawdust and associated activities. Sawmill debris has blanketed good salmon rivers in Scotland and northern Europe also.

The Decline of the Great Lakes Fishery

Modification of lakes and rivers for navigation often has unexpected consequences. Development of the St. Lawrence waterway provided a nearly 3000-km connection between the Great Lakes and the north Atlantic Ocean. Opening of the first Welland Canal in about 1830 and a second canal 100 years later allowed the invasion of the sea lamprey, *Petromyzon mari-*

nus, and the alewife, *Alosa pseudoharengus,* into the upper Great Lakes. Both have caused serious fish management problems and are taking many years to attain a level of balance in their new environment. Sudden mass mortalities of the alewife have littered beaches with tons of rotting fish, and its population explosion has made this species the most abundant fish in Lake Michigan in terms of both weight and numbers (Fig. 14-10*b*).

The case of the sea lamprey deserves further mention. Although they existed in Lake Ontario for many years, Niagara Falls prevented their passage into lakes Erie, Huron, Michigan, and Superior until completion of the Welland Canal. They began their invasion of lakes Huron and Michigan in the 1930s, and in 1946 they were found in Lake Superior (Fig. 14-10*b*). The sea lamprey is a very effective predator, about 30 cm long and 3 to 4 cm in diameter. The lamprey feeds by attaching to large fish with its suckerlike mouth and extracting the fish's internal fluids. A single individual will consume almost 9 kg of fish in its lifetime. The appearance of the lamprey was coincident with the decline of the salmonid fishery in lakes Michigan, Huron, and Superior (Fig. 14-10*a*).

The story of the decline of several major fish species in the Great Lakes provides an excellent example of the complex structure of aquatic ecosystems. More complete explanations have been given by Baldwin (1964), Smith (1968), Beeton (1969), and Christie (1974).

The major commercial fish in the Great Lakes were all members of the salmonid family (Table 11-9). They include lake trout, *Salvelinus namaycush,* and an inshore fish, *Coregonus clupeaformis,* usually called whitefish. The smaller deepwater coregonids are collectively called cisco but include one species, *C. artedii,* referred to as a lake herring. The lake trout, a large fish which grows to about 50 cm length, was a mainstay of the century-old commercial fishery in the Great Lakes until the 1940s. The very rapid decline of this species

occurred more or less simultaneously in all five lakes in the decade 1945–1955 (Fig. 14-10 *a,c*). The decline was initially attributed entirely to predation by the sea lamprey. However, the decline occurred in Lake Michigan at a time when lampreys were relatively uncommon (Fig. 14-10 *a,b*) and in Lake Ontario where lampreys had always been present (Fig. 14-10 *c,d*). Obviously, factors in addition to lamprey predation were involved.

In the Great Lakes, fishing pressure was concentrated on the larger fishes. Unfortunately, these were the same size class preferred by the sea lamprey. Lampreys do not usually kill larger fish by occasional parasitic feeding, but mortality is severe if smaller individuals are attacked or larger fish are repeatedly attacked. The lake trout is unusual in that it becomes sexually mature at 6 to 7 years of age, which is later than most fish of this type. Thus both commercial fishing and the lamprey were exploiting the breeding population. Commercial fishing had already placed a heavy stress on the breeding populations of trout, and the extra predation due to the lamprey's introduction tipped the balance toward collapse of the fish stock as the lamprey population exploded after the initial introduction. The trout might have better survived lamprey predation without fishing pressure since they have coexisted for centuries in Lake Ontario.

Due to the numbers of intermediate-sized but sexually immature trout and a possible increase in fishing effort or efficiency, fish catches remained constant in the Great Lakes (Table 14-1) while the numbers of reproductive adults probably declined rapidly. Eventually, with the almost complete loss of the breeding stock, the fishery suddenly collapsed and failed to recover (Fig. 14-10 *a,c*). Though an occasional successful year can restock the population as in the case of perch in Windermere (Fig. 14-7), there is a point where there are insufficient adults remaining to influence future stocks. Since the decline of the lake trout, exten-

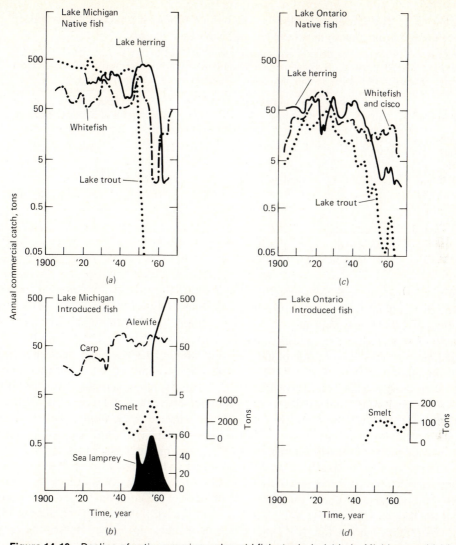

Figure 14-10   Decline of native premium salmonid fish stocks in (a) Lake Michigan and (c) Lake Ontario following commercial fishing and (b), (d) introduction of exotic fish. Lake Ontario always had populations of alewife and sea lampreys (no data on introduced carp), but these species only gained access to the upper Great Lakes in 1920–1930. Smelt were introduced at about the same time. Note the rapid decline of the largest salmonids, which are variously called lake trout, whitefish, cisco, and herring. The combined pressure of heavy fishing of both breeding and nonbreeding stock and lamprey predation on the breeding stock decimated trout and whitefish in all the Great Lakes. The enormous biomass of smelt (note scale change) probably reduced populations of another salmonid, locally called lake herring, by competition for food. Note recovery of whitefish in Michigan when the lamprey numbers were decreased by poisoning. Also note reduction in smelt numbers in Michigan but not in Ontario. The rapid increase in alewife and smelt is probably due to the loss of the large fish predators like lake trout. Note increase in carp fishery in Michigan as the other fisheries collapsed. (*Modified and redrawn from Beeton, 1969; and Christie, 1974.*)

Table 14-1   The Sudden Decline in Catches of Premium Species (Lake Trout, Whitefish, Larger Cisco) in All the Great Lakes Combined*

The catches now are considerably less than in the 1900–1920 period and contain large proportions of smaller fish such as yellow perch, rainbow smelt, and alewife, as well as smaller cisco. These fish cost more to catch and sell for less than the premium species.

Decade	Premium species as a % of total	Total catch, thousands of tons per decade
1900–1909	77	77
1910–1919	77	61
1920–1929	80	54
1930–1939	78	61
1940–1949	73	54
1950–1959	67	52
1960–1969	45	52

*Modified from Christie (1974).

sive electrocution of adults and poisoning of lamprey larvae in breeding streams has reduced lampreys (Fig. 14-10*b*). There may also have been a gradual buildup of natural predators and diseases which have established a degree of natural lamprey population control (Fig. 14-10*b*). In addition, since commercial fishing for lake trout has been banned and artificial stocking has been carried out, the populations have shown some recovery.

A decline in commercial fisheries for whitefish, *Coregonus clupeaformis*, lake herring, *C. artedii*, and smaller *Coregonus* species, called *cisco* (Fig. 14-10*c*) occurred at about the same time as that of the lake trout (Fig. 14-10*a,c*). All these fishes are of the salmon family (Table 11-9). In this case a mixture of overfishing and lamprey predation combined with other environmental pressures may have been responsible for their decline.

Rainbow smelt, *Osmerus mordax*, were introduced into the Great Lakes from Crystal Lake, Michigan, in the 1920 to 1930s, and huge stocks rapidly developed in the shallower areas of lakes Michigan and Huron. The commercial catches of 2000 to 4000 tons annually in the late 1950s in Lake Michigan compare to yields of only 500 tons for trout, whitefish, and lake

herring (Fig. 14-10*a,b*). Smelt are planktivorous feeders which, because of their enormous numbers, almost certainly compete for food with the young of native fish. In most lakes, including Lake Michigan, the smelt contributed to the decline of lake herring stocks (Fig. 14-10*a,b*). The smelt may have had little direct effect on whitefish since large populations of both species coexisted in some of the other Great Lakes (not shown in Fig. 14-10).

The rapid increase in the alewife population in Lake Michigan (Fig. 14-10*b*) could also have been responsible for the decline of the herring because both compete for a similar zooplankton food source. Once again the decline of the herring fishery in Lake Ontario (Fig. 14-10*c*), where the alewife has always been present, shows the crucial role of overfishing salmonid stocks already pressured by competition and lamprey predation (Regier and Loftus, 1972). The alewife increase in Lake Michigan occurred between 1950 and 1960. The rise is probably real but exaggerated by the relative lack of fishing for alewife before the mid-1950s (Fig. 14-10*b*). It is interesting to speculate why the alewife was not very abundant prior to this time. The small alewife occupied a niche in the Great Lakes' fish population structure because it spends much of its life in open water. Alewives increased as the larger deepwater predators such as lake trout and burbot, *Lota lota,* declined, but now are faced with predation from introduced Pacific salmon.

The demise of the whitefish and cisco (Fig. 14-10*a, c*) must be attributed mostly to a mixture of overfishing and lamprey predation. In the case of the whitefish the introduction of carp (Fig. 14-10*b*) may have accelerated the decline by physical interference rather than resource competition. Whitefish are more littoral-zone species than the open-water ciscos, and the presence of large numbers of bottom-grubbing carp may have interfered with whitefish breeding.

Finally, mention should be made of the fishing strategies used in the Great Lakes until all but native American commercial fishing was banned and replaced by sport fishing in the 1970s. When stocks of one large species declined in the 1950s, more effort was put into catching the remaining stocks of large fish as well as initiation of fishing for smaller fish. Some idea of the direct effects of fishing is shown for a coregonid, the blackfin cisco, *C. nigripinnis,* which was virtually exterminated by overfishing long before the entry of lampreys. The loss of the lake sturgeon to excessive fishing was mentioned previously.

The several successive species of cisco were overfished. Starting with the largest, *C. artedii* and *C. nigripinnis,* both fish reaching 40 cm, the fishery progressed down to the smallest, the bloater, *C. hoyi,* which is about 20 cm long. In addition the fishery switched in part from a profitable high-quality product for human consumption to a less valuable fish-meal industry which utilizes carp and alewife as well as indigenous small fish.

Other factors influencing the general decline of the Great Lakes fisheries include pollution, harbor construction, dredging of the rivers and bays used for spawning, and the intentional introduction of a variety of exotic trout and salmon.

Exotic Predators, Fish Kills, and Pollution

Unexpected fisheries management problems have occurred in a variety of freshwaters around the world. In Central America the intentional introduction of the piscivorous sunfishes, largemouth bass, and the black crappie has raised havoc with the native fishery in Guatemala's Lake Atitlan. A similar introduction of an Amazon River cichlid, the peacock bass, to Gatun Lake in Panama (Zaret and Paine, 1973) has altered the trophic structure of the entire system (Fig. 14-11). An unexpected side effect has been an increase in mosquitoes which carry malaria. The loss of mosquito larvae predators, most notably the topminnow *Gambusia,* to predation by the peacock bass may be responsible for a general increase in mosquitoes.

Water quality remains of prime importance in the management and conservation of fish resources. A common cause of changes in water quality is the increase in nutrients which cause eutrophication (Chap. 18). Eutrophication in its early stages is almost certain to increase fish production, especially in flowing waters, while at the same time altering the fish species composition to what may be less desirable. One of the most serious effects of eutrophication in lakes is its effect on depleting oxygen in the hypolimnion. This may result in the loss of deepwater spawning grounds and the elimination of some benthic invertebrates utilized for food. Rivers, streams and estuaries often show fewer undesirable effects of increased productivity (Chaps. 16, 17), but once again less-desirable fishes may predominate as eutrophication progresses.

Fish kills may also result from excessive eutrophication where bacterial and plant respiration depletes dissolved oxygen during a dark or windless period in summer or under a cover of winter ice and snow. Inflow of water with low dissolved oxygen from peat swamps may also produce winter kills. The cause of

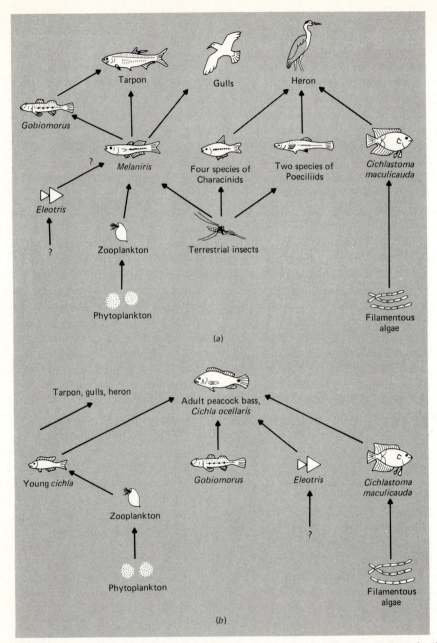

Figure 14-11   Changes in the food web of a tropical reservoir following the introduction of a large exotic predator, the peacock bass, *Cichla ocellaris*. (*a*) The postulated system prior to *Cichla* introductions with about a dozen fish species and several higher predators including gulls (terns), heron, kingfishers, and tarpon. (*b*) The situation after *Cichla* introductions. Eleven important fish species were drastically reduced and only one or two, including *Cichlastoma*, increased. The role of some common small fish such as *Eleotris* is unknown. Losses shown in populations of the higher predators are derived from semiquantative observations. (*Redrawn and modified from Zaret and Paine, 1973.*)

these kills was found by fisheries biologists who were among the first to make extensive use of the Winkler method for measuring the oxygen dissolved in water. Unfortunately, too much reliance on oxygen levels alone as evidence for pollution has developed, and occasionally an industry has successfully defended itself from pollution charges by directing attention to this variable while ignoring the direct toxic effects of chemical pollutants.

Fish kills resulting from industrial pollution are sufficiently dramatic to raise a public uproar against the polluter. Reports of fish kills in the United States have greatly increased in recent years. Chemical pollution or low oxygen levels may destroy the fish populations or form a block through which they cannot migrate. The high biological oxygen demand of sewage has at times dropped dissolved oxygen levels, which stops the upstream migration of salmon in rivers. Oxygen reduction is often more difficult for fish to overcome than thermal effects. Oxygen levels below 5 mg liter^{-1} can block migration of coho salmon even though they were undeterred by river temperatures of 23°C, which is well above the normal limits for this fish. The River Thames in England (Chap. 20) and the River Clyde in Scotland had a variety of fishes which formerly migrated through the main river and estuary. By blocking migration, pollution can have an effect upstream as well as downstream from the source of pollution (Maitland, 1974).

Aquatic organisms may concentrate some types of pollutants within their bodies without necessarily being affected. Although human beings are remarkably resistant to many potential poisons, they face increasing levels of dangerous substances concentrated by favorite food organisms. Pacific salmon, which were introduced in the Great Lakes for sports fishing, have at times been considered a health risk if regularly consumed because of high concentrations of chlorinated hydrocarbons such as polychlorinated biphenyls (PCBs). DDT, DDD, and PCBs are sometimes concentrated through successive levels of the food chain to the point where the large predatory fish contain hazardous levels in their fat. Similarly, methyl mercury derived from extensive use of metallic mercury in a variety of industries may contaminate the aquatic food chain.

An example of the concentration of chlorinated hydrocarbons in the aquatic food chain is provided by Clear Lake, California. The high fertility of this lake supports a very productive sport and commercial fishery. The lake has been the site of numerous attempts to control the Clear Lake gnat, *Chaoborus astictopus*, with insecticides. Predatory fish such as largemouthed black bass and catfish attained high levels of DDD and its derivatives in their fat (Hunt and Bischoff, 1960). The western grebe *Aechmophorus occidentalis*, presumably feeding on these contaminated fish, declined catastrophically and took years to recover after DDD use ceased. Although the fishes seemed little affected, the older individuals among the predatory species, particularly catfish, showed a high pesticide level in their tissues. The Clear Lake example served to alert the public to the dangers of persistent pesticides (Carson, 1962; Rudd, 1964).

Waste heat under some circumstances may increase growth rates of some fish but may be lethal to others. The most pronounced effects of thermal pollution occur in rivers where power plants utilize a large fraction of the flow for once-through cooling. A good example is the Wabash River, a typical large Mississippi drainage river, where up to two-thirds of the river volume is used for cooling in the autumn low-flow period. The fish are dominated by gizzard shad which are vegetation and detritus feeders; also important are carp, catfish, three species of redhorse, and the sauger, a relative of the walleye pike. The large number of species commonly present (about 20) allows a type of composite diversity index to be used to show effects after power station construction. Laboratory studies predict the effects of various temperatures on individual species (Gammon et al., 1981). Diversity, density, and biomass of the fish community decreased where the river temperature rose to 28 to 32°C near the outfall. Even at these high temperatures there was no direct mortality of the large adult carp, channel catfish, sauger, or redhorse, although the latter two selected lower temperature if available. After dilution the river temperature is elevated only 1 to 4°C for about 6 km downstream, but even this small change can affect the fishes. The composite index, which is based on numbers, biomass, and diversity, showed a decrease below the power plants.

Management approaches to overcoming fisheries problems include stream improvement through removal of natural blocks which prevent migration, construction of fish ladders, the development of gravel spawning stretches below reservoirs or in tributary streams, and in the last several decades, an increasing reliance on fish hatcheries below dams to supplement or replace natural reproduction. Unfortunately, fish ladders usually do not perform up to expectations, and adult fish may even require truck transport around dams. Although adult salmonids may be taken through a dam by elevator fish ladders or truck, the young when migrating downstream frequently are unable to find their way through the still waters of the reservoir or

are killed in the turbines. Air narcosis (a form of diver's bends) can also kill salmonid fish below high dams when water is discharged over the top of the dam. Both oxygen and nitrogen contribute to this effect, and while as little as 7 percent supersaturation may cause narcosis, measured values to 43 percent are not uncommon below dams.

Many lakes throughout the midwest and western United States are crowded with enormous numbers of tiny centrarchids in which inter- and intra-specific competition restricts growth. Water-level control in lakes or reservoirs can reduce fish reproduction and increase the size of individuals. The water level can be lowered in the spring as a means of stranding fish nests. Other methods including destratification are discussed in Chap. 20.

New Reservoirs

New reservoirs typically go through an early period of high fish production soon after they are filled and then decline to a much lower level of productivity (Table 14-2). This high initial productivity is due to flooding of terrestrial plants and nutrient-rich soil. It appears to reflect high initial nutrient content and is followed by a decline in basic fertility as nutrients are lost by outflow of the nutrient-rich hypolimnetic water and by sedimentation. An unfavorable species composition may also limit the

Table 14-2   Decline in Productivity in the Surface Waters of a New Reservoir, Dworshak Reservoir, North Idaho, in the First 3 Years after Filling*

The lake is 80 km long. Numbers are expressed as a percent of 1972 values except for photosynthesis, where 1973 is the base year. Little change in levels of NO_3^- or soluble PO_4^{3-} occurred.

	1972	1973	1974
Phytoplankton total numbers	100	2–70	9
Photosynthesis	...	100	28
Zooplankton:			
Copepods	100	40	30
Cladocerans	100	70	46

*Modified from Falter (1977).

harvest of fish from a reservoir. The lake formed by Kariba Dam on the Zambesi River in Africa may not produce the protein formerly harvested from the now-flooded valley (Harding, 1966), although recent data are more encouraging (Fig. 14-12). The Niger River in Nigeria has been dammed to form Kainji Lake (Table 2-1), which has large fluctuations in water level. Preimpoundment studies of fisheries in the river and the associated swamps were compared with postimpoundment catches 2 years after damming. Although the populations have not yet stabilized, a rise in predatory fishes in the reservoir was attributed to an increase in small forage fish (clupeids). Cichlids of the genus *Tilapia* which were not numerous in the river increased, but below the dam the total catch declined as detritivores lost their food supply (Blake, 1977).

The use of reservoir water for power and irrigation may lower fish production by drying out spawning and food-producing shallow areas just at the time they contribute most to the fertility of the system. Reservoir drawdown affects the shallow warm areas which contain most of the higher aquatic plants and the plankton and benthos. In the shallows they receive an energy subsidy from the wind which provides top-to-bottom mixing not possible in the deeper areas of the reservoir. In the tropics *Tilapia*, which are favored human food, may increase with the proliferation of attached and floating aquatic plants.

In Lake Baikal, the world's deepest and volumetrically largest lake, the establishment of a hydroelectric dam raised the water level and placed spawning areas below optimum depth for the most important coregonid present. The influence of industrial complexes on the major tributary, the Selenga River, and the cellulose plant on the South Basin may soon cause deleterious effects even in this enormous volume of water. The lumbering activities in the watershed already mentioned may also have de-

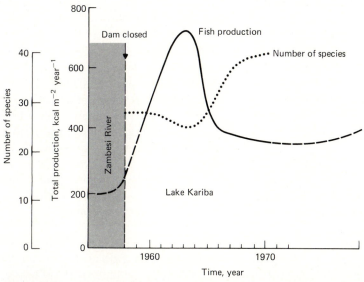

Figure 14-12 Fish production and species numbers in the Zambesi River before the creation of Lake Kariba and the changes that occurred afterwards. Note the rapid rise in production after dam closure and its similarly rapid fall to about pre-impoundment levels after a few years (see also Table 14-2). (*Modified from Balon and Coche, 1974.*)

stroyed spawning gravels by depositing forest debris in streams once utilized for breeding. New management strategies for the Baikal basin are badly needed.

Fish Stocking

Hatching and stocking fish have long been standard management practices throughout the world. This may be a suitable practice where angler demands are high but is less satisfactory than enhancing conditions for natural reproduction. Exotic species have been introduced on a worldwide basis and are at best a mixed blessing. The carp and the mosquito fish, *Gambusia,* are large and small fishes which have displaced native species in many places but which also survive where many of the native species cannot. Rare or endangered species may be eliminated by the introduction of an exotic that rapidly takes over the habitat. Exotics such as the Asiatic grass carp which has been introduced into central United States for weed control may be limited in their expansion by their inability to breed in lakes. This species has a requirement for high river flows at a constant temperature for successful breeding. However, once in streams grass carp are likely to outcompete many endemic associates, and are currently banned from introduction into many states. In Hungary, where it was introduced to control macrophyte growth in shallow weedy lakes, grass carp have proved ineffective and difficult to harvest.

In an ultraoligotrophic lake such as Lake Tahoe, fish production only amounts to a few tenths of a kilogram of fish per hectare. The fish populations are now largely exotic with lake trout, locally known as mackinaw, occupying the deeper reaches following their introduction from the Great Lakes (Fig. 14-5a). The rainbow and brown trout occupy the near-surface layers, and the pelagic zone is sometimes dominated by kokanee. All these species have been introduced.

A form of fish management still in its infancy is the development of hybrid species whose growth characteristics may be superior to the native forms. The centrarchids (sunfish family) cross frequently in nature and may be hybridized artificially to produce races which have a higher growth rate than their parents. Generally speaking, the hybrids have lower fecundity due to imbalanced sex ratios and will disappear from the population with time if they are not artificially bred.

Another means of fish management is the alteration of the structure of the aquatic food chain through introduction of new food items. One promising technique utilizes crustacean detritivores as fish food. The remarkable growth of rainbow trout in some British Columbia lakes has been in part credited to the abundance of the detritus-consuming amphipod *Gammarus.* On the other hand, introduction of the amphipod *Hyallela aztea* to the alpine Rae Lakes in California did not benefit the fishery. The sad example of *Mysis* introduction into Lake Tahoe has already been mentioned. It was assumed that *Mysis* was a detritivore, not the voracious predator on zooplankton that it appears to be in this lake.

The limnologist may be of particular assistance to the fisheries biologist through controlled fertilization of infertile waters. With the increase in eutrophication of most natural waters, this is more likely to be a tool in aquaculture than in general fish management. However, in certain environments, trace-element deficiency can be relieved at minor cost with a resulting increase in the production of trout (Fig. 10-13). As more is known about the structure of aquatic food chains, it may be possible to tailor fertilization to enhance specific portions of the food chain which go directly into fish production. Fish managers and limnologists should try to maintain stocks of native fish; new introductions of exotic fish or invertebrates should be made with caution, and then only after very careful study.

FURTHER READINGS

Allen, K. R. 1951. "The Horokiwi Steam. A Study of a Trout Population." *Bull. Mar. N. Z. Fish.,* **10:**1–238.

Balon, E. K., and A. G. Coche (eds.). 1974. "Lake Kariba: A Man-Made Tropical Ecosystem in Central Africa." *Monogr. Biol.,* vol. 24. 767 pp.

Barbour, C. D., and J. H. Brown. 1974. "Fish Species Diversity in Lakes." *Am. Nat.,* **108:**473–489.

Christie, W. J. 1974. "Changes in the Fish Species Composition of the Great Lakes." *J. Fish. Res. Board Can.,* **31:**827–854.

Fryer, G., and T. D. Isles. 1972. *The Cichlid Fishes of the Great Lakes of Africa: Their Biology and Evolution.* Oliver & Boyd, Edinburgh. 641 pp.

Lagler, K. F., J. E. Bardach, R. R. Miller, and D. R. M. Passino. 1977. *Ichthyology.* 2nd ed. Wiley, New York. 506 pp.

Magnuson, J. J. 1962. "An Analysis of Aggressive Behavior, Growth and Competition for Food and Space in Medaka, *Oryzias latipes* Pisces (*Cyprinodontidae*)." *Can. J. Zool.,* **40:**313–363.

Marshall, N. B. 1966. *The Life of Fishes.* Universe Books, New York. 402 pp.

Moyle, P. B. 1976. *Inland Fishes of California.* University of California Press, Berkeley. 405 pp.

Rawson, D. S. 1952. "Mean Depth and the Fish Production of Large Lakes." *Ecology,* **33:**513–521.

Ricker, W. E. 1975. *Computation and Interpretation of Biological Statistics of Fish Populations.* Department of Environment Fisheries and Marine Service, Ottawa, Canada. 382 pp.

Chapter 15

Food-Chain Dynamics

OVERVIEW

Energy and nutrients are transferred from their original sources through successive trophic levels by phytosynthesis, bacterial decomposition, or the feeding of herbivorous and carnivorous animals. Energy is a convenient measure of this transfer, but organisms also require a balanced diet of essential elements or preformed organic molecules such as amino acids and vitamins. The nutritional value per calorie of food may vary widely, so energy-flow diagrams must be carefully interpreted. Most of the energy and many nutrients acquired by one trophic level are lost as heat or excretion with only a small amount retained for growth. The efficiency of these transfers is best known for energy and varies from 2 to 40 percent, depending for the most part on the age and biological complexity of the organism concerned. The most generalized concept of biomass or energy transfer is the *biological pyramid*. Like a pyramid it is broad at the base where there is a large plant biomass, has herbivores in the middle, and is small at the top where there are relatively few higher carnivores.

Many aquatic organisms are versatile in their choice of food. Their omnivorous diets make it difficult to measure transfers of energy or nutrients between trophic levels. In fact, the term *trophic level* is best thought of as an idealized concept. The varied choices of food are best described in a *food web* which is usually derived from qualitative measurement of the contents of animal guts or knowledge of the nutrients required by algae or other plants. When the amounts and types of food utilized by each organism are known, a *dynamic food web* can be constructed where the importance of each component of the web can be shown. It is hard to actually make all the necessary measurements needed for constructing a simple food web, and a complete dynamic food web is even more difficult.

An alternative to studying the whole food web is to measure changes in subcompartments of the system. Stable or radioactive isotopes are frequently used to trace the flow of nutrients or biomass over short distances within the food web. The use of oxygen concentrations or carbon 14 uptake to measure primary productivity is a good example. The rate of transfer of essential substances can be measured by using additions of limiting nutrients to pure cultures or natural populations of bacteria and algae. These studies enable the researcher to find biochemical constants such as the maximum growth rate ($\hat{\mu}$) and to measure the plant's ability to take up nutrients at low levels (K_s). The constants vary with species and are useful in explaining the seasonal phytoplankton cycles as well as predicting changes if the nutrient levels are altered. The theory behind these biochemical concepts is stretched to or beyond its limits when applied to whole lakes, streams, or estuaries.

INTRODUCTION

Less is known about the dynamics of the food chain than is known about such other factors as light, temperature, or the distribution of the various species of organisms. Much of the difficulty in food-chain studies stems from our inability to extrapolate accurately from laboratory studies to natural plant growth or animal feeding rates. In the laboratory large animals like fish can rapidly consume a variety of prey. In contrast, under more natural conditions the choice of prey is limited and feeding rates are considerably lower than observed in the laboratory. In this chapter we discuss general physiological and biochemical factors such as enzyme growth constants and the percentage efficiency of energy transfer between trophic levels. This approach extends the classic method which relies on analysis of gut contents or decrease in prey species numbers through grazing or predation. Food chains, when reduced to simplified flow diagrams, provide a view of the dynamic structure of the aquatic ecosystem (Chap. 2).

MEASUREMENTS

At the base of the food chain, photosynthesis is measured in relation to light, nutrients, and a variety of other environmental factors. The various elements present in plant tissue are easily measured and show the nutrients available to the next trophic level. Further, they may provide clues as to the nutrient factors limiting plant growth. Another approach to plant dynamics is to measure the plant's response to additional nutrients in growth experiments. For aquatic animals the number and health of eggs and young indicate the food and energy transferred to the next generation. To study food webs limnologists use bomb calorimetry, whole-organism nutrient content, and nutrient-uptake kinetics. These investigations may involve radioactive and stable isotopes as tracers. Details of the most commonly used techniques are given in the IBP Manuals 8 (Golterman et al., 1978) and 12 (Vollenweider, 1969), Strickland and Parsons (1972), and the latest APHA volume. Individuals or whole populations can be combusted in a bomb calorimeter and the energy measured is then used to determine the energy flow. This approach taken alone is limited since the limnologist is as much concerned with the nutritive value of the food as with its absolute energy content.

THE REGULATION OF NUTRIENT AND ENERGY FLOW

Before attempting food-chain studies, year-round measurements of the major groups of organisms are needed. In this way a *biological pyramid* lumping organisms into general functional groups such as carnivores, herbivores, detritivores, etc., can be constructed (Fig. 15-1). More-refined measurements splitting organisms into particular functional groups such as mayfly-eating fish, algal-scraping insects or snails, heterotrophic bacteria, etc., enable a *food web* to be drawn (Fig. 15-2). Finally, a knowledge of the energy and nutrients passed between each small functional group will allow a *dynamic food web* to be constructed (Figs. 15-3 and 15-4).

Biological pyramids are derived from measurements of the numbers or biomass of organisms present at each season of the year: primary producers form the bottom level (P), herbivores or grazers are included in the second level (H), primary carnivores, grazing on small herbivores make up the next level (C_1), and secondary or tertiary carnivores, if present, form the apex (C_2–C_4). Figure 15-1 shows summer and winter biological pyramids for a typical temperate lake. Normally numbers of organisms or biomass decrease as one moves up to successive trophic levels. Biomass is perhaps the best

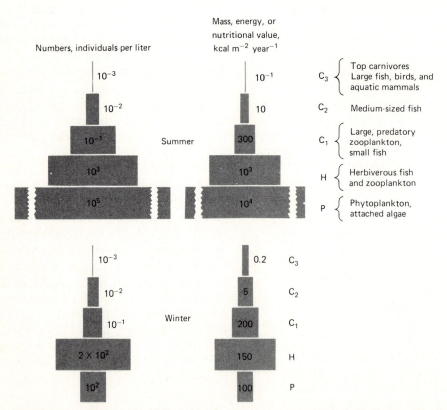

Figure 15-1 Biological pyramids of numbers or biomass, energy, and food value for a typical temperate lake in summer and winter. Note that only values in summer make a real pyramid. Also note the thicker upper part of the pyramid for mass or food in winter. This is due to the death of most lower organisms in winter while the larger organisms like fish live on stored food reserves. The quantitative values are for illustrative purposes only but are of the correct order of magnitude for some lakes.

unit to use for biological pyramids but numbers are also important. In winter the pyramid may be inverted, with only a few primary producers relative to grazers and carnivores. This pyramid is made stable because of the relatively high growth rate of the few algae present in relation to zooplankton grazing. Algal production, which is a nearly temperature-independent photochemical process, is less affected by low temperature than is the rate of animal feeding. During cold periods most small cladocerans and rotifers form some kind of resting stage and are absent from the plankton. Those remaining in the open water are the larger copepods and their young, which feed slowly due to the lower temperatures. It should be remembered that in contrast to the land, most of the world's aquatic net primary production occurs in the cold waters of upwelling areas and polar or subpolar seas, not in the tropics. Some tropical lakes have very high gross primary production, but because of elevated respiration rates, net production is often greatly reduced.

The next stage in analysis of the ecosystem is to break down each trophic level by measurement of food eaten in order to construct a food web. A detailed collection and identification of organisms is made together with an analysis of the gut contents of the herbivores and carnivores. This tedious analysis is not usually accomplished for all trophic levels in any one aquatic ecosystem. Figure 15-2 shows a typical food web developed from studies on the Truckee River in the mountains of California. Samples for analysis of benthic invertebrates and attached algae were taken simultaneously on the same natural rocks. The algae and invertebrates were counted and the guts of several individuals of each of the common grazing insects were removed and examined to find which algae were being consumed. Fish were also examined for their choice of prey.

In this example (Fig. 15-2) the individual species have been lumped into genera or feeding classes, which simplifies construction of the food web. Inevitably there are too few individuals of some species to obtain a reliable estimate of their food. It is not a serious drawback since most aquatic animals are omnivorous, and closely related species usually eat similar food. This is the logic behind the functional classification of the structure of stream animal communities proposed by Cummins (1973), which is discussed in Chap. 16. Food webs sometimes show food preference (Table 15-1). For example, in the Truckee River the uncommon pennate diatom *Cocconeis* was frequently eaten by net-spinning caddis fly larvae. In contrast, the common pennate diatom *Nitzschia* was eaten less frequently than would be expected if the animals were not selective (Table 15-1). Similarly, rainbow trout preferentially selected caddisflies (Fig. 15-2).

The key to biological success is efficient transfer of suitable types of energy and nutrients from one trophic level to the next. The energy used in hunting, avoidance of predation, or uptake of very dilute nutrients against a large gradient is lost for growth and reproduction. Although efficiency varies considerably, the approximation that only 10 percent of energy is transferred from one trophic level to another is a good starting point for discussion. The maximum efficiency of energy transfer in young, actively feeding animals is about 30 to 40 percent and is probably limited by their ability to convert food protein to body protein. Due to the expenditure of reproductive energy, the overall efficiency of transfer from adult to the adult of the next generation is considerably lower. For mammals the average value is only 2 to 5 percent, but it may range as high as 20 to 25 percent for fast-growing zooplankton. Much of the difference is due to the necessity of maintaining a constant body temperature in mammals, which is not needed by the poikilothermic fish and invertebrates.

A major use of energy in all adult organisms

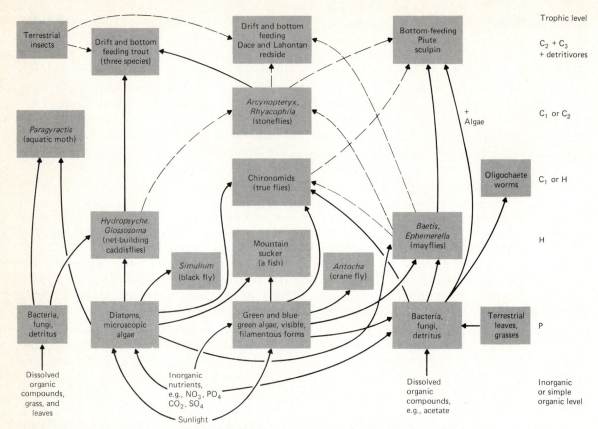

Figure 15-2 Qualitative food web for the Truckee River, California. Solid lines indicate measured pathways. Broken lines are assumed pathways derived from the other studies in adjacent waters. Note the omnivorous feeders (e.g. dace, trout, sculpin) using more than one trophic level. Most herbivores prefer microscopic diatoms to large filamentous green and blue-green algae.

is to produce and disperse young. Any mechanism which reduces this energy requirement but maintains the same survival rate of young is advantageous. For small aquatic organisms the energy of turbulent eddies (Chap. 5) enables rapid dispersal of young.

In general the higher the level in the food chain the greater the energy and nutrient content of the food items consumed. Thus fish eating smaller fish or zooplankton gain more calories per unit weight of food than do herbivorous fish or zooplankton which extract energy from plant and detrital material. The ef-

ficiency of energy transfer is thus understandably higher as one proceeds up the food chain. Several assumptions about the efficiency of energy transfer are needed to evaluate any food chain, and, unfortunately, few direct verifications of these efficiencies are currently available.

A study of the energy flow in a pond in Georgia was carried out by Welch (1967). This pond is used for sport fishing and is managed to increase the harvest of sunfish and bass. The management produces a more efficient system for producing centrarchids than a natural lake,

Table 15-1 Ingestion of Different Benthic Diatoms by Common Benthic Insects in a California Stream*

+ = taken in preference; 0 = no preference shown; − = under utilized; total = rating of the group. The diatoms were scraped from the same rocks and gravel area from which the insects were collected. The differences shown could be due to active selection by the insects in the case of the grazers or to a differential drifting of the algae in the case of the filterers.

Insect herbivore	Genus of diatom eaten						
	Gomphonema	Navicula	Cymbella	Cocconeis	Epithemia	Diatoma	Nitzschia
Glossosoma	+	−	0	+	0	−	−
Baetis	0	−	0	+	0	−	−
Heptagenia	−	0	−	+	0	−	−
Ephemerella	−	0	0	+	0	−	−
Chironomidae	+	+	+	−	+	0	−
Total	0	−1	0	+3	+1	−4	−5

*Unpublished data from A.J. Horne and P.W. Johnson.

and results in a simple food web (Fig. 15-3). In this artificial system the efficiency of energy conversion between trophic levels is easy to calculate. The values obtained are given in Table 15-2 and show that in this simplified situation, conversion may exceed 10 percent between trophic levels and generally increases from light to algae to herbivores to primary carnivores to top predator (bass). Depending on how it is calculated, energy transfer to higher carnivores may be quite inefficient. We will return to this problem later when a more-complex, natural food web is discussed.

The example chosen to illustrate a dynamic food chain was measured during IBP studies on a lowland river, the Thames in England (Fig. 15-4). Because of the relatively slow flows, large size, and high nutrient loadings, this river has a community structure which combines major elements of both lakes and rivers. It has an important plankton community, as occurs in lakes, but also has the large allochthonous contribution typical of flowing waters (Chap. 16).

Figure 15-4 illustrates the complex interactions between nutrients, bacteria, and the various trophic levels of algal primary produc-

ers, grazing insects, and other invertebrates, and finally the primary and secondary carnivores. It should be remembered that the term trophic level is an abstract concept useful in recognizing interactions and patterns existing within food webs. Organisms may spend part of their time at one trophic level and part at another. For example, some nominally carnivorous chironomid midge larvae feed partially on small animals but also use plant material when other food is unavailable. The many links in the food web allow most herbivores and carnivores to utilize more than one food source. Obviously if there are only a few links in the food chain, overall efficiency will be greater than if there are many. Considerable energy loss occurs at each transfer, and aquaculturists derive highest production from maintaining the shortest possible food chain.

In the Thames, light provides 7×10^5 cal m^{-2} year^{-1} for plant growth which is mostly used by phytoplankton. In turn the primary producers in the river provide about 1900 cal m^{-2} year^{-1} for direct use by various kinds of herbivores. The diet of herbivores is supplemented by allochthonous detritus from tree leaves, aquatic

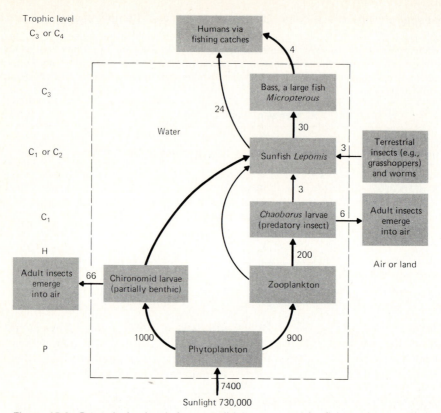

Figure 15-3 Dynamic food web for a small lake managed for fish production. The management simplifies the food web by reducing the habitat for competing animals or plants. Values shown are in kilocalories per square meter per year uncorrected for losses in respiration or assimilation. Heavy lines indicate the most important pathways. Note that even in this system the omnivorous feeding habits of the sunfish and people make a calculation of trophic-level efficiency difficult (see Table 15-2). Also note that a large part of the higher trophic-level production was lost to the air as adult chironomid flies. (*Redrawn from Odum, 1971, who originally modified data from Welch, 1967.*)

macrophytes, and aufwuchs which provide only 460 cal m^{-2} year^{-1}. Thus there is an efficiency of conversion of light to plant material in the river of roughly 0.3 percent. This is a typical value for most plants. Table 15-3 indicates that the efficiency of conversion in the Thames is highest for the upper trophic levels.

As mentioned earlier it is difficult to be precise about energy-transfer efficiency. Depending on how one partitions the potential food of the higher trophic levels, efficiencies can be made to range from 2 to 24 percent (Table 15-3). For example, if chironomid larvae act as carnivores their efficiency would be 24 percent, but if they consume only plant material their efficiency would be only 2 percent. Since no analysis of gut contents was made, actual efficiency is unknown, but plant material is often consumed by chironomid larvae (Table 15-1).

All major interrelations are obvious in energy flowcharts because the energy transfer between each compartment is measured (Fig. 15-4). For

Table 15-2 **Energy Conversion in Simplified Lake System Managed for Fish Production**

	Annual production, kcal m^{-2}	% efficiency of transfer
Light	730,000	—
P : phytoplankton	7,400	1
H : zooplankton + chironomids	1,900	26
C$_1$: *Chaoborus* larvae + sunfish	550	29
C$_2$: sunfish (on *Chaoborus*)	3 ⎫	
C$_3$: bass + people (on sunfish)	54 ⎬	11
C$_4$: people (on bass)	4 ⎭	

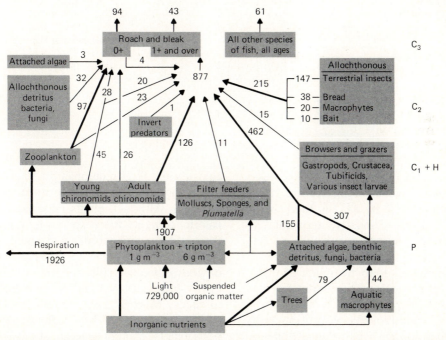

Figure 15-4 Dynamic food web for a natural system. Energy flowchart for the River Thames below Kennet mouth. In general, primary producers are shown at the bottom, invertebrate animals in the center, and fish at the top, but to avoid complex networks of arrows sources of attached algae, detritus, and allochthonous materials are shown in two places. Heavy arrows indicate the largest channels of energy flow. Note the twin flow of energy to fish from low quality attached algae and high-quality animal food from terrestrial insects and adult chironomids. Energy imput from dissolved organic matter was not measured directly. (*Redrawn from Mann et al., 1972.*)

Table 15-3   Various Ways of Looking at Food-Chain Energy Conversion in the River Thames, below the Entry of the River Kennet*

Note that C_2 fish must acquire most of this high-quality food from allochthonous sources since the river cannot provide enough to maintain measured growth. This food is mostly terrestrial insects, worms, and chironomids from surrounding ponds. Also, note that the four cyprinid fish common in this river are quite omnivorous and take food from several trophic levels depending on season, prey availability, and age class of the fish.

Trophic level	Annual production, kcal m^{-2}	% efficiency of transfer†
Light	729,000	
P:		
Plants, mostly phytoplankton	2,369	0.33
H:		
Herbivores/grazers, filtering organisms	191	8.1
H + C_1:		
Young chironomids (primary carnivores),	45	2–24 (average 11)
		0.1
Leeches	1.7	~0.1
Allochthonous fish food, terrestrial insects, etc.	215	n.c
Adult chironomids	152	6–64 (average 29)
H + C_1 + C_2:		
Fish	198	19

*See Fig. 15-4 for energy food webs.
†Annual production of trophic group per annual production of food.
 n.c. = not calculated since food source is unmeasured terrestrial material.

example, in relation to phytoplankton, terrestrial plant detritus and macrophytes are relatively unimportant to the base of this food chain in the River Thames. Similarly, zooplankton are a major energy source for fish such as the roach. Older minnows like bleak use periphyton and benthic diatoms for much of their energy but chironomids and terrestrial insects also supplement this diet. These insects, although of lower calorific value than plants, are high-quality food with a well-balanced amino acid content.

The subject of food quality brings us to the final section of food-chain analysis. As indicated for bleak in the Thames, as well as in most lakes, rivers, and oceans, the consideration of energy transfer alone oversimplifies organisms and food relationships. Energy utilization is largely determined by nutritive value of the food. Although plants may have a similar caloric content, there are large variations in their protein, phosphorus, potassium, and other mineral content (Table 15-4). Some plants are obviously better-quality food than others. Animals show less variability in their protein to energy ratio. The differences in overall nutritive usefulness of any particular prey cautions against making generalizations from measurement of only energy or a single element such as carbon, nitrogen, or phosphorus. The digestibility of food items is also important, since certain items such as crayfish require almost as much energy to digest as they yield to the fish that consume them.

Table 15-4 Quantities of Selected Chemical Substances and Caloric Contents of Several Species of Aquatic Macrophytes*

Values for all constituents except tannins were obtained for samples from a single lake and should be subject to a minimum of between-site variation.

Species	Protein†	Non-cell wall constituents	Tannins	Ash (minerals)	Phosphorus	Potassium	Caloric content, kcal g⁻¹ dry wt.
				% dry weight			
Typha latifolia	4.0	34.7	2.1	7.5	0.14	2.65	4.26
Hydrotrida carolinensis	10.5	60.6	2.5	...	...	...	4.06
Brasenia schreberi	10.9	70.4	11.8	7.6	0.14	0.99	4.03
Utricularia inflata	11.4	59.3	...	14.0	0.12	1.98	4.02
Nelumbo lutea	12.1	56.9	9.2	8.8	0.19	2.27	4.23
Myriophyllum heterophyllum	13.5	63.0	3.2	12.2	0.16	1.25	3.96
Eleocharis acicularis	14.1	...	2.0	11.2	0.24	2.86	4.26
Najas guadalupensis	14.4	...	1.4	12.8	0.15	3.49	3.92
Nymphaea odorata	14.6	58.1	15.0	8.1	0.18	1.28	4.18
Ceratophyllum demersum	17.1	53.1	1.9	14.9	0.26	4.01	3.91
Nuphar advena	21.6	68.5	6.5	10.6	0.40	1.88	4.32

*Modified from Boyd and Goodyear (1971).

†Protein—sum of amino acids

GROWTH KINETICS AND ISOTOPIC TRACERS

These methods provide an alternative means of estimating the flux of energy and nutrients in food webs. Biochemical techniques are best-suited to small parts of the food web, especially at lower trophic levels, since upper levels can be described by gut contents. The phytoplankton ecologist needs to know how algae respond to different nutrient concentrations. In most lakes and rivers nutrients are highest in spring and then decrease until the autumn when they increase again. In addition summer rain, zooplankton and fish excretion, or wastewater discharge change the flux of nutrients. The limiting-factor concept, or law of the minimum, is often called *Leibig's law*. As applied in ecology it states that under steady-state conditions only one factor can limit growth. In aquatic systems a whole spectrum of factors may limit growth in the course of a day or a season (see Chap. 10). At any time plants are most limited in their growth by a single factor which may be light, a nutrient, or their intrinsic maximum growth rate. When light or nutrients fall to limiting levels, those plants which require the lowest concentrations are most likely to succeed.

Growth Kinetics

One approach, which uses a *substrate-growth curve,* is now widely applied to microbial kinetics. The technique involves measuring the growth of natural or laboratory cultures of phytoplankton or bacteria at several different nutrient concentrations. After incubation the differences in growth between treatments is apparent. If all nutrients are added at the start of the test, it is called a *batch culture.* If nutrients are added continuously and with the overflow volume removed, it is called a *chemostat* or *continuous-flow culture.* Growth (μ) can be measured in different ways using, for example, cell counts, chlorophyll *a,* or ^{14}C uptake. The

nutrient or substrate S added at various levels must be growth-limiting, and all other nutrients, light, and temperature should be maintained at optimum levels. The limiting nutrient must be added so that the growth requirement is met at the highest substrate levels where the maximum, nutrient-independent growth rate ($\hat{\mu}$) will be achieved. A wide range of additions is needed to cover responses between low levels and saturation. Under ideal conditions a progressive increase in growth will be produced with increasing nutrient levels (Fig. 15-5). The equation describing such growth is

$$\mu = \frac{\hat{\mu}\,S}{K_s + S}$$

where μ = measured growth rate
$\quad\ S$ = substrate (nutrient) concentration
$\quad\ \hat{\mu}$ = maximum growth rate
$\quad\ K_s$ = half saturation constant, i.e., the substrate concentration at which half the maximum growth rate occurs (see Fig. 15-5)

Note that $\hat{\mu}$ and K_s are, in principle, fixed intrinsic biochemical properties of the plant species and are constant, while μ and S can change with environmental conditions. The equation also assumes a steady-state condition in the system which can be achieved with continuous-flow laboratory cultures but not with batch cultures. This method will usually work with most organisms and is easy to carry out. In its simplest form the method requires only a few flasks and a routine growth-measuring technique. Pioneering work by Eppley and Thomas (1969) and others on algal, inorganic nutrient kinetics, and by Wright and Hobbie (1966) on organic, bacterial kinetics are now common in studies of phytoplankton (Fig. 15-7).

The usefulness of the method is shown by reference to examples which illustrate a common effect in lakes. As nutrients wax and wane between spring and fall, phytoplankton with different $\hat{\mu}$ and K_s will respond differentially.

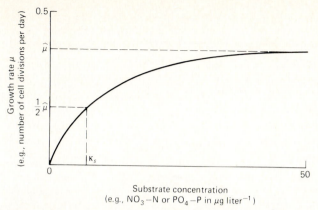

Figure 15-5 The theoretical nutrient-uptake growth-rate curve (Substrate-growth or Monod curve). Growth proceeds to $\hat{\mu}$. At $^1/_2\hat{\mu}$ the dotted lines show how κ_s is derived. In practice, the equation is rewritten as $S/\mu = (1/\hat{\mu})\ S + \kappa_s/\hat{\mu}$ to give a straight-line graph of the type $y = mx + c$. S/μ is then plotted against S and κ_s obtained from the intercept $\kappa_s/\hat{\mu}$ since $\hat{\mu}$ is known. This plot is more accurate than any other transformation. (*Dowd and Riggs, 1965.*)

Three examples are shown which cover most possibilities. For two competing species of phytoplankton (referred to as species 1, species 2, etc., in each example) these are:

a. $\hat{\mu}_1 = \hat{\mu}_2 = 0.4$ cell divisions per day
   $K_{s,1} < K_{s,2}$
   where $K_{s,1} = 5\ \mu$g liter^{-1} substrate
   $K_{s,2} = 15\ \mu$g liter^{-1} substrate

b. $\hat{\mu}_3 > \hat{\mu}_2$
   where $\hat{\mu}_2 = 0.4$ cell divisions per day
   $\hat{\mu}_3 = 0.6$ cell divisions per day
   $K_{s,3} = K_{s,2} = 15\ \mu$g liter^{-1} substrate

c. $\hat{\mu}_4 > \hat{\mu}_2$
   where $\hat{\mu}_2 = 0.4$ cell divisions per day
   $\hat{\mu}_4 = 0.5$ cell divisions per day
   $K_{s,4} > K_{s,2}$
   where $K_{s,4} = 24\ \mu$g liter^{-1} substrate
   $K_{s,2} = 15\ \mu$g liter^{-1} substrate

These conditions are plotted in Fig. 15-6. Consider the question, which species out of the pair will eventually become dominant at 10 and 50 μg liter^{-1} of substrate? The graphs show that for case *a*, species 1 outcompetes species 2 at lower nutrient levels but that at higher levels they are equal competitors. For case *b*, species 3 always outgrows species 2. For case *c*, the

most interesting one, species 2 wins at low substrate levels but at about 40 μg liter^{-1} species 4 starts to be dominant. Examples for laboratory-pure cultures of a blue-green alga and a diatom are shown in Fig. 15-7.

Prediction of Phytoplankton Seasonal Cycles

Because $\hat{\mu}$ and K_s are essentially constant cellular attributes of an algal species or at least a genetic race within a species, we can predict the outcome of changes in any nutrient concentration for any algal species if $\hat{\mu}$ and K_s are known. Diatoms have a high maximum growth rate and a high K_s relative to many algae and therefore require high concentrations of nutrients for growth. Therefore, diatoms bloom in spring and fall when high nutrient concentrations occur (Fig. 12-3). In summer when nutrients are low a fast growth rate is less advantageous than the ability to take up nutrients at low concentrations (low K_s). Blue-green and flagellated algae owe their success in part to their ability to grow slowly at low nutrient levels. Of course many other factors such as grazing, parasitism, sinking, and active swimming modify the final outcome of phytoplankton competition during the year (Chap. 12).

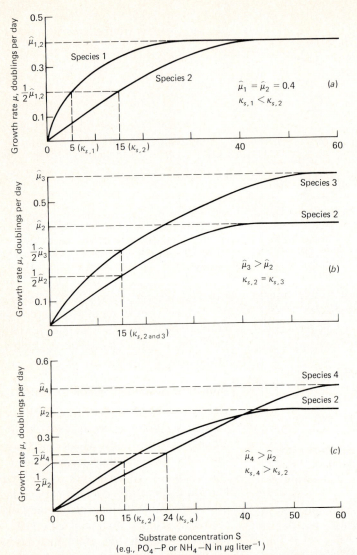

Figure 15-6 Three idealized examples of competition between algae with different half-saturation constants (see text for more details). Note the different results in overall growth at 10 and 50 μg liter^{-1} substrate for the different species. Note that phosphate or ammonia is shown as the growth-limiting substance, but in real systems, the limiting factor could also be NO_3, SiO_2, Fe, CO_2, or light. Note that at approximately 50 μg liter^{-1} the enzyme systems are fully saturated and growth rate predominates over uptake kinetics. Note that in the bottom graph species dominance changes with substrate concentration. This type of competition occurs each year as nutrient levels decrease in spring.

The results of interalgal competition can only be predicted if differences in the maximum growth rate $\hat{\mu}$ or K_s are large and constant and the same factor remains limiting (Tables 8-3, 9-2, 10-2; Figs. 15-6, 15-7). In many lakes, however, a whole spectrum of factors may limit

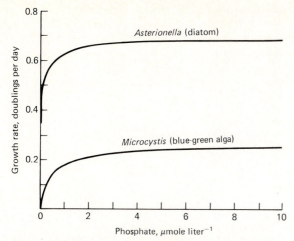

Figure 15-7 The substrate-growth curves for laboratory cultures of two common planktonic algae. Note that the diatom easily outgrows the blue-green alga under the conditions of this experiment; i.e., temperatures are moderate (20°C), silica is plentiful, and light is quite high (100 fc). At half this illumination or with low silica, neither the blue-green alga nor the diatom would have any distinct advantage. (*Redrawn from Holm and Armstrong, 1981.*)

growth during a single day as well as through the year. For example, phytoplankton may start the season with an ample, well-balanced supply of nutrients but deplete one more rapidly than the others. This *resource competition* has been tested experimentally for two algae, *Asterionella formosa* and *Cyclotella meneghiniana*. Both of these diatoms are common in lakes and have a similar $\hat{\mu}$. The two species have different growth kinetics, with *A. formosa* outgrowing *C. meneghiniana* when phosphate is low. When silica is low, however, *C. meneghiniana* outgrows *A. formosa*. As the ratio of available nutrients, Si/P, changes from more than 90 to less than 90, laboratory growth studies predict that *Asterionella* will be outcompeted by *Cyclotella* (Fig. 15-8). Resource competition of this sort explains the observed diatom community structure in Lake Michigan (Fig. 15-8), where ratios of Si/P in summer range from less than 1 near inflowing rivers to greater than 100 in the pelagic waters.

Growth rates and resource competition result directly from changes in physical variables such as light, temperature, and pH. The availability of many substances, particularly CO_2, is also influenced indirectly by pH (Fig. 7-2). Carbon dioxide is most available at low pH and is the only carbon fraction used by many plants. Lowering the pH experimentally in lake enclosures has been demonstrated to alter the species composition of phytoplankton (Shapiro, 1973). In the controls where pH was 8–9 blue-green algae were dominant, possibly because they can use bicarbonate as a carbon source. Blue-green algae also have a lower K_s for CO_2 than do the other phytoplankton. In the experimental enclosures with lower pH, green algae and diatoms increased while blue-green algae decreased. This change could be due either to an inability of greens and diatoms to use HCO_3^- or because they are favored with a high K_s for CO_2. Another interesting possibility is the change in microbial predators with pH since some viruses may only be able to lyse blue-green algae under acid conditions.

Isotopic Tracers

The use of radioactive or stable isotopes of biologically important elements to trace the pathway of nutrients from one trophic level to another is still in its infancy. ^{14}C is widely used to measure photosynthesis, and ^{15}N and ^{32}P are often used at one trophic level to determine uptake of N_2, NH_4, NO_3, or PO_4. Their use is only essential if changes in natural concentrations cannot be measured accurately. An entirely different use is to label one trophic level and to follow the tracer directly all the way through several other trophic levels. In this way there is no doubt of the origin and amounts of each element transferred. Unfortunately carbon, the most abundant element of organic material, cannot be used this way due to its loss in respiration and likelihood of recycling as CO_2. Recycling can also be a problem for other tracer elements.

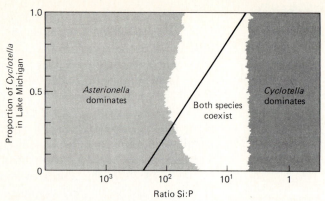

Figure 15-8 Effect of changes in the ratio of silicate to phosphate on the outcome of competition between two diatoms, *Asterionella formosa* and *Cyclotella meneghiniana*. The shaded
areas represent ratios at which one or another species should dominate as ascertained by kinetic studies on pure cultures of each species and by long-term laboratory experiments on mixtures
of the two. The black line shows measured proportions of *Cyclotella* vs. Si:P ratios in Lake
Michigan. The agreement between laboratory prediction and field data is imperfect, especially at
low Si:P ratios. Considering all the other factors which contribute to algal growth (Chap. 12),
prediction from kinetic parameters alone is quite good. (*Redrawn and modified from Tilman,
1978, Kilham and Tilman, 1979.*)

When $^{15}N_2$ gas is added to a natural mixed assemblage of mosses, grasses, and the nitrogen-
fixing alga *Nostoc*, the gas can only be fixed
into combined nitrogen by *Nostoc*. However,
experiments using this community growing in
damp hollows in coastal areas showed that the
grasses and mosses became labeled with ^{15}N
(Stewart, 1967; Horne, 1972). Obviously ni-
trogenous material from *Nostoc* had been lost
by excretion or decay and had been used by the
other plants. This method can be used to show
nitrogen flow through several trophic levels,
especially as ^{15}N is a stable, nonradioactive
isotope with no radiation hazard (Preston,
Stewart and Reynolds, 1980). Whole-ecosys-
tem studies often require large amounts of
label and a high degree of sensitivity of detection for the diluted isotope.

Other isotopes present particular problems.
The useful radioactive isotopes ^{32}P or ^{13}N
decay too rapidly for long-term experiments,
and other elements such as ^{99}Mo or ^{33}S are
needed in such small amounts or have such
short half-lives that they are unreliable tracers

for food web studies. The dangers associated
with isotopes, especially those which emit
strong radiation, preclude large-scale use in nature. Recent government attempts to produce
large quantities of cheap nonradioactive isotopes such as ^{15}N and ^{13}C for scientific research
provide hope for more quantitative studies of
food chains.

PRIMARY PRODUCTIVITY

Measurements of photosynthesis or the
primary productivity of the ecosystem are essential in food-chain studies. The
daily and seasonal carbon flow in the system
forms the base of the annual food pyramid and
can be used to estimate the maximum production at higher trophic levels. The *sustainable
yield* of a fishery can be estimated by using only
measurements of primary production (Steele,
1974). In addition to longer-term effects,
changes in photosynthetic rates may show the
almost instantaneous effect of a nutrient, toxicant, or physical change in the environment.

The advantage of primary productivity assays is that they directly measure effects before they are masked by adaptation or dilution in the food chain. Cell counts, chlorophyll *a* levels, ATP, total organic or inorganic carbon, pH, or oxygen can be used to measure changes in plant growth. Most biomass estimates are relatively insensitive except where plant growth is dense. In practice, changes in oxygen and uptake of the radioisotope ^{14}C are the most widely used methods (Figs. 15-9, 15-10, 18-7).

The light-and-dark-bottle method of measuring community metabolism by changes in dissolved oxygen has been very valuable in es-

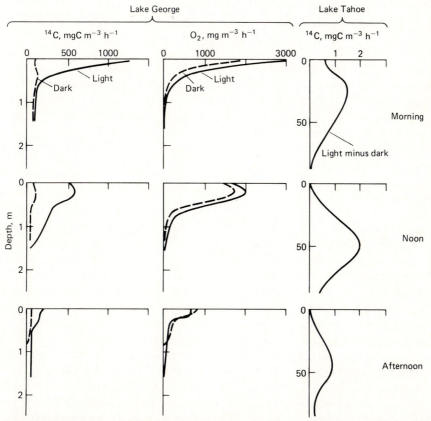

Figures 15-9 Diurnal changes in gross primary productivity with depth for two contrasting lakes, productive Lake George, Uganda, and unproductive Lake Tahoe. Primary productivity was measured with both the ^{14}C and the O_2 light- and dark-bottle methods in Lake George, but only the ^{14}C method is sensitive enough for Tahoe. In the productive tropical lake, respiration (dark-bottle results) almost equals photosynthesis, producing high gross but low net productivity values. Note that despite the different methods and differences in absolute gross production of over 3 orders of magnitude, the daily and depth patterns are similar. Productivity peaks near the surface in the early morning, a subsurface peak may be produced around noon, and a smaller morning-type peak is produced in the afternoon. These lakes were well-illuminated with continuous sunshine at the time of measurement. In climates with more rain, wind, and cloud cover in summer, these patterns may be less obvious (see Fig. 15-10). (*From Ganf and Horne, 1975; Tilzer and Horne, 1979.*)

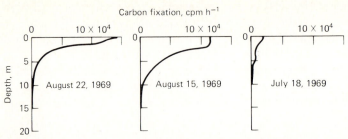

Figure 15-10   Variations in primary productivity as indicated by counts per minute (cpm) for after-noon incubations in the South Basin of Windermere, England. British summers are often wet and cloudy, and no appreciable surface inhibition was observed (in contrast with sunny areas such as shown in Figs. 15-9; 18-7; and 19-9.

timating net and gross production in moderately eutrophic waters. Both respiration (dark bottle) and photosynthesis (net = light bottle, gross = light + dark bottle O_2 changes) can be measured, so the oxygen method should be used when possible. This method involves measuring the oxygen change that occurs in both light and opaque bottles during an in situ incubation of several hours (Fig. 7-1a). The time period of incubation can be very short in highly productive waters, but requires as much as 24 h in less-productive ones. In productive waters loss of carbon dioxide or increases in oxygen can be measured directly in the water column of lakes to give an approximate measue of photosynthesis.

For unproductive waters, or for short incubations, the sensitivity of the $^{14}CO_2$ method using the light-and-dark-bottle technique is required. The isotope is added in the form of sodium carbonate or bicarbonate, which rapidly disperses through the inorganic carbon pool (Fig. 7-3). The addition of labeled carbon increases total carbon by an insignificant amount. After a short incubation, the water is filtered and the radioactivity contained in the phytoplankton is measured with a Geiger-Müller counter or a scintillation counter [for details on primary production, see Strickland and Parsons, 1972; IBP Manual 12 (Vollenweider, 1969); or Goldman, 1960, 1963, 1968a, b]. The radioactivity (disintegrations per minute or dpm) of the filter as

well as that added initially and the total inorganic carbon level in the water ($\Sigma^{12}C_i$) are needed to calculate the primary productivity, which can be expressed by the equation:

$$^{12}C \text{ fixed} = \frac{(^{14}C, \text{ light dpm}) - (^{14}C, \text{ dark dpm})}{^{14}C, \text{ dpm added}}$$

$$\times \Sigma \; ^{12}C_i \times BA \times 1.06$$

where BA = bottle aliquot correction factor

1.06 = isotope discrimination factor, since the heavier isotope ^{14}C is taken up approximately 6 pecent more slowly than ^{12}C

Rates of primary productivity for a variety of lakes and estuaries are given in Figs. 7-1b, 18-7, 19-8, 19-10b, 20-6, and Table 17-1.

Both the oxygen and ^{14}C methods have been very useful, especially in eutrophication studies (Fig. 20-6). There are a few drawbacks, mostly due to the enclosure of a small volume of water in the incubation bottle. The major problems in ^{14}C studies are due to nonphotosynthetic dark uptake and the release of soluble extracellular products of photosynthesis (ECPP) to the medium. Dark uptake is related to respiration and is subtracted from light uptake. In some oligotrophic waters dark uptake, particularly in winter, may exceed light uptake. Under these circumstances dark uptake is usually ignored. ECPP pass through the filter but may amount to as much as 95 percent of total production (An-

Table 15-5 Diel Variations in Extracellular Release of Photosynthetic Products (ECPP) by Phytoplankton in Contrasting Lakes, Very Unproductive Lake Tahoe (Secchi Disk ~ 30 m) and Productive Lake George, Uganda.*

The values for Lake George are given for several depths on one day and those for Tahoe for various days throughout the growth period. Note the regular increase in surface ECPP in Lake George as the day progresses. In Tahoe, note the low ECPP during the most favorable growth period (June) and the high releases at night. For both lakes, ECPP is an important component of primary production which is directly usable by bacteria. Units are in percent of total carbon fixation.

Lake, date	Depth	Morning		Noon	Afternoon		Night	Day average
		0700–1000	1100–1230	1300–1400	1500–1600	1700–1900		
Eutrophic Lake George, Uganda, March 26, 1968	0–8 cm	7	20	27	40	34		26
	15–23 cm	9	13	18	27	37		21
	30–40 cm	23	20	22	44	57		33
	50–60 cm	45	29	36	50	45		41
	150 cm	54	45	24	29	26		36
	Depth av.	28	25	25	38	40		31
		0600–1000		1000–1400	1400–1800		1800–0600	
Oligotrophic Lake Tahoe, 1975:	April	14		12	15		42	17
	June	6		8	5		18	7
	July	13		15	13		13	16
	Sept.	11		9	10		45	10

*Modified from Ganf and Horne (1975); Tilzer and Horne (1979); Goldman (unpublished).

derson and Zeutschel, 1970). Since the measurement of ECPP requires tedious laboratory work, it is fortunate that such high losses are rare. About 10 percent is typical. Some soluble ECPP values are shown in Table 15-5. Since carbon is respired as well as fixed, there is some loss and recycling of ^{14}C by the plants. Despite some technical problems, the oxygen and ^{14}C methods remain the best estimates of primary productivity. For example, gross photosynthesis estimated from several different methods run simultaneously are in reasonable agreement (Goldman, Tunzi, and Armstrong, 1969; Ganf and Horne, 1975).

Because primary productivity measurements require much time and effort, experimental measurements can be combined with mathematical modeling. This is especially useful where there are many similar lakes in an area, as in Canada, Minnesota, and Wisconsin. Photosynthesis can be modeled with an equation developed by Talling (1970):

$$\Sigma_a = n \frac{P_{max}}{K_e} F\left(\frac{I'_0}{I_k}\right)$$

where Σ_a = photosynthesis or carbon fixed per unit area

n = population density (e.g., in mg chlorophyll a m^{-3})

P_{max} = photosynthetic rate at light saturation which must be determined empirically using a lake water sample

K_e = average light extinction coefficient (see Chap. 3)

I'_0 = incident light intensity

I_k = light intensity for P_{max}

F = function of the ratio I'_0/I_k

Using this model, only one measurement of photosynthesis and chlorophyll may be required since the other values remain relatively constant. Unfortunately, there are many problems with the model due to algal patchiness (Figs. 5-3, 12-13, 12-14, 12-15, and Plate 6b).

Changes in P_{max} with chlorophyll, variations in the physiology of the algae, as well as horizontal variations in K_e further reduce the model's usefulness. Nevertheless, the method provides a useful approximation of photosynthesis if a complete study is impractical, and has stimulated further theoretical studies of light and photosynthesis (Vollenweider, 1965; Rodhe, 1965; Margalef, 1965; Tilzer et al., 1975).

Although the rivers and estuaries discussed in the next two chapters are very different from lakes, a knowledge of their basic structure (Chap. 2) combined with a knowledge of the food-chain dynamics expressed in this chapter greatly simplifies understanding of their organization. We feel strongly that even limnologists who will work exclusively with lakes can benefit greatly from a better knowledge of flowing waters and estuaries.

FURTHER READINGS

Butler, E. I., E. D. S. Corner, and S. M. Marshall. 1969. "On the Nutrition and Metabolism of Zooplankton. VI. Feeding Efficiency of *Calanus* in Terms of Nitrogen and Posphorus." *J. Mar. Biol. Assoc. U. K.,* **49:**977–1001.

Goldman, C. R. 1969. "Photosynthetic Efficiency and Diversity of a Natural Phytoplankton Population in Castle Lake, California." Proceedings IBP/PP Technical Meeting, *Prediction and Measurement of Photosynthetic Productivity,* pp. 507–517.

Ivlev, V. S. 1963. "On the Utilization of Food by Plankton-Eating Fishes." *Fish. Res. Board Can.* (Translation Series No. 447. 17 pp.)

Jassby, A. D., and C. R. Goldman. 1974. "Loss Rates from a Lake Phytoplankton Community." *Limnol. Oceanogr.,* **19:**618–627.

Mann, K. H., R. H. Britton, A. Kowalczewski, T. J. Lack, C. P. Mathews, and I. McDonald. 1972. "Productivity and Energy Flow at All Trophic Levels in the River Thames, England," pp. 579–596. In Z. Kajak and A. Hillbricht-Ilkowska (eds.), *Productivity Problems of Freshwater,*

PWN (Polish Scientific Publishers), Warsaw-Kraków.

Mullin, M. M., and P. M. Evans. 1974. "The Use of a Deep Tank in Plankton Ecology. II. Efficiency of a Planktonic Food Chain." *Limnol. Oceanogr.,* **19:**902–911.

Steele, J. H. 1974. *The Structure of Marine Ecosystems.* Blackwell Scientific Publications, Oxford. 128 pp.

Strickland, J. D. H., O. Holm-Hansen, R. W. Eppley, and R. J. Linn. 1969. "The Use of a Deep Tank in Plankton Ecology. I. Studies of the Growth and Composition of Phytoplankton Crops at Low Nutrient Levels." *Limnol. Oceanogr.,* **14:**23–34.

Teal, J. M. 1957. "Community Metabolism in a Temperate Cold Spring." *Ecol. Monogr.,* **27:**283–302.

Winberg, G. G. 1970. "Energy Flow in the Aquatic Ecological System." *Pol. Arch. Hydrobiol.,* **17:**11–19.

Chapter 16

Streams and Rivers

OVERVIEW

The flowing waters of streams and rivers are more closely linked to their watersheds than are the relatively stationary waters of lakes. The productivity of streams is often dependent on terrestrial leaves, grasses, and other debris. This *allochthonous* material may contribute most of the food and energy to small, shaded forest streams. Where sunlight can penetrate, *autochthonous* production from attached algae, higher aquatic plants, and mosses in the stream forms the base of the food chain. Unlike lakes, benthic invertebrates, especially insect larvae, constitute the bulk of the invertebrate fauna. True plankton are almost absent in streams and are common only in deep slow-moving stretches of rivers. All biota in streams are influenced by the unidirectional current. Flowing waters are referred to as the *lotic* environment as contrasted to the standing or *lentic* habitat of ponds and lakes. The lotic habitat is subdivided into two zones: the cool, shallow, and often stony-bottomed stream and the warmer, deeper, silty-bedded river. Streams consist of shallow gravel *riffles* separated by deeper *pools* which contain some organic debris. Rivers are larger and deeper and usually lack riffles and pools. Most studies have been made in the shallow streams; ecological investigations of large rivers are scarce.

The type of water motion is the single most important environmental feature since it controls the physical structure of the streambed. This, in turn, influences the amount of benthic biomass, since fist-size stones provide the most favorable habitat for stream algae and invertebrates. The greater the *discharge* (volume/time), the greater the *current* (distance/time) and also the amount of suspended material that is transported. *Spates,* or *floods,* are events of major importance in the lotic environment. Normal spring floods can uproot and kill large numbers of organisms, while summer spates can completely denude streams of benthic biota. The annual change in stream temperatures is 10 to 20°C, similar to that of lakes, but diel changes are greater than in lakes. Although large rivers do not change in temperature very much on a daily basis, a small unshaded stream may heat up 10°C in a few hours on a hot summer's day and cool by the same amount at night. Temperature changes caused by clearance of streamside trees and bushes alter the benthic species present. This, in addition to erosion control, is a good reason for leaving buffer strips along the edges of streams when harvesting timber or clearing for permanent agricultural use.

The chemical composition of streams is affected by their irregular discharges. Most streams and rivers have a maximum discharge during winter rains or from spring snowmelt. Particulates are scoured from the riverbed only during highest discharges. These carry almost all the annual flux of nutrients like phosphorus and iron which are closely associated with particulate matter. High discharges also transport much of the annual load of soluble nutrients such as nitrate and silica. Streams fed by snow or glacial melt show large diel variations in discharge. Streams fed by springs have more constant nutrient concentration and discharge.

The rock type predominating in the drainage basin affects the vegetation and hence the quality of allochthonous matter from the watershed. Soft- or acid-water streams occur on very hard or acidic rocks such as metamorphosed volcanics or sandstones. The terrestrial vegetation is composed of *Sphagnum,* other mosses and coniferous forests which have relatively low productivity and produce limited allochthonous material. This type of watershed also releases few nutrients. Hard-water streams, often called *chalk streams* in Britain, are formed on more easily broken-down rocks and are usually surrounded by highly productive deciduous trees or grasslands. Such watersheds release both allochthonous material and ample nutrients for high autochthonous production. Fungi and bacteria process dead allochthonous and autochthonous organic material, collectively called *detritus,* which is then eaten by other stream dwellers. This microbial decomposition requires ample supplies of inorganic nutrients, particularly nitrogen, to break down detrital organic carbon. The hard-water stream provides both more nutritious leaf detritus and more inorganic nutrients for faster decomposition than the soft-water stream. Only the coating of bacteria and fungi on most detritus ingested by stream dwellers is digested. This coating is much more nutritious than the debris itself, which may cycle several times through animal guts before its disappearance. *Drift* is composed of detritus, algae, and small invertebrates which are washed along by the current. Many benthic invertebrates feed by filtering drift or by grazing on algae and detritus. Attached algae and macrophytes occur whenever light, nutrients, and substrata are adequate. Predatory invertebrates, fish, and other vertebrate predators complete the lotic food chain.

INTRODUCTION

Streams can never be considered by themselves since the role of adjacent land is always crucial. The stream, river, and watershed can greatly modify the chemistry of the water before it reaches its terminus in a lake or ocean. The distinction between lakes and streams is encompassed by their formal names *lentic* (calm, e.g., lakes and ponds) and *lotic* (flowing, e.g., rivers). As we saw in Chap. 5, lake waters are by no means still but the slowest currents in rivers are about the maximum found in lakes. The faster currents and frequent dependence on allochthonous (i.e., outside the system) primary production is a major factor distinguishing rivers from lakes.

The main part of the river through the lowland plains is subject to more human pollution than the upstream waters, and one example is dealt with in detail in Chap. 20. Most towns, industries, and farmers have historically regarded rivers as dumping grounds for waste; only recently have healthy rivers become common in industrial areas.

The distinction between streams and rivers is similar to that between ponds and lakes, which are ill-defined words in common usage. Nevertheless, it is useful to separate streams and rivers so long as the definitions are used cautiously. In common English, streams are small and often have steep gradients, while rivers are large and usually have low gradients at lower elevations. Despite appearances the average current in rivers is more rapid than in small tributary streams.

MEASUREMENT

The physical and chemical limnology of streams or rivers is usually measured by the same methods as those used for lakes. Discharge in small streams may be measured by direct collection at a culvert or estimation of flow over a V-notch weir. *Discharge* is water flow expressed as volume per unit time (e.g., m^3 s^{-1}) and *current* is distance traveled per unit time (e.g., cm s^{-1}). For larger rivers, discharge is estimated from current measurements. A current meter with revolving cups is appropriate, but its use in streams only a few centimeters deep requires a miniature meter to ensure that the cups are completely submerged. The speed of the current is conventionally measured at five equally spaced sites across the stream, with the meter held two-fifths of the way up from the bottom. Accurate calculation of the whole stream discharge is then made using special tables. In very shallow water or between rocks the current can be determined with a *pitot tube,* a piece of capillary whose end is held into

the flow and the current speed then determined by differential pressure. The height of fluid in the capillary is proportional to the current.

The entire microscopic attached community of streams is termed *aufwuchs*. This community of bacteria, fungi, algae, and small animals can grow luxuriantly almost year-round. The aufwuchs community is collected by scraping rocks, aquatic plant surfaces, or submerged logs. Artificial substrates of almost any size or material may be placed in streams to study aufwuchs accumulation. Large stream plants, generally flowering angiosperms, are collected by cutting or uprooting. The extent of streambed plant cover can be estimated with transects or photography, as in terrestrial ecological studies.

Lotic benthos in water over 1 m deep can be obtained either by diving or with mechanical samplers. However, virtually all studies have been carried out in water less than $\frac{1}{2}$m deep, and most techniques are designed for shallow waters. The most widely used device is the *Surber sampler*. This samples a 1-ft square (about 930 cm^2). Benthic organisms dislodged by disturbances in the square are trapped in an extended downstream net (Fig. 16-1). Small rocks are turned over by hand in the water to release benthos, while larger rocks and those with cracks and holes are removed for the retrieval of invertebrates on shore. There are several drawbacks to the Surber sampler. Small organisms, including most algae, are lost through the large mesh, and deep benthos in the gravel are not collected. The Surber sampler is also restricted to shallow waters which do not flow over the top of the net. This drawback may be overcome by using a taller system, for example a Hess sampler, which can be employed in water up to 1 m in depth.

The collection of fish and other free-swimming organisms is discussed in Chaps. 13 and 14. A classification of streams based on 13 easily measured variables, such as width, substrate, and maximum and minimum temperatures, has been proposed by Pennak (1971). This system

Figure 16-1 (a) The Surber sampler as used for collecting benthic invertebrates in a stream. Water flow is from right to left. (b) The entire sampler on land showing the rigid brass square which is pressed into the substrate when under water.

is very useful for comparing streams in different regions.

THE STREAM

Streams are zones where a rapid flow of shallow water produces a shearing stress on the streambed, resulting in a rocky or gravel substratum covered by fully oxygenated water. Where the geological strata provides sediment, lowland streams have muddy beds. For purposes of definition, the stream waters usually have an average monthly mean temperature much less than 20°C, except in some tropical areas. Many streams are shaded by trees which reduce solar heating. Deeper *pools* of relatively slow-moving water are separated by *riffles*, which are areas of shallow turbulent water passing

through or over stones or gravel of a fairly uniform size (Fig. 16-2). Intermediate areas of moderate current often found in larger streams and rivers are termed *runs*. Riffles make excellent sites for stream surveys. They are actually the "larder" of the lotic environment, containing the majority of the stream's benthic invertebrates (Fig. 16-3). In deeper rivers where riffles and runs are present, a few studies have been made by diving. The biological and physical structure at these depths is apparently similar to that in small streams. Stones in sunlit riffles are often covered with the algae and mosses constituting the main in situ primary producers of the stream. In contrast, the pools, which usually cover several times the area of the riffles, have a distinct, usually less dense biota living among a mixture of stones and fine-grained sediments. A characteristic feature of

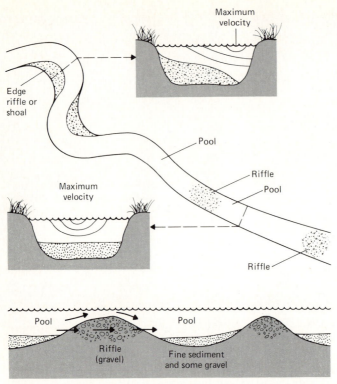

Figure 16-2 Diagrammatic representation of water flow and the major structural components of a stream. Note the alternating pools and riffles. A major feature of the riffle is that water flows through the gravel as well as over it. This enables fish eggs and small benthic invertebrates to obtain the oxygen they need while being protected from predation by larger organisms such as fish.

pools is the accumulation of decaying terrestrial debris. Pools and riffles support higher plants called *aquatic macrophytes,* but rarely will the whole streambed be covered (Gessner, 1955, 1959).

THE RIVER

River waters are deep and generally sufficiently turbid to prevent light penetration to the riverbed. This, together with rapid current in some transparent rivers such as the Brooks River in Alaska, restricts the growth of aquatic macrophytes. These become important only at the edge where they provide fish breeding sites and refuges for smaller organisms. Au-

tochthonous primary production is carried out by true phytoplankton, typically centric diatoms, and by blue-green algae in warm climates. In contrast to streams, water in rivers may reach mean monthly temperatures in excess of 20°C. Shading by vegetation is usually small due to the larger size of rivers. Deep rivers generally show only a small bottom slope. Shear stress on the bed is low because shear decreases with increase in discharge. This condition produces muddy, debris-laden sediments, and riffles are usually absent except below waterfalls. As mentioned previously the average current, as well as discharge, becomes greater as the overall size increases so that river waters move faster than streams, despite the difference

Figure 16-3 The alternation of pools and riffles in the Truckee River, California. Riffles usually occupy a smaller area than pools but are more productive.

in slope. The proportionately smaller fraction of water in frictional contact with the bed in rivers and the generally smoother bed partially explain the differences in flow.

The water in rivers may occasionally be less than fully oxygenated when natural or artificial organic loads are high. Leaf fall in autumn, waste discharges, or ice cover may cause this depletion. In the Salt Fork River of central Illinois, sewage discharge during a low flow period one winter promoted a bloom of phytoplankton under the ice. A snowfall then reduced photosynthesis to the point where respiration of the algae decreased oxygen levels and a major fish kill occurred.

THE LOTIC ENVIRONMENT

Discharge

Annual patterns of stream flow determine many of the physical and biological properties of lotic systems. It is dependent on rainfall, catchment geology, area, bed slope, and dam control by vegetation, beavers, and, most importantly, human beings. For detailed discussions of river flow, physics of stream water-channel interactions, and turbulence, readers are referred to major texts such as those by Leopold et al. (1964), Morisawa (1968), and Hynes (1972). Just as in lakes, water movements in streams are spatially nonuniform. Currents vary from highest just below the surface near the stream center to lowest at the streambed and along the bank (Fig. 16-2). This distribution is the basis for the current-measuring methods discussed earlier. There are areas of high and low flow around submerged objects. Bends produce both turbulent and quiet areas.

Seasonal and daily variations in discharge are important in river and stream ecology. For example, high mountain streams show their greatest discharge during midafternoon snow melts in late spring and lowest discharges after freezing at night (Fig. 16-4). Most important from

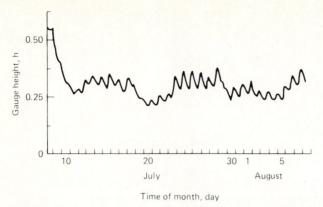

Figure 16-4   Daily changes in stream discharge in a stream fed by snowmelt. The relationship between gauge height h and discharge Q is given by $Q = h^b$, which indicates an exponential increase in Q with increase in h. Thus the changes in discharge in the above example would be much more extreme than they appear from the normally measured stream depth.

both the biological and physical points of view are the very high discharges known as floods, or *spates*. When in flood, scouring and bank-eroding effects are much larger than during normal discharge. Most biota avoid spates either by migrating to calm backwaters or by having life cycles which are terrestrial or aerial at these times. When floods occur at unusual times, the fauna may be severely depleted and require several years to recover. Some larger organisms, particularly the salmon, often choose to ascend rivers during periods of high discharge, presumably because there are few shallow water barriers. Spates in streams carry large amounts of suspended material (Fig. 16-5) which is temporarily deposited downstream in lakes, rivers, or flood plains. Severe floods are detrimental to smaller biota if they leave only inhospitable large rocks, but may benefit fish migration by removing major obstacles. Smaller floods can remove excessive silt and produce better environments for fish eggs, benthic invertebrates, and algal production. In rivers floods are generally less destructive, stirring the already silty bottom and spreading it over the flood plain. This cannot occur if the river has been canalized, levied, or dammed. In areas such as the lower Nile, floods were a positive

benefit, adding fertile mud to the fields. This process has now been greatly reduced by the Aswan High Dam. Since phosphate and most metals are generally transported as particles, flood transport is important to the phosphorus budgets of unpolluted lakes (Fig. 16-6). Nitrogen, which is largely transported in dissolved forms, is more affected by the volume of water discharged than by current (Table 16-1, Chap. 8).

Temperature

The temperature of most streams is lowest in the uplands and becomes gradually warmer in the lower reaches. Even arctic rivers are warm in the lowland; for example, the Brooks River, Alaska, approaches 15°C in summer. High mean temperatures in many rivers restrict the spread of some organisms. Streams may show daily temperature variations of up to 10°C, which will normally be greatest in level, rocky, unshaded zones (Fig. 16-7). These large diel changes are normally found above the tree line or when the stream shrinks to the center of its channel. Clearance of streamside vegetation increases water temperature and, together with the loss of tree root habitat, can cause dramatic

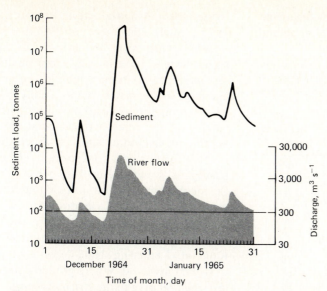

Figure 16-5 Relationship between river discharge and sediment load for the Eel River, northern California. Note that most sediment is carried in a few days of peak river flows. This river, like the Yellow River in China, carries much more sediment than most rivers, partially due to natural erosion at the headwaters. The general effect is similar in most rivers. (*Redrawn from Waananen et al., 1970; and Beaumont, 1975.*)

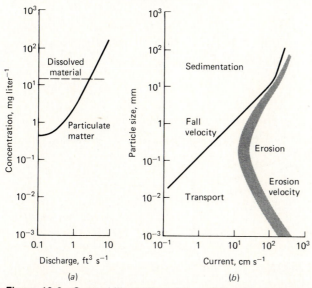

Figure 16-6 General illustration of the relationship between discharge, current, and the concentrations and transport of soluble and particulate materials. (*a*) A small stream in a heavily wooded part of northeastern United States in which most particulate matter is carried only at the highest discharges. This is shown in detail for another stream in Fig. 16-9. (*Modified from Likens et al., 1977.*) (*b*) A general figure for all flowing waters, showing erosion and deposition of a uniform material such as a sandy bank. Note that the size of the particle, as well as water current, controls how easily it will be eroded. (*Modified from Morisawa, 1968.*)

Table 16-1 Gross Annual Outflow of Dissolved and Particulate Matter in a Small Woodland Stream*

Figures are given as a percent for the total of that element. Note that most phosphorus and all iron moves in the particulate form, while nitrogen and silicon move mostly as dissolved forms.

Element	Particulate fraction, %	Dissolved fraction, %
P	63	37
N	3	97
Si	26	74
Fe	100	0
S	0.2	99.8
C	32	68
Na	3	97
K	22	78
Ca	2	98
Mg	6	94
Cl	0	100
Al	41	59

*Modified from Likens et al. (1977).

reductions in fish production (Ringler and Hall, 1975). Such clearances of the riverside, or *riparian,* vegetation are now illegal in many areas. Rivers, by contrast, show smaller daily temperature variations since their large volume and large heat capacity act as a buffer. Average stream temperature is not necessarily the same as that of the microhabitats occupied by the benthic invertebrates and fish which may hide in the gravel or near cold springs. The effect of shading is shown in Fig. 16-8, which illustrates the increase in temperature resulting from a few meters of unshaded meadow, followed by a decrease in temperature when the stream reenters a shady zone. The amount of cooling found after entering shade depends on the type of stream involved. Where groundwater inflows are small relative to the stream volume, back radiation and conduction to the streambed produce the cooling (Burton and Likens, 1973). Where groundwater or side stream inflows are

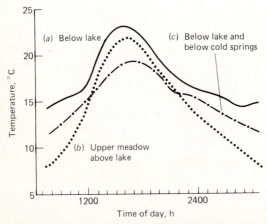

Figure 16-7 Daily variation in average summer temperatures for three reaches of Martis Creek, California, a small, partially shaded mountain stream. Note the typical large diel variation of more than 10°C in all reaches. A lake in the same area would show only 1 or 2° diel change in temperature. Surface discharges from a lake to the stream heats up the section (*a*) below the lake relative to an unshaded meadow section (*b*) above the lake. Cold springs reduce temperature (*c*) to below that initially present prior to lake heating. (*Redrawn from McLaren et al., 1979.*)

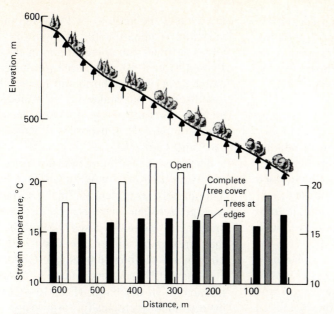

Figure 16-8 Effects of shade on the temperature of a small stream in midsummer at midday on a sunny day. Black bars represent stream temperatures under dense tree cover, open bars show temperature in unshaded reaches, and hatched bars show effect with trees in a strip only 10 m wide on each side of the stream. Arrows show where temperatures were taken. Note the rapid increase and decrease of up to 5° C in less than 50 m. That the decrease is due to the immediate shading effect is shown by the similar temperature reduction given by full forest and the small buffer strip only 10 m wide. On cloudy days the fluctuation in temperature was reduced to 1.5°C. The effect is mostly due to the heat stored in streambed rocks. The trees by this stream in the Hubbard Brook ecosystem were cut as part of an experiment. Normally such a regular pattern of shade and open water would not occur. (*From Burton and Likens, 1973.*)

large, they exert the major cooling effect (Brown, 1970).

The amount of radiant heat retained by a stream can be expressed by the equation

$$\Delta T = \frac{HA}{Q}$$

where ΔT = temperature rise, °F
H = heat flux, btu ft^{-2} min^{-1}
A = area, ft^2
Q = discharge, ft^3 s^{-1}

This equation applies to streams with gravelly beds which are highly permeable to water. The heat of both water and gravel is distributed by water mining. In streams with rock beds, some solar radiation is lost by conduction to the earth and less heating of the water occurs. Similarly, rocky streams lose more heat than gravelly ones which increases the likelihood that *anchor ice* will form in winter in rocky streams (see Chap. 19).

Cool water released from the bottom of reservoirs can be deleterious to the growth of fish or rice crops. On the other hand, if well-oxygenated, these releases of cool water are beneficial for salmon spawning or trout fisheries in warm climates. Shady streams are poor environments for primary production, although temperature does not limit algal photosynthesis directly. Once cleared of shade, primary production may increase, but not always usefully, since unpalatable benthic algae may coat the streambed. It is wise stream management practice to leave a buffer strip of original vegetation about 30 m wide. This vegetation reduces erosion and prevents temperature increases while preserving the original benthos. (Erman et al., 1977).

Nutrients

The chemistry of rivers and streams in relation to the distribution of biota is often overshadowed by physical considerations. As mentioned previously some nutrients such as nitrate or silicate in rivers are largely present in soluble form while others like phosphate are associated with the particulate phase (Table 16-1). For example, in the Hubbard Brook ecosystem, a wooded area in New Hampshire, 97 percent of the nitrogen and 74 percent of the silicon was carried in soluble form as nitrate or silicate. In contrast only 37 percent of phosphorus was transported as soluble phosphate. These proportions vary with season, climate, discharge,

floods, and the geology of the watershed. This variation is apparent for phosphorus in Ward Creek, a small subalpine stream in the western United States. Here, soluble phosphate is the major form of phosphorus during conditions of low discharge but makes up only a few percent at high discharge (Fig. 16-9a). The effect of floods on particle transport is indicated in Figs. 16-5 and 16-6.

The seasonal variation in major nutrients during a year is shown for Ward Creek in Fig. 16-9. This pattern is typical for many streams with a large seasonal increase in discharge from snowmelt. Extreme cases where the stream may freeze in winter are dealt with in the chapter on regional limnology (Chap. 19). In streams fed by snowmelt there is a large daily fluctuation in discharge (Fig. 16-4). Although little-studied ecologically, large tropical rivers such as the Amazon, Nile, Niger, and Congo also undergo enormous seasonal fluctuations in discharge and nutrient transport.

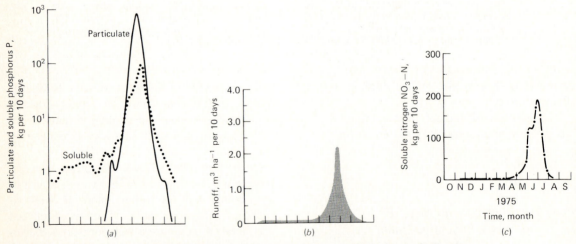

Figure 16-9   Flux of nitrogen and phosphorus relative to water flow in Ward Creek, California, a subalpine stream fed mostly by snowmelt. Most phosphorus (a) moves in the particulate form and nitrogen as soluble nitrate, as is the case for most streams (Table 16-1). The spring-summer melt (b) in June increases the flux of particulate phosphorus (note log scale) much more than nitrate. In this example in 1975 particulate phosphorus carried during the high melt discharges increases 10,000 times, but nitrate and soluble phosphate increases by only 100 to 200 times. This difference is due to high discharges which carry particulate material eroded from banks or resuspended in pools while soluble nitrate is carried in more constant concentrations during all discharges (see Fig. 16-6). (*Redrawn from Leonard et al., 1979.*)

Reduced variation in discharge may occur in lowland streams. For example, Wraxall Brook, a chalk stream in southern England, shows fairly constant nitrate levels and discharge throughout the year (Fig. 16-10). Nitrate levels are probably more stable under natural conditions since peaks may be due to the influx of fertilizer from adjacent fields. One reason for the constant level of nutrients in Wraxall Brook is its spring source. The discharge of springs to chalk streams often increases very soon after a rain. This results from increased head pressure, but the water actually discharged may have accumulated years earlier. Water in the underground reservoir has a constant nutrient content compared to surface water (Table 16-2). The beneficial effect of springs on stream biota in alpine and polar regions is discussed in Chap. 19. The buffering effect of underground reservoirs provides the constant temperature and chemical composition of springs which is of worldwide economic importance to the brewing, whiskey, and bottled water industries.

Year-to-year variations in nutrients in most streams (Fig. 16-11) are more pronounced than in lakes since in lakes the large volume dilutes the effect of inflows. Even in undisturbed systems the causes of year-to-year changes in nutrients in streams and rivers is not well-under-

Table 16-2 Annual Means and Ranges in the Levels of Nitrate in Hollybush Spring, a Chalk Spring*

| Year | Nitrate concentration, mg liter^{-1} | | No. of samples |
	Mean	Range	
1968	5.36	5.11–5.96	8
1969	6.00	5.10–6.96	52
1970	5.54	4.96–6.10	52
1971	5.47	4.72–7.48	52
1972	5.14	4.61–5.56	51
1973	5.27	4.56–5.74	51

*From Casey (1977).

stood. A combination of many factors in the watershed controls nutrient flux. In the case of nitrate, the sources and sinks of nitrogen in the watershed are dominated by levels of nitrate and ammonia in rainfall, biological nitrogen fixation and denitrification, as well as freezing and thawing of the soils, natural fires and erosion. and the amount of nitrogen recycled by vegetation or held in the humus layer. Some of these effects are shown for the Hubbard Brook ecosystem in Fig. 16-11. Here a gradual increase in nitrate in rainfall has apparently been responsible for an increase in nitrate levels in the brook. There is, however, no direct correla-

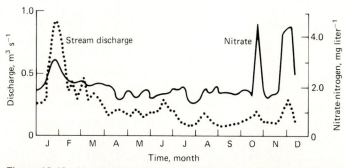

Figure 16-10 Levels of nitrate and water discharge for Wraxall Brook, a small chalk stream partially fed by springs. Note the relatively small seasonal changes in nitrate and, to a lesser extent, in discharge. The levels of nitrate in this stream are higher than in most lakes where 0.5 to 1.0 mg liter^{-1} NO$_3$-N is the usual maximum (see Tables 8-2, 19-1, 19-3 and Figs. 8-7 and 12-10). (*Redrawn from Casey, 1977.*)

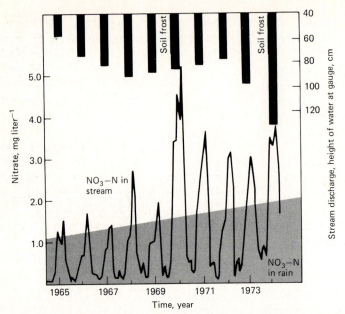

Figure 16-11   Year-to-year variation in seasonal patterns of nitrate concentration in Hubbard Brook, a small creek in a heavily wooded watershed in the northeast United States, as related to nitrate in rainfall and in the stream. The nitrate increases more or less proportionately with discharge (black bars at the top). The increase in nitrate in rainfall (shaded area at bottom) over the 20 years apparently shows up as an overall trend but does not account for the high values in 1969–1970 or 1973–1974. These unusually high levels of nitrate from the watershed might be due to nitrate released by severe frosts which increased the freeze-thaw cycle in the surface soils. (*Redrawn and modified from Likens et al., 1977.*)

tion between stream nitrate levels and the measured inflows of nitrate during any single year. The two largest peaks of nitrate are related to years when unusually severe frosts occurred when the soil was not insulated with a blanket of snow. This caused freezing and thawing of the upper soil which is believed to have increased release of nitrate to the stream (Fig. 16-11).

In almost all cases the soil horizon containing humus is an important reservoir of nitrogen for the watershed and can temporarily reduce and even reverse the flow of nitrogen. The lack of any correlation between nitrate in 12 rivers in England and nitrate added as fertilizer is shown in Table 16-3. At best only half and usually only one-third of nitrogen added to

agricultural land finds its way into crops. The rest is lost to groundwater, rivers, denitrified to nitrogen gas, or held in the humus.

Some nitrogen is taken up by plant growth in streams and rivers, and this buffering effect could interfere with correlations such as those shown in Table 16-3. However, in many streams and rivers nitrate is often so abundant (Table 16-4) that biological uptake removes only a small fraction (Casey, 1977). This is not the case in some alpine and arid-zone streams where nitrogen is in short supply and biological uptake may greatly reduce its concentration. Phosphorus probably shows similar large modification by the watershed, but due to the high affinity of phosphate for soils (Chap. 9), little is ever released to streams unless soil erosion

Table 16-3 Effect of the Watershed on Flows of Nitrogen into Rivers in England*

Most values are for the period 1953–1967. The nitrate purchased for fertilization has been correlated with nitrate levels in the rivers concerned. Note the negative, positive, and lack of correlations in the various watersheds. Although much of the nitrate in these rivers is almost certainly derived from fertilizers, there is not often a direct correlation even over quite long time periods (see also Fig. 16-11).

River	Significant ($P < 0.1$) correlation coefficients
Dee	−0.89
Rother	+0.90
Severn	+0.75
Stour	+0.71
Devon	+0.52
Wensum	+0.52
Twelve other rivers including the Thames	No significant correlation

*Modified from Tomlinson (1970).

Table 16-4 Nutrient Concentrations in Rivers and Streams

The highest nitrate, ammonia, and soluble phosphate values are found in rivers polluted with agricultural drainage and sewage. The unpolluted maximum concentrations are probably nearer 5.0 mg liter^{-1} for NO_3-N and 0.01 mg liter^{-1} for PO_4-P. Minimum values for nitrate and maximum values for phosphate and sulfate are usually found in arid areas such as east Africa or the southwest United States. Maximum nitrate and minimum phosphate concentrations are often found in temperate areas such as northern Europe or the northeastern United States. Values for rivers are world averages and examples are from most major watersheds (see tables in Chaps. 8–10). Values for streams are from a less extensive data base and are only illustrative.

Nutrient	Typical value, mg liter^{-1}	Range
Rivers		
Nitrate (NO_3)	1.0	0.003– 7.0
Ammonia (NH_4)	0.05	0.005– 10.0
Silica (SiO_2)	9	1.4 − 35
Soluble phosphate (PO_4)	0.1	0.001− 1.0
Sulfate (SO_4)	20	0.4 −290
Streams		
Nitrate (NO_3)	0.5	0.003− 5
Ammonia (NH_4)	0.04	0.005− 1.0
Silica (SiO_2)	6	0.5 − 20
Soluble phosphate (PO_4)	0.002	0.001− 0.013
Sulfate (SO_4)	8	0.3 − 20

occurs. Soluble phosphate concentrations in unpolluted rivers are usually less than 0.01 mg liter^{-1} PO_4-P and often only 0.001 mg liter^{-1}.

The biomass of animals and plants will usually increase downstream from large influxes of nutrients from agricultural or domestic waste discharges. This increase in stream fertility can rarely be ascribed to the effect of any single nutrient. The lower trophic levels, primary producers and saprophytes, are less affected by low nutrient concentrations in rivers and streams than in lakes. A major reason for movement in lake algae is to "stretch" the diffusion shell described in Chap. 12 and allow more rapid access to scarce nutrients. Adapting this idea to stream plants and animals, it can be seen that running water gives a permanently stretched diffusion shell so that even nutrients in low concentrations are available for plants. For example, the most rapidly moving lake phytoplankton cover less than 5 m h^{-1} and most

travel much more slowly. A similar cell as part of a filament attached to a rock in a stream with water current of about 50 cm s^{-1} would experience more than 300 times the water contact of the most rapidly moving lake plankton.

Rivers and streams are more akin to lake edges with regard to sediment-water interface ratios and thus derive more nutrients from the sediment. Although some nutrients in streams are often present in very small concentrations (Table 16-4), deficiencies are not as obvious as

those in lakes. Rivers are often rich in accumulated plant nutrients (Table 16-4), due to
the decreased nutrient uptake as water progresses downstream. The turbid water and
smaller bed to volume ratio are unfavorable for
high plant growth rate.

Over the centuries riverbanks have hosted
considerable human development and municipal, industrial, and agricultural wastes have
increased nutrient loading (Oglesby et al.,
1972). Grossly polluted rivers such as the
Rhine in Europe and Cuyahoga River in North
America remain disgraceful evidence of this
phenomenon. Here, nutrients are superabundant but plant and animal growth is restricted
by light limitation from turbid water or toxicants such as phenols, chromium, cyanide, or
foaming agents (Fig. 16-12).

We should now consider what creates soft-
and hard-water streams and how this affects
allochthonous and autochthonous additions.
Table 16-5 lists the factors involved and demonstrates the indirect but overwhelming influ-

ence of the character of the watershed on lotic
systems. This situation contrasts strongly to
lakes, which maintain a large buffering effect by
virtue of their slowly exchanging volume. Invertebrates and fish are generally more abundant in hard waters than they are in soft waters.
This is because hard waters are usually rich in
other nutrients as well as the calcium, magnesium, carbonate, sulfate, and other ions which
give them the designation "hard." It is reasonable to assume that hard and soft correspond
to nutrient-rich and nutrient-poor. We already
showed that nutrient levels in streams are less
important to primary producers than in lakes.
Although basic fertility is important, the greater
quantity of allochthonous plant debris found in
hard-water streams is largely responsible for
their high productivity.

Most benthic invertebrates gain little nutritive value from leaves until they have un-
dergone considerable microbial modification.
The changes are similar to those which occur in
composting or silage formation and require con-

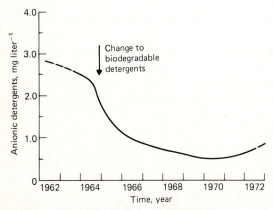

Figure 16-12 Decrease in foam-producing toxic detergents in a lowland river, River Don,
England, due to a change from domestic hard (i.e., nonbiodegradable) detergents to biodegrad-
able detergents. Industrial use continues. This change reduced a problem at weirs or waterfalls
where the detergent stabilized foam several feet thick which occasionally blew onto adjacent
roads and stopped traffic! One of us (Horne) remembers kayaking on this river through foam so
thick that only the head and shoulders were visible. Although foaming is much reduced, the
phosphate released in biodegradation can cause eutrophication downstream if nitrogen is also
present in abundance (see Chap. 20). In the 1950s toxic detergents and other waste eliminated
virtually all higher forms of life as well as most of the dissolved oxygen from many rivers in industrialized countries. (see Figs 20-13, 20-14). (*Redrawn from Toms, 1975.*)

Table 16-5 The Factors Involved in the Production of Hard-Water Streams, Which Are Often Alkaline and Nutrient-Enriched, as Compared with Soft-Water Streams, Which Are Usually Acid and Nutrient-Depleted

Factor	Soft/Acid	Hard/Alkaline
Watershed	Sandstone, sedimentary, or meta-morphosed rock with slow weathering rates. Little chemical breakdown to buffer the rainfall which is acidic. Stream waters produced are poor in calcium and are acidic.	Alkaline sedimentary rock (e.g., limestone or chalk) which is rapidly weathered by rain to yield stream water rich in calcium and slightly alkaline.
Vegetation of watershed	Acid-tolerant plants (e.g., *Sphagnum*) where waterlogged and coniferous forest soil result in slow decomposition of litter by microbes. In the tropics, e.g., the Amazon basin, ion exchange is tightly controlled by efficient recycling so that nutrient washout is slow.	Deciduous woodlands typically have higher input of nutrients from rock weathering and a greater loss of organic|matter to streams. Microbial decomposition in the leaf litter is rapid.

siderable supplies of inorganic nutrients, particularly nitrogen (Kaushik and Hynes, 1971). Dead leaves and dry grasses lose much of their protein-nitrogen before they fall. In addition decomposition in streams usually occurs slowly.

There are some benthic invertebrates which are particularly dependent on a single element. Mollusks and some crustaceans, for example, require calcium for their shells. Despite this need, low calcium levels in soft-water streams do not appear to limit these organisms since their growth and thus their annual demand for calcium is low (Young, 1975). In soft or acidic waters such as high mountain lakes in granitic basins, reduced shell thickness may result from low calcium concentrations.

The Food Chain

Phytoplankton and zooplankton are minor components of undisturbed lotic ecosystems. This is not the case where the currents are much reduced by dams or locks such as on the Missis-sippi (United States), Thames (England), or Dnieper (Russia). Here, true phytoplankton occur because the residence time of the water is longer than the time required for cell division (Fig. 15-4). In streams in the semiarid areas of northwest America or the treeless uplands of northern Europe and Asia, autochthonous primary production plays a major role as it does in lakes. The main primary producers are *attached algae*. Pennate diatoms like *Nitzschia* and *Gomphonema* are more common than centric forms, but filamentous green algae like *Ulothrix* and *Cladophora* are more common than single-cell forms. Blue-green algae like the filamentous *Phormidium* and the gelatinous blobs of brown *Nostoc* are very common in some streams. For convenience, small, usually unicellular attached algae, normally diatoms, are called *microphytes*. The larger green and blue-green algae with some chain-forming diatoms are easily visible with the naked eye and are called *mesophytes*. This distinguishes them from higher aquatic plants, or *macrophytes* (see also Chap. 11). A typical distribution of micro-, meso-, and macrophytes is shown for a subalpine river in Fig. 16-17. In small streams deciduous riparian vegetation may shade out plant growth except in early spring (Fig. 16-14). Thus abundant growth of plants is usually confined to large, shallow rivers or streams above the tree line.

Macrophytes are often uncommon in turbid rivers due to lack of light or in rapid-flowing ones due to lack of suitable substrate for rooting. Many rivers contain a good trout population which depends on allochthonous material and microphytes. In contrast, the highly prized trout chalk streams in England contain dense masses of macrophytes such as *Ranunculus*, the water crowfoot (Fig. 16-13). Enormous numbers of periphyton, diatoms, and small animals are found in the shelter of these dense growths of aquatic plants. Much of the food chain may be based on the existence of detritus-producing macrophytes. Attached algae growing on rocks, logs, leaves, and stems of larger aquatic plants may form the main diet of grazing benthic invertebrates. In contrast, higher plants

in streams are seldom directly eaten although exceptions occur since macrophytes are consumed by manatee and grass carp (white amur). Dinoflagellates and ciliates are rare, presumably since their vigorous swimming ability is of no use in turbulent, shallow water.

The major distinction between the energy and nutrient budgets of lakes and woodland streams is the dominating role of the associated forest. Algae and higher aquatic plants are sparse in these shaded, shallow streams. Much of the research in streams has been done in heavily wooded areas, and this tends to overemphasize the role of allochthonous production and underestimate autochthonous primary production (Minshall, 1978). The typical lotic biomass or energy pyramid is truncated and consists mostly of (1) allochthonous detritus and/or in situ primary production, (2) benthic invertebrates, (3) fish and an assortment of sparse vertebrate predators (see Chap. 11). The stream metabolism in shaded, woodland streams is dominated by heterotrophic organisms, principally fungi, bacteria, and small

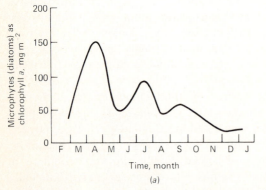

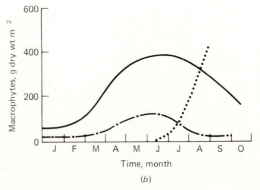

Figure 16-13 Seasonal variation in microphyte diatoms, expressed as chlorophyll *a*, and macrophytes in the chalk stream ecosystem of Brere Stream, England. Note the large and typically irregular seasonal variations in microphyte chlorophyll on the flint rocks (*a*). This contrasts with the regular rise of three macrophytes (*b*). The differences are probably due to lack of direct grazing by benthic invertebrates on macrophytes and continual grazing on the microphyte algae. In part (*b*) the solid line = water buttercup or crowfoot (*Ranunculus penicillatus*), the dotted line = water-cress (*Rorippa nasturtium-aquaticum*), the dashed line = another water buttercup or crowfoot (*R. calcareus*). (*Redrawn and modified from Westlake et al., 1972; and Marker 1976a.*)

invertebrates. The overall community thus uses more oxygen than it produces. This is well-illustrated in New Hope Creek which flows through deciduous woodland in North [illegible] United States. An obvious oxygen debt is [illegible] when the tree cover is complete, but [illegible] when trees are without leaves there [illegible] balance between oxygen production and [illegible] ration (Fig. 16-14). In large rivers and unshaded streams, there is a more even balance between autotrophy and heterotrophy.

The amount of energy supplied by the surrounding land is variable, but nearly all is allochthonous in temperate woodland and tropical forest streams (Fig. 16-14). R[illegible] source, leaves or aquatic plant remains are collectively referred to as *detritus,* and its role is vital in understanding aquatic systems. In temperate regions, fallen leaves provide a great deal of food during the autumn and winter, while [illegible]sses contribute most in spring and [illegible]tritus recycled from one grazer to [illegible] important food source in summer. [illegible] noted, only a small amount of leaf litter is eaten directly. Leaves are usually modified on the surface by microorganisms and shredded by invertebrates. Some stoneflies, amphipods, and crayfish bite or tear whole leaves and ingest the fragments, many of which are then dropped or are partially digested and

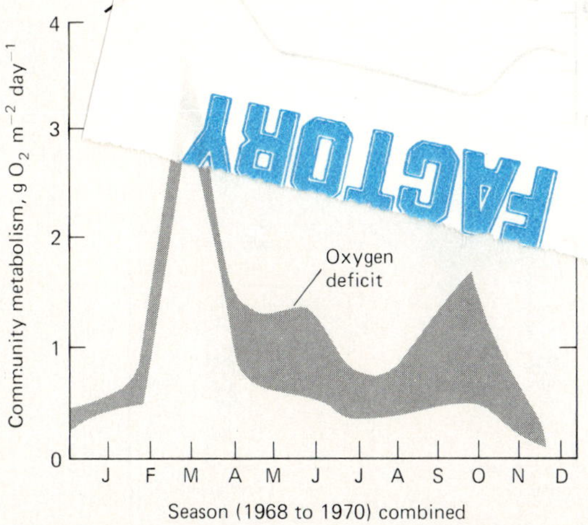

Figure 16-14 Metabolism of New Hope Creek, North Carolina, a woodland stream in the eastern central United States. For most of the year this type of stream is heterotrophic; i.e., community respiration is greater than photosynthetic oxygen production. The black area represents the oxygen deficit or degree of heterotrophy. The upper boundary of the black area represents respiration and the lower boundary photosynthesis. The resulting deficit must be made up by instream aeration. This heterotrophy is due to the lack of light (upper curve) for photosynthesis when the trees leaf and block out light to this small stream which is overshadowed by the riparian vegetation. Rivers and streams with less vegetation in dry or alpine areas do not show as much heterotrophy. (*Redrawn and modified from Hynes, 1975.*)

defecated (Fig. 16-15, and Cummins et al., 1973; Cummins, 1974; Mathews and Kowalcezwski, 1969). There are many microbes which possess special enzymes to aid in plant decomposition. Cellulase, which breaks down the abundant cellulose of the plant cell wall, is probably the most common. Just as in the terrestrial system, fungi play a major role in the initial breakdown of plant material. Some aquatic fungi, for example the Hyphomycetes, have unique triradiate spores which are adapted to adhere to leaves and twigs even in swirling stream waters. These fungi also possess low-temperature enzyme systems which are adapted to utilize their food source at its most abundant period. Only fungi can break down

plant structural components like lignin; the thick fungal coating over leaf fragments is a nutritious part of detritus. The proportions of bacteria or fungi taking part in the modification of detritus is highly variable.

Bacteria appear to play a role similar to fungi in detritus formation in streams. Suspended bacteria are common in streams and constitute a major food of blackfly larvae (Simuliidae). However, the decay of plant material in water is always slow relative to that on land. For example, leaves or logs stored in lakes by beavers for their winter feed or by human beings for commercial timber show little decay because fungi do not grow without good supplies of oxygen. Thus the low solubility of this gas in water

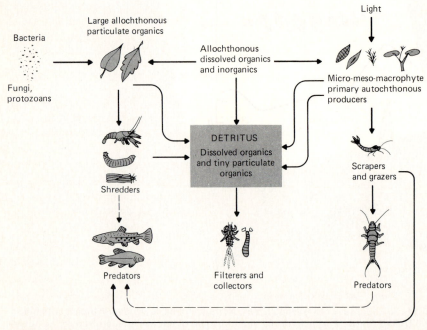

Figure 16-15 Diagrammatic representation of the functional roles of benthic invertebrates in streams. The crucial role of detritus is shown by its position at the center of the diagram. Fish, larger stoneflies, and some other carnivores are almost the only organisms not directly involved with detritus. The detritus pool contains fungi, bacteria, and small protozoans which continually convert the rather inedible cellulose and lignin of detritus into food by using nitrogen and phosphorus dissolved in the water. For further details of the functional roles of the organisms see Table 11-8. (*Modified from Cummins, 1974.*)

(see Chap. 7) slows decomposition. Mention should be made here of *Sphaerotilus,* the so-called *sewage fungus,* which is not a fungus at all but a bacterium which utilizes dissolved organic matter. *Sphaerotilus* often grows as masses of long whitish streamers where sewage is present. Pennate diatoms may cover the surface of the bacterial threads, imparting a rusty-brown color. *Sphaerotilus* may also occur where the leachate from forest soils provides sufficient dissolved organic carbon.

In addition to leaves and other plant detritus, there is also a considerable living allochthonous contribution from the land. Caterpillars, flies, earthworms, crickets, and other insects fall from overhanging vegetation or, during rains, are washed from the soil to be consumed by fish and larger invertebrates. In the tropics the constant supply of fresh, nutritious blossoms and fruits is considered by Fittkau (1964) to be a major allochthonous food source. Once allochthonous or autochthonous material has been partially decomposed and broken into small bits, it forms true detritus which recycles through the guts of many organisms with only a small fraction being utilized at each passage. It is probable that the coat of bacteria and fungi is the major nutrient source on these "dinner plates" of detritus.

Carnivores and Herbivores

Invertebrates dominate the stream benthos. Even casual observation of rocks and logs in a stream reveals caddisfly larvae and snails, and examination of the underside of stones shows a great variety of insect larvae in various stages of development. Apart from insects, only oligochaetes, nematode worms, crustaceans, and mollusks are abundant in the stream benthos. For extensive discussions of the role of benthic invertebrates in running water, readers are referred to Hynes (1972) and Macan (1974). In typical stream riffles, the species vary but the organisms appear rather similar all over the world due to the process of convergent evolution. Among the largest organisms are the insect orders Trichoptera (caddisflies), Plecoptera (stoneflies), the Ephemeroptera (mayflies), Odonata (dragonflies), oligochaete worms, and some snails (Mollusca). On rocks and branches in swifter currents, such as waterfalls, or the raceways of dams, the larvae of blackflies (Simuliidae) are often abundant and may be the only invertebrates present.

The annual changes in the resident benthic invertebrate population are complicated by *drift* (Fig. 16-16) and by hatching and emergence of those species which have a flying adult stage. These include the majority of the benthic insects such as stoneflies, mayflies, and caddisflies but not the various worms and mollusks. As with lake zooplankton (Chap. 13), benthic invertebrates in streams may complete several, one, or only part of a generation each year. In addition some benthic organisms may require a true resting stage or diapause to complete their life cycle, while others spend many months as inconspicuous eggs. The net result is that insect numbers in rivers appear highly variable through time. Their populations can be estimated if their life cycles of hatching, egg laying, and resting stages are known.

Modifications due to natural or human causes can change the relative numbers of invertebrates (Hynes, 1960). The prediction of effects of stream changes from a knowledge of the ecology of the benthic dwellers is a major challenge to modern limnologists. For example, an increase in Simuliidae larvae, which carry the tropical disease, river blindness, may result from the creation of fast, shallow waters in the raceway of a dam. The presence of the disease often requires public health agencies to use pesticides which may kill the entire benthic insect population and reduce fish production. One alternative strategy used in African reservoirs is to allow the raceways to dry occasionally and kill the blackfly larvae.

The key to understanding the ecology of

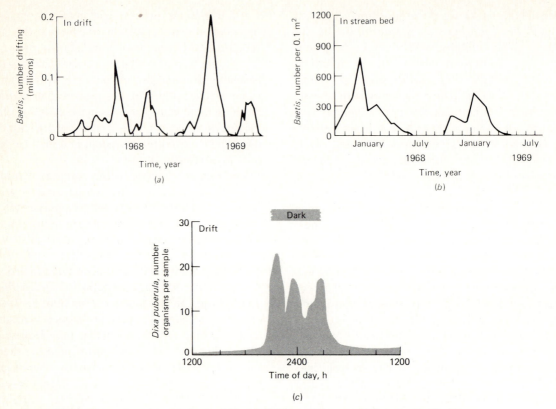

Figure 16-16 Examples of invertebrate drift. (*a*) Drifting of the mayfly larvae, *Baetis bicaudatus*, compared with (*b*) numbers in the streambed in the Temple Fork of the Logan River, Utah. This stream is less than 4 m wide at the sampling station shown. The drift is a major food for fish. (*Redrawn and modified from Pearson and Kramer, 1972.*) (*c*) Diel variation in the drift of the larvae of the midge *Dixa puberula* in Wilfin Beck, a stony stream in the English Lake District. Note the rapid onset of drift as night falls. This reduces predation by fish which depend on sight to detect their food. (*Redrawn from Elliott and Tullett, 1977.*)

streams is knowledge of the *food web* (Figs. 15-2, 15-4). As in lakes and on land, these webs are constructed from actual measurements of individual biomasses of stream organisms and measures of transfer between trophic levels. These values are interlinked by known or assumed predator-prey relationships.

The feeding methods of benthic invertebrates are *grazing*, catching *drift*, and direct *predation*. Together with *shredding* organisms mentioned earlier, these groups constitute a set of functional divisions of the benthic invertebrate community (Fig. 16-15). Grazing of attached algae,

bacteria and fungi, and small animals like rotifers is readily observed for snails but is also common in benthic insect groups such as stoneflies and mayflies. Grazing on pennate diatoms may be supplemented by direct consumption of mosses. As mentioned earlier, higher aquatic plants are not normally eaten directly, but the rich covering of periphyton on their leaves and stems provides a nutritious food source for various grazing animals. Stem-mining insect larvae such as chironomids are an exception and exist entirely on macrophyte tissues.

Drift

Drift, which contains many small living organisms and pieces of detritus, is the major food of many stream invertebrates. Drift includes clumps of bacteria, algae, detritus, other aquatic and terrestrial invertebrates, and fragments of vegetation. In many cases, particularly in fishery-related contexts, drift may refer only to free-floating invertebrates. Some organisms have specialized nets to ensnare drift. The caddisfly *Hydropsyche* spins a remarkable funnel-shaped net for collecting drift. Most stream dwellers, however, use the bristles on their legs or mouthparts as filters and scrape up and eat the particles they collect. The collection of drift food by most stream insects is not simply a matter of remaining stationary and waiting, since diurnal variation in drift and exposure to predation complicate feeding.

Much drift consists of small insect larvae which can only move large distances to new habitats by releasing their hold on the substrate and allowing the current to move them. A good example is shown by a mountain stream in Utah (Fig. 16-16*a, b*). The quantities of total drifting material are sometimes astonishingly large, and millions of individual animals may pass one spot in a single night in a large stream. Invertebrate drift normally occurs at night (Fig. 16-16*c*), probably to avoid predation by sight-dependent fish. The living invertebrate component of the drift represents only a small fraction of the total benthos. Many insects gradually drift downstream during their lifetimes and when they emerge for a brief aerial existence, fly upstream to lay their eggs. Mayflies (Ephemeroptera, from *ephemeral,* meaning "short-lived") and some other insects may live only a matter of hours after emergence, which limits their flight range. The creation of a reservoir can block upstream migration of the egg-bearing females, which may reduce the productivity of headwaters. However, the abundant plankton from the reservoir may increase the production of a few filter-feeding species found below the dam. Floods scour the streambed and increase drift enormously. An obvious effect of reservoir building is to decrease the size and effect of floods and resulting scour. Likewise, channelization and levee construction will increase the scouring effects of floods. Insects may be unable to recolonize their original habitats.

Most predation occurs in daylight, since fish and carnivorous insects hunt mainly by sight. The result is that herbivores feed near or under stones and emerge to drift or feed in more exposed places at night. During daylight, many herbivores stay under stones to avoid predators. Some insects possess special structural adaptations to prevent them from being dislodged by the current and becoming edible drift. The flattened bodies of stoneflies and mayflies together with hooks at the ends of the legs are adaptations to swift water. The Simuliidae (blackflies) are particularly good at holding on and they are usually abundant in torrential sites such as waterfalls or rapids. Few predators can feed effectively in such rapid water.

Fish are an important component of most stream biota. They are also specially adapted for the currents normally encountered in their chosen habitat. Both stream and lake fish are considered in Chap. 14. The distribution of fish is influenced by physical habitat and the presence of other animals and plants, particularly those used for food and shelter. Fish may be territorial in streams and rivers and often remain in one small reach for most of their lives.

The distribution of several trophic levels, including fish, is given for a western mountain stream in Fig. 16-17. The distribution of six fish species can be seen to be basically unrelated to possible food sources such as microphytes, mesophytes, higher plants, or benthic invertebrates but does appear to be related to temperature. Two of the three trout species are found in the cooler water downstream from

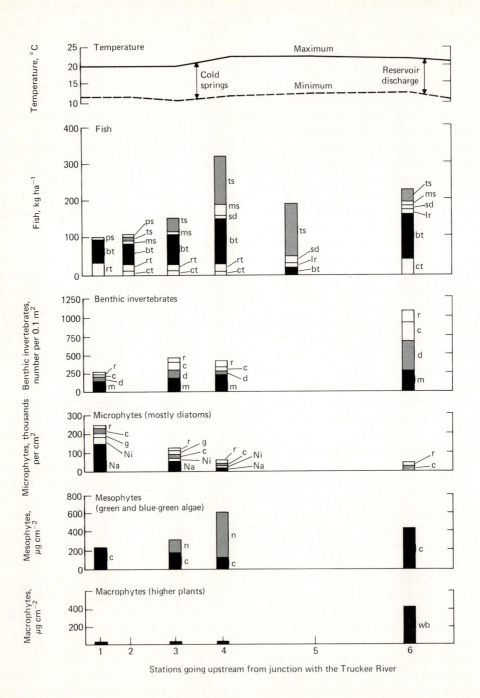

Stations going upstream from junction with the Truckee River

cold springs, and the minnows (Cyprinidae) and suckers (Catostomidae) dominate just below the warm surface outflow of a shallow reservoir. Food does not seem to be a limiting factor here, but in other streams there may be more direct relationships between food supply and fish distribution. In larger streams and rivers collection becomes more difficult at all trophic levels, and few precise conclusions can be drawn. Instead, a general ecological role can be ascribed to each functional group of fish in the stream (Fig. 14-5) in a similar fashion to that done for invertebrates (Fig. 16-15). The indirect dependence of cyprinid fish such as carp on macrophytes is similar to the indirect dependence of trout and sculpins on microphyte diatoms and mosses. The relationship between these widely separated trophic levels is through the food web (Fig. 15-2) and depends on the intermediate role of the benthic invertebrates which constitute the major food supply for these fish.

Where the river first encounters the influence of the sea, the limnologist finds the physical boundary of the science. Estuaries are common waters for limnologists and marine biologists alike, and this seaward limit of the river provides the basis for the next chapter.

Figure 16-17 Longitudinal distribution of animals and plants relative to temperature in Martis Creek, a small mountain stream in California. There are cold springs between stations 3 and 4 and warm surface discharges from a reservoir above station 6. For diel variation in this stream see Fig. 16-7. Black bars represent the most common organisms in each taxonomic grouping, e.g., mayflies for benthic invertebrates, brown trout in the fish group. Shaded bars are the second-most common. Letters indicate the actual organisms present (see key below). Note the predominance of brown trout in the cooler sections and the Tahoe sucker in the warmer sections. This corresponds to their known temperature preferences. There is no obvious relationship between fish and their food (benthic invertebrates for trout and other carnivores and algae for the others). Mayflies are little changed downstream, but chironomids (Diptera) dominate below the reservoir where their food of detritus and lake planktonic algae is washed downstream. The microphytes in general, especially the small pennate diatom *Navicula*, are not favored by this reservoir discharge. Microphytes have been replaced as the primary autochthonous producers below the reservoir by the mesophyte green filamentous alga *Cladophora*, the emergent macrophyte *Ranunculus*, the water buttercup, and submerged plants such as the pond weed *Elodia* (not shown here). The mesophyte *Nostoc* favors the faster-flowing middle sections of this creek, while the other common mesophyte, *Cladophora*, prefers the slower upper reaches in common with the habitat preferences of the macrophytes. The large contribution of the irregularly distributed mesophytes and macrophytes to the detritus pool is obvious at station 6. Fish have been plotted as weight, but if those which were too small to catch using conventional electrofishing (Chap. 14) were accounted for, smaller fish such as Lahontan redside would be more apparent.

Key. Fish: bt = brown trout *Salmo trutta;* rt = rainbow trout *S. gairdnerii;* ct = cutthroat trout *S. clarkii;* ts = Tahoe sucker *Catostomus tahoensis;* ms = mountain sucker *C. platyrhynchus;* sd = speckled dace (minnow) *Rhinichthys osculus;* lr = Lahontan redside (minnow) *Richardsonius egregius;* ps = Piute sculpin *Cottus beldingi.* Benthic invertebrates: m = mayflies (Ephemeroptera); d = true flies or Diptera; c = caddisflies (Trichoptera); r = other types. Microalgae, mostly diatoms: Na = *Navicula;* Ni = *Nitzschia;* c = *Cymbella;* g = Gomphonema; r = other types. Mesophytes (green and blue-green algae): c = *Cladophora;* n = *Nostoc.* Macrophytes (higher aquatic plants): wb = water buttercup *Ranunculus.* (Redrawn and modified from Horne et al., 1979.)

FURTHER READINGS

Blum, J. L. 1956. "The Ecology of River Algae." *Bot. Rev.*, **22**:291–341.

Cummins, K. W., R. C. Petersen, F. O. Howard, J. C. Wuycheck, and V. I. Holt. 1973. "The Utilization of Leaf Litter by Stream Detritivores." *Ecology,* **54**:336–345.

Hynes, H. B. N. 1972. *The Ecology of Running Waters.* University of Toronto Press, Toronto. 555 pp.

———, 1975. "The Stream and Its Valley." *Verh. Int. Ver. Limnol.,* **19**:1–15.

Kaushik, N. K. and H. B. N. Hynes. 1971. "The Fate of Dead Leaves That Fall Into Streams." *Arch. Hydrobiol.,* **68**:465–515.

Likens, G. E., F. H. Bormann, N. M. Johnson, D. W. Fisher, and R. S. Pierce. 1970. "Effects of Forest Cutting and Herbicide Treatment on Nutrient Budgets in the Hubbard Brook Watershed Ecosystem." *Ecol. Monogr.,* **40**:23–47.

Minshall, G. W. 1978. "Autotrophy in Stream Ecosystems." *Bioscience,* **28**:767–771.

Moore, J. W. 1976. "Seasonal Succession of Algae in Rivers. I. Examples from the Avon, a Large Slow-flowing River." *J. Phycol.,* **12**:342–349.

Mordukai-Boltovskoi, Ph.D (ed.). 1979. "The River Volga and Its Life." *Monogr. Biol.,* vol. 33. 473 pp.

Omernik, J. M. 1976. "The Influence of Land Use on Stream Nutrient Levels." U.S. EPA-600/3-76-014. Corvallis, Ore. 68 pp. plus appendix.

Pennak, R. W. 1971. "Towards a Classification of Lotic Habitats." *Hydrobiologia,* **38**:321–334.

Pfeifer, R. F., and W. F. McDiffett. 1975. "Some Factors Affecting Primary Productivity of Stream Riffle Communities." *Arch. Hydrobiol.,* **75**:306–317.

Swale, E. M. F. 1969. "Phytoplankton in Two English Rivers." *J. Ecol.,* **57**:1–23.

Ward, J. V., and J. A. Stanford (eds.). 1979. *The Ecology of Regulated Streams.* Plenum, New York. 398 pp.

Whitton, B. A. (ed.). 1975. *Studies in Ecology,* vol. 2: *River Ecology.* University of California Press, Berkeley. 725 pp.

Chapter 17

Estuaries

OVERVIEW

Estuaries are among the most complex bodies of water commonly encountered by limnologists, and they are also the most productive. Estuaries have fewer plant or animal species than fresh or sea waters, but may have very large numbers of individuals of the species present. The dominant features of an estuary are variable salinity, a salt wedge or interface between salt and fresh water, and often large areas of shallow, turbid water overlying mud flats and salt marshes. A workable definition of an estuary is a partially enclosed body of water of variable salinity, with a freshwater inflow at one end and sea water introduced by tidal action at the other.

Nutrients are supplied to the estuary from the sea, the inflowing river(s), and adjacent marshes. Tiny silt particles flocculate at the saltwater-freshwater interface to form larger clumps which settle more rapidly and serve as food for filter-feeding animals. More detritus is produced in estuaries than in lakes but some is lost to the ocean. Tidal motions provide an energy subsidy for the estuarine ecosystem by bringing food to sessile organisms. Large, productive salt marshes may develop in the waterlogged salty soil between high and low tide levels in temperate climates. In the tropics, mangrove swamps are the equivalent of these salt marshes. The shallow well-mixed waters of most estuaries are often ideal sites for high rates of photosynthesis and secondary production. The various habitats, mud flats, deeper water, and salt marshes or mangrove swamps, effectively divide the biota into size and age classes. Juvenile stages of many commercially important fish and crustaceans depend on the shallow, protected estuary. The adults spawn at sea and the eggs and larvae may be carried into the estuary where they spend most of their early life. Other seawater species return directly to the estuaries to spawn. This anadromous habit is especially well-developed in the warm-temperate Atlantic and Gulf coast estuaries of the United States. In the colder seas of the north temperate regions and the Arctic, there is a strongly anadromous group of fish that spawn in fresh water and are only briefly in estuaries.

INTRODUCTION

Although not as frequently studied by limnologists as lakes and rivers, estuaries provide a great challenge. Tidal fluctuation and variation in salinity complicate both sampling and the interpretation of measurements. The study of water movement in estuaries must take into account at least two different salinity layers as well as the temperature-caused density layers found in lakes. The shallow nutrient-rich muds of estuaries can produce rapid changes in the chemical composition of overlying waters. In addition, the biota of estuaries is often quite low in diversity but extremely high in number of individuals. The large number of interacting physical (marsh, mud flat, channels, open water, shoreline) and biological components (benthic micro- and macroalgae, zoobenthos, salt marsh vegetation, as well as phytoplankton, zooplankton, and nekton), make the overall food web very complex.

Despite the difficulties inherent in studying them, estuaries are an extremely interesting habitat and produce valuable fish and shellfish. Unfortunately, this seafood harvest is not often fully realized because estuaries tend to be the sites of large port cities which affect the biota through dredging and industrial or domestic pollution.

MEASUREMENTS

Sampling in estuaries involves a combination of methods already described for lakes and rivers.

Sediment, benthos, and plankton can be collected by standard limnological methods. Corrosion is a major factor in saline water, and steel clips or instruments must be replaced by brass or plastic components. Since estuaries are tidal, water currents are sufficient to measure using relatively insensitive cup-type flowmeters. In deeper, open areas more sensitive ocean or lake-type instruments may be required. Salinity is easily measured by conductivity using portable salinometers. Salinity, the dissolved solids present in a kilogram of filtered seawater after oxidation of nutrients, is usually reported in parts per thousand, written as ppt, or $^o/_{oo}$. Sodium chloride dominates the overall total dissolved solids. Full-strength seawater has a salinity of about 35 ppt. Salinity falls gradually from the sea to the upstream limit of the estuary, which is considered fresh water at about 0.5 ppt. It should be noted that seawater has a different ionic composition than water from saline lakes. These lakes are derived from the evaporation of fresh water (Chap. 19) and have relatively greater quantities of calcium, magnesium, and sulfate.

In most estuaries considerable areas of mud, which are often cut by deeper channels, are exposed at low tides. A shallow-draft, propellerless "jet" boat is best for sampling the mud-flat zones at high tide but is usually unstable for rough open water where a larger boat is needed. In addition, sampling the mud flats at low tide requires special footwear such as mud skis or snowshoe-type "mud striders" to prevent sinking.

VARIABLE SALINITY AND THE SALT "WEDGE"

Variable salinity is the most characteristic feature of estuaries. Salinity at one place changes daily with tides and usually changes dramatically with the seasons. The head of the estuary is the least saline, but during the low flows of summer, almost full-strength seawater may penetrate to this zone. Conversely, in winter, floods of fresh water may reach the mouth of the estuary. Control of river flows by dams usually damps out these extreme natural fluctuations. Where fresh and salt water mix the resulting mixture is called *brackish* water. The length of the estuary, if defined as the zone of brackish water, will vary depending on the volume of river flow relative to the size of the estuary. In an extreme case, the Amazon River in Brazil, the flow is so large that seawater never penetrates upriver. In the days of the early European explorers, ships were able to take on fresh drinking water well out of sight of land from the enormous discharge of this river.

In West Africa the heavy inland rains periodically increase river flow so that the salinity of the coastal waters of the Gulf of Guinea are reduced to 27 ppt some 50 km offshore. In this case, and in the similar case of the inshore waters of the Gulf of Mexico, the whole coastal zone takes on some aspects of an estuary. At the other extreme, where a small stream flows into a large sunken valley (Chap. 18), the estuary may be only a few hundred meters long. The rest of the estuary will then be an arm of the sea mostly enclosed by land. Thus the relative dominance or equilibrium of fresh and salt water is a variable factor which establishes the unique characteristics of different estuaries.

To survive variable salinities or brackish water, organisms must be euryhaline (i.e., tolerant of a wide salinity range) and equipped with special physiological mechanisms to eliminate salt or excessive fresh water depending on the outside salinity. This osmotic regulation is needed because the body fluids of all organisms are similar. Marine fish and other organisms which live in salt water are hypotonic because they contain less salt than their surroundings and must continually gain water and excrete salt. Freshwater organisms are hypertonic and must continuously conserve their salt by excreting water from which the salt has been reabsorbed in the kidney. The constantly varying salinity of estuarine waters requires both

types of osmotic regulation in the same animal. Such a physiological and anatomical flexibility is unusual. Anadromous and catadromous fish, such as salmon or eels, must convert their osmoregulation from hypotonic to hypertonic, or vice versa, as they migrate between salt and fresh water. Animals must remain for a period above or below the estuary while their osmoregulator functions are reorganized. Most estuarine inhabitants cannot osmoregulate very well but survive because their cells can tolerate changes in internal salinity which would kill most other organisms. Only a few species have evolved this osmotic tolerance. Adaptation to changing salinities, rather than adaptation to either low- or high-salinity, characterizes the estuarine inhabitant. Few species of bacteria, fungi, algae, higher plants, invertebrates, or fish can live in brackish water. Thus, the flora and fauna of estuaries tend to be poor in species (Figs. 17-1, 17-2), mostly of marine origin, with almost no insects.

Where salt water and fresh water meet in relatively sheltered waters, the denser salt water becomes overlain with the lighter river water. The salinity-density effect dominates over any temperature-density effect. The result is called a *salt wedge* and is a permanent feature of most estuaries (Fig. 17-3). The salt wedge is not fixed in place but moves down the estuary with high winter flows and up the estuary during the period of low summer discharge. Horizontal and vertical profiles of a salt wedge are shown in Fig. 17-4. The wedge is modified by the degree of wind mixing and the shape of bottom topography. The saltwater-freshwater interface can be almost vertical, as occurs with the large tidal ranges and open exposure of the Chesapeake Bay or the Thames estuary. Alternatively, large freshwater flows, such as are found in the Mississippi delta, cause a pronounced horizontal wedge. Because of the irregular cross section of most estuaries the deep salt wedge(s) in the major channel(s) will be up-

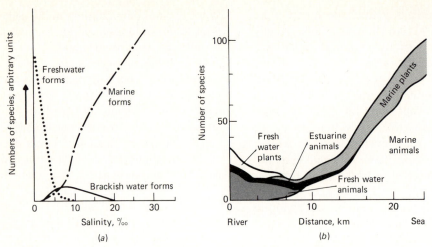

Figure 17-1 Diagrammatic representations of the low species diversity in estuaries relative to adjacent marine and fresh waters. (*a*) Simplified diagram. (*Modified from Barnes, 1974; and Remane and Schlieper, 1971.*) (*b*) Numbers of species of animals and plants along the Tees Estuary, England. (*Modified from Alexander, Southgate, and Bassendale, 1935.*)

stream of smaller salt wedges in the shallower regions. The term salt wedge is only an approximation of the complex set of salt density gradients, or *picnoclines*, actually present.

As the fresh water moves seaward, there is a corresponding counterflow of saline bottom water upstream. The amount of countercurrent is in part dependent on how much mixing occurs continuously at the salt wedge interface and the relative strength of the freshwater and tidal flows (Fig. 17-4). The upstream movement of bottom water is vital for the retention and recycling of organic detritus. The transport of young larvae, especially crabs and oysters, from the open sea to the upper parts of an estuary is also dependent on this countercurrent along the bottom.

The saltwater-freshwater interface is more than just an interesting physical phenomenon since it provides an excellent site for the precipitation, or *flocculation*, of organic and inorganic particles. A common technique in chemistry is to add a salt to a solution to precipitate an otherwise soluble compound. This "salting out" occurs continuously at the salt wedge interface.

River silt particles of 4- to 60-μm diameter flocculate into much larger clumps when the fresh water meets the mass of cations in seawater. Flocs may further sorb inorganic and organic compounds such as phosphates and carbonaceous nutrients. The resulting particles fall to the bed of the estuary and may form mounds visible to divers. The particles are often supposed to be held in a "nutrient trap" by the counterflow currents in the salt wedge (Fig. 17-4), although there is some dispute as to the importance of this phenomenon. If the flocculated particles remain for more than a few hours near the salt wedge, bacteria are able to grow on them and provide a concentrated food source for both planktonic and benthic animals. The zone of highest benthic animal production is often associated with the salt wedge. Since the wedge moves up and down the length of the estuary, planktonic organisms and fish also benefit.

Over the seasons the average net accumulation of organic detritus from flocculation and sinking of silt or dead grass and leaves has been estimated to average about 2 mm year^{-1}. Both

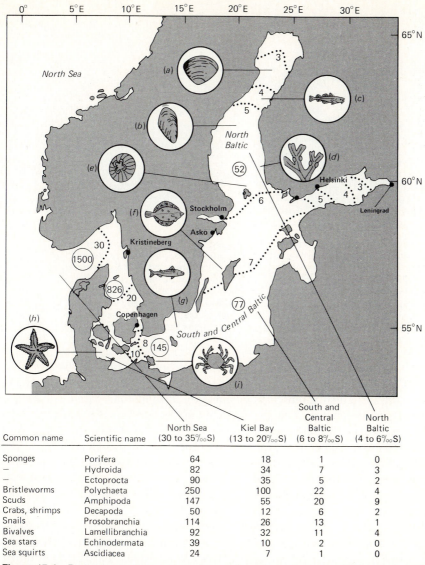

Common name	Scientific name	North Sea (30 to 35‰S)	Kiel Bay (13 to 20‰S)	South and Central Baltic (6 to 8‰S)	North Baltic (4 to 6‰S)
Sponges	Porifera	64	18	1	0
—	Hydroida	82	34	7	3
—	Ectoprocta	90	35	5	2
Bristleworms	Polychaeta	250	100	22	4
Scuds	Amphipoda	147	55	20	9
Crabs, shrimps	Decapoda	50	12	6	2
Snails	Prosobranchia	114	26	13	1
Bivalves	Lamellibranchia	92	32	11	4
Sea stars	Echinodermata	39	10	2	0
Sea squirts	Ascidiacea	24	7	1	0

Figure 17-2 Reduction in the number of macrofaunal species in brackish waters. In the Baltic Sea estuary salinity ranges from less than 3 ppt to full-strength seawater (dotted lines with numbers indicate isohalines). There are only 52 macrofaunal species in the North Baltic, but 1500 are found in the North Sea (numbers in circles).

Letters in parentheses are distribution limits for some common marine species: (*a*) *Macoma baltica*, (*b*) *Mytilus edulis*, (*c*) cod, (*d*) the brown macroalga, *Fucus vesiculosus*, (*e*) the "jellyfish," *Aurelia aurita*, (*f*) plaice, (*g*) mackerel, (*h*) *Asterias rubens*, (*i*) *Carcinus meanas*.

The table shows the general distribution of some individual groups of organisms. (*Redrawn and modified from Jansson and Wulff, 1977; Barnes, 1974; and Remane and Schlieper, 1971.*)

(a)   *(b)*

Figure 17-3 (a) The salt wedge in San Francisco Bay as outlined by inflowing turbid water during high winter discharges of the Sacramento River. The white coloration is the muddy fresh water which here reaches the open ocean. The South Bay, having no direct river inflow, remains more saline, darker here. The gradation in color from dark in the open ocean to very light upstream corresponds to decreases in surface salinity (Fig. 17-4). (*Photograph from 20,000 m by NASA.*) (*b*) The salt wedge in winter as it appears looking seaward from a nearby hillside opposite the entrance to the sea. The brown fresh water lies in a band isolating more saline water on either side. As these photographs show, salt wedges are more complex than most diagrams can indicate.

the marine and fresh waters contribute to this rich source of food for the common detritus-feeding organisms such as flatfish, crabs, oysters, clams, and polychaete worms. The detritus also serves as a refuge from predation and as a buffer to salinity changes. While overlying water may vary from almost fresh to salty, the interstitial water shows less variation (Fig. 17-5). Even very soluble compounds like nitrates are only slowly exchanged from the interstitial mud to the overlying water. The slow exchange permits some organisms like marine clams to survive occasional floods of fresh water. Estuarine mud is resuspended by the tide until it is lost to the sea or settles amongst salt-tolerant vegetation.

EUTROPHIC ESTUARIES

Estuaries are among the most productive systems in the world (Table 17-1). The reason for this is best represented in terms of physical and chemical energy subsidies as described by E.P. Odum (1971). The energy subsidy in estuaries may be usefully compared to lakes as follows: (1) Tidal action brings nutrients and food to the organism, saving the energy which would otherwise be expended in searching or capture; (2) the mixing of large quantities of seawater with fresh water causes material to flocculate into particles which become the food supply for smaller animals; (3) the shallow water and dark mud exposed at low tide is easily heated by the sun, increasing nutrient recycling through bacterial decay and accelerating the growth of benthic animals; (4) estuaries are usually shallow and gain much of their food from allochthonous sources such as mangroves or salt marshes. These nutrients tend to be recycled by tidal action rather than lost to the sediments as commonly occurs in stratified lakes; and (5) the mere presence of the adjacent sea tends to reduce extremes in temperature.

Although many estuaries are very eutrophic, some are unproductive: for example, meromictic fjords, where the sediments act as a nutrient trap (Chap. 18), or estuaries in polar climates subject to the scouring action of ice.

The back-and-forth motion of the tide is an economical way to create and maintain a food supply. In lakes animals and phytoplankton must swim or sink to find new food sources or nutrients (see Chaps. 12 and 13). In rivers nu-

trients are supplied by the current to fixed organisms, but the biota is in constant danger of being swept away (Chap. 16). Alternatively, the food supply may pass without being captured. Only in the estuary are nutrients and plankton moved back and forth and only slowly passed out into deep water. It is thus not surprising that many fish and crustaceans either spawn or

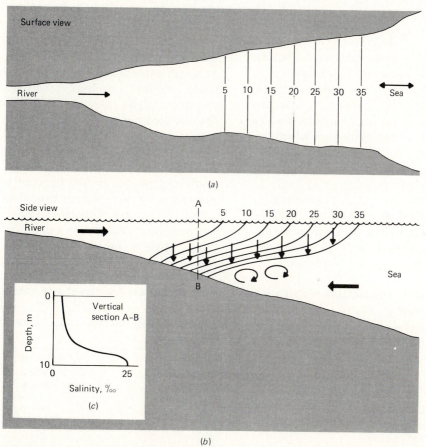

Figure 17-4 Typical salt wedge of an estuary in diagrammatic form. Isohalines indicate salinity in parts per thousand (a) at the surface or (b) with depth. Note the upstream flow of dense salt water counterbalancing the seaward flow of river water. The thin vertical and circular arrows indicate precipitation and sedimentation of particles in this brackish-water zone. (c) The very sharp density gradient, or picnocline, which is much stronger than any thermocline, resists vertical mixing by the wind except in very shallow areas such as mud flats exposed at low tide. There are other types of salt wedges possible, and isohaline profiles can be more or less vertical than those shown in part (b). The same estuary may appear very different throughout the year (see Fig. 17-14 for examples from Chesapeake Bay). (*Modified from Barnes, 1974.*)

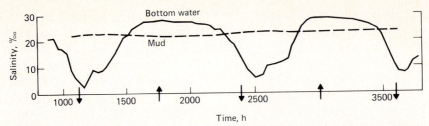

Figure 17-5 Changes in the salinity of estuarine bottom water and in the interstitial waters of the underlying mud over the tidal cycle. ↑ = high tide, ↓ = low tide. Note that although the salinity of the bottom water ranges from almost fresh to nearly full-strength seawater, that of the mud remains constant at about 22"/₀₀. The mud thus provides a refuge from osmotic pressure fluctuations for many small or burrowing organisms. (*Modified from Mangelsdorf, 1967.*)

at least spend their juvenile stages in estuaries. Most estuaries have a gentle wave action near the shore due to protection by land or vegetation which dampens wave action. In contrast the open seacoast or exposed beaches of large lakes are subject to pounding surf which reduces productivity.

Marshes, mangroves, and other aquatic vegetation such as the sea grass, *Zostera,* provide an abundance of autochthonous organic matter in the estuary. In the open water high rates of autochthonous inputs may occur from photosynthesis by planktonic and semiplanktonic algae and macroaglae or seaweeds (Fig. 17-6). Although estuaries are often turbid with suspended silt, their shallow depth usually permits planktonic algae adequate light for growth. Thick-walled pennate diatoms play a more important role in estuarine production than in most lakes. These heavy diatoms would be lost through the thermocline of most lakes, but are easily resuspended by the frequent top-to-bottom mixing of the water column in an estuary.

The efficient use of the high primary production by higher trophic levels is what makes estuaries such a valuable resource to people.

Table 17-1   Productivity of Estuaries as Compared to Other Aquatic and Terrestrial Areas*

Net carbon production was converted to gross energy values, where necessary, by multiplying by 20.

Site	Gross primary production, kcal m^{-2}y^{-1}	Total for the world, 10^{16} kcal y^{-1}
Estuaries and reefs	20,000	4
Tropical forest	20,000	29
Fertilized farmland	12,000	4.8
Eutrophic lakes	10,000	
Unfertilized farmland	8,000	3.9
Coastal upwellings	6,000	0.2
Grassland	2,500	10.5
Oligotrophic lakes	1,000	
Open oceans	1,000	32.6
Desert and tundra	200	0.8

*Modified from Odum (1971), with additions from Goldman (1981) and Goldman and Wetzel (1963).

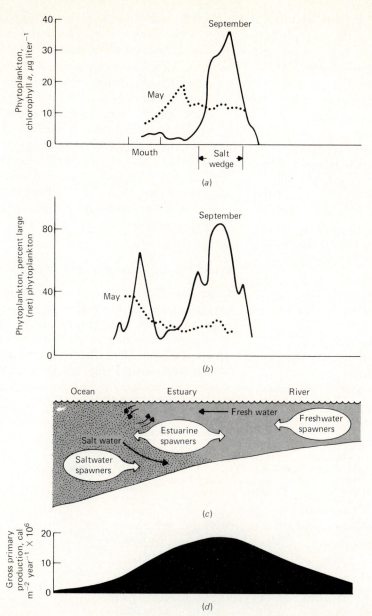

Figure 17-6 Spatial and temporal variation in phytoplankton biomass and productivity in relation to fish reproduction. Many fish spawn so that their young are carried by the water currents into the salt-wedge area where phytoplankton production is greatest. The young fish feed on zooplankton which also congregate where phytoplankton are abundant. The general applicability of this concept is shown by the fact that parts (a) and (b) are from San Francisco Bay, while (c) and (d) are based on the Hudson River which flows through New York. (*Modified from Cloern, 1979; and Hall, 1977.*)

Channels in salt marshes and shallow mud flats are poor habitats for most of the larger fish, which could quickly decimate populations of young fish, shrimps, and other plankton that live there (Fig. 17-7). Protection from large predators together with the constant supply of food from algal and marsh plant debris makes them an excellent refuge for the eggs or juvenile stages of most of the major *forage, sport,* and *commercial* fish and shellfish. It is a pity that these intertidal mud flats and marshes are sometimes considered unsightly. They are frequently filled as sites for airports, oil refineries, factories, garbage dumps, and marinas. Such developments frequently destroy the base of the food chain upon which the fishery depends.

THE BIOTA

In estuaries the physical compartmentalization of the biota in the muds, shallows, marshes, or mangroves is more pronounced than in large lakes (Fig. 17-7), where plankton often dominate the system. Constant tidal flooding wa-

terlogs the soil between high and low tide, and only a few groups of specialized plants have evolved to tolerate the extremes of the habitat. In places like southwestern England or eastern Canada tidal fluctuations may exceed 10 m, which results in huge areas covered with one or two species of short reeds or grasses (Fig. 17-8). These plant beds are intersected by an extensive network of steep-sided muddy drainage channels. Great care should be taken in such salt marshes, which can be crossed safely only at low tide. One of the authors (AJH) and his grandfather narrowly escaped with their lives from a Devon estuary when trapped by a fast-moving tide.

Marsh Plants, Seaweeds, and Phytoplankton

Common plants such as *Spartina,* the cord grass, or *Carex,* a sedge (Chap. 11), provide a major source of primary production which is usually cycled into the estuarine food chain as debris after the plants have died (Figs. 17-9, 17-10). The vertical sides of the reeds and the salt-marsh channels provide an excellent site for attached algae, particularly diatoms,

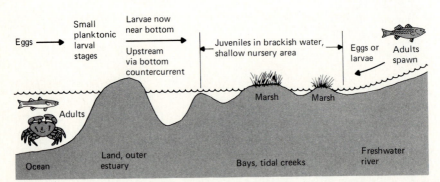

Figure 17-7 The use of the several physical compartments of the estuary as a nursery ground for fish and shellfish. Estuarine fish, shrimps, mollusks, and crabs often move to the constant salinity of the open sea to breed and lay their eggs. Young fish and the several larval stages of crustaceans drift around in the ocean plankton. In the estuary larvae drop to the bottom and move up the estuary to the shallow, marshy areas on the bottom countercurrent (see Fig. 17-4). Some fish move to the freshwater rivers to breed and some breed in the outer estuary. All utilize the very high primary production, high temperatures, and protection from predation afforded by the estuarine nursery grounds to grow rapidly during their first year. (*From an idea by Odum, 1971.*)

Figure 17-8 Complex arrangement of channels, mud flats, and open water in a tidal salt marsh on the Bradley River, Alaska. Large stands of one species of reed cover most of the marsh. The central portion shows the main river channel and mud banks. The many physical compartments provide refuges for small fish and shellfish.

Figure 17-9 Edge of a small reed bed on Chesapeake Bay. The colonization of the hard clay by these plants increases the diversity of habitat and helps to make estuaries one of the most highly productive ecosystems (see Table 17-1).

green, and blue-green algae which are grazed directly by snails and small animals or are eaten as detritus when they become dislodged or die. The net productivity of this marsh-mud-higher plant–algae system is unusually high when compared with other aquatic and terrestrial environments (Table 17-1).

Most of the larger plant inhabitants of salt marshes are long-lived perennials which may survive for up to 50 years. They usually become established on bare ground from seeds or small pieces of living plant. Their tough underground stems spread vegetatively but often the aboveground portions will die back in winter. The marsh maintains much of its physical integrity since the decay rate of marsh plants is slow relative to that of animal matter (Fig. 17-11). A strong winter structure is essential if the marsh is to avoid destruction of its

sediment base by winter storms. In turn the tall stems provide year-round cover for birds and small mammals.

As the sediment builds up around the stems of the pioneer marsh plants such as *Spartina*, other plants with less tolerance for saline water begin to colonize the marsh. This leads to a very obvious vertical zonation of marsh growth. Mud flats occupy the lowest level, *Salicornia* and *Spartina* the next, reeds such as *Juncus* and *Scirpus* the next, and eventually salt-tolerant grasses and shrubs at the higher levels. The lowest zones are often only a few centimeters different in elevation from the upper ones and can easily change due to natural fluctuations in rainfall and tidal action. Marsh zonation is particularly amenable to mapping by remote sensing using the characteristic infrared signature of the major marsh plant types. Use is made of this

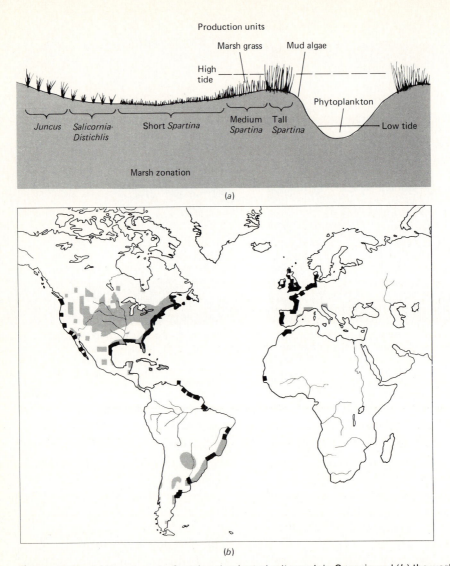

Figure 17-10   (a) Zonation of a *Spartina*-dominated salt marsh in Georgia and (b) the worldwide distribution of this genus of cordgrass (black = complex of marine species, shaded = freshwater species). Note the different ecotypes of the same species, *S. alterniflora*, in the Georgia marsh. There are short, tall, upper tidal, and lower tidal forms, each with its own productivity, associated epiphytes, and animal communities. Similar *Spartina* communities are found throughout the temperate areas of the world and in some parts of the tropics. Mangrove trees replace grasses and reeds in many tropical estuaries (e.g., Figs. 17-19, 17-21), and there are virtually no estuaries in desert areas such as the northwest coast of Africa or the west coast of South America (*Modified from Odum, 1971; and Chapman, 1977.*)

method in defining changes due to human activities, especially filling, waste disposal, freshwater diversion, and dredging.

Seaweeds or macroalgae are a common sight in estuaries, usually as patches of green sea lettuce, *Ulva*, on the mud flats, or as brown or red

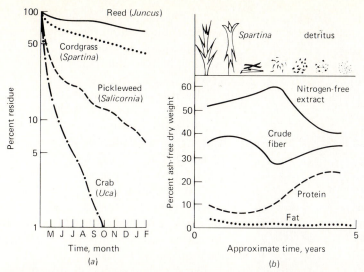

Figure 17-11 Decomposition of salt-marsh plants and animals. (*a*) The very slow degradation of estuarine grasses and reeds compared to the rapid decay of animal remains such as those of crabs. (*b*) The changes in various nutritive components of *Spartina* as it decays. The colonization by bacteria and fungi as the detritus becomes smaller results in the typical increase in protein which makes detritus such a good food source. (*Modified from Odum and de la Cruz, 1967.*)

seaweed strands on rocks or piers. Their large size and rapid growth rate makes them important contributors to the overall autochthonous production of some estuaries. Estuarine macroalgae are difficult to measure quantitatively, so there is much less data on their rates of production than are available for the phytoplankton. In San Francisco Bay a distinct population of the 1-m-long red alga *Gigartina*, estimated at 100 to 1000 tons, moves within a section of the Bay during the summer. (Fig. 17-12).

Large meadowlike areas of specialized higher plants often occur below low tide. The sea grasses, examples of which are *Zostera* and *Ruppia* in cooler waters and *Thalassia, Poseidonia*, and other genera in tropical waters, are common and bear an extensive crop of nutritious aufwuchs. Their annual die-off is a major contributor to the detritus pool. Benthic samples from the deep ocean trenches are sometimes dominated by partially decayed sea-grass fronds washed in from shallow areas.

The seeds and stems of these plants, some of which depend intermittently on fresh water, are important in the diet of some estuarine animals, especially the large flocks of geese and ducks which migrate annually along the coasts of the major continents. Large marshes are sometimes managed to provide the most suitable plant foods for migrating waterfowl. The seeds of the alkali bulrush, for example, provide 90 percent of the winter food of Suisun Marsh ducks migrating through the Pacific flyway.

The phytoplankton in estuaries may show spring and fall blooms which are similar to those found in nearby shallow lakes (Fig. 12-3). In some estuaries, however, there may be a single large bloom which occurs in autumn or spring. Two examples from temperate estuaries are shown in Fig. 17-13. The very turbid water in many estuaries together with river and tidal effects may account for this single peak of production. The large benthic and grazer populations may also prevent an accumulation of planktonic algae. Unlike zooplankton, benthic

Figure 17-12 A common estuarine macroalga or seaweed, the red alga *Gigartina*. These large algae grow in shallow, muddy water attached to rocks or small pieces of shell, and in about 4 years reach full size. They provide sites for the attachment of eggs for fish such as Pacific herring. (*Photograph by A. Nanomura.*)

of the phytoplankton production, biomass, and nutrients throughtout the estuary. Variations in phytoplankton biomass may be large, and a maximum near the salt wedge has been recorded (Fig. 17-6).

The types of planktonic, benthic, and attached microalgae (i.e., excluding seaweeds or macroalgae) in estuaries is somewhat different from that in adjacent freshwater lakes or streams. The general estuarine rule of larger populations of fewer species applies. Diatoms are the dominant form, especially in cooler waters (Fig. 17-13), and green algae (Chlorophyceae) are often reduced to a few flagellated and attached forms. At times these green algae may produce dense blooms.

Dinoflagellates and ciliates commonly form red tides in estuaries. Under specific hydrographic conditions their daily vertical migrations between seaward-flowing upper layers and upstream-flowing deeper layers maintain their position in the estuary. An example of dinoflagellate migration has been described recently in Chesapeake Bay (Fig. 17-14), and more spectacular examples are known from tropical waters (Seliger et al., 1970).

The importance to people of dinoflagellates in estuaries is related to their role in paralytic shellfish poisoning (PSP). Toxin production by dinoflagellates assumes special significance because estuaries are major sites for both natural growth and mariculture of clams, mussels, and oysters. Almost every year people die from PSP after eating shellfish. Although most toxic dinoflagellates seem to originate from the open sea, they occasionally penetrate estuaries, often with fatal results. Some genera, such as *Pyrmnesium,* are genuine estuarine forms.

A characteristic feature of most estuaries is their shallow depth and well-mixed muddy waters. Under this vigorous mixing regime, phytoplankton and benthic microalgae are more difficult to distinguish than in lakes or the ocean. For example, in San Francisco Bay ($\bar{z} = 6$ m) a major diatom species, *Melosira*

filter feeders such as mussels persist over winter and need only the spring increase in temperature to accelerate their feeding rates.

As in lakes, the onset of estuarine algal growth is probably due to the increase in light in late winter or spring (Fig. 12-3). Cessation of estuarine phytoplankton blooms is sometimes due to nutrient depletion, but the effect is often not as clear-cut as in lakes.

The spatial heterogeneity of estuarine biota is much greater than in lakes. This patchiness hinders interpretation of the seasonal dynamics

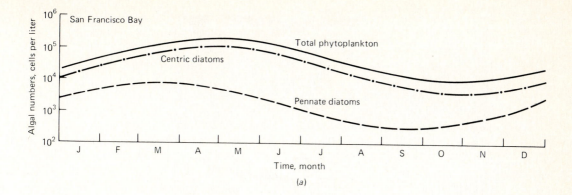

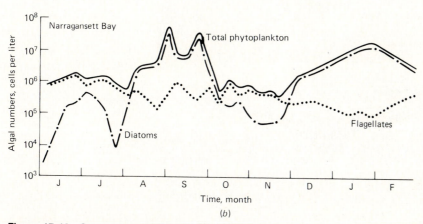

Figure 17-13 Seasonal variations of major types of estuarine phytoplankton. The phytoplankton is dominated by centric diatoms and, to a lesser extent, pennate diatoms. Diatoms usually comprise the majority of the phytoplankton in estuaries, although flagellates including dinoflagellates are occasionally abundant (see Fig. 17-14). (a) San Pablo Bay, a part of the San Francisco Bay estuary. (*Redrawn and modified from Storrs et al., 1966.*) (b) Lower Narragansett Bay, a large estuary on the east coast of the United States. (*Redrawn from Smayda, 1957.*)

moniliformis (Fig. 11-4), is present in small chains in the plankton or on the sediments in spring. However, it is a species commonly attached throughout the year to rocks and pilings.

Some interstitial pennate diatoms remain in the mud or sand during high tides and emerge only when the substrate is exposed at low tide. These algae have an internal clock which tells them when daylight and low tide coincide. A beautiful golden-brown sheen of diatoms can appear in a few minutes as the tide recedes. The advantage of this strategy is that filter-feeding

predators like clams feed only when the substrate is submerged. The diatoms thus avoid exposure to predation.

Zooplankton and Benthic Animals

Zooplankton and algae show a similar close relationship between benthic and planktonic states. They also seem to follow the maxim of few species but an abundance of individuals. Some forms, such as the copepod *Acartia* which is common in the east and west coast estuaries of the United States, are mostly plank-

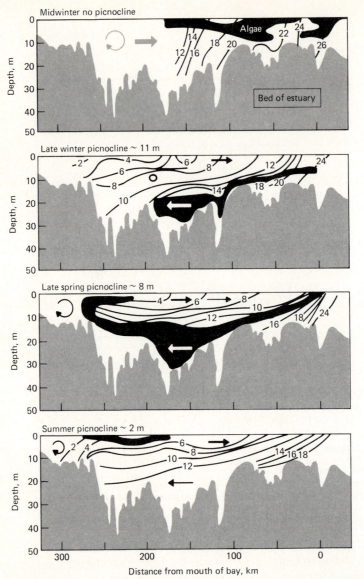

Figure 17-14   Growth and blooms of a red tide dinoflagellate in an estuary; Chesapeake Bay. Black areas represent relatively high concentrations of *Prorocentrum mariae-lebouriae*; isohalines are every 2 ppt. Arrows indicate direction of water movement. Maximum densities range from 6×10^5 individuals per liter in winter to 14×10^6 during the surface red tides in summer. This motile phytoplankter is present near the surface at the mouth of the estuary in midwinter when density stratification is weak and net freshwater outflow at any one depth is low. When snowmelt and rainfall increase in late winter and spring, strong density stratification is established, fresh surface water flows rapidly seaward, and the more saline countercurrent moves quickly upstream. *Prorocentrum* moves upstream in the deep channels below the major picnocline and grows slowly in the nutrient-rich but dimly lit bottom waters. In late spring the alga is mixed vertically in the shallow, almost-fresh waters of the uppermost estuary. The alga remains at the surface due to its positive phototaxis (see Chap. 12) and forms dense surface blooms or red tides in high summer. It is then gradually transported seaward in the surface layer and mixed vertically as stratification ends in autumn. (*Modified from Tyler and Seliger, 1978.*)

tonic (Fig. 17-15). Others such as the crustaceans *Corophium* and *Neomysis* are at least partially benthic, perhaps because of the abundance of diatoms and detritus at the mud-water interface. Estuaries, like the open ocean, have no pelagic insects, and more copepods than cladocerans in their planktonic crustacean fauna than are found in fresh waters. Zoobenthos is more important in estuaries than lakes.

Shallow estuaries with vigorously moving bottom water are physically similar to streams or lake littoral zones. Much of the harvest of the estuary is dependent on the huge populations of polychaete worms, clams, oysters, and crabs living in or on the sediment. Benthic shrimp such as *Crangon* or the polychaete *Nereis* feed on debris in the mud and provide food for shorebirds at low tide or larger fish during high tide.

The *infauna* contains most of the true estuarine benthic fauna and is dominated by relatively sessile types. This sediment habitat reduces both bird and fish predation as well as salinity variation. The infauna of estuaries has fewer species than are found in the ocean sediments, but these are often present in large numbers. This is illustrated in San Francisco Bay (Fig. 17-16), but similar examples occur in other estuaries, for example, the Baltic Sea (Fig. 17-2) or the Elbe (Fig. 17-17).

Unlike the plankton or the nekton, the infauna can often avoid unfavorable temperatures or salinities by burrowing into the more stable sediment environment (Fig. 17-5). Reduction of contact with changing salinities is probably more important for infaunal survival than is special osmotic adaptation to salinity variations.

Most studies on estuarine benthos concern the macroinfauna, usually those which will not pass through a mesh of 0.5 mm. The majority of the macroinfauna is directly dependent on the mixture of littoral algae and sunken or flocculated autochthonous and allochthonous material which microbial action transforms to detritus. Some polychaete worms can burrow directly through the accumulated detritus, eating as they go, but others form characteristic tubes and pump detritus-filled water to feed. A similar pumping mechanism is used by most bivalves which may live deep in the sediment, on the sur-

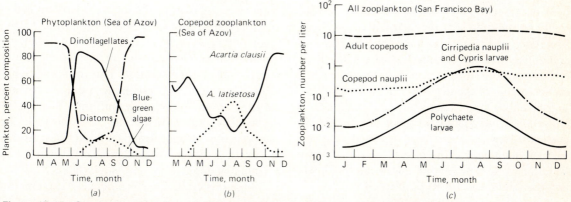

Figure 17-15 Seasonal cycles of zooplankton and some phytoplankton in two estuaries. (*b*) *Acartia* spp., copepods common in many eastern American estuaries, may show peaks in summer or winter. This may be related to (*a*) phytoplankton food. (*c*) Estuarine benthos produces many larval stages (e.g., Polychaetes and barnacles) which have no counterpart in the freshwater zooplankton. (*Redrawn from Zenkevich, 1963 and Storrs et al., 1966.*)

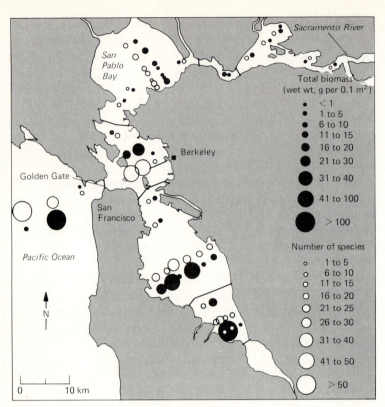

Figure 17-16 Species and biomass distributions for the benthic infaunal biomass from 0.1-m² sediment samples screened to collect organisms larger than 1.0 mm. Note the relatively few species present in most areas of San Francisco Bay which, like many estuaries, has several different types of sediment. In many places one species makes up most of the biomass (see also Fig. 17-17). (*Modified from Nichols, 1979.*)

face, or attached to rocks. Some snails graze directly on the mud surface and bivalves such as *Macoma,* the vacuum-cleaner clam, use a siphonlike tube to suck mud from around their buried position. Sedentary polychaetes use tentacles to collect the detritus. Finally, other estuarine dwellers such as the amphipod *Corophium* and some crabs pick out individual particles. Further discussions of estuarine macrofaunal feeding and the role of detritus are to be found in R. S. K. Barnes' short book (1974), and the larger text by R. C. Newell (1970).

The benthos in estuaries faces several problems during reproduction. Since the net flow of water is seaward, the sedentary parents must get their eggs or motile larvae upstream if dispersal from the home site is required. In the case of many species this is accomplished by using the bottom countercurrent which moves upstream on the flood tide. The youngest planktonic larvae of the common American oyster *Crassostrea virginica* are uniformly distributed vertically, but older larvae actively seek the more saline deeper water on flood tides. In this way they progressively move upstream. The young of many fish and shellfish such as the Dungeness crab, *Cancer magister,* also employ the bottom countercurrent to move upstream, apparently also using the higher salinity as a guide. In Chesapeake Bay larvae theoretically

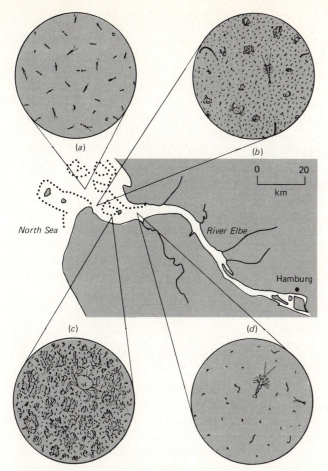

Figure 17-17 The infauna, or bottom-dwelling invertebrates, of four places in the Elbe Estuary, Germany. Dotted lines enclose shallow mud or sand banks. This estuary has a long history of scientific research (see Chap. 2). The organisms are brought up in a 0.1-m² bottom dredge after screening through a 1-mm mesh sieve. (a) Fine sand substrate dominated by 63 mysid shrimps (*Gastrosaccus spinifer*), five *Haustorius arenarius*, one *Bathyporeia tenuipes*, one small shrimp (*Crangon crangon*), five cockles (*Cardium edule*), and one Spionoid. (b) Soft mud substrate dominated by 240 mussels (*Mytilus edulis*) and 140 amphipods (*Corophium volutator*). Also present were 22 amphipods (*Gammarus zaddachi*), one mysid shrimp (*Neomysis vulgaris*), one *C. crangon*, four crabs (*Carcinus maenas*), 12 clams (*Petricola pholadiformis*), and 36 bristle worms (*Nereis succinea*). (c) Dominance of 2035 *M. edulis* and 450 *G. zaddachi* in a navigation channel where currents are swift and sediment is scoured out. The only other organisms present were 10 *C. maenas*. (d) In the fine loam and sand only a small number of several species were present: 30 *Bathyporeia* spp., seven polychaete worms (*Heteromastus filiformis*), one *C. crangon*, two *H. arenarius*, one *G. zaddachi*, and one *N. vulgaris*. (*After Caspers, 1951.*)

could be transported upstream almost 10 km day^{-1} but it is doubtful if such efficient use of the currents is ever made.

Fisheries

The high quality and abundance of fish in and near estuaries is a result of the food and protection for young that estuaries provide (Figs. 17-6, 17-7, 17-11, Table 17-1). The ease of fishing in estuaries, relative to the open ocean, makes them very valuable when commercially important fish are present. Most of the ocean's harvest is caught close to shore and many of the fish in the catch depend on adjacent estuaries. For example, the Gulf Coast between Pascagoula, Mississippi, and Port Arthur, Texas, supplies about 20 percent of the total U.S. marine and freshwater fisheries catch. The Gulf catch is dominated by a fish called the menhaden (*Brevoortia* spp.), which is considered an estuarine species. Gunter (1967) estimates that 97.5 percent of the total commercial fish and shellfish catch of the Gulf states is directly dependent on estuaries at some stage in their life cycle. Similar dependence is probably true in many areas of the world's oceans. Five out of six most important commercial fish and most sport fish in the United States are considered estuarine.

Large sport fish like the striped bass (*Morone saxatilis*) require fresh water for spawning and estuarine marsh backwaters during the first few months after hatching. The Pacific herring, *Clupea harengus pallasii*, lay their eggs on estuarine grasses or seaweeds. Still others like the oyster *Crassostrea virginica* spend their whole life in and around the mud flats. Many other fish such as the Atlantic herring, *C. h. harengus*, spawn offshore in the constant conditions of the open ocean, but masses of their young migrate to the more fertile brackish waters of the estuary. Fish such as the menhaden, the grey mullet, *Mugil cephalus*, and most harvest shrimp (e.g., *Penaeus setiferus*) breed similarly. Commercially important shellfish like the

Dungeness crab and blue crabs, *Callinectes sapidus*, are also born at sea but grow to almost full size in the estuary (Fig. 17-7).

Much of the need for the constant, if infertile, conditions of the open ocean stems from the relative youth of estuaries. It is probable that more advanced organisms such as fish and crabs have had insufficient time to evolve physiological mechanisms for breeding in the brackish waters of the estuary. A few million years ago there were probably very few estuaries because most were recently formed by drowning river valleys when the last ice age terminated (Chap. 18). Careful investigation of estuarine species shows that they can survive and breed in full-strength seawater off the coast. Because most estuarine organisms have marine origins, the occasional freshwater flushes during heavy rains maintain low diversity in estuaries compared to associated seas or rivers (Fig. 17-1*b*). Grey mullet swim far out to sea to spawn due to the high sensitivity of the eggs and larvae to changes in salinity.

Birds are too often ignored by limnologists but in estuaries their importance is obvious. Vast flocks of geese, ducks, swans, and waders take refuge on the estuarine mud flats in winter. The recreational value of hunting waterfowl has been important in the protection and management of estuarine marshes. Aside from their sporting importance, large flocks of birds feeding on one or two types of prey can substantially influence the population structure of estuarine mollusks, polychaetes, and sea grasses. For example, wading birds have been estimated to reduce populations of the common clam *Macoma* by 4 to 20 percent annually. Large ducks and geese may each eat 3000 snails during a single low tide, but have a small effect on the vast shellfish resources of productive estuaries. The principal food of the brant goose (*Branta bernicla*) in estuaries is *Zostera marina;* reclamation projects that destroy the eel grass beds have decreased the bird's numbers. In a reversal of the usual roles the bivalve, *Ischadium*, occasionally traps the toes and presumably drowns its major predator in San Francisco Bay the Clapper Rail, *Rallus longirostris* (Morris et al., 1980).

TROPICAL ESTUARIES: MANGROVE SWAMPS

In hot climates the typical estuarine grass or reed-dominated salt marsh of the temperate zone is replaced by low trees collectively called *mangroves*. The word may originate from *mangue*, a mixture of the Malay, Spanish, French, or Portuguese for "wood" and "grove", English for "a small wood." Several genera dominate mangrove swamps but *Rhizophora*, *Avicennia*, and *Bruguiera* are common representatives. Mangrove swamps cover huge areas of the lowland tropics (Figs. 17-18, 17-19) and *Avicennia*, a cold-tolerant species, ranges as far north as Florida. Mangroves can tolerate salinities from full-strength seawater to fresh water and dissolved oxygen levels as low as zero.

Their ecological role is very similar to a salt marsh. Mangrove swamps develop in wa-

Figure 17-18 Large mangrove swamp on the Indian Ocean coast in tropical east Africa. The estuary is totally undisturbed and shows the typical estuarine dense growth of emergent plants which occurs right up to the water's edge (compare with San Francisco Bay, Fig. 17-3). The different species of mangrove (Fig. 17-21) are indicated by differing shades of gray in this photograph.

terlogged silt from upriver which is continuously deposited and enriched by flocculation at the freshwater-saltwater interface. Mangroves grow rapidly and support large growths of attached animals and plants on their root system. These are exposed at each tide (Fig. 17-20) and are the most obvious feature of mangrove swamps at the water level. The roots form masses of stiltlike branches which grow out away from the main trunk and can penetrate anoxic muds to obtain water and nutrients. The upper part of the roots actually supplies the necessary oxygen for the respiring root hairs below. Some mangrove species have special root modifications called *pneumatophores* to assist in root oxygenation. Shallow water and the fencelike root system prevent most large predators from entering the area enclosed by the trees. As in other marshes, the resulting refuge is an excellent rearing area for juvenile fish, shellfish, and reptiles. Crabs and their usually conically topped burrows are the most obvious animal feature of mangrove swamps at low tide, and as many as a dozen species may be present. Like the plants, the various crab species are zoned and are characteristic of certain seawater levels and water tables.

As in salt marshes, mangrove swamps show distinct floral zonation as one progresses inland. A typical transect through an Australian mangrove swamp relating plant types to tidal height is shown in Fig. 17-21.

Mangrove swamps are disappearing as they are cut down for charcoal or are drained and filled to provide industrial or residential sites. Oil exploration and production on major river deltas in the tropics, such as the Niger in Africa, threaten vast areas of mangroves important for future protein production.

In many tropical countries, mangrove swamps are far more biologically productive than nearby oceans or the often-arid land even a few hundred yards inland from high tide (Table 17-1). It would be much better to build behind existing swamps and cut only necessary access

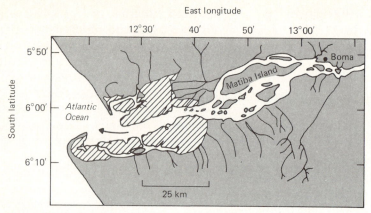

Figure 17-19   Estuary of the Congo (Zaire) River in west Africa showing mangrove swamps (hatched areas). (*Redrawn from Eisma and Van Bennekom, 1978.*)

channels through the mangroves. In the long run mangrove swamps are worth far more as a natural aquatic farm for fish, shrimp, and crocodiles than as sterile docks or factory sites.

Figure 17-20   The stiltlike roots of mangroves provide protection for small animals and a substrate for attachment of aufwuchs. This picture is from Florida near Miami Beach.

FURTHER READINGS

American Fisheries Society. 1966. *A Symposium on Estuarine Fisheries.* R. F. Smith (Chairman). Special publication *No. 3.* 154 pp.

Barnes, R. S. K. 1974. *Estuarine Biology.* E. Arnold, London, and Crane, Russak and Co., New York. Studies in Biology no. 49, 76 pp.

Chapman, V. J. (ed.). 1977. *Ecosystems of the World,* vol. 1: *Wet Coastal Ecosystems.* Elsevier Press, New York. 428 pp.

Coull, B. C. (ed.). 1977. *Ecology of Marine Benthos.* The University of South Carolina Press, Columbia. 467 pp.

Gunter, G. 1961. "Some Relations of Estuarine Organisms to Salinity." *Limnol. Oceanogr.,* **6:**182–190.

Hedgpeth, J. W. 1957. "Estuaries and Lagoons II. Biological Aspects." In "Treatise on Marine Ecology and Paleoecology." *Geol. Soc. Am. Mem. 67,* **1:**693–729.

Hildebrand, S. F., and W. C. Schroeder. 1928. Reprinted 1972. "Fishes of Chesapeake Bay." *U.S. Fish Wildl. Serv., Fish. Bull.,* ser. IV, vol. 53, pt. 1. Smithsonian Institution, Washington. 388 pp.

Jefferies, R. L., and A. J. Davy (eds.). 1979.

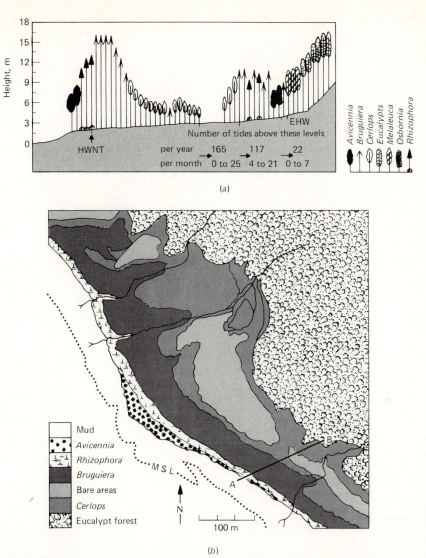

Figure 17-21 Vegetation changes on Magnetic Island, an Australian mangrove swamp, as one moves away from the sea. The amount of tidal exposure controls the species of mangroves present. (a) EHW = extreme high water, HWNT = high water at neap (lowest) tide. Note the trees (drawn to scale). (b) Map of mangroves on the western shore of Magnetic Island. Notice the distinctly zoned mangrove area and the presence of a bare area within the thickets of *Cerlops*. The landward fringe is so narrow that it does not show on this scale. The heavy line at lower right (b) is a line of transect A-B for (a). MSL = mean sea level. (*Slightly modified from Macnae, 1967.*)

Ecological Processes in Coastal Environments. Blackwell Press, Oxford. 684 pp.

Jørgensen, C. B. 1966. *Biology of Suspension Feeding.* Pergamon, New York. 357 pp.

Lauff, G. H. (ed.). 1967. *Estuaries.* American Association for the Advancement of Science, Washington. Publication 83. 757 pp. A multiauthor book with physical, chemical, and biological sections.

Livingstone, R. J. (ed.). 1979. *Ecological Processes in Coastal and Marine Systems. Marine Science,* vol. 10. Plenum, New York. 548 pp.

Olansson, E., and I. Cato. 1980. *Chemistry and Biogeochemistry of Estuaries.* Wiley-Interscience. New York. 452 pp.

Perkins, E. J. 1974. *Biology of Estuaries and Coastal Waters.* Academic, New York, 678 pp.

Ranwell, D. S. 1972. *Ecology of Salt Marshes and Dunes.* Chapman & Hall, London. 258 pp.

Chapter 18

Origin of Lakes and Estuaries, Eutrophication, and Paleolimnology

OVERVIEW

The geological events which created the millions of water bodies throughout the world are the basis of lake classification. The bulk of natural lakes and estuaries were formed by *tectonic, volcanic,* or *glacial* forces. The rest resulted from landslides, river action, wind, meteorites, and the activities of animals, including people. Most lakes in the world were created by glaciers. Some of the oldest and deepest lakes have a tectonic origin, and some of the clearest and most recent are volcanic.

Lake succession from creation to destruction occurs slowly for all but a few small lakes. Some lakes quickly fill up with sediments, while others are eventually destroyed by the same catastrophic processes which created them. Once the lake is created a biotic community becomes established. The lake progresses quickly to some trophic equilibrium which then changes slowly but may oscillate back and forth as the climate varies. If the supply of nutrients is high, the lake is defined as *eutrophic*. If nutrients are in short supply, the lake is considered *oligotrophic*. Eutrophic lakes are usually productive, shallow water bodies with considerable fluctuations in surface and benthic oxygen levels. Oligotrophic lakes are unproductive, often deep, with very transparent waters which are usually fully saturated with dissolved oxygen. It is important to recognize that the terms oligotrophic and eutrophic are rather subjective terms and should only be used for lakes within one general lake district. A eutrophic lake in one area may be better classified as mesotrophic or even oligotrophic in another region. Estuaries and rivers are generally more productive than lakes. These typically more fertile environments may also be classified as oligotrophic or eutrophic in relation to others in their district.

Although lakes are geologically ephemeral, a few are at least several million years old. They are reservoirs of history in the sense that they record in their sediments a record of what has transpired on their watersheds as well as in their waters. The *paleolimnologist* reconstructs this history by analyzing the plant and animal remains preserved in the sediments. Dating of sediments may be accomplished with isotope techniques, the most common of which uses the naturally occurring radioactive carbon 14. Geologic events may be recorded by sediment discontinuities, distinctive layers of volcanic ash, or unusual and identifiable sediment types. Present in some lakes are alternate layers of dark and light sediment called *varves* which show winter and spring sediment inflows. Pollen grains, plant fragments, organic carbon, pigments, diatom frustules, and remains of some zooplankton, zoobenthos, and fish species in the sediment provide an indication of previous lake fertility.

ORIGIN OF LAKES

The origin of lakes provides one of the methods used for classification of lakes, streams, and estuaries and enables comparisons to be made between units which are of similar origin but which may be of very different character. Furthermore, the particular origin of a lake basin, stream course, or estuary is important in determining its morphometry (shape), which was the first element of our consideration of structure (Chap. 2).

Lakes, streams, and estuaries have definite life spans and are ephemeral in the geologic sense. They are born through gradual or catastrophic geological events and will usually be rejuvenated or become extinct through similar processes. Because a lake's existence is finite, its geologic origin can be traced, its rate of "aging" measured, and its eventual disappear-

ance predicted. The process may nevertheless take a very long time. Lake Baikal in eastern Russia, for example, has existed for 25 million years since at least the Tertiary Period, yet still remains the world's deepest lake (Fig. 18-1).

Most basins are created by natural events such as slow glacial activity or gradual deformations of the earth's crust, while a few result from rapid catastrophic geologic events such as earthquakes, landslides, or volcanic eruptions. Occasionally a lake will vanish as quickly as it was created. Such a lake existed in Alaska by virtue of a volcanic dam in the Valley of Ten Thousand Smokes, created by the Katmai eruption of 1913. The lake was so short-lived that it was not named before its volcanic dam gave way and the lake emptied in a great flood.

The lake basins formed in a particular area are generally created by common natural events and thus are similar in appearance. It has therefore been convenient to group lakes into *lake districts* on the basis of the geologic event responsible for their basin formation. Although lakes within a given lake district have similar characteristics, they also have distinct differences from one another (Chap. 19). A good example is the Great Lakes of North America. Although the entire Great Lake system was formed by recent glacial activity, lakes Superior

Figure 18-1 Part of the steep, rocky shores of Lake Baikal, Siberia, a graben lake which is the oldest and deepest lake in the world. (*See also Kozhov, 1963.*)

and Erie differ greatly from one another on the basis of water characteristics, morphometry, and biological productivity. G. Evelyn Hutchinson noted that, "it is this diversity in unity that gives the peculiar fascination to limnology." Lakes may be compared both within districts and between districts.

Just as taxonomy has been employed for the purpose of comparing the relationships among organisms, lakes, likewise, have been subjected to systems of classification for comparative purposes. Hutchinson classified lakes based on the natural agent creating the basin, and a simplified version is presented here in categories of (1) tectonic agents, (2) volcanic agents, (3) glacially formed lakes, and (4) others which include landslides, solution, river excavation, wind, natural coastline activities, organic accumulation, animal behavior, and meteoritic impact. Some of these types of lakes are shown diagrammatically in Fig. 18-2 and Plates 1 to 8.

Tectonically Formed Lakes

Tectonic basins are those formed by deep earth crustal movements, with the exception of volcanism. Most lakes formed tectonically are a result of faulting. They may be one of two types, those associated with single faults and those associated with multiple faults, or *grabens* (Fig. 18-2a). Single-fault basins result from depressions brought about through tilting, while grabens are relatively large depressed areas located between adjacent faults. An example of a single-fault lake is Abert Lake in Oregon. Graben-fault basins are well-exemplified by Lake Tahoe (Plate 1b), Lake Baikal in Siberia (Fig. 18-1), Lake Biwa, Japan, and the huge rift lakes of central Africa. Most old, deep lakes, and many of the world's largest lakes were formed by tectonic action. The long chain of rift lakes in Africa, lakes Malawi, Tanganyika, Edward, Mobutu Sese Seko, formerly Albert, and Turkana, and the now marine Red Sea were all formed by one great fault during or before the Pliocene era, about 1 to 10

million years ago. However, not all tectonic lakes are deep and relatively narrow like Baikal. The size of the lake depends on the magnitude of the faulting and the amount of silting over the years since formation. The two smaller arms of Clear Lake, California, for example, are grabens but are only a few kilometers long and 10 to 12 m deep (Plate 6a).

A second type of tectonic basin is that arising from the uplift of portions of the sea floor. The shallow lakes of Florida are believed to have been formed in this way. One of these, Lake Okeechobee, with an area of 1840 km², possesses the second largest surface area of any freshwater lake in the United States. Droughts and water diversions can greatly reduce the size of Lake Okeechobee due to its very shallow depth. The huge sub-Saharan Lake Chad also changes shape drastically with each season's rains. The Caspian Sea and the Asian Sea of Aral were created by uplift of the sea bottom or closing of the Tethys Sea by plate tectonics. These basins called *relic* lakes, retain their original marine structure. They still possess characteristics of their original environment.

A third type of tectonic basin is that formed by the tilting or upwarping of the earth's crust in such a way as to cause the reversal of the existing drainage system. A good example of this is in east Africa where a reversal of the flow of the Kafu River north of Lake Victoria caused a flooding of the valleys in this region, creating Kioga Lake.

A fourth variation of the tectonic basin is a large basin created through crustal warping. Lake Victoria, Africa, and Thurston Lake, California, are examples. They differ from the previous type in that instead of resulting from the flooding of a valley they resulted from the formation of a basin by the uptilting of the land surrounding the basin. Variations on the same theme are the basins produced by local subsidence or landslides following earthquakes. Reelfoot Lake was formed in 1811 along the Mississippi River in Tennessee by subsidence

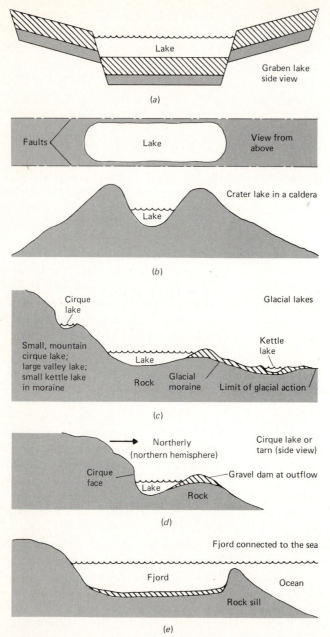

Figure 18-2 Sketches of the origins and present situations of the most common types of lakes. (a) A tectonically created graben lake formed by the sinking of land along faults. Examples include almost all the world's large, deep, old lakes. (b) A crater lake in a caldera inside a volcanic cone. Large and small examples are found in older volcanic regions throughout the world (see also Plates 1a,3). (c), (d), (e) Glacial lakes and a marine fjord formed by the scouring action of glaciers in the recent geological past. These types of lake are very common in temperate areas and in mountains (see also Plate 8 and Fig. 18-5).

from a great earthquake. Natural examples can be found in the southeastern United States. Lakes caused by subsidence over old coal mines are common in Britain. A landsliide created Slide Lake, Wyoming, in 1925 (Fig. 18-3).

Lake Turkana, the northern-most of the African Rift Valley lakes, exemplifies a lake partially formed by a tectonically dammed syncline, or downwarping, while Fahlensee in Switzerland was formed by the damming of a valley by an anticline, or upwarping. Lake Titicaca in the Andes is an example of a lake formed by the uplifting of the surrounding plains.

Volcanically Formed Lakes

Lake basins created through volcanic activity are found in many parts of the world. Perhaps the best-known type is one formed constructively and known as a *crater lake*. The lake is created when the crater of a volcano remains intact and later becomes filled with water. Examples are the lake in Crater Butte in Mount

Lassen National Park, California, or Lake Mahega, Uganda (Fig. 18-4). Crater lakes formed volcanically may be of three types: those existing in the unaltered cone of a previous volcano, those formed in the depression resulting from a volcanic explosion, and those resulting from the collapse of the rim of a volcano. Collapsed cones which form the largest basins are called *calderas* (Fig. 18-2b). The collapse occurs after the magma has been ejected, leaving the rim of the cone unsupported. A prime example of a caldera lake is Crater Lake, Oregon, which is only about 10 km across but over 600 m deep (Plate 1a). The Eifel district of Germany contains numerous lakes formed by explosions at considerable depths below the earth's surface resulting in low rims but deep basins. Examples of these are also found in the Auvergne district of France, as well as at the base of Mount Ruwenzori in central Africa. Big Soda Lake in Fallon, Nevada, is another example of this type of lake. Lakes may also develop from basins formed in the lava due to differential cooling following an eruption. Lake Mývatn in Iceland and Yellow-

Figure 18-3 A lake created by large-scale earth movement. Slide Lake, Wyoming, was formed about 60 years ago when part of a hillside (center, marked by light-colored bare area) slid across the river in the valley. A dam was formed (right) which, although partially eroded in the first few years, has stabilized to form a large lake.

Figure 18-4 A small crater lake, Lake Mahega, Uganda. This is one of many shallow, alkaline lakes in this region. These lakes may appear brown, red, purple, or black and have Secchi disk depths of only a few centimeters. Thus not all crater lakes are as transparent and deep as Crater Lake, Oregon (Plate 1a).

stone Lake in Yellowstone National Park owe their origins to this process.

Another type of lake associated with volcanic action but not with craters are those created by the damming of a valley by lava. Snag Lake in Lassen National Park, California, parts of Clear Lake, California, and many of the Japanese lakes have been created in this manner. A valley may also be dammed by obstruction caused by the appearance of the volcano itself. Lake Kivu in central Africa is an example of this type.

Glacially Formed Lakes

Glacial activity probably constitutes the most important of the lake-creating forces, at least in the last few millenia. Many small lakes were formed during the Pleistocene era when glaciers covered much of the earth.

Most lakes in North America, northern Europe, and Asia, and equivalent areas in the southern hemisphere are either of glacial origin or have been substantially modified by glacial action. The famous lake districts of the Laurentian Great Lakes, the Wisconsin lakes, the English and Scottish lake districts, and the Scandinavian and Alpine lakes are all dominated by lakes of glacial origin. Less well known are those of southern New Zealand, the Himalayas, and the Antarctic coastal areas.

Basins which have been formed by the gouging action of a moving glacier are referred to as *glacial rock basins* (Fig. 18-2c). There are several types, one of which is the ice-scour lake formed by valley glaciers in their slow movement downslope. Rocks frozen into the ice form effective chisels for eroding the valley floor. Glacial rock basins are particularly numerous in the mountains of Norway, southern Sweden, and Scotland. In California, ice-scour basins are found in Devil's Basin southwest of Lake Tahoe and at the head of the south fork of the San Joaquin River.

A second and very common type of glacial-rock-basin lake is the *cirque lake* (Fig. 18-2d).

The french term *cirque* is otherwise used to mean "semicircle" or "amphitheater." This type of lake is usually found at the head of glaciated valleys where the valley abuts against the steep slope of a mountain range. Cirques are usually deepest on the uphill side and shallow near the outlet downhill where they may be dammed by a moraine of glacial debris. The exact process involved in creating the deep gouging which forms cirques is not entirely understood, but the most popular theory is that the rock is eroded away from the glacier, aided by a continual freezing and thawing activity which fractures the rock. Most of the weight of the glacier is adjacent to the rock face and applies its greatest eroding force below the cirque face. Castle Lake in California is an excellent example of a cirque basin with the deepest portion adjacent to the cirque wall (Fig. 18-5, Plate 8a).

If a series of valley-rock-basin lakes occur connected by streams, they are called *paternoster* lakes. They derive their name from the rosary bead–like arrangement of several lakes in a linear series (Plates 2a and 2b). Numerous examples of these can be found in the moun-

Figure 18-5   A typical glacially formed subalpine cirque lake, Castle Lake, California. The steep rock wall, the cirque face, here hidden from view, rises from the deepest part of the lake (see Plate 8a). Outflow across a shallow lip of rock and old glacial moraine occurs at the side of the lake opposite the cirque face (center).

tains of North America, and a six- to seven-lake chain exists in Glacier National Park, Montana. The Sierra Nevada, the eastern side of the Continental Divide in Colorado, and the Bighorn Mountains of Wyoming also contain examples of paternoster lakes.

Fjords and *piedmont* lakes are both types of glacial-rock-basin lakes. Fjords occur along glaciated coastlines and are typically deep and steep-sided. Hornindalsvatn, one of the fjord lakes on the west coast of Norway, is the deepest lake in Europe at 514 m, ranking ninth on a worldwide scale. Fjords are generally associated with Norway where, indeed, some of the best are found, but other fine examples occur in Scotland, New Zealand, and North America. A limnologically important feature of marine fjords is the presence of a partial dam of rock or moraine near the mouth which prevents full circulation with the ocean (Fig. 18-2*e*). *Piedmont* lakes are formed by a descending glacier and are frequently large and deep. These lakes are typically located at the foot of mountains. Good examples of piedmont lakes such as Lago di Garda and Lago Maggiore occur in the Italian Alps. Another example is Lake Wakatipu on New Zealand's South Island where a glacier excavated the basin to 69 m below sea level.

Some lakes are formed by the damming of streams by glacial moraines, which are the rocks and debris left behind after the glacier recedes. Big Cedar Lake, Washington, is an example of a lake formed between two terminal moraines which mark the farthest advance of a glacier. In the United States probably the best examples are the well-studied southeastern Wisconsin lakes: Green Lake, Lake Winnebago, and Lake Mendota. It is sometimes difficult to distinguish between lakes held by moraine dams and those in the irregularities of the material under a glacier, which is called the *ground moraine*. Heron Lake, Minnesota, is an example of a *drift basin lake* formed within a ground moraine. Melting of blocks of ice

trapped in glacial moraine is thought to form kettle lakes (Fig. 18-2*c*). *Kettle lakes* characteristically have very steep sides and are sometimes meromictic because of their small surface to volume ratio and because the wind fetch is small. Numerous kettle lakes occur in the central United States. Waldon Pond in Massachusetts is an example of a double kettle, as are the intensively studied Linsey ponds in Connecticut.

Cryogenic lakes, also known as *thaw* or *thermokarst* lakes, are formed in areas or perennially frozen ground of permafrost. If an area of ground is somehow exposed so as to induce melting, this area will not refreeze and may fill with meltwater. The depression is increased through subsequent melting at the frost-water interface. Examples of these may be found in Alaska and Siberia. Still another form of cryogenic lake is to be found on existing glaciers. These may be formed through differential melting when dust or debris absorbs solar radiation and melts out a pocket. We observed a number of these in the Antarctic.

Lakes with Miscellaneous Origins

Solution lakes are those formed by the dissolution of soluble rock by water percolating through it. Limestone is particularly susceptible to such action. The best examples of this are found in Florida and the Karst region along the Dalmation coast of the Adriatic Sea. Others are found in the European Alps and the Balkan Peninsula. Rock salt or halite (NaCl) and gypsum ($CaSO_4 \cdot 2H_2O$) may also dissolve away to produce solution basins.

Fluviatile lakes are lakes produced by the action of running water. They can be created through destructive processes such as excavation or obstructive processes such as damming. Those known as *plunge-pool* lakes were originally created by erosion at the base of waterfalls. *Fluviatile-dam lakes* are created by the deposition of sediment by a tributary across the main river. An example of this is Tulare Lake,

California, formed through damming a portion of the San Joaquin Valley by sediment deposited by the King's River. Where deposition of sediment by the main river causes obstruction of the lateral valley, a *lateral lake* is formed.

Several different types of small lakes are formed in the flood plains of rivers. One of the most familiar is the *oxbow lake,* formed when the loop of a meandering river is cut off by silt deposition. Examples can be found in the flood plains of almost any river in the world. Where a former river channel has moved from sediment deposition at a bend, *scroll lakes* may form. Examples of these can be seen along the Mississippi and Minnesota rivers and the flood plain of the Paraná River, Argentina (Fig. 2-4*b*).

Flood-plain lakes are connected to the river during high floods and provide a protected habitat for eggs and young of many river fish. Some South American fish migrate hundreds of kilometers from the swamps at the head of the river to breed in the flood plain lakes. Some river species in Africa make lateral migrations during high water to feed and breed in adjacent flood plains.

Beavers and people are two important lake builders. Results of the beavers' activities are to be seen throughout North America and on a few European rivers. Reservoirs have been built on small rivers for several thousand years in Sri Lanka and Egypt although all large dams are recent. Lakes may also form accidentally when mines, borrow pits, gravel diggings or quarries are abandoned. Kimberly Lake in South Africa, formed in the old DeBeers diamond mine, is probably the deepest artifical lake in the world.

Recently huge lakes have been created throughout the world for power generation, irrigation, and flood control (Fig. 18-6). Most of the world's major rivers are now dammed, and those remaining are constantly threatened with enclosure despite frequent associated problems. Reservoir limnology has now become a major challenge for limnologists. The various

Figure 18-6   People are now the main agents creating lakes, usually for power, irrigation, or flood control. This picture shows the dam of one of the largest reservoirs, Lake Kariba on the River Zambesi between Zambia and Zimbabwe. The lake created has an area of over 2300 km². An even larger reservoir downstream is under construction. Reservoirs of this size rival the largest natural lakes.

environmental problems associated with impounded waters are considered in Chap. 20 and effects on river fisheries in Chaps. 14 and 16.

ORIGINS OF ESTUARIES

Geological events which turned rivers into lakes will create estuaries near the ocean. Most estuaries originate from the two processes of glacial melt which produce *drowned river valleys,* and glacial scour which creates *fjords,* but some are produced by *tectonic processes* (Prichard, 1967). Biologists will wish to add a fourth process, the estuaries of large river deltas, which are of lesser concern to the physical limnologist.

Glaciation and the associated ice ages have created the largest number of lakes in the world and produced fjords as well as drowned river valleys. The sea level has risen as much as 400 m since the climax of the last ice age, and the meltwater has flooded many of the large river valleys created during the ice age. Most estuaries in coastal plains are of this type. Perhaps the best-studied one in the United

States is the huge Chesapeake Bay on the Atlantic, but similar estuaries occur on many major rivers throughout the world. The River Thames in Britain (Chap. 20) is a well-studied drowned-valley estuary.

Where offshore barriers such as sandspits or small islands have been created in the mouth of a drowned river valley, the system is then described as a *bar-built* estuary. The major difference is the restriction of water circulation with the sea since the estuary is then almost completely surrounded by land. Many of the complex estuaries in the Gulf of Mexico are of this type, but small, often temporary examples can be found among other estuaries throughout the world.

Where large earth movements have caused depression in the earth's crust, tectonically formed estuaries may result. San Francisco Bay was created about 10,000 years ago when the sea broke through a low point in the Coast Range at the Golden Gate to flood a tectonically depressed, marshy freshwater area. The bay continues to sink at a measurable rate. Similar examples can be found in volcanically active coastlines throughout the world.

Rivers carrying large quantities of silt, like the Mississippi or the Nile, build out their estuaries into the sea rather than being flooded. The very shallow, frequently changing deltas of these large rivers are often highly productive areas and serve as nursery areas supporting the estuary and adjacent marine food chains (Chap. 17).

LAKE SUCCESSION AND EUTROPHICATION

Ecologists in the early part of this century were fascinated with questions of biological succession. They were concerned with changes in animal and plant communities that occur with time, and it is not surprising that there has been considerable effort to categorize lakes and streams on this basis. In this section we will consider how lakes do, in fact, change with time and how eutrophication, particularly *cultural eutrophication*, affects this process.

The trophic state may range from *eutrophic* (literally, "well-nourished") and *mesotrophic* ("moderately nourished") to *oligotrophic* ("little-nourished"). Eutrophication results from increases in such essential plant nutrients as nitrogen, phosphorus, iron, and carbon. Eutrophic lakes have good supplies of all essential plant nutrients from nutrient-rich inflows and recycling. Special physiological adaptations such as nitrogen fixation or alkaline phosphatase production may help compensate for shortages of one or more essential nutrients (Chaps. 8, 9). In contrast, oligotrophic lakes are poorly supplied with nutrients. The general characteristics of oligotrophic and eutrophic lakes are summarized in relative terms in Table 18-1.

Much confusion has occurred over the use of conceptual terms like eutrophic and oligotrophic which are not easily quantified. Comparative limnology (Chap. 19) provides some resolution of this problem. The concept of oligotrophic-eutrophic is applicable only to lakes within one lake district or possibly groups of lake districts. For example in the Great Lakes of North America (Chap. 19), Lake Superior is considered oligotrophic and Lake Erie extremely eutrophic. By the standards of many European lakes, Lake Erie might be considered only mesotrophic since many of the world's lakes have much higher plant growth. If applied to western North America the Great Lakes would occupy the center of a much larger trophic range. For example, lakes Tahoe, Crater, and Waldo are much more oligotrophic than Lake Superior, and Clear Lake, California, is a great deal more eutrophic than Lake Erie. There are a variety of reasons for these differences but there are no completely satisfactory generalities. The terms eutrophic and oligotrophic should be used only relative to other lakes within a single lake district or restricted to very general usage. Examples of relative trophic status and associated physical, chemical, and biological variables are given in Chap. 19 for several lake districts (Tables 19-1 to 19-4).

No matter how a lake basin originated, whether dug as a pit for the excavation of minerals or scraped out of the landscape by a glacier, the lake will show succession. This includes colonization by both plants and animals and a slow but progressive change in water chemistry. In arid areas evaporation gradually increases the salt concentration. In new lakes change during the first few decades may be very rapid whatever the initial trophic state of the lake. Over a longer time, some lakes pass through different trophic states, beginning with the lower fertility or oligotrophy and gradually arriving at a moderately productive or mesotrophic state. Particularly in lakes formed by glacial action, eroded sediment may arrive in sufficient quantities to change the depth of the lake significantly. This, when combined with rising fertility from the increased drainage area to lake volume ratio, may result in a highly productive or eutrophic state. However, if the now-shallow lake is exposed to wind action, the suspended sediments may so reduce light penetration in the muddy water that less plant growth occurs than previously. Technically the lake will be eutrophic but actually it will be relatively unproductive. If erosion is severe and the original basin shallow enough, the lake will eventually fill in completely. First, perhaps, it will pass through a *dystrophic* or bog stage, where humic acids from leaching in the soil give the water a characteristic yellow-brown or humic color. Eventually, invasion around the margins by rooted aquatic macrophytes chokes the aquatic habitat with plant growth. The lake is extinct when it no longer contains water. Formerly glaciated mountains contain innumerable meadows and forests which in their time were mountain lakes.

Ideas about eutrophication were initially based on the very obvious successions seen in small mountain lakes which eventually became meadows when they were completely filled. The observed changes seemed to follow well-understood successions of terrestrial plants which are thought to reach a *climax,* a stable state where no further change takes place unless the climate changes or a large fire occurs. The succession of a European oak forest from grassland to birch and willow trees and finally to a climax mature oak woodland is an example. As mentioned earlier (Chap. 12), the main vegetation in lakes, the phytoplankton, is not necessarily specialized for a particular lake type. Most species are quite generalized and are found in lakes with a variety of trophic conditions, although their relative abundance often reflects their trophic status (Tables 19-2, 19-4).

Succession and climax in lakes is not as well-based theoretically as is terrestrial succession, which is also being reevaluated. In fact, consideration of lakes throughout the world leads to the conclusion that the idea of oligotrophic to mesotrophic to eutrophic to extinction is only one of several possible routes of lake evolution. Lakes are created and destroyed by geologic or climatic events. For many lakes, their lifetime will be more likely determined by these catastrophic events and not by gradual sedimentation or inflow of nutrients. Certainly for most large deep lakes including the Great Lakes, Lake Baikal, or Crater Lake, Oregon, it is more likely that they will remain oligotrophic until they are destroyed by some catastrophic process such as the one that created them. With human assistance, a more rapid change in trophic status can occur.

In the geologic sense all lakes, like the mountains and plains which surround them, are ephemeral. The speed with which lakes change with time is, however, very much dependent upon the size and nature of their drainage area, upon the average depth of the original basin, and upon fluctuations in local climate, especially temperature, and erosive rainfall. Paleolimnological evidence from the study of deep lakes undisturbed by human beings suggests that a lake may change from an oligotrophic to a eutrophic state but then reverse and become oligotrophic again (Fig. 18-10). This cycle may occur several times during the existence of the lake and is related to climatic changes in the region. However, vegetation may greatly modify the influence of climatic factors.

Geologic events like landslides, volcanic eruptions, and faulting can greatly alter the progress of eutrophication. The geologic origin of the lake basin often controls the lake's future trophic state since it largely determines the depth of the basin and the size of the watershed relative to it. Lakes in deep grabens such as Tanganyika, Baikal, and Tahoe would take millions of years to fill with sediments. For example, Lake Washington, Seattle, with an unusually high annual sedimentation rate of 3 to 5 mm over the past few decades, would take about 20,000 years to fill. A lake in a volcanic cone usually has a small watershed for erosion and little prospect for rapid evolution. At the other end of the time scale, oxbows and scroll lakes which are associated with rivers may be created or disappear on an annual basis depending upon the height of the spring flood.

The reason that eutrophication is of such great interest to the limnologist is because it alters the physical, chemical, and biological character of natural waters. The distribution of temperature and oxygen (Chap. 7) is helpful in classifying lakes as to their trophic status and has been used to compare eutrophic and oligotrophic lakes (Table 18-1). As eutrophication proceeds, nutrients increase and autochthonous organic matter production rises. The general distribution of primary production in the water column is illustrated in Fig. 18-7, where permanently frozen Lake Vanda and monomictic Lake Tahoe, which are oligotrophic lakes, are compared with mesotrophic Castle Lake, eutrophic Clear Lake, and very eutrophic Lake George, Uganda. The higher production per unit volume is particularly evident. It should be

noted that if the productivity is measured per unit area of surface, as is normal in terrestrial studies, there is often less difference between oligotrophic and eutrophic lakes. The long photic zone of the clearer lakes partially compensates for their low production per unit volume (Fig. 18-7). The higher production per unit of surface area of eutrophic lakes is evident from (Fig. 18-7).

In general, *oligotrophic* lakes have steep sides and relatively small drainage areas. This is a broad generality since some extremely oligotrophic lakes are no more than shallow pans on the granitic slopes of the Precambrian shield or in the high mountains. All oligotrophic lakes are characterized by low nutrient and organic matter levels and usually maintain high hypolimnetic and benthic oxygen levels throughout the year whether they stratify or not. There is good light penetration through the clear waters (Fig. 3-6b), and, with the exception of some alpine lakes, there is little or no rooted vegetation. Phytoplankton density is low and production is limited at higher trophic levels. In contrast, *eutrophic* lakes are apt to be shallow with gradually sloping basins and a large drainage area to lake surface area ratio (Table 18-1). Nutrient and organic matter levels are high, and in temperate lakes hypolimnetic oxygen is likely to be depleted both in summer and under ice cover. In tropical eutrophic lakes oxygen levels near the mud are low year-round. Because of the high production of algae, light penetration is limited (Fig. 3-5a), but there is often rooted and emergent vegetation around the lake margin. In some eutrophic lakes phytoplankton growth is sufficiently dense to shade out most macrophytes even around the lake edges. Perhaps the most obvious feature of eutrophic lakes is the high standing crop at all levels of the food chain. In particular, eutrophic lakes usually show good fish production per unit of surface area.

It is paradoxical that an oligotrophic lake may sometimes have more attached, nonrooted algal growth than a eutrophic one. Recent stud-

Table 18-1 How Eutrophic Lakes Can Be Distinguished from Oligotrophic Ones

Most of the very productive lakes in the world are shallow and unstratified, while most of the very unproductive ones are very deep. The terms oligotrophic and eutrophic are best applied to lakes within one lake district or climatic region. WL ratio = ratio of watershed area to lake area.

Factor	Oligotrophic (unproductive)	Eutrophic Productive	Eutrophic Less productive
Nutrients	Low levels and low supply rates of at least one major nutrient (e.g., nitrogen, phosphorus, silica)	High supply rates and often high winter levels of all major and minor nutrients	Often high levels of nutrients year-round
O_2	Does not vary much from saturation in epilimnion or hypolimnion ($100 \pm 10\%$)	Great variation from saturation. Depression in hypolimnion ($0-100\%$) and mostly supersaturation in epilimnion ($80-250\%$)	Similar to oligotrophic.
Biota	Low densities and yields of phyto-plankton and zooplankton, zoo-benthos and fish	High densities and yields of phyto-plankton and zooplankton, zoo-benthos and fish	Similar to oligotrophic.
Light	Transparent water, deep light penetration, often to below thermocline. Secchi depth 8–40 m	Water not very transparent, light penetration relatively low, often not reaching thermocline or lake bed. Secchi depth 0.1–2 m	a. Water often cloudy; low light penetration due to peat fragments or humic acids (*dystrophic* lake) or to suspended sediments. b. Water clear but acid, pH < 4 (*acidotrophic* lake).
Basin shape and watershed	Lakes deep and steep-sided. Undisturbed, rocky, or unproductive watershed. WL ratio low (e.g., 1:1).	Lakes shallow with gently sloping sides. Often unstratified. Cultivated, disturbed, or naturally fertile watershed. WL ratio high (e.g., 100:1).	Lakes usually small and shallow. Watershed with peat wetlands, coniferous forest or easily eroded soils. Acid volcanic springs, acid rain or muddy inflows. WL ratio variable.

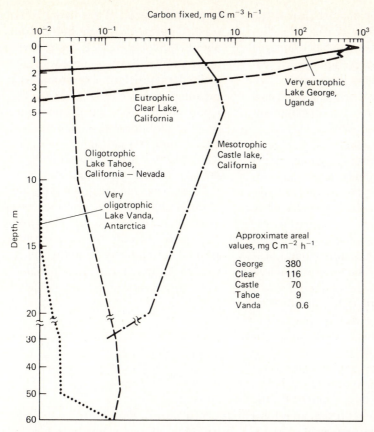

Figure 18-7 Primary production in five contrasting lakes. Very oligotrophic Lake Vanda, Antarctica, oligotrophic Lake Tahoe, mesotrophic Castle Lake, California, eutrophic Clear Lake, California, and very eutrophic Lake George, Uganda. Note log scale for primary production. Production per unit volume is very different between the lakes, but on an areal basis there is less difference due to the deeper photic zones of the more transparent lakes. It is difficult for any organism to utilize this higher per area productivity since it is spread so thinly in transparent lakes. Areal values are from literature values; curves are on specific, not necessarily average, days. (*Redrawn from Goldman, 1968a, b; Goldman and Wetzel, 1963; Ganf and Horne, 1975; and Tilzer and Horne, 1979.*)

ies using special equipment such as a submarine show growth in an oligotrophic lake down to 160 m, well below the level for conventional scuba diving. The deep photic zone of oligotrophic lakes and scarcity of grazers allows slowly growing periphyton to accumulate in the relatively nutrient-rich hypolimnetic waters. On the underside of ledges, benthic grazers such as crayfish may be ineffective in removing attached algae, leaving large greenish stands on the otherwise bare cliffs.

EUTROPHICATION IN FLOWING WATERS

Estuaries and rivers should be included by limnologists in discussions of eutrophication. The destruction of the Thames and its restoration (Chap. 20) provide an example of the limitations of applying the trophic concept. Estuaries are normally much more productive and eutrophic than lakes, due to tidal action. This energy subsidy may effectively stir up nutrients in shallow mud flats, flush salt marshes, and provide for the salting-out of particles at the saltwater-freshwater interface (Chap. 17). Nutrient accumulation from extensive drainages coupled with the tidal energy input cause estuaries to be among the most productive environments on earth. They are of great value as a food resource if their fertility is properly managed and if pollution is controlled.

With higher rates of production than lakes, estuaries such as those found draining most temperate or tropical lowlands, are highly eutrophic. Chesapeake Bay in the northeast United States or the mouth of the Purari River in New Guinea are examples of eutrophic estuaries which should be maintained to sustain their valuable fisheries. High-latitude and fjord-type estuaries are amongst the least productive estuaries but marine fjords are more productive than their freshwater counterparts as shellfish production attests. Thus estuaries, like lakes, can be classified as eutrophic or oligotrophic on a relative basis.

Eutrophication in rivers and streams is in many ways similar to that found in estuaries. Once again an energy subsidy imparted by the water flow can result in higher production for a given nutrient concentration than is usually encountered in lakes. One can occasionally find extensive attached algal growth even in the most pristine alpine streams where nutrient levels are always low. Additions of nutrients will often further increase the growth of aufwuchs. The choking of some rivers with long strands of the green alga *Cladophora* can be a major eutrophication problem in temperate rivers (Lund, 1972).

CULTURAL EUTROPHICATION

On the basis of sediment profiles, Hutchinson concluded that more-or-less steady states of eutrophication exist for considerable periods of time. For example a small lake along a highway, the Via Cassia north of Rome, became eu-

trophic as a result of the deforestation of its watershed during the height of the Roman Empire about 2000 years ago. Examination of the sediment profile down to about 25,000 years indicated a consistent, low productivity until the building of the road (Fig. 18-8). Linsey Pond, Connecticut, was originally an oligotrophic kettle lake and showed a long period of trophic equilibrium in a eutrophic state. A return to a less eutrophic state apparently occurred during post-European settler soil erosion.

The changes in phosphorus in the sediments of culturally eutrophic and then restored Lake Washington (Chap 20) are shown in Fig. 18-9. The 1958 sediment core contained large amounts of phosphorus in the upper few centimeters but much less prior to waste discharge. The lower concentrations now being deposited

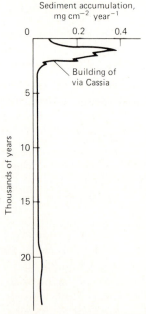

Figure 18-8 Accumulation of sediment in the Lago di Monterosi, Italy, over time measured by ^{14}C dating. Note large increase when the road Via Cassia was constructed and subsequent recovery as the soil erosion decreased afterward. (*Units corrected from Cowgill and Hutchinson, 1970.*)

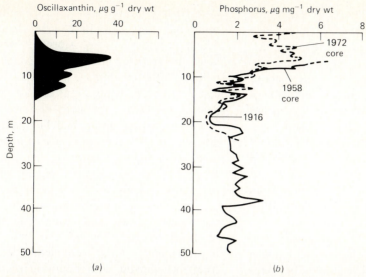

Figure 18-9 Changes in sediment content in Lake Washington, Seattle, before and after its restoration from a eutrophic state (see also Chap. 20). The core has been adjusted vertically to coincide with a known marker event in the lake in 1916. (a) The rise and fall in oscillaxanthin content which derives from changes in the abundance of the blue-green alga *Oscillatoria agardhii*, a major component of the plankton when the lake was eutrophic. (b) The lower phosphorus content before waste discharge in 1941 and its rise until waste diversion in 1968. The rise and fall of oscillaxanthin coincides with the changes in phosphorus. (*Redrawn from Edmondson, 1974b; Griffiths and Edmondson, 1975.*)

are clearly illustrated in the 1972 core, a few years after waste diversion. Other examples are discussed in this chapter in the section on paleolimnology (Figs. 18-10 and 18-11). These changes Hutchinson termed *cultural eutrophication,* an idea which has become widely accepted in the limnological literature. The cultural eutrophication of the Great Lakes is general public knowledge (Chap. 19).

Cultural eutrophication differs from natural eutrophication in that it is greatly accelerated in the sense of geologic time. As the term implies, people are the causative agents through enrichment of natural waters in various ways. Only in the last hundred years have community sewage treatment plants become more the rule than the exception in the developed nations. Today most sewage treatment plants settle out and break down particulate matter but discharge the major nutrient load to the receiving water. It is

perhaps one of the ironies of our time that lakes, rivers, and estuaries may have been better protected by the wide dispersal of waste treated in individual cesspools or septic tanks than by its collection in a single large outfall which flows directly into a lake or river (Chap 20). Certainly municipal sewage, along with agricultural drainage, is one of the oldest causes of cultural eutrophication. Most waste water is rich in nutrients and, unless chlorinated, serves as an immediate stimulant for the growth of algae and higher aquatic plants. Some more-elaborate treatment plants are now in existence which reduce phosphorus, and a handful also reduce nitrogen and use activated charcoal to absorb organic contaminants. This advanced treatment is extremely expensive and energy-demanding; recently, authorities have been reconsidering use of the soil mantle, particularly for phosphorus removal (Chap 9). In sunny climates large ponds can be designed to remove all types of nutrients in a most economical fashion.

The use of detergents high in phosphate has stimulated cultural eutrophication in some previously phosphorus-deficient lakes, since about half the phosphorus in sewage effluent is derived from detergents. Intensive agriculture is another source of nutrients for aquatic systems. Invariably some of the fertilizers applied are washed off by rain or during irrigation. This washout is particularly evident when nitrogenous fertilizer is applied to frozen soils in spring. During the thaw, most is washed off directly to the lake or river (Fig. 16-11).

Industrial wastes may be stimulatory and create cultural eutrophication or toxic, causing *cultural oligotrophication*. The stimulatory ones include the inorganic products of fertilizer factories and the organic waste residues of food processing plants. Long strands of the bacteria *Sphaerotilus,* which use organic carbon as a growth substrate, are a common and unpleasant sight in rivers downstream from canneries and pulp mills. Another source of nutrients is sediment transport to lakes and streams during land clearing for road and house building. These developments frequently promote growths of green filamentous algae, and the turbid waters make fishing difficult. Although more fish may be supported by the increased fertility, the species composition is often less desirable to the angler.

Clear-cutting, a prevalent form of lumbering throughout western North America and in many tropical forests, is particularly destructive to spawning streams and contributes to eutrophication of waterways (Fig. 6-2). The roads used for removing the lumber and the careless use of heavy equipment such as bulldozers in the forest can create enormous erosion. The worst results occur when no buffer strip of riparian vegetation is left between the cut zone and the river, and where roads cross the streambed. The lack of protection of the soil from rain during storms causes heavy sedimen-

tation in any receiving waters. The simple act of removing the vegetation releases nutrients which previously were stored in the humus layer and recycled continually through the vegetation. Although damage may be short-lived, it can destroy spawning areas and, if continued, destroy the fishery for many preferred species.

In the chapter on applied limnology (Chap. 20), we discuss the cultural eutrophication of lakes Washington, Shagawa, and Tahoe, the River Thames, and the subsequent restoration programs. Attempts are now being made on the massive task of reducing the eutrophication of the Great Lakes, and Lake Erie has been reported to be showing the first signs of improvement. Since eutrophication is so frequently linked with pollution or excessive fertilization, the point may be missed that eutrophic environments are also more productive at higher trophic levels than oligotrophic ones. In general, managed ecosystems are maintained at an optimal level of productivity (Fig. 15-3). If carp, bass, or tilapia are the end product, the fisheries manager is well-advised to maintain highly eutrophic, warm-water conditions while at the same time guard against oxygen depletion. However, if trout are to be cultured, cool, more oligotrophic waters are necessary.

Lake succession proceeds very slowly in the geologic sense and can reverse itself (Fig. 18-10). However, the discharge of sewage, excess fertilizer, and eroded sediments into lakes and rivers has imposed an accelerated rate of succession upon many inland and coastal waters of the world. Since soil is a slowly renewed resource, and the cost of fertilizers rises with that of energy, it is important that ecologists recommend conservation of both soils and nutrients. In this way, the basic fertility of the land will not all eventually end up in the sediments of our lakes, estuaries, and coastal waters.

Water itself is often too valuable to be utilized only for the dilution of domestic and industrial wastes and sediments. Wasting ni-

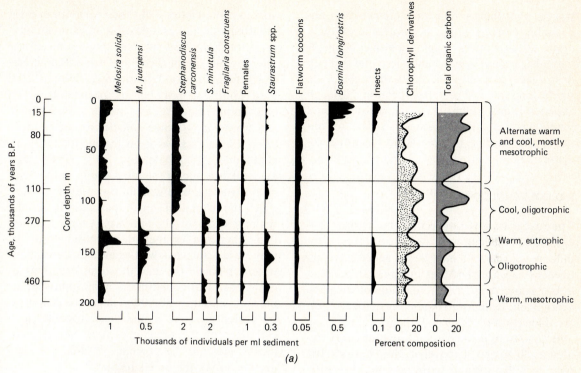

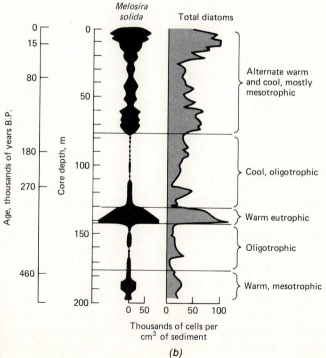

Figure 18-10 The paleolimnological record of the graben Lake Biwa, Japan, over the past half-million years. Two hundred meters of sediment have accumulated over this period. The lake has changed shape considerably over time. This core was taken beneath the present lake. (a) The changes in phytoplankton, zooplankton, and benthos with time and change in climate and trophic state as ice ages came and went. (b) The oscillations in the lake's trophic state from oligotrophic to eutrophic and back as indicated by changes in the most common diatoms. (*Redrawn from Kawanabe, 1978; Fuji, 1978; Mori, 1974; and Horie, 1981.*)

trogen, phosphorus, and trace elements on undesirable eutrophication becomes increasingly tragic as the world's populations rise. Neglecting this aspect of applied limnology has already destroyed waterways and lakes throughout the world. The reversal of eutrophication, although expensive and often slow, is possible, and limnologists should constantly strive to improve the quality of the world's limited resources.

PALEOLIMNOLOGY

The reader is now aware that the natural process of eutrophication or its reversal may span millions of years or may be so accelerated by events on a watershed that changes can occur in a few years. A special area of limnological investigation, called *paleolimnology,* focuses on the dimension of time.

Botanists, in their search for evolutionary evidence, long ago discovered that besides the remains of large plants in the geological records they could also identify specific vegetation by its characteristic pollen (Wright et al., 1963). Pollen grains are proteinaceous but resist decay. They exist indefinitely in marsh deposits as well as in the sediments of lakes. The paleolimnologist can use pollen to reconstruct the ancient plant communities that were present in the watershed of a particular lake and thus determine what climatic conditions existed.

Pollen is prepared for identification by soaking the sediments in a very strong base, such as sodium hydroxide. This dissolves any accumulation of diatoms and renders the preparation ready for pollen analysis. Because of their siliceous frustule (Fig. 10-2), diatoms are among the best-preserved fossils in the lake sediment. In contrast to the pollen, they are extraordinarily heat- and acid-resistant and can be cleaned of their organic matter and associated detritus by boiling in acid or burning at red heat in a muffle furnace. Their intricate structure makes them ideal for taxonomic identification,

and a knowledge of their environmental requirements makes it possible to estimate temperature, light, and nutrient conditions during the previous life of the lake. Other organic material such as wood fragments, charcoal, the hard chitinous remains of zooplankton and chironomids, scales, vertebrae or spines of fish, and freshwater sponge spicules are often identified. Pioneering work on the identification of cladoceran fragments by Frey (1964) has evolved into one of the most useful methods of reconstructing lake trophic histories.

At the chemical level, particularly in the realm of organics, the options are almost unlimited. Pigments, or their degradation products, may be measured as well as organic carbon, nitrogen, phosphorus, and the complete range of other chemical constituents. Chlorophylls dating back as far as 30,000 years were described by Vallentyne (1960), and these, together with estimates of the organic content of the sediments, indicate the presence of different algal groups during the lake's history.

One of the most difficult, and often most expensive, aspects of paleolimnological investigation involves dating the various strata. In young glacial lakes, such as those created 10,000 years ago in North America, the sediment consists entirely of the accumulation since glaciation. In older glacial lakes, and lakes of different origins, estimation of total age is more difficult. *Varves* are alternate light and dark bands in the sediment. The light portion is from inorganic accumulation during winter and early spring runoff, while the dark band represents organic deposition during the main period of plant growth. If varving occurs, the year-by-year sedimentation can be estimated by counting alternate dark and light bands. Geological events often provide important and recognizable discontinuities in the sediment profile. Among the most certain methods for aging particular strata in lakes is the presence of volcanic ash from known eruptions. The Katmai eruption of 1913 left an indelible layer of ash in many Alaskan lakes, and the recent Mt. St. Helens eruption in Washington will provide a similar mark in lakes along the path of the ash plume. The type of glass in the volcanic ash tends to be a specific signature for particular eruptions, and can be used for dating sediments over the entire area of ash fallout. As in the ocean, the magnetic orientation of iron-containing sediments may also be utilized since there is a growing knowledge of the magnetic changes which have occurred with time. X-ray examination of sediment cores for lead has also been accomplished with interesting results, for example, the pinpointing of the year 1916 in Fig. 18-9 (Edmondson and Allison, 1970).

In mesotrophic Castle Lake it has been possible to reconstruct the vegetation from an examination of a few meters of sediment which has accumulated in the 10,000 years since the lake was created in a glacial cirque. The bottom of the core is composed of glacial flour. This rapidly blends into layers of increasing organic content and greater accumulations of pollen grains from the various conifers and deciduous trees that grew on the watershed. Yellow pine replaced the lodgepole pine and alder trees eventually became abundant. In eutrophic Clear Lake, California, the sediments are very deep and at least 250,000 years old in the main basin. The large, round basin has an extremely uniform particle composition in the upper 166 m of sediment. From this one can infer that the depth of the lake and erosion in the watershed has been rather constant. In contrast, cores from the two smaller basins, when examined for fish remains and pollen grains, showed that these sections of the lake had varied between shallow swamps with extensive lily beds and their present much deeper state. The major paleolimnological events in Clear Lake concern volcanic and tectonic activity which alternately isolated the small basins of the lake from the main lake basin. The shallow main basin has apparently sunk continually since its formation. This has produced a paradoxical paleolimnological record which indicates that this oldest of North American lakes has always been roughly the same depth!

Some of the most extensive paleolimnological studies have been carried out on Lake Biwa, Japan, where 1000 m of sediments, accumulated over several million years, have been examined (Horie, 1981). Among the most interesting results is the study of microfossils of zooplankton and zoobenthos over the last 500,000 years in 200 m of sediment (Fig. 18-10a). In this sediment, the remains of several species of Cladocera and flatworm cocoons could be distinguished. Most noticeable were the intermittent presence of insects and the presence of *Bosmina* only in the last 100,000 years. In contrast, turbellarian flatworms showed little change over half a million years (Fig. 18-10a). Among the most sensitive indicators of change are the phytoplankton (Tables 19-2, 19-4). In Lake Biwa large centric diatoms such as *Melosira* or the green alga *Staurastrum* fluctuated considerably while pennate diatoms including *Fragilaria* remained fairly constant (Fig. 18-10a). The changes in biota can be correlated to the changes in climate and trophic level over at least half a million years. Some diatoms, for example, *Melosira solida,* were only abundant during eutrophic phases and were scarce when the climate was cool and the lake perhaps more oligotrophic. The total numbers of diatom frustules also demonstrate these changes (Fig. 18-10b). We can conclude from this long record that eutrophication is not always a one-way process but one which may reverse as climates change. Lake restoration takes advantage of this fact to reverse cultural eutrophication (Chap. 20).

From the early days, people have been attracted to the shores of lakes which provided protection, food, and easy transport. These lake dwellers left an archeological record in the sediments beneath their wood and reed houses, which often extended out into the lake on lightweight pilings. The history of the English

Lake District from the time of the stone age "beaker people" to modern times is preserved in the lake sediments. The early residents traveled along high ground and burned or cut the forests, reducing tree pollen in the sediments. The pollen record shows the pine's replacement by grassland and acidic sedge and heather moors. Later, an agrarian society drained the lowland birch and alder swamps and replaced them by grasses for grazing sheep. All this is clearly recognizable in the pollen and charcoal content of the Lake Windermere sediments (Fig. 18-11). Even where trees are absent, grasses, weeds like the ragweed, and cultivated cereals may be used to reconstruct agricultural activities.

Since evolution remains a unifying theme throughout much of biology, reconstructing the plant and animal communities remains a fascinating area of investigation. The oldest lake, Baikal, which after 25 million years now con-tains over a thousand endemic plants and animals, can provide important insights into the evolutionary process. The intellectual value and practical aspects of the application of the time dimension to lake studies provides fascinating opportunities.

FURTHER READINGS

Cushing, E. J., and H. E. Wright, Jr. (eds.). 1967. *Quaternary Paleoecology*. Yale University Press, New Haven, Conn. 433 pp.

Gorham, E., J. W. G. Lund, J. E. Sanger, and W. E. Dean, Jr. 1974. "Some Relationships between Algal Standing Crop, Water Chemistry, and Sediment Chemistry in the English Lakes." *Limnol. Oceanogr.*, **19**:601–617.

Horie, S. (ed.). 1974. *Paleolimnology of Lake Biwa and the Japanese Pleistocene*. Otsu Hydrobiology Station, Kyoto University, Japan. 288 pp.

Hutchinson, G. E. 1957. *A Treatise on Limnology*, vol. 1: *Geography, Physics and Chemistry*, "Ori-

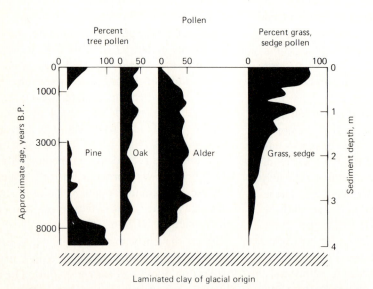

Figure 18-11   The pollen record in the sediments of Windermere, given as a percentage of total tree or grass and sedge pollen. The record begins about 10,000 years before present (B.P.) with laminated clays laid down as the glaciers retreated. Pines died out or were removed, and grassland increased about 8000 years B.P. Drainage of swamps reduced the number of alders and further expanded grasslands about 800 years ago. (*Modified from Pearsall, 1950.*)

gin of Lake Basins," chap. 1, pp. 1–163. Wiley, New York.

———, and U. M. Cowgill. 1973. "The Waters of Merom: A Study of Lake Huleh. III. The Major Chemical Constituents of a 54 m Core." *Arch. Hydrobiol.*, **72**:145–185.

Lund, J.W.G. 1972. "Eutrophication." *Proc. R. Soc. London, Ser. B*, **180**:371–382.

Manny, B. A., R. G. Wetzel, and R. E. Bailey. 1978. "Paleolimnological Sedimentation of Organic Carbon, Nitrogen, Phosphorus, Fossil Pigments, Pollen and Diatoms in a Hypereutrophic, Hardwater Lake: A Case History of Eutrophication." *Pol. Arch. Hydrobiol.*, **25**:243–267.

National Academy of Sciences. 1969. *Eutrophication: Causes, Consequences, Correctives.* Washington. 611 pp. Includes lakes, rivers, and estuaries.

Parma, S. 1980. "The History of the Eutrophication Concept and the Eutrophication in the Netherlands." *Hydrobiol. Bull.* (Amsterdam), **14**:5–11.

Vollenweider, R. W. 1968. *Scientific Fundamentals of the Eutrophication of Lakes and Flowing Waters, with Particular Reference to Nitrogen and Phsophorus as Factors in Eutrophication.* Tech. Rept. No. DAS/CSJ/68.27. Organization for Economic Cooperation and Development, Paris. 159 pp.

Chapter 19

Comparative and Regional Limnology

OVERVIEW

Comparisons of different lakes within one geographical area or *lake district* is one of the most efficient ways to study limnology. Comparing conclusions reached concerning one lake district with those from other lake districts often enables generalities to be drawn about such difficult limnological problems as eutrophication. Fertility is after all relative, and the degree of eutrophication can only be evaluated in terms of other lakes. What is considered eutrophic in one lake district may even be oligotrophic in another. Finally, comparison between very different types of lakes or lake districts, such as temperate versus semiarid, polar versus tropical, or saline versus fresh water brings to light principles which might otherwise be missed. In this chapter we consider the comparative limnology of two lake districts, the well-studied English Lake District, and the huge, difficult-to-study, Laurentian Great Lakes of North America.

Warm-water tropical lakes can be compared with cold-water alpine or polar lakes. The seasonal constancy of the flora and fauna of many tropical lakes contrasts with the very pronounced seasonal effects in polar and alpine lakes. High temperatures provide stable density stratification in tropical lakes, sometimes on a daily basis, but unfrozen cold lakes (less than 10°C) remain unstratified. The two effects of water temperature and degree of mixing combine to produce phytoplankton populations dominated by large diatoms in cold water and large colonies of blue-green algae in warm-water systems. The role of light as an inhibitor of photosynthesis is pronounced in both polar and alpine lakes. As a result, in shallow waters extensive benthic algal mats develop in polar regions and submerged macrophytes coated with epiphyton often dominate alpine lakes. Phytoplankton production in these lakes is usually less important than production by attached plants.

Biological diversity of plants and animals decreases rapidly as one moves from fresh to saline lakes, and fish and macrophytes are entirely absent in the saltiest lakes. Low species diversity in saline lakes and estuaries is indicative of the physiological stress produced by variable osmotic conditions. As a result of adaptation to high salinity, there is more similarity between the biota of saline lakes throughout the world than is found among freshwater lakes.

In contrast to lakes, streams have a characteristic biota the world over. Only a few streams share such special environmental conditions that they prevent the development of the typical array of organisms. Examples include polar or alpine streams and hot springs. In cold climates ice formation on the streambed and *frazil ice* in the water can be a major problem for living organisms. Where the temperature in a stream is constant due to spring inflow, an oasis of luxuriant plant growth is likely to persist throughout the year.

INTRODUCTION

Possibly the most efficient method to advance our knowledge of limnology is to utilize comparative studies of different types of lakes within the same geographical area. Limnology in these *lake districts* (Chaps. 1, 18) has the advantage of considering lakes in areas of generally similar climate, soil, and vegetation. Individual lakes within a lake district share these common denominators but may be shallow or deep, large or small, or of very different fertility. In any one district, the effects of particular environmental factors such as wind shelter, the introduction of a particular fish, or change in nutrient levels can be best studied. Great limnologists have been associated with lake districts,

for example, Lund and Macan in the English lakes, Ohle and Ruttner in German and Austrian lakes, Birge and Juday with the Wisconsin lakes, Beadle and Talling with East African lakes, Rawson with Canadian lakes, and Welch and Eggleton with Michigan lakes. The major drawbacks of comparative limnology are the long time and comparatively high cost required before conclusions are reached.

MEASUREMENT

Gathering sufficient samples from several lakes involves a considerable physical effort and a good study design. Whenever possible, identical methods should be used in each lake. However, care must be taken to avoid cross-contamination which may occur if the same phytoplankton or zooplankton nets are used. Separate nets should be used for each lake, and other collection devices should be carefully washed after use.

Relatively simple experimental designs are essential. In comparative limnology it is certain that data gaps will occur if complex apparatus is required or if sampling time is insufficient for the inevitable delays from bad weather and equipment failure. The lack of certain kinds of information in comparative limnology is in part compensated for by the long data runs assembled. Very rapid sampling techniques have been used for simultaneous or synoptic sampling. Helicopters for surface sampling have been employed in the Great Lakes, the estuaries of San Francisco Bay, and the Elbe in Germany, and in the Environmental Protection Agency's survey of 600 lakes in the United States. Remote sensing from satellites or aircraft is possible, but at present, lack of resolution and verification of reflectance signatures limits its application. The classic method still employed in multilake studies is a coordinated effort by many scientists working in a particular lake district. The creation of the Experimental Lakes Area in Canada, where up to 1000 lakes could be sampled, is an example (*J. Fish Res. Bd. Canada* **28**: 123–301. Special Issue).

LAKES IN THE TEMPERATE ZONES

These areas contain most of the world's extensively studied lakes. The lake districts in the temperate zone were formed or modified by glacial action (Chap. 18) and are often situated in partially forested areas, with moderate rainfall most of the year. They often show phosphorus, nitrogen, and silicon limitation. Of the many possible examples, we will discuss in detail only the English Lake District and the Laurentian Great Lakes.

The English Lake District

In this area of northwestern England about 18 lakes radiate out from a low mountain range (Fig. 19-1, Plate 8*b*). Most seldom freeze and are monomictic, a few are dimictic, and some are polymictic. They show a wide variety of

physical, chemical, and biological characteristics, and their most important features are summarized in Tables 19-1 and 19-2. The proximity of these lakes to each other and their small size have made them an ideal subject for intensive study over many years.

A feature of limnological studies on animals and plants in the English Lake District has been the abundance of some species in only a few lakes (Table 19-2). This occurs despite apparently favorable conditions for these species in other lakes in the area. For example, the zooplankton copepods *Cyclops* and *Mesocyclops* are found together in only 3 out of the 18 lakes (Chap. 13). The thermal stratification responsible for the onset of spring blooms of the diatom *Asterionella* occurs earlier in shallow lakes of the English Lake District because shallow lakes stratify more rapidly than deep ones (Table 19-1). The stratification prevents the algae in the epilimnion from being mixed into the dark, deeper waters which would reduce their growth rate (Chaps. 5, 12).

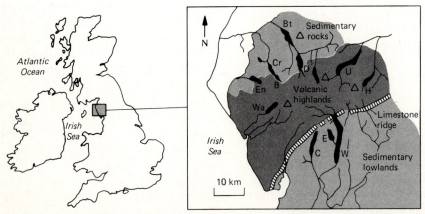

Figure 19-1 Fifteen of the lakes of the English Lake District (black areas) arrayed like spokes in a wheel around the central mountains. All the lakes were formed as the most recent glacial period ended over 10,000 years ago. W = Windermere, E = Esthwaite Water, Wa = Wastwater, En = Ennerdale Water, C = Coniston Lake, B = Buttermere, H = Haweswater, U = Ullswater, Cr = Crummock Water, D = Derwent Water, Bt = Bassenthwaite Water. The volcanic origin of the central massif contrasts with the sedimentary deposits of slate and sandstone in which most of the lakes actually lie. The central rocks are quite barren and produce inflows low in nutrients, while the lower terrain contains softer rocks, more vegetation, and inflows with higher nutrients (see also Chap. 16). (*Modified from Mitchell, 1956.*)

Table 19-1 Characteristics of Selected Lakes in the English Lake District.[*]

All are monomictic except for Blelham Tarn and Esthwaite Water which are occasionally dimictic and Bassenthwaite Water which is polymictic (P).

Lake	L, km B, km A, km^2	Depth z_{max} ($\bar{z}$), m	Drainage basin area, km^2	Thermocline depth, m	Typical max (min) surface temp., °C	Approximate period of thermal stratification	Winter (summer) nutrients, μg liter^{-1} NO$_3$-N	PO$_4$-P	SiO$_2$
Windermere	17 1.5 15	67 (24)	231	5–20(N) 5–20(S)	17 (3–5)	May–end Nov.	300 (100)	3.0 (0.5)	600 (200)
Wastwater	4.8 0.8 2.9	79 (41)	49	15–20	15 (<4)	June–Dec.	>140 (~100)		
Ennerdale	3.8 0.9 2.9	45 (19)	44	15–25	19 (<4)	May–Oct.	>140 (50)		
Esthwaite	2.5 0.6 1.0	16 (6.4)	14	9–12	20 (2)	Apr.–mid-Sept.	400 (100)	2–4 (0.5)	1000 (100)
Blelham	0.8 0.3 0.11	15 (6.6)	2.3	8–10	20 (3.5)	Apr.–end Sept.	600 (200)	4 (0.5)	1000 (200)
Bassenthwaite	6.2 1.2 5.4	21 (5.5)	238	P	21 (<4)	Irregular	~250		

[*]From Macan (1970); Heron (1961).

Table 19-2 The Lakes of the English Lake District as Ranked by Their Biological Differences*

Windermere in many ways behaves like two lakes, an oligotrophic North Basin and a mesotrophic South Basin. In this table, average values for both basins are listed. For convenience, lakes Thirlmere and Haweswater have been omitted since they have been much enlarged in size by dam construction just before and during the last 80 years, so some data are not comparable. Note how the 1920s trophic classification based on percent rocky bottom, calcium or bicarbonate is supported by the distribution of phytoplankton, zooplankton, and zoobenthos.

Lake	Modern relative trophic status	1960 basin population	Secchi disk, m	Physical, % rocky and shallow bottom	Chemical, mg liter^{-1}		Phytoplankton		Zooplankton *Cyclops leuckarti*†	Zoobenthos *Lymnaea palustris*†
					Ca	HCO$_3^-$	*Asterionella*†	*Staurastrum*†		
Wastwater	O	50	9.0	73	2.4	3.2	·	d	—	—
Ennerdale	O	32	8.3	66	2.2	3.5	·	d	—	—
Buttermere	O	32	8.0	50	2.1	2.6	·	·	—	—
Crummock	O	200	8.0	47	2.1	2.9	·	·	—	—
Derwentwater	M	720	5.5	33	4.5	5.4	d	·	+	+
Bassenthwaite	E	5,900	2.2	29	5.3	10.0	d	·	+	+++
Coniston	M	1,400	5.4	27	6.1	10.8	n	—	—	+++
Windermere	M	13,300	5.5	28	6.0	10.3	d	·	+	++
Ullswater	M	782	5.4	28	5.7	12.7	d	·	—	++
Esthwaite	E	1,200	3.1	12	8.3	18.3	d	·	+	+

*From original texts by Pearsall and from Macan (1970).

†*Asterionella* is a common diatom in the spring bloom, *Staurastrum* is a desmid (green alga). *C. leuckarti* is a copepod zooplankter, and *L. palustris* is the "marsh" snail which inhabits the littoral zone.

Key: · = present but not dominant; d = dominant; — = absent; + = present; n = no data.

One of the oldest and best-known studies in the English Lakes concerns classification of the waters by trophic level. In the 1920s, the great aquatic botanist Pearsall ranked the lakes by the type and abundance of algae. He also indicated the amount of agriculture in the watershed and the percentage of shallow, rocky, and thus unproductive, littoral zone. His list is given in Table 19-2, which has been revised to include more modern classification. Pearsall's work was done when limnology was a little-known discipline, and the terms eutrophic and oligotrophic were not used as they are today. However, his rankings do not differ greatly from the way these lakes are ranked using modern information.

The terms eutrophic and oligotrophic have been shown by more than half a century of collection to be better applied to lakes in one district than to lakes between districts. For example, Esthwaite Water is considered a eutrophic lake in the English Lake District (Fig. 11-2, Table 19-2). It has anoxic sediments in summer, and occasionally some parts of the lake surface are covered by a thin film of the blue-green alga *Aphanizomenon flos-aquae*. These are similar conditions to those found in Clear Lake, California, which lies at the eutrophic end of the spectrum of California lakes, but there the similarity ends. Chlorophyll levels reach 300 μg liter^{-1} in Clear Lake but only about 10 μg liter^{-1} in Esthwaite. This English lake supports a good trout fishery, while the turbid California lake has many carp and contains an excellent warm-water bass and crappie fishery. Without these comparisons, the terms eutrophic and oligotrophic cannot be understood.

Using biological records gathered over many years, some species of animals and plants appear to confirm Pearsall's original rankings. *Asterionella formosa* is, at times, dominant in the mesotrophic and eutrophic lakes, while species of *Staurastrum* dominate only in the most oligotrophic ones. The zooplankter *Cyclops* and the snail *Lymnaea* are for the most part confined to

the eutrophic and mesotrophic lakes (Table 19-2). The validity of rankings of this sort should be confirmed over a wide range of physical, chemical, and biological variables to be of real value in interpreting cultural eutrophication (Chap. 18). When a lake changes its rank or does not appear to be properly ranked, there may be good grounds for suspecting pollution. The aim of lake restoration (Chap. 20) is to return the lake to its original rank. Not all variables will immediately return to their former levels. For example, in Lake Washington, North America, nutrient levels rapidly returned to prepollution levels after waste diversion. However, many years were required before the supposed prepollution ratio of diatoms to blue-green algae was reestablished.

The Laurentian Great Lakes

These five lakes constitute the largest unfrozen mass of fresh water anywhere in the world (Fig. 19-2). The Great Lakes, in contrast to the English Lake District, present severe sampling problems for the limnologist. They are too large and stormy to sample easily in summer and are extremely difficult to sample during the very cold windy winters. Large oceanographic-type vessels are necessary to sample these very large lakes. Despite their importance, much less is known of their limnology than is known of many smaller, more easily accessible lakes near institutes or university campuses.

Lake Superior is the largest and deepest of the Great Lakes. It is monomictic and does not become completely ice-covered even though the winters are very cold. The other lakes are also monomictic and have partial winter ice cover except for Lake Erie, which may freeze over completely during the coldest years. The water motion in all the Great Lakes resembles that found in the oceans except that the lakes are all holomictic and, of course, nontidal.

Despite their huge size, the Great Lakes are only about 10,000 years old, similar to most of the English Lake District lakes, which also

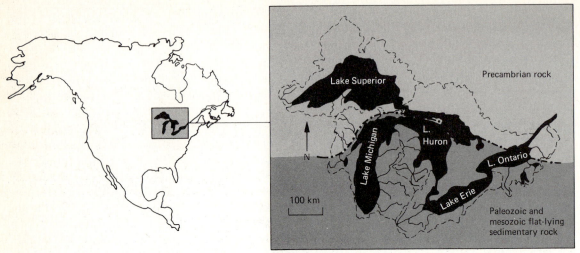

Figure 19-2   The Laurentian Great Lakes of North America formed 10,000 to 15,000 years ago. This lake district contains more freshwater than any other and consists of five major lakes. Note the small size of the drainage basins (indicated by dashed lines) relative to the size of the lakes. This tends to reduce nutrient inflow but gives most of the lakes a long residence time (see Table 19-3). The underlying rock strata is of two types (–·–· marks the boundary). The northerly Precambrian rock is not easily weathered, is overlain by extensive coniferous forests, and releases few nutrients to its drainage system. The southern rock is composed of more fertile sedimentary material and was overlain with some deciduous forest before the recent introduction of extensive agriculture to the area. This zone releases more nutrients to the drainage system.

became free of glacial ice at the end of the Pleistocene Era. Because of their youth, the Great Lakes do not possess a large list of endemic species such as is found in 25-million-year-old Lake Baikal, Siberia. As with *Cyclops abyssorum* in the English Lake District, cold-water relic species such as *Mysis relicta* still survive in some parts of the Great Lakes.

A summary of information on the Great Lakes is given in Tables 19-3 and 19-4. The large size of the lakes is obvious, but also important is the ratio of lake area to drainage basin area. The Great Lakes have a much smaller surface area to drainage basin ratio, about 1:2, than most lakes. Partially for this reason, the water inflow relative to lake volume is small and the lakes are low in dissolved materials. The hydrologic residence time for the larger, upstream lakes, Superior and Michigan, is centuries rather than years (Table 19-3).

Climatic variations or human-influenced nu-

trient inflows take a long time to change large lakes with long residence times. The smaller, more-frequently flushed lakes such as Erie and Ontario respond more rapidly. A good example of the chemical accumulation presently occurring is shown in Fig. 19-3. Most waste from residential, industrial, and agricultural pollution goes into lakes Erie, Michigan, and Ontario. Superior and Huron have few large cities on their shores, and their drainage basins are too cool and rocky for intensive agriculture. The increases shown in Fig. 19-3 were deduced from only small portions of the lakes, but are thought to be typical of the whole.

Biologically dynamic substances such as nitrate, phosphate, or silicate have shown local increases and decreases because they are rapidly used by the phytoplankton and attached algae. Nutrients introduced into Lake Michigan along the Milwaukee-Chicago side of the lake take months or years to reach the center or

Table 19-3 Physical and Chemical Characteristics of Pelagic Waters of the Laurentian Great Lakes*

	L, km B, km A, km²	z_{max} ($\bar{Z}$), m	Drainage basin area, km² × 10³	Thermocline depth, m	Max/min summer (winter) temp, °C	Hydraulic residence time, years	Approximate period of thermal stratification	Winter (summer) surface nutrients, μg liter⁻¹ — $NO_3-N + NH_4-N$	PO_4-P	SiO_2
Superior	560 256 82,000	406 (149)	125	10–30	14 (0.5)	184	Aug.–Dec.	280 (220)	0.5 (0.5)	2200 (2000)
Michigan	490 188 58,000	281 (85)	118	10–15	18–20 (<4)	104	July–Dec.	300 (130)	6 (5)	1300 (700)
Huron	330 292 60,000	228 (59)	128	15–30	18.5 (<4)	21	End June –Oct. or Nov.	260 (180)	0.5 (0.5)	1400 (800)
Erie†	385 91 26,000	w: 13(7.3) c : 24(18) e: 70(24)	59	w: p c: 14–20 e: 30	24 (<4)	w:0.13 c:1.7 e:0.85 all:3	mid-June– Nov.	w: 640 (80) c : 140 (20) e : 180 (20)	23 (2) 7 (1) 7 (1)	1300 (60) 350 (30) 300 (30)
Ontario	309 85 20,000	244 (86)	70	15–20	20.5 (<4)	8	End June– Nov.	280 (40)	14 (1)	400 (100)

*From Schelske and Roth (1973); Dobson et al. (1974); Ragotzkie (1974); Bennett (1978).
†For Lake Erie, w =western; c = central; e = eastern, p = polymictic.

Table 19-4 The Laurentian Great Lakes Ranked by Their Biological and Chemical Differences*

Note that, as in the English Lake District, the trophic classification based on transparency (Secchi disk) or chlorophyll a is also supported by the distribution of phytoplankton, zooplankton, and zoobenthos.

Lake	Local trophic classification	1970 basin population, millions	Secchi disk, m	Summer surface chlorophyll, μg liter^{-1}	Chemical mg liter^{-1} Ca^{2+}	HCO$_3{}^-$	Tabellaria + Asterionella	Aphanizomenon + Microcystis	Diaptomus silicis	Cyclops vernalis	Pontoporeia
							Phytoplankton		Zooplankton		Zoobenthos
Superior	O	0.9	11.3	0.9	14	51	d	r	d	p	c
Michigan	O	7.5	4.8	1.3	37	130	d	r	c	c	d
Huron	O–M	2.0	5–9	1.8	26	200	c	p	c	c	d
Ontario	M	6.1	—	5	42	—	—	—	p	—	—
Erie, all	—	12	—	4.3	—	—	—	—	—	—	—
East	M	—	4.4	5.5	38	—	p	p	p	r	r
Central	E	—	—			—					
West	E	—	2.0	11	32	—	p	d	p	r	r

*From Beeton and Chandler (1963); Patalas (1971); Scheiske and Roth (1973).

Key: d = dominant; c = common; p = present in smaller amounts; r = rare or absent; — = no data; O = oligotrophic; M = mesotrophic; E = eutrophic.

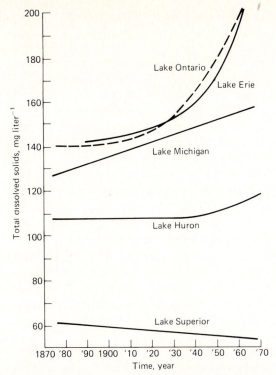

Figure 19-3 Changes in total dissolved solids (TDS), which are mostly calcium and magnesium sulfates, in the Laurentian Great Lakes. Note the large change in the "initial" TDS in 1870 and the rapid increases in the smaller lakes Ontario and Erie relative to Michigan. The downward trend in Superior and upward trend in Huron may not be significant. (*Redrawn from Beeton, 1969.*)

upper sections of the lake. Polluted inflows from the Detroit-Toledo area into Lake Erie pass through the shallow, polymictic western basin before they penetrate to the thermally stratified basins (Table 19-3). In contrast, nutrients in Clear Lake, California, circulate around the lake in about a week (Fig. 5-4).

As with many temperate lakes, phytoplankton production throughout the summer appears most limited by supplies of phosphorus and silicon. Nitrogen limitation is less important in the Great Lakes but occurs in eutrophic inshore waters and bays like the polluted Green Bay of Lake Michigan and the western basin of Lake

Erie. The phytoplankton of the Great Lakes is dominated by diatoms, particularly the colonial pennate forms *Fragilaria, Tabellaria, Asterionella,* the solitary pennate *Synedra* or the chain-forming centric genus *Melosira.* Blue-green algae as single filaments or small colonies of *Lyngbya* and *Oscillatoria,* which are typical of unproductive lakes, are also widespread. In the most productive areas such as western Lake Erie, Saginaw Bay, or Green Bay, the large colonial blue-green algae *Aphanizomenon, Anabaena* and *Microcystis,* characteristic of eutrophic conditions, are abundant (Table 19-4).

The zooplankton in the deep, colder lakes are mostly calanoid copepods, for example, species of *Diaptomus.* Cyclopoid copepods, like *Cyclops* and the cladoceran zooplankters *Daphnia* and *Bosmina,* increase in importance in the more productive, warmer lakes. Rotifers and cladocerans tend to dominate the most productive, eutrophic lake basins such as western Lake Erie. An unusually high number of limnetic zooplankton coexist in the Great Lakes. The large numbers of crustaceans are not entirely due to the lakes' large size since this diversity is not found in large tectonic lakes. High species diversity in the Great Lakes may be due to the greater turbulence found in their cold waters.

The benthos of lakes Superior and Michigan contain small populations of "relic" forms left over from the previous ice age. The shrimp *Mysis* and the amphipod *Pontoporeia* together with some species of cold-water midge larvae have survived in cold deep waters where oxygen is abundant and their major predators are scarce. In the benthos of the more productive parts of the lakes where oxygen is lower, *Pontoporeia* has been replaced by chironomid larvae. A similar gradual replacement of the "relic" copepod *Limnocalanus* by *Bosmina* or *Daphnia* has also occurred in the more productive sections and bays of some of the Great Lakes. The fish of the Great Lakes, which have been considerably modified by human activities, are discussed in Chap. 14.

The examples given for the English Lake District and Great Lakes serve to illustrate the value of comparing lakes within the same general region. Long-term studies enable the investigator to both classify and better understand changes occurring within particular lakes if they are compared with the most similar lakes in the same district.

LAKES IN TROPICAL OR ARID CLIMATES

The major characteristic of this group of lakes is their continually high temperature. For most of the year temperatures are above 25°C, and the lowest values are approximately the highest found in temperate lakes. Lakes in tropical and arid zones range from shallow, very eutrophic waters to the deep, less-productive ones. In this section we will concentrate on Lake George, Uganda.

Limnological methods in tropical lakes are similar to those generally employed at other latitudes, but some aspects deserve particular attention. First, tropical lakes and their drainage basins have health hazards not usually experienced in temperate or polar lakes. In addition to minor problems of infected scratches or intestinal disorders, there are a variety of serious vector-borne diseases such as malaria, bilharzia (Schistosomiasis), and river blindness. Limnologists should get their full set of vaccinations and take antimalarial pills regularly! Simple wiping off of water splashes near any snail habitat will help prevent bilharzia. Second, the indigenous animals are often dangerous. More than one limnologist has been attacked by crocodiles, tossed by buffalos, or attacked by the surprisingly aggressive hippopotami.

Lake George, Uganda

This remote, large shallow lake is situated astride the equator, 915 m above sea level ($A = 250$ km^2, $\bar{z} = 2.4$ m, residence time $= 0.36$ year). The lake is polymictic and is situated in a large open plain near the Ruwenzori Mountains in the Western Great Rift Valley of Africa. Gross productivity is high with very dense algal crops and large populations of zooplankton, zoobenthos, and fish. Papyrus swamps are common on the tributaries of African lakes and cover 60 km^2 adjacent to Lake George.

A year-round water supply comes from the Ruwenzori Mountains as well as from local runoff during the two wet seasons characteristic of this area. Incident solar energy varies annually by only 16 percent from the mean of 1700 J cm^{-2} day^{-1}. There is no significant cultural eutrophication of the basin by the sparse population of commercial *Tilapia* fishermen and subsistence farmers (Fig. 19-4). Lake George provides a baseline for eutrophication studies in tropical lakes. Because of the constant water inflow and its location at the equator, there is probably less seasonality in this lake than anywhere else in the world. The constancy of the environment allows observation of a lake that has probably been at near steady-state conditions for at least a thousand years. In some ways Lake George, Uganda, is a huge long-term natural experiment where normal lake seasonality does not exist. Unlike most remote lakes, it has been thoroughly studied for

Figure 19-4 Subsistence-level farming and fishing typical of the drainage basin of Lake George, Uganda. Wooden dugout canoes are now often powered by outboard motors. Note the dense papyrus stand on the lake edge.

several years by a team of international and Ugandan limnologists largely sponsored by the Royal Society of Britain as part of the International Biological Programme (*Proc. R. Soc. London, Ser. B*, 1973, **184**:229–298).

The remarkable year-round constancy of the lake is shown in Fig. 19-5*a*. Nutrients, phytoplankton, chlorophyll *a*, and zooplankton are virtually constant. The small changes which occur seasonally are minute. Detailed studies of the individual species present show that with few exceptions this seasonal constancy is typical of each component of the plankton and is not due to species replacement maintaining constant density (Ganf, 1974; Burgis, 1971). Lake George may be contrasted to temperate Loch Leven in Scotland which is also a productive polymictic lake (Fig. 19-5*b*). The annual change in zooplankton in Loch Leven is enormous when compared to the slight fluctuations found in Lake George. In contrast to seasonal patterns, daily patterns show very large changes (Ganf and Horne, 1975). One of the most dramatic is the daily change in dissolved

oxygen from less than 100 to 230 percent saturation. Similar changes occur in temperature, chlorophyll, and carbon and nitrogen fixation (Figs. 19-6, 7-1).

An important conclusion can be drawn from the seasonal constancy and diurnal variability. The farther we move from temperate lakes toward the equator, the more daily cycles tend to dominate the lake's ecology. Phytoplankton in lakes normally have a maximum growth rate of two or three divisions per week and zooplankton grow even slower. In temperate climates the major environmental changes occur more slowly than planktonic division rates. Thus, the plankton respond most to seasonal changes. In tropical lakes plankton experience the largest environmental changes every 24 h and must adapt to those rather than the smaller seasonal changes. Any group of species sufficiently adaptable to grow well during a 24-h period will tend to dominate the lake (Round, 1971; Ganf, 1974). This is borne out in Lake George where the colonial blue-green algae *Microcystis aeruginosa* is the most abundant

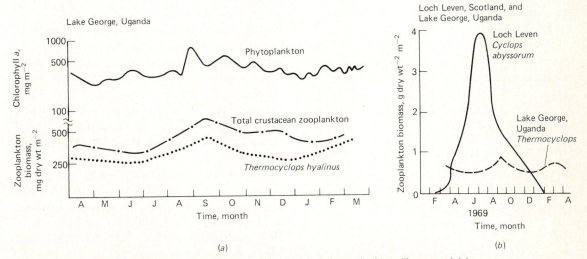

(a)   (b)

Figure 19-5 Seasonal changes in phytoplankton (log scale) and zooplankton (linear scale) in (*a*) tropical Lake George Uganda, and (*b*) a comparable temperate lake. Note the absence of seasonal cycles in the tropical lake. *Cyclops* make up 99 percent of the zooplankton in the temperate lake (*From Burgis, 1974; Ganf and Viner, 1973; Bindloss et al., 1972*).

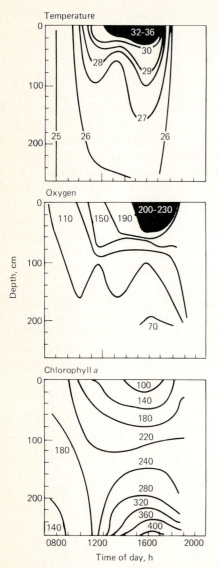

Figure 19-6 Diurnal changes in temperature, percentage oxygen saturation, and chlorophyll *a* with depth in shallow, eutrophic, tropical Lake George, Uganda. Oxygen isopleths are every 40 percent oxygen saturation, chlorophyll *a* every 40 μg liter⁻¹, temperature every 1°C. (*Redrawn from Ganf and Horne, 1975.*)

Phytoplankton in Lake George comprise 99 percent of the total plant and animal plankton biomass even when the planktonic stages of the abundant midge *Chaoborus* are included (Fig. 19-5). The major feature of this blue-green dominance is its year-round constancy although the absolute quantity may be equaled by algal blooms in eutrophic temperate lakes.

A population of about 3000 hippopotami live in the lake, graze on surrounding vegetation at night, and digest and excrete in the lake during the day. The waste from this animal population is equivalent to a city of about 20,000. Although this would appear to be a large enrichment for a lake of this size, analysis of nitrogen shows that hippopotami wastes contribute only a few percent to the lake's annual nitrogen inflows (Table 19-5). This is similar to the low percentage of limiting nitrogen contributed by people in eutrophic Clear Lake, California (Horne and Goldman, 1972). In these warm, shallow lakes, the natural fertility of the watershed and the efficient nutrient recycling is the dominant factor.

The outstanding physical feature of Lake George is its high temperature, 25 to 36°C (Fig. 19-6). This has a direct effect on species composition and biological production. The high temperature produces a strong temporary stratification if the day is sunny. Blue-green algae thrive at temperatures above 20°C, while many other algae do not. In addition, the colonial gas-vacuolate species of blue-greens are sufficiently buoyant to regulate their position in the water column at optimal depth (Figs. 19-6, 12-13).

Tropical ecosystems are usually deficient in nitrogen (Odum and Pigeon, 1970). In Lake George, about one-third the nitrogen used for algal growth is fixed from the atmosphere by blue-green algae (Table 19-5 and Horne and Viner, 1971; Burgis et al., 1973).

The zooplankton in Lake George is dominated by the raptorial cyclopoid adults and copepodites which feed on colonies of

lake plankton at all times of year. Similarly, the copepod *Thermocyclops hyalinus* dominates the zooplankton population in the lake at all times (Burgis, 1971).

Table 19-5 Nitrogen Balance Sheet for Lake George, Uganda*

Stream inflow is a minimum value and does not include some storm runoff which could not be measured adequately. Note the large role played by nitrogen fixation in this very productive lake.

Annual gain, metric tons		Annual loss, metric tons	
Nitrogen fixation	1280	Denitrification	?
Stream inflow	322	Stream plankton outflow	3180
Rainfall	277	Sedimentation	655
Hippopotami excretion	76–99	Commercial fish export	50–75

*Modified from Horne and Viner (1971).

Microcystis. Filter-feeding cladocerans like *Daphnia* are uncommon because of a lack of small algae as food. This situation is not necessarily common to all tropical lakes since many deeper ones, like Lake Tanganyika, apparently have low populations of all types of algae.

The major fish species in Lake George are two herbivorous cyclids, *Haplochromis nigripinnis* and *Tilapia nilotica.* These and the dominant zooplankton *Thermocyclops hyalinus* are all dependent on *Microcystis* and other large planktonic blue-green algae for food. This contrasts with temperate lakes where diatoms and flagellates are more important for herbivores. Other tropical lakes where blue-greens are abundant may have a food chain similar to Lake George. The mechanism employed by *Tilapia nilotica* to digest blue-greens is to increase the acidity of the stomach in order to reduce the pH to about 1.4 (Fig. 14-3). This acidity is sufficient to rupture algal cells and allow digestion of otherwise unavailable blue-green algae. Stomach acids in other fish lower pH only to about 2, which is 6 times less acid than the *Tilapia* stomach (Moriarty, 1973).

Extensive free-floating growths of large aquatic plants are a feature of Lake George common to many tropical lakes, but unfamiliar to most temperate limnologists. In Lake George, as in many tropical reservoirs, *Pistia,* the water lettuce, covers parts of the lake so thickly that it is impossible to drive a boat through (Plate 4*a*). In the huge subtropical reservoir Lake Kariba, as much as one-third of the water surface may be covered by *Salvinia,* the water fern. These large plants provide shelter for many smaller organisms and disease vectors. They also block the oxygen transfer across the lake surface and greatly increase transpirational loss of water. The plants eventually pile up on the downwind shores in huge rotting masses (Mitchell, 1969). Where lake levels in the tropics do not fluctuate very much, floating macrophytes are common. A major design criteria for tropical reservoirs may be to ensure sufficient water level fluctuations to strand the young plants before they become a serious nuisance for boating, fishing, and power generation.

No discussion of tropical lakes is complete without mention of papyrus swamps. Like many tropical lakes, Lake George has a huge papyrus swamp (*Cyperus papyrus*) through which the major inflowing rivers must flow (Fig. 19-7). The main effect of these swamps is to slow down the seasonal heavy rains and filter out particulates. Tropical streams are usually brown with silt during rains, but after passage through a papyrus swamp, the water is so transparent that emergent plants like lilies can grow up from the bottom.

POLAR AND HIGH ALPINE LAKES

These lakes are distinguished by constant low water temperatures for all or most of the year. Large annual changes in biota are caused by the huge seasonal variations in daily total incident light. The maximum daily radiation on mountains or near the poles is greater in midsummer than that experienced by tropical lakes.

Polar and high alpine lakes are usually transparent and unproductive, with low standing crops of plankton. Much of the plant biomass is often found in the large benthic green alga *Chara* or in a dense benthic felt of blue-green algae which harbors many small invertebrates. Higher plants with a covering of aufwuchs may also be more important than the plankton, especially in some alpine lakes. The small, unmodified rock and clay drainage basins of these lakes reduce annual nutrient loadings so that it may take years to achieve even a moderately thick algal felt.

In this section we will discuss two types of unpolluted lakes: coastal lakes in Antarctica where the temperature rarely exceeds 10°C and a lake high in the European Alps.

Limnological methods are similar to those employed in other areas, but the extreme cold and remote situations of polar and alpine lakes require some special attention. Immersion in polar waters will kill most people in a few minutes, so a companion with a rope tied to a sampling limnologist is advisable. Whiteouts of blowing snow can often reduce visibility to zero in these areas, so snow-cave survival techniques should be as familiar to the polar and alpine limnologist as snowshoes and wide cross-country skis.

There are two well-studied similar sets of lakes in Antarctica (Goldman, Mason and Wood, 1972; Heywood, 1972; Horne, 1972). The edges of the Antarctic continent and adjacent islands contain many small lakes which are influenced by the surrounding ocean and are often called *maritime lakes*. Other large lakes in the interior Antarctic desert such as amictic Lake Vanda and Lake Bonney were mentioned previously (Fig. 4-3).

Antarctic maritime lakes are usually quite small and shallow ($A < 10$ ha, $z_{max} < 6$ m) and are often very shallow ($z_{max} < 1$ m). These small lakes do not dry up because of their low temperature and the high humidity of the coastal zone. In common with Arctic and alpine lakes, maritime Antarctic lakes are usually

(a)

(b)

Figure 19-7 Papyrus stands in tropical lakes. (a) An aerial view of an arm of Lake Victoria, Uganda and (b) the way in which papyrus swamp blocks small inflowing rivers to create large swamps. Part (b) is the Mpanga River flowing into Lake George, Uganda. The papyrus stands about 3 m high and in a small boat, we have taken an entire afternoon and made only a few yards progress through the dense root system.

oligotrophic in the strict sense that inflowing nutrient supplies are low. Eutrophication can occur when the lakes are loaded with nutrients from sea birds, seals, or people.

Also, like most Arctic and alpine lakes the maritime Antarctic lakes usually have only small populations of phytoplankton or zooplankton. The algae are often not truly planktonic but appear in the open water after being washed in from the snow or dislodged from the bottom. Unlike many sparse phytoplankton populations, Antarctic phytoplankton are unable to photosynthesize at high rates to compensate for small numbers. Goldman, Mason, and Wood (1963) provided an explanation for this phenomenon which is now generally accepted for all lake and ocean surface waters. They discovered that even at sea level there was a severe light inhibition of photosynthesis for most of the day during the Antarctic summer. Light inhibition was always more pronounced at noon so photosynthesis was inverse to incident light energy (Fig. 19-9). As we shall see, for deeper alpine lakes this is not a simple function of light but also results from long day length.

Figures 19-8 and 19-9 show the effect of varying levels of natural sunlight on carbon fixation in phytoplankton populations in two contrasting Antarctic lakes. In transparent Algal Lake, summer photosynthesis by the sparse phytoplankton population was low at noon when light was highest and maximal during the low-light period around midnight. In nutrient-enriched Skua Lake overall photosynthesis was higher but showed approximately the same pattern. In the Antarctic fall, a more normal pattern of maximum productivity near noon occurred in both lakes (Fig. 19-8). The reason for this change in daily maximum is *light inhibition of photosynthesis*. The lower section of Fig. 19-8 shows the very high sunlight levels present for much of the day during the short Antarctic summer. In contrast, levels are much lower in the fall, which is only a month later. This effect is shown most clearly in photosynthetic efficiency (Fig. 19-9).

Light inhibition of photosynthesis occurs in the upper waters of almost all lakes and seas, but overall production per unit area of lake surface may not be reduced since the production maxima can move deeper in the water column during high light. This is impossible in the very shallow Antarctic lakes as well as in shallow lakes in the Arctic and alpine regions. However, the lost production is compensated for by extensive growths of benthic algae and mosses which form thick felts on the lake beds (Fogg and Horne, 1970; Goldman et al., 1972).

The algal felts are mostly filamentous blue-green algae and dominate the stony sections of the lake beds for several reasons. Blue-green algae are usually the first to colonize exposed surfaces such as are left by ice scour. The upper filaments produce protective carotenoid pigments which act as a screen against inhibiting levels of sunlight. Digging through an Antarctic lake's benthic felt may show a yellow-orange layer with a blue-green layer below. Some species of blue-greens appear more tolerant of high light and consequently are more dominant near the surface.

The benthic felts of algae and moss provide a major habitat for Antarctic lake animals. Large populations with relatively few species are typical due to the isolation of this continent from the rest of the world's land masses. In alpine and Arctic lakes the felts are more diverse. Protozoans, rotifers, and crustacean copepods and cladocerans spend parts of their lives in amongst the felt where temperatures are often several degrees higher than the water above. The increase in temperature is due to light energy being absorbed by the felt in these transparent, shallow lakes.

Fish and all but one flying insect are absent in the Antarctic due to the difficulty of colonizing this remote region. Fish are present or have been introduced to most of the larger alpine and some Arctic lakes.

Alpine lakes are found in most high mountains. They are cold, often deep, and ice-free for much of the summer. Most limnologists have seen alpine lakes, but the difficulty in working there year-round often prevents the acquisition of first-hand knowledge. The example dis-

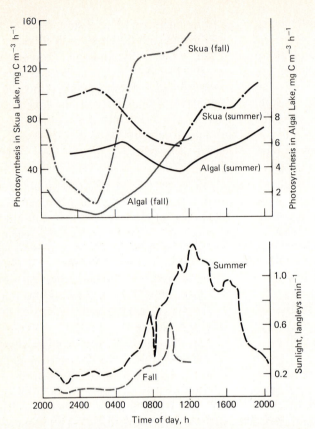

Figure 19-8 Diel changes in photosynthesis in typical transparent Algal Lake and in bird-fertilized Skua Lake, in midsummer and fall (January 9–10 and February 17–18). Note that maximum production occurs at midnight and minimum at noon in summer but normal patterns occur with a maximum at noon in fall. The light reaches saturation at about 0.3 langley, which is about 20 percent of the incident light at that season. Above that level, light actually inhibits photosynthesis. (*Redrawn from Goldman et al., 1963 and 1972.*)

cussed here is Vorderer Finstertaler See in the European Alps which is situated at an elevation of 2237 m. Thus the lake receives very intense sunlight in summer although day length is shorter than at an equivalent season in polar regions. The lake water remains below 10°C in summer. In common with most polar or alpine systems, the structure of the lake ecosystem is simple. Dinoflagellates are the only common phytoplankton group, and the zooplankton community contains two rotifers and one copepod. Two species of salmonid fish make up the

nekton in this lake (Pechlaner et al., 1972). Mosses, blue-green algae, and epiphytic diatoms occur on the lake bed where sufficient light penetrate. These plants form a felt inhabited by nematodes, oligochaetes, ostracods, and chironomid larvae.

Light inhibition of photosynthesis occurs in the surface waters but, unlike the shallow polar ponds, the lake is sufficiently deep for motile algae to avoid damaging light intensities. Maximum photosynthesis in winter occurs under the ice at low light levels near the ice surface

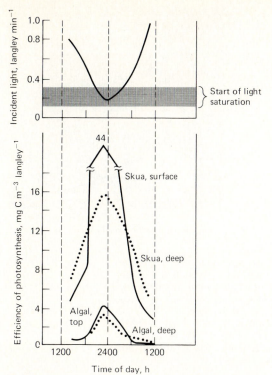

Figure 19-9 Inhibition of photosynthesis in midsummer (January 15, 1962) for a normal (Algal Lake) and a turbid bird-fertilized (Skua Lake) lake. Note maximum efficiency at lowest light due to photosynthetic inhibition by high sunlight during the daytime. These lakes are very shallow, but the inhibition is reduced even in the "deep" samples about 50 cm below the surface. ly = langleys, a unit of light = calories per square centimeter. A skua is a predatory sea bird, *Stercorarius* spp. (*Modified from Goldman, Mason, and Wood, 1963.*)

ter phytoplankton are abundant but photosynthesize little since they are pre-adapted to the summer conditions of very high light intensities and long days. Under winter ice the motile dinoflagellates can remain at optimal light levels, grow slowly, and gradually become adapted to the dim light. Their photosynthetic efficiency rises as adaptation occurs, but their winter biomass tends to remain small (Fig. 19-10a). The low biomass could also be due to winter grazing by the rotifers *Keratella hiemalis* and *Polyarthra dolichoptera*, but under-ice temperatures are always low, so grazing rates are small. The ice and snow cover take most of the spring to melt due to the high latent heat of water (Chap. 4). The eventual melt in late spring exposes the dark-adapted algae to full summer sun. Phototactic avoidance of inhibiting light intensities (Fig. 19-10) by the phytoplankton can produce maximum photosynthetic efficiency at this time (Fig. 19-10b). This response indicates that only light was limiting growth in the dark-adapted algae. Lakes all over the world show this kind of response in spring, which has been clearly demonstrated in the English Lake District (Lund, 1949, 1950).

As summer progresses the phytoplankton efficiency decreases as biomass increases, probably due to nutrient deficiency. Other experiments indicate that Vorderer Finstertaler See is phosphorus-limited in summer and fall. In general, different nutrients limit algal growth in other alpine and polar lakes since nitrogen, phosphorus, and trace metals can all stimulate growth (Goldman, 1960; Kalff, 1967, 1971).

SALT LAKES

There are many salt lakes throughout the world despite their desert locations. There is almost as much saline lake water (104×10^3 km³) as fresh lake water (125×10^3 km³, Table 3-1). When climates become drier or geological events change drainage basins, the annual flows

(Fig. 19-10a). In spring as the snow melts and the ice becomes more transparent, the photosynthetic maximum moves to 5–15 m. Under ice-free conditions the maximum is still deeper (Fig. 19-10a). The dominant dinoflagellates *Gymnodinium* and *Glenodinium* are highly motile and phototactic and make possible the downward movement of maximum photosynthesis (Fig. 19-10). The adaptation of the phytoplankton to light under ice and in the open water of Vorderer Finstertaler See is illustrated diagrammatically in Fig. 19-10b. In early win-

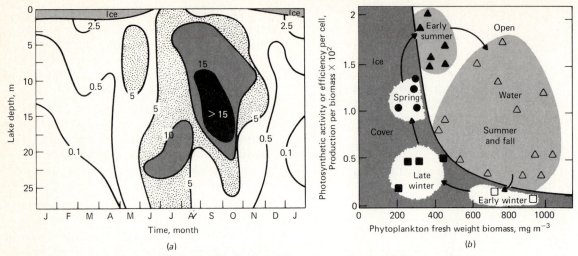

Figure 19-10 Changes in photosynthesis with depth and season in an alpine lake, Vorderer Finstertaler See. (a) Note the maximum photosynthesis near the surface under ice but much deeper down in summer due to light adaptation and active avoidance of high light by the dominant phytoplankter, the dinoflagellate *Gymnodinium* (see text). Values for isopleths are milligrams of carbon fixed per cubic meter per day. (b) How photosynthesis per algal cell changes with season. In winter and spring, light levels limit production, while nutrients limit production in summer and fall when the water surface is free of ice (see text for further explanation). (■, ●, ▲ = dark-adapted cells in late winter, spring, early summer; △, □ = light-adapted cells in summer and winter (*Modified from Pechlaner et al., 1972.*)

into a lake may be greatly reduced. The lake may then cease to have a significant outflow and become a terminal or sink lake. The salts from inflowing streams are concentrated by evaporation and are no longer flushed out through outflow. Eventually the lake may dry up completely. In dry climates with high evaporation, the lakes gradually become salty. In the Pleistocene Era two vast lakes, Bonneville and Lahontan, covered much of the southwestern United States. The rise of the Sierra Nevada cut off most of the rain-bearing clouds from the Pacific and now only Great Salt Lake, Utah, and Lakes Walker and Pyramid in Nevada remain as saline relics of wetter times. Other good examples of salt lakes are found in areas of drier climate such as Australia, southwest and northwest United States, South America, east Africa, Antarctica, Russia, and the dry northern side of the Himalayas.

Salt lakes vary from those with almost fresh-waters in dry temperate climates to viscous organic soups covered with seasonal crusts of salt such as those in tropical east Africa. Lakes are considered to be salt lakes when they contain more than 3 ppt salinity (Williams, 1978) but they can reach 350 ppt which is the approximate level of saturation. Seasonal rains vary the salinity and cause severe osmotic stresses for organisms adapted to life in this brine solution. One of the most tolerant of salinity change is the brine shrimp *Artemia salina*, which has become the focus of recent aquacultural exploitation for fish food.

Most salt lakes are derived from the evaporation of fresh water and not from seawater. Therefore the ionic composition of most salt lakes is usually very different from the ocean. They can, in fact, be classified by the relative dominance of the anions chloride, sulfate, or bicarbonate-carbonate (Hutchinson, 1957). For example, the Great Salt Lake in Utah is influenced by sodium chloride from nearby salt beds. High salinity alters several important physical processes. Salt lakes stratify easily and require more wind energy to destroy thermal stratification than fresh water at the same temperatures. For example, Cole (1968) measured a spectacular temperature gradient of $0.4°C$ cm^{-1} in a shallow Algerian pond. This is about 100 times greater than that found in typical freshwater lakes and results from the fact that saline water is more viscous and its temperature-density curve has a very steep slope (Fig. 19-11).

Another important physical feature of salt water is the depression of the freezing point. At high salinity, this may amount to several degrees Celsius and is sufficient to prevent some salt lakes from freezing. Mono Lake, California (elevation 1979 m) remains monomictic although nearby lakes freeze over in winter. An example of an amictic salt lake is Lake Vanda, in the dry Taylor Valley of the Antarctic (Figs. 4-3, 19-12). It is permanently ice-covered and has a warm saline bottom layer (monimolimnion). Another effect of high salinity is its influence on reducing evaporation relative to a freshwater lake in a similar climate. Harbeck (1955) calculated that evaporation of fresh water in deserts may exceed 10 m year^{-1}, but a very saline lake may lose as little as 3 m.

Ecologists have been attracted to salt lakes for the same reasons that they are attracted to polar and alpine lakes. These lakes typically share a low diversity of organisms which makes food-chain analysis easier. As the salinity increases, diversity decreases (Table 19-6). Even on a worldwide basis the diversity of the saline biota is much less than in fresh water. Ciliates, foraminiferans, spirochaetes, gastropod snails, oligochaete worms, crustaceans such as anostracans, copepods, cladocerans, amphipods and isopods, chironomids, ephydrid flies, and some mosquitoes are found (Williams, 1981; Chap 11). At the lowest trophic level macrophytes such as *Ruppia*, the widgon grass,

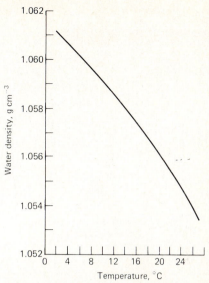

Figure 19-11 Change in density of saline Mono Lake, California (salinity ~ 70 ppt) with temperature. The curve is steeper than for pure water, and there is no density inflection at 4°C, the maximum density for pure freshwater (*Redrawn from Mason, 1967.*)

the *Halobacteria* group, and several phytoplankton including *Dunaliella* and the blue-green algae *Spirulina* are often dominant.

Table 19-6 Numbers of Animal Species Recorded within Various Salinity Ranges for the Saline Lakes in Victoria, Australia*

Note the rapid decline in species diversity as salinity rises above brackish water levels. A similar effect is shown in estuaries (Figs. 17-1, 17-2).

Salinity, ppt	No. of species
1–10	71
10–100	36
100–200	8
200–300	1
>300	0

*From Williams (1978).

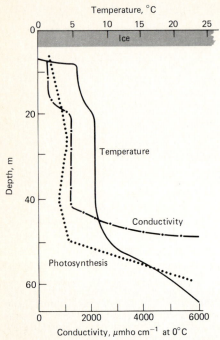

Figure 19-12 Conductivity (salinity), photosynthesis and temperature profiles in amictic, desert Lake Vanda, Antarctica. Note how the salinity stablizes the warm deeper water (*Modified from Goldman, Mason, and Hobbie, 1967.*)

The productivity of saline lakes is often high based on a few very abundant species. Few carnivorous zooplankton occur and, because of the osmotic stress, fish are often totally absent. An exception is the tiny desert pupfish (*Cyprinodon* spp.) which is very salt-tolerant. One species has been found in salinities as high as 140 ppt (Simpson and Gunther, 1956). Some other fish can be acclimated to quite high salt levels, but there is usually a requirement of lower salinity water for successful breeding. A sport fishery for corvina, which may grow to a length of 75 cm, has been maintained for several years in the increasingly saline Salton Sea. This large lake was formed accidentally by the diversion of the Colorado River in 1905 and has salinities of up to 60 ppt.

The lower tropic levels in salt lakes have very limited diversity. For example, in Mono Lake,

California (salinity = 70 ppt, $A = 200$ km^2, $z = 19$ m), there are only six phytoplankton species, several protozoans, two rotifers, and one brine shrimp (Mason, 1967). Many saline lakes are small and shallow and periodically dry up completely. This is certain to reduce diversity by providing fewer niches! Evidence for the importance of large size and depth as well as low salinity in promoting higher diversity is found in Pyramid Lake, Nevada. It has an area of 446 km^2, is deep ($z_{max} = 100$ m), and a salinity of 3.8 ppt. This lake contains 31 phytoplankton genera, seven zooplankton, and three rotifers (Galat et al., 1981).

Comparative limnological studies on animals in the Australian salt lakes have revealed three faunal assemblages: a halobiont group ($\sim 50-100$ ppt salinity), a halophylic (salt-loving) group ($\sim 10-60$ ppt), and a salt-tolerant freshwater group ($< 1-20$ ppt).

During the drying phases of lakes, many common species die out or become rare as salinity increases. The drying up of the large lake, Lake Chilwa, Malawi, Africa, is apparently a cyclic process and has occurred several times in the last few hundred years.

Blue-green algae, especially attached forms such as *Schizothrix* and planktonic forms like *Anabaena* and *Nodularia*, are abundant in many salt lakes. One of the most spectacular events in moderately saline lakes is the bloom of the filamentous blue-green alga *Nodularia* (Fig. 19-13). This genus is characteristically the most visible species in summer and may cover most of Pyramid and Walker lakes in Nevada as well as much of the brackish Baltic Sea in Europe. Also abundant and characteristic of saline lakes are the brine flies and midges, some of which bite. These insects are often neglected by limnologists since they occupy the interface between open water and the dry shore, but are a major food supply for some of the shore birds which frequent saline lakes. They were even utilized as food by native Americans who found adult and larvae of the brine fly, *Hydropyrus*,

Figure 19-13 *Nodularia* blooms in saline Pyramid Lake, Nevada. Note the patchy distribution of *Nodularia,* the nitrogen-fixing blue-green alga, which is restricted to brackish water (salinity > 5 ppt) in saline lakes and estuaries.

along the foam-covered edge of Mono Lake, California.

Although fish may be absent or play a subordinate role in saline lakes, birds and reptiles do not. Most of the world populations of roseate flamingos depend on saline lakes in east Africa for food and breeding sites. These birds build mound-like mud nests to allow for temporary flooding. The hot climate and treacherous salt crust prevents animal predators from crossing the drying lakes to reach these nests. The well-known upside-down position of the head during flamingo feeding facilitates filtering out the large filaments of blue-green algae and zooplankton (Fig. 11-13).

SPECIAL STREAM ENVIRONMENTS

Unlike lakes, streams and rivers are permanent features of the landscape although their courses and flows constantly change. Because of their continuous existence, streams contain very similar looking but taxonomically distinct organisms (Chap. 16). Rivers and their head-waters are less geographically isolated than lakes since many lake organisms cannot survive even quite short stretches of river passage. Streams have great similarity no matter where they occur, but a few share special environmental features which prevent the usual array of organisms from developing. These include those at very high latitudes and altitudes as well as hot and cold springs. In all these cases, the specialized limited fauna and flora last only a short distance along the course before representatives of the more typical river community appear.

Some cold-water streams rarely exceed 1° C even in summer and in winter may even freeze solid. However, low temperatures in themselves do not reduce the diversity or activity of stream biota since stream invertebrates may be cold-adapted. Stonefly larvae, for example, often show their maximum growth during winter. A few insects survive very cold periods as eggs or in a diapause phase. Even in frozen streams the temperatures in the deeper gravels are usually above freezing and this zone acts as a refuge. The main restriction for most insects is that warmer conditions are required for the aerial mating of adults. Suitable weather may occur for only a few days in the short polar and alpine summer.

The main problem for stream organisms in winter is probably ice formation; as air cools, hoar frost forms by condensation on vegetation along the stream edges. Ice crystals drop into the water and form nuclei for *frazil ice*. The effect looks like a snowstorm in the water. Frazil ice accumulates in large spikey masses on any solid object and forms ice caverns which disguise the familiar rocks and pools. After some time the ice becomes firmly attached as *anchor ice* which dams up and diverts water out of the streambed. Fish and most insects become disoriented by the lack of recognizable objects or flow patterns in the ice caverns and may wander into the diverted channels. Next morning when the ice dam melts and the water returns to its normal channel, large numbers of fish and insects may become stranded and die. In some high Sierra Nevada streams in the United States the effects of ice dams are thought to be the major cause of mortality in young trout and overwintering insects (Needham and Jones, 1959).

Even without the dangers of anchor ice, the large changes in daily radiant energy in summer make cold streams an extremely variable habitat. The diel changes in flow for various types of streams fed by snowmelt are shown in Fig. 19-14. The effect in small streams is a change in

flow of 10 to 20 percent day^{-1}, which is equivalent to a daily flood, and flood scouring remains one of the most important factors in stream ecology (Chap. 16).

Large diel temperature changes occur in small streams above the tree line, and some Arctic streams may rise from 4 to 15°C between dawn and midafternoon. This increase rules out the presence of obligate cryophylic forms which grow only at low temperatures. Most cold-stream species are not specially adapted to low temperatures (Fig. 19-15). Despite the very different temperatures, from 1 to

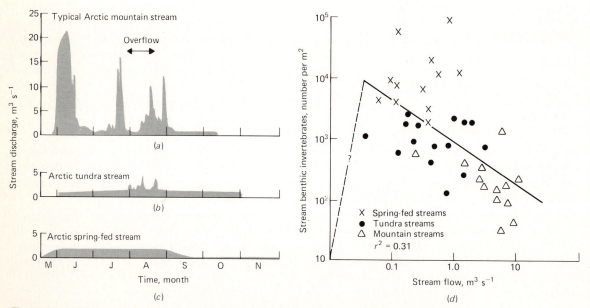

Figure 19-14   Stream flow and its relationship to stream animals in various types of cold-water streams in the Arctic. (a) The mountain stream has a highly variable discharge in summer, (b) the tundra stream is less variable, and (c) the spring-fed stream has a constant-discharge. Mountain and tundra streams usually freeze solid in winter, but spring-fed streams maintain their flow at low levels. Summer temperatures reach about 8°C for all types of streams. (d) Note the low benthic invertebrate populations in the continually fluctuating mountain streams relative to large numbers in the constant "Arctic oasis" of the spring-fed stream. Also note that flow characteristics only explain about one-third ($r^2 = 0.31$) of the differences in animals between streams. Much of the rest of the variation is probably due to the moderating effects of constant temperature and lack of winter freezing in the spring-fed stream (*Redrawn from Craig and McCart, 1975.*)

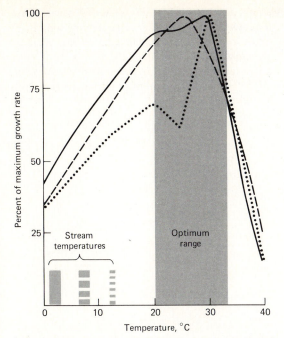

Figure 19-15 Growth of various types of stream algae at their normal and elevated temperatures. Solid line is the very common thallate (platelike) green alga, *Protococcus*. Dashed line is for two filamentous blue-green algae, *Oscillatoria* and *Phormidium*. Dotted line is the filamentous green alga *Zygnema*. Note that the optimal temperatures for all three groups are 20–30°C but their actual growing temperatures in the streams are 1°C (solid black, *Protococcus*), 7°C (thick dashes, blue-green algae), 12°C (thin dashes, *Zygnema*). (*Modified from Mosser and Brock, 1976.*)

mountains and in the Arctic. The bright green color of the year-round moss-dominated flora near the spring outlet contrasts sharply with the winter snow. These small open areas have been called the "oases of the Arctic." Unless the stream is frozen solid, the normal complement of stream organisms occurs within a short distance from the spring source. As with snowmelt streams there are no relic or endemic species present, and the only difference from more typical temperate streams is the small number of species found.

FURTHER READINGS

Brock, T. D. 1967. "Life at high temperatures." *Science.*, **158**:1012–1019.

Carpelan, L. H. 1958. "The Salton Sea. Physical and Chemical Characteristics." *Limnol. Oceanogr.*, **3**:373–397.

Castenholz, R. W., and C. E. Wickstrom. 1975. "Thermal streams." pp. 264–285. In B. A. Whitton (ed.), *River Ecology*. Univ. California Press, Berkeley.

Cole, G. A. 1968. "Desert Limnology." In G. W. Brown, Jr. (ed.), *Desert Biology*. Academic, New York.

Goldman, C. R. 1960. " Primary Productivity and Limiting Factors in Three Lakes of the Alaskan Peninsula." *Ecol. Monogr.*, **30**:207–270.

Heywood, R. B. 1972. "Antarctic Limnology: A Review." *Br. Antarct. Surv. Bull.*, No. 29:35–65.

Macan, T. T. 1970. *Biological Studies of the English Lakes.* American Elsevier, New York. 260 pp.

Schelske, C. L., and J. C. Roth. 1973. "Limnological Survey of Lakes Michigan, Superior, Huron, and Erie." *Great Lakes Res. Div., Univ. Mich., Publ. 17.* 108 pp.

Wetzel, R. G. 1964. "A Comparative Study of the Primary Productivity of Higher Aquatic Plants, Periphyton and Phytoplankton in a Large, Shallow Lake" *Int. Rev. Ges. Hydrobiol. Hydrogr.*, **49**:1–61.

Williams, W. D. 1972. "The Uniqueness of Salt Lake Ecosystems," pp. 349–361. In Z. Kajak and

12°C, at which three algae grow in Montana and Wyoming, their optimal temperatures for growth in the laboratory are between about 20 and 30°C. The same general temperature relationships are to be found with bacteria, fungi, lichens, insects, and fish. The chironomid *Diamesa* and the colonial alga *Palmella* are among the only species which seem to be strictly confined to very cold waters and are found within a few meters of glacier ice.

A similar effect is found for cold, but constant-temperature, spring-fed streams in high

A. Hillbricht-Ilkowska (eds.), *Productivity Problems of Freshwaters*. PWN (Polish Scientific Publishers), Warsaw-Krakow.
—— 1981. "Inland Salt Lakes: An Introduction." In W. D. Williams (ed.), *Salt Lakes: Proceedings of an International Symposium on Athalassic (Inland) Salt Lakes*. Junk Publishers, The Hague.

Various authors. 1973. "Physical, Chemical, Phytoplankton, Zooplankton, Zoobenthos and Fisheries in Lake George, Uganda." *Proc. R. Soc. London, Ser. B*, **184**:235–346.
Various authors, 1974. *J. Fish. Res. Board Can.* **31**:689–854. (On Laurentian Great Lakes physics, chemistry, biology.)

Chapter 20

Applied Limnology

OVERVIEW

This chapter presents some of the uses of limnology. We focus on several well-documented examples of rational lake management with wide application. With limnology at a stage in between youth and maturity, the best-understood systems are inevitably those dealing with the simplest levels of organization —nutrients and phytoplankton. The best examples are found in the manipulation of medium-sized lakes—small enough for adequate sampling but large enough to minimize edge effects. It can only be hoped at this stage that sensible management of whole food chains will be possible in the future. Until then it would be wise to cease the more speculative lake manipulations involving additions of "miracle" chemicals, fish, or invertebrates until we can better anticipate the effects of their introductions. This chapter evaluates examples of the restoration of already polluted eutrophic lakes, such as the straightforward waste diversion from Lake Washington and phosphorus removal from wastewater entering Shagawa Lake, to more involved waste diversion schemes, including dredging in Lake Trummen, Sweden. The case of eutrophication control in deep ultraoligotrophic Lake Tahoe, still threatened by nutrient enrichment from waste inflows and sediment, is discussed as an example of prevention rather than cure. Methods which have limited application for pollution control, such as flushing, aeration or mixing, and algal or macrophyte harvesting, are also discussed. Finally, the restoration of the Thames and older remedial measures such as copper sulfate for algal control are considered in light of the development of resistance and new findings on copper toxicity.

INTRODUCTION

Twenty years ago almost all limnologists had academic positions. Today there are probably more limnologists employed by consulting engineering firms. Applied work requires fast response to difficult questions and the results have often lacked the rigor of the old methods. The challenge to the new limnologist is to be both fast and correct.

The rationale behind lake or stream management is commonly some consensus of public opinion concerning the current versus desired trophic state of the water body in question. The actions taken may range from an artificial increase in nutrients to promote fish production, to a decrease in nutrients or addition of toxicants to decrease nuisance plant growth to improve the appearance of culturally eutrophic waters. Management policies dealing with nutrients will not be successful unless based on an understanding of the processes responsible for eutrophication (See Chap. 18). Another common practice is alteration of an existing food web by heavy stocking for a put-and-take fishery.

The importance of a carefully defined, realistic approach to aquatic resource management is perhaps the single most important factor in the successful manipulation and restoration of lake, stream, and estuarine ecosystems. Management is best planned as a long-term continuing process which normally requires the expenditure of rather large sums of public or private funds. Although the larger lakes are owned by the state and tax revenues can be used in their management, many involve the political jurisdictions of more than one county, state, or country. Unfortunately, the political decision makers almost always require quick solutions, necessitating compromises and shortcuts by limnologists.

In the most developed countries, increases in both population and leisure time over the last century have created public pressure to manage a variety of lakes and streams for public recreation. Despite a degree of incompatibility of use, these water bodies are required to meet multipurpose recreation needs, from appreciation of their beauty to fishing, duck hunting, swimming, and boating. In contrast, in most developing countries aquatic resource management is primarily concerned with food production. Exploitation of the natural fisheries, or occasionally large lake animals such as hippopotami, crocodiles, or alligators, is becoming more widespread. The prospects for successfully developing aquaculture over a broader range than the fish farming of mullet, trout, and carp require an understanding of aquatic food chains as well as new approaches in biological and engineering technology (see Chap. 15).

In addition, aquatic ecosystems have been, and will continue to be, the ultimate sink for soluble waste. Despite this the quantitive assessment of waste discharges on the biota of receiving waters is a neglected branch of limnology. To the question, "How much treatment is enough?", we must fit the treatment much better to anticipated effects on the biota. If we do not, the present costly muddle of overtreatment here and undertreatment there will ensure the loss of beneficial use of many lakes, estuaries, and rivers.

Although it would be desirable to leave many beautiful lakes completely alone, it is unrealistic to assume that any lakes are now completely isolated from pollution, if only from the atmosphere. Even in the remote Antarctic, the ice has recorded a steady increase in lead content from atmospheric transport. The lead has been present since the first smelting over 2000 years ago and has shown a particularly dramatic increase which began when tetraethyl lead was first used in gasoline (Patterson and Salvia, 1968). In northern Europe, acid rain has resulted from sulfur pollutants derived from the combustion of coal. Even the most remote rockbound lakes and streams receive an increased supply of major and minor nutrients from aerosols and other volatile fallout from industry and agriculture. As lakes increase in recreational value with population growth, their careful management becomes essential. Even Lake Baikal in remote Soviet Siberia has seen a steadily increasing number of tourist visitors each year, and at least one recreational camp has been established on its shore.

Before any management is attempted, a number of factors should be considered. It is essential to first establish exactly what are the desired goals of management. To meet these goals, limnological measurements over a number of years are desirable to include flood, drought, and normal years, with sufficient sampling to develop a nutrient budget. The managers must also secure sufficient popular and political support for the project to be successful. Management should be both environmentally sound and ecologically reversible to allow for unforeseen negative effects of the selected strategy. The project should realistically evaluate the initial capital cost, long-term maintenance costs, and any secondary impacts of the management scheme.

MEASUREMENTS

Normal climatic cycles, especially the frequency of extreme events, are as important in limnology as in terrestrial ecology. In small lakes a very heavy flushing rain at the beginning of spring can easily decimate the phytoplankton bloom. Similarly, an extreme summer calm can create oxygen depletion, fish kills, and a visual nuisance from an otherwise unnoticed blue-green algal population. In the lotic environment, a summer flood may scour most benthic fauna from the streambed. In other cases unusual human activities can temporarily disturb the annual cycle. For example, a rather inaccessible small English lake had been studied for several years when, for no apparent reason, a totally different diatom flora appeared for one season. As it turned out, a passing farm tractor had accidentally dropped a sack of nitrogen fertilizer into the pond which caused the change. In many smaller lakes similar, less-obvious changes in nutrient and toxicant loading are likely to occur. Such variations will rarely result in wrong management decisions if the data from several years are available.

Having established the duration of the premanipulatory study and the management goals, what then should be measured? Judgment as to relative importance as well as economy of time and money are the deciding factors in selecting the parameters to measure. Fortunately the most important variables to be measured may not require particularly complicated equipment or analysis. Important measurements such as transparency, dissolved oxygen, or primary productivity can easily be made. In this chapter we discuss the kind of observations which are necessary for management. These may not be strictly quantitative since the phrase "good fishing" does not merely reflect a high catch rate but also the species composition and the actual fishing conditions (Chap. 14).

The Secchi disk, which measures lake transparency, is a useful and inexpensive method to assess water quality, particularly algae and suspended sediments (Fig. 20-1). However, the seasonal variations in water transparency prior to the pollution of lakes are usually unknown, and a level of transparency must be arbitrarily established as a goal for lake restoration. For example, a Secchi disk depth of 50 cm indicates a severe nuisance condition for most lakes, but a fourfold increase to 2 m would provide satisfactory transparency for most observers, while a fourfold change in Secchi depth from, say, 2 to 8 m may not be readily apparent to the casual observer. Plotting transparency and algal abundance measured as chlorophyll a produces a hyperbolic tangent similar to that found for algal growth and nutrients (Fig. 15-5). Only

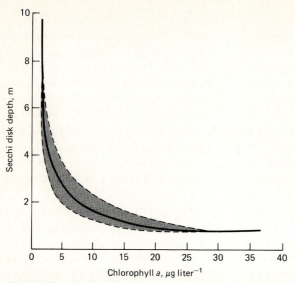

Figure 20-1 Generalized relationship between acetone extracts of chlorophyll a from phytoplankton and Secchi disk depth in lakes. The shaded area shows the variability to be expected with different lakes at different seasons. This relationship will be poor if algae are present in large clumps (Fig. 12-2). Water color or turbidity from suspended sediments may also distort this relationship.

over a limited range do small changes in algal concentration produce large changes in transparency (Fig. 20-1).

A complicating factor in some eutrophic lakes is the presence of chlorophyll contained in large clumps of blue-green algae up to a few centimeters in diameter. Other phytoplankters are dispersed as individuals or short filaments. Clumps produce higher than expected transparency for any given chlorophyll level. Under such conditions, the Secchi disk may seriously underestimate nuisance conditions.

In very clear lakes, considerable eutrophication may occur before any change in Secchi depth is noted. In some cases, increased growth of attached algae at the edges may be the most obvious change. This has occurred in the clear littoral waters of oligotrophic Lake Tahoe (Goldman, 1981). Even though the Secchi disk is often a good indicator of lake eutrophication, no one measure is likely to establish the necessary criteria for lake restoration. This chapter discusses several measurements used to evaluate the progress of lake restoration. These include Secchi disk, chlorophyll, attached algae, primary productivity, oxygen, and changes in fish, zoobenthos, and zooplankton populations.

Dunst et al. (1974) have listed 17 types of potential methods for lake restoration which involve both lake and watershed management strategies. The best-known examples of watershed management concern sewage and other wastewater treatment or diversion, although reforestation may also be important. Diversion usually sends the problem elsewhere, yet has been the most widely used method of curbing wastewater inputs. Studies of wastewater diversion from lakes, such as in Lake Mendota, Lake Washington, Lake Tahoe, and Lake Zurich, are important because they illustrate the causes and effects of nutrient loading. First, diversion can prove that human pollution was responsible for much of the nuisance. Second, diversion shows that the gross features of eutrophication are, to a large extent, reversible in most lakes. Third, long-term inputs from old sewage leach fields, sediments, or other land disposal may negate for many years the beneficial effects of major sewage diversion especially in shallow lakes.

CASE STUDIES

Lake Washington: Waste Diversion

Lake Washington, Washington, United States, is a large monomictic coastal lake of the fjord type ($A = 88$ km^2, $\bar{z} = 33$ m, $z_{max} = 62$ m). The lessons learned from its pollution and restoration are of general application. In addition, the limnological information needed to interpret the effect of lake management on Lake Washington over the last 40 years is available (Edmondson, 1972a, b).

Early in this century, Lake Washington was a mesotrophic to oligotrophic lake with a phytoplankton mainly composed of diatoms and a minimum Secchi disk depth of 3 to 4 m. By the mid-1950s, the phytoplankton was dominated by the filamentous blue-green alga *Oscillatoria* and the maximum Secchi disk depth had dropped to about 1 m. The loss of water clarity within memory of lakeshore residents and unsightly floating mats of blue-green algae produced the necessary public demand for a lake cleanup program.

The question first asked was, what caused the deterioration?, and then, what could be done to restore the lake? The likely causes were either sewage or land drainage. Sewage inflow from nearby towns was treated and then disposed of in the lake; land drainage, chiefly discharge from agricultural, logging, and other activities, also entered the lake. Three solutions were suggested: (1) it was possible to divert treated sewage out of the drainage basin; (2) legislation could be implemented to change agricultural, logging, and residential development practices in the drainage basin; (3) the problem could be ignored as unimportant except to a minority of property owners near the lake. A fourth solution, not technically possible in 1960, would be to drastically reduce either phosphate or nitrogen inflows by advanced wastewater treatment. The removal of sewage phosphorus could be carried out simultaneously with a prohibition of phosphate-containing detergents (see Shagawa Lake).

Lake restoration procedures were carried out, largely due to the presence at Lake Washington of an eminent limnologist, Prof. W. T. Edmondson of the University of Washington. It is unlikely, even as late as 1970, that most consulting engineering firms had sufficient limnological experience to solve the problem. It was important to differentiate between the point-source inflows of sewage and the diffuse inflows which included natural runoff and pollutants from logging. At this time the lake was receiving some 24,000 m^3 day^{-1} of secondary-treated effluent from about 70,000 people. Sewage, therefore, seemed potentially responsible for the deterioration of the lake. Secondary-treated effluent contains 20–30 mg liter^{-1} NH$_4$-N and 5–10 mg liter^{-1} PO$_4$-P. A lake nutrient budget in 1960 showed that roughly half the phosphorus input, but only 15 percent of the annual nitrogen income to the lake, was derived from sewage. Thus sewage diversion was the only alternative which would have any chance of substantially reducing the nutrient loading of the system. Further deterioration had occurred by 1964 when point sources of sewage were shown to be contributing almost 75 percent of the annual phosphorus income. A good correlation existed between the early spring concentration of phosphorus in the surface water and the crop of algae produced 3 months later. Similar correlations between early spring silicon and maximum crops of the diatom *Asterionella* had been made in Windermere some years previously (see Chap. 12).

There was much controversy over which was the limiting nutrient for algal growth in Lake Washington. However, in most lakes polluted by sewage, it is unimportant which nutrient eventually becomes limiting because it is obvious that too much has been added of all the essential elements. In Lake Washington, after many years of waste inflows, nitrogen usually limited growth in summer because so much more phosphorus than nitrogen had been added. Treated domestic sewage has a ratio of N:P of 3:1 by weight, while plants use approximately an N:P of 10:1 by weight. A good correlation between spring phosphorus and summer phytoplankton crops and a lack of a similar correlation with nitrogen suggested that diversion of inflowing phosphorus alone would eliminate algal nuisance problems. At the time when a political decision had to be made there was only limited scientific evidence for phosphorus control and sewage diversion would only remove a small amount of nitrogen. The remedy suggested by Edmondson involved courageous scientific intuition and considerable risk to a well-established scientific reputation. This uncertainty is often inherent in applied limnology.

A major drawback with almost all previous successful lake reclamation schemes is a lack of scientific information upon which to generalize future solutions. Although the problem of a particular lake may have been solved, no real idea of the cause was known. Thus the next lake problem had to be tackled afresh. Lake Washington and Lake Zurich in Switzerland are among the first restoration examples where a general theory of lake management was tested. The dramatic effects of the Lake Washington waste diversion are given in Fig. 20-2, which shows the remarkable increase in water transparency following soon after the total diversion of sewage. Figure 20-2 illustrates another feature which is at first puzzling, namely, that when only 30 to 40 percent of the waste had been diverted, no change in Secchi depth was noted. This was because mean epilimnial chlorophyll a was at first unaffected and later only fell from 40 to approximately 20 μg liter^{-1}. Figure 20-1 demonstrates that even large changes in concentration of chlorophyll a above 20 μg liter^{-1} do not have much of an effect on Secchi depth. Whether the lake would have shown

any noticeable improvement for recreation if only half the sewage phosphorus had been diverted is uncertain (Fig. 20-2). It is likely that any increase in transparency would have taken much longer.

The effects of diversion of phosphate-phosphorus, the most important limiting nutrient for Lake Washington, are given in Fig. 20-3 where nitrate is also shown. Diversion decreased phosphorus but not nitrogen, as expected. This whole-lake experiment has enabled lake managers to predict with greater confidence the effects of wastewater diversion in other lakes.

Shagawa Lake: Advanced Wastewater Treatment

If phosphorus were the only element which controlled algal growth in Lake Washington, total sewage diversion would not have been essential. The same results could have been achieved with phosphorus removal from sewage without diversion. Diversion was employed because of the proximity of Puget Sound to Lake Washington. In inland areas like northern Minnesota, where Shagawa Lake is situated, diversion is often impractical.

The phosphorus present in treated domestic wastewater was also implicated as the prime factor responsible for algal blooms in culturally eutrophic Shagawa Lake. In this case, an experimental sewage treatment plant was designed and built to remove over 99 percent of wastewater phosphorus using a method of chemical precipitation. Shagawa Lake ($A \approx 1000$ ha, $\bar{z} \approx 6$ m, $z_{\max} \approx 14$ m) is smaller, shallower, and has a shorter hydrologic residence time (8 months) than Lake Washington (3 years). Shagawa Lake is typical of many dimictic lakes which have developed nuisance algal blooms over the last century. There were large spring peaks of diatoms (*Asterionella* and *Synedra*) and green algae followed by blue-green algal blooms in late summer (*Aphanizomenon* and *Anabaena*). These blooms were in marked

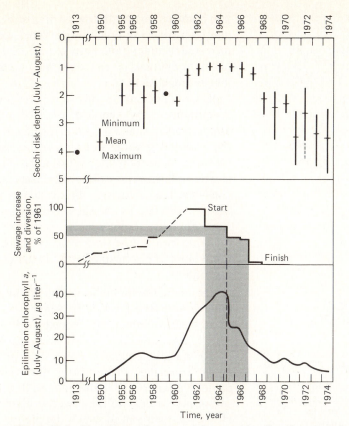

Figure 20-2 The effects of sewage inflow and its diversion on Lake Washington as measured by water transparency (Secchi disk) and epilimnetic chlorophyll *a*. Note the rapid drop in chlorophyll and subsequent increase in Secchi disk depth after diversion of about 50 percent of the wastewater. Note also that diversion of about 35 percent only halted the deterioration of some aspects of water quality (i.e., Secchi depth, but not chlorophyll *a*). Dots indicate single readings. (*Replotted from Edmondson, 1972a, b, with additional information courtesy of W.T. Edmondson.*)

contrast to conditions in nearby oligotrophic lakes which presumably were similar to Shagawa Lake's original state. Almost all pollution appeared to come from one source, sewage from the town of Ely. This was a simple problem, at least in concept. Main inflow to the lake, apart from wastewater, was a river draining the adjacent oligotrophic Burntside Lake, which made the restoration of Shagawa Lake more feasible. In many lakes, the multiplicity of lesser nutrient inflows complicates the before-and-after picture so important in a demonstration project. An annual nutrient budget es-

tablished that about 80 percent of the inflowing phosphorus but only 25 percent of the nitrogen came from the domestic wastewater (Malueg et al., 1973).

Bioassay experiments indicated that addition of phosphorus to the lake water produced the greatest increase in algal growth, and a similar effect was produced by secondary-treated effluent (Powers et al., 1972). A special, well-controlled treatment plant was built, and an extensive limnological study was carried out as a demonstration project for the U.S. Environmental Protection Agency.

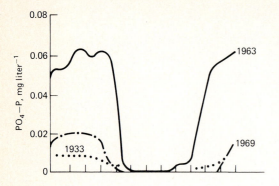

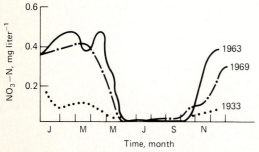

Figure 20-3 Changes in phosphate and nitrate from 1933 (little waste discharge) to 1963 (maximum waste discharge) to 1969 (all waste diverted). Ammonia was not measured, but could be expected to be approximately 0.1 mg liter^{-1} (1963) and 10-20 μg liter^{-1} (1933 and 1969). In addition, nitrogen fixation, which was not measured, probably contributed between 0.5 and 10 percent of annual inflowing N (see Horne, 1977). (*Modified from Edmondson, 1972b.*)

The waste-treatment plant began operation in the period January–April 1973, and the results on water quality are shown in Fig. 20-4. Virtually all sewage phosphorus was removed, reducing the annual inflow by approximately 80 percent, which is close to the 75 percent removed from Lake Washington. The algal levels and transparency recorded for the first year were actually just below the normal year-to-year variation for the pretreatment period. After 2 years large decreases did occur in annual mean orthophosphate and chlorophyll *a*. Transparency as measured by Secchi disk depth increased, but there was little change in the available nitrogen. Because Shagawa Lake

has both a short hydrologic and a short phosphorus residence time, more rapid lake improvement might have been expected. If phosphorus removal is to work for lakes with longer residence times, such as the Great Lakes, the reasons for this slower response are important. In the simplest theories, nutrients are thought to spend the same residence time in the lake as the water does. In reality, summer releases of accumulated phosphate from the anoxic sediments constitute a diffuse source not considered in calculating the residence time. In the case of relatively shallow Shagawa Lake in summer, *internal loading*, i.e., the rate of PO$_4$-P recharge from the sediments, may be as important a source of phosphorus as is the gross annual inflow.

The annual epilimnetic variations in chlorophyll *a* and PO$_4$ for typical years before and after waste treatment are shown in Fig. 20-5. The figure indicates a dramatic decrease in PO$_4$-P but little change in green algae or diatom crops in spring. Presumably, the winter accumulation of PO$_4$-P is adequate to develop the spring bloom despite the lowered waste input. However, the point at which the pre- and posttreatment PO$_4$-P levels are closest is in August (Fig. 20-5) due to sediment PO$_4$-P releases. At least part of these releases is due to macrophytes "mining" the sediment with their roots and releasing PO$_4$ into the water from their leaves. Such a phenomenon has been well-documented for the sea grass beds in coastal Alaska (McRoy et al., 1972). Unfortunately, blue-green algal nuisance blooms and peak recreation coincide in midsummer. Since the major blue-green algae present can fix nitrogen to supply much of their nitrogen needs, the release of PO$_4$-P and possibly iron from the sediments provides all that is needed to grow a sizable algal bloom. As already noted, average chlorophyll *a* levels must be lowered below 20 μg liter^{-1} for differences in transparency to be obvious to the recreationist (Fig. 20-1).

Shagawa Lake's restoration program demon-

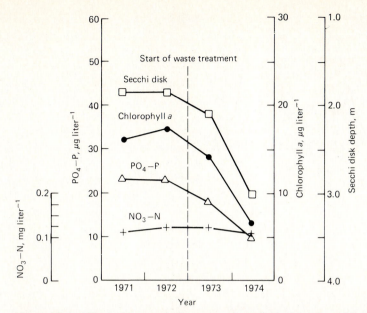

Figure 20-4 Effect of removal of sewage phosphorus on lake nutrients and algal growth in Shagawa Lake, Minnesota. The rest of the wastewater was still discharged to the lake. Levels given are annual means for the epilimnion (0–5 m). Note declines in chlorophyll and phosphorus, increase in water clarity, but no change in nitrate. (*Redrawn from Malueg et al., 1973, with additions from Malueg by personal communication.*)

strates a number of important points. The lake's annual limiting nutrient concentrations reached predicted levels. In shallow lakes of this dimictic type, the annual average concentrations are of limited value for predicting the rate of supply of nutrients during midsummer. Internal loading of major nutrients from the sediments must be considered in lake restoration projects. Although the percentage of phosphorus removed from lakes Shagawa and Washington was similar, the percentage of water in contact with the bottom mud was much greater in Shagawa Lake and slowed restoration. Eventually, internal loading should diminish as nutrients are buried or locked into refractory compounds. The time required is at least a decade and may be even hundreds of years for shallow lakes with a thick layer of polluted sediments.

An important facet of the removal of phosphate-phosphorus from wastewater is the removal of most soluble metals by the high pH used for precipitation. Although this may not often be important, it should be noted that the correlation of phosphorus removal and lowered biomass could in part be due to removal of iron, manganese, or other minor nutrients.

Lake Tahoe: Preventative Measures

In general, most lakes which suffer from pollution are not large, deep high-mountain lakes. Lake Tahoe, located near the crest of the Sierra, is one of the deepest lakes in the world ($z_{max} = 501$ m). It is so beautiful (Plate 1 *b*) that it has attracted a large, year-round population with a recreation-based economy of gambling, downhill skiing, boating, and fishing. Sewage disposal and land erosion associated with development of the Tahoe basin have resulted in increased flow of nutrients into the lake (Goldman, 1981). Because of the extreme dilution in the 156 km³ of lake water (Table 2-1) and active plant uptake, there has not yet been a measur-

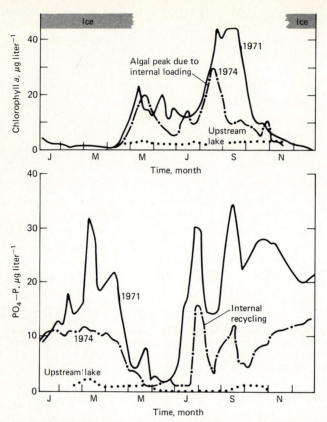

Figure 20-5 Shagawa Lake. Effect of waste treatment including PO_4 removal on seasonal changes in phytoplankton and epilmnetic PO_4-P. Treatment started January 1973. Due to releases from anoxic sediments in August–September 1974, an unexpected fall blue-green algal bloom was produced even after 2 years of treatment. The type of sediment release shown here is called *internal loading* (Larson et al., 1981). It is the main reason for the lack of total success of lake restoration projects in relatively shallow stratified or unstratified lakes such as Shagawa Lake ($\bar{z} = 6m$), and failure of diversion in Lake Trummen. In contrast deep fjordlike Lake Washington ($\bar{z} = 33m$) recovered rapidly (Fig. 20-2). (*Redrawn from Malueg et al., 1973.*)

able increase in nutrient concentration. The lake's remarkable clarity (Secchi readings to 40 m) is slowly declining over much of the lake, and attached algae around the lake's margins are visual evidence of increased nutrient accumulation.

There was, as at Lake Washington, a fortunate coincidence of basic limnological studies with the period when decisions on sewage treatment strategy were being made. The various available options were, first, to continue existing practices of secondary treatment and near-

lake disposal of effluent from mixed septic tanks and primary treated sewage. It was even believed by some that such a large, deep, and cold lake could dilute all nutrients added to an insignificant level. A second alternative was to pipe the waste flows to the lake bottom, because such a deep lake was thought by some consulting engineers and geologists to be meromictic. Another alternative was simply to continue the current practice of secondary treatment but reuse the wastewater by spraying it on the land and forests of the lake basin. A fourth proposal was to completely sewer the watershed and treat wastes to the most advanced level possible, and then dispose of the effluent in the lake. Finally there was diversion of secondary-treated waste outside the basin as was done for Lake Washington. Waste treatment would then be combined with a land use plan to prevent uncontrolled development from further increasing erosion of the basin.

Nutrient-addition bioassays were devised using $^{14}CO_2$ uptake as a measure of phytoplankton responses to highly treated wastewater (Goldman and Carter, 1965). These studies showed that because of the long (approximately 700 years) residence time of the water in Lake Tahoe, even very small nutrient additions would continue to have a stimulatory effect and cause the lake's eutrophication (Fig. 20-6). Another bioassay technique used the organic

^{14}C-labeled acetate to demonstrate how bacteria were stimulated by inflows of eroded sediments (Paerl, 1973). Through this experimental approach, the lake was shown to be extremely sensitive to nutrients and to be unable to absorb treated wastewater and increased sediment inflows from the watershed without progressive eutrophication.

Further limnological study of this deep lake found it to be monomictic during most years. This conclusion was based on oxygen measurements indicating no depletion with depth and was used as an argument against deep-lake sewage disposal since nearly isothermal cold waters have little resistance to vertical mixing by the wind. Eventually, studies using the nitrocline between the low nitrate-nitrogen of the photic zone and the slightly higher nitrate-nitrogen levels below confirmed the earlier conclusion

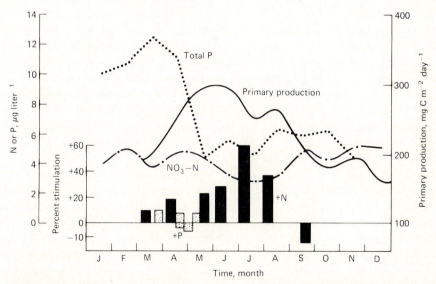

Figure 20-6 Response of Lake Tahoe water to additions of very small amounts of nitrate and phosphate (2 to 10 μg liter^{-1}). Also shown are mean monthly averages of nitrate, total phosphorus, and primary production for the 0–105 m water column in 1978. Dotted bars refer to PO_4-P additions, black bars to NO_3-N additions. Note the stimulation by nitrogen in this nitrogen-limited subalpine lake which contrasts with the phosphorus stimulation in some other systems (see Fig. 20-4). As in most oligotrophic systems, where all nutrients are scarce, additions of nitrogen, phosphorus, iron, silicon and trace metals produce the largest increases in growth. (*Redrawn and modified from Goldman, 1981.*)

based on oxygen and temperature (Paerl et al., 1975). The measurement of a nitrocline required extremely careful measurements of nitrate-nitrogen at trace levels (2 to 30 μg liter^{-1}) with an error of only ± 2 μg liter^{-1}.

There is sound hydrologic and economic sense in reusing wastewater, especially within a lake basin. This is usually accomplished by spraying secondarily treated wastewater, rich in nitrogen and phosphorus, over agricultural land or golf courses. These are normally irrigated with nutrient-poor ground or river water to which nitrogen and phosphorus fertilizers need to be added. After use in growing crops, the excess water which is then partially nutrient-stripped rejoins the natural drainage, thus retaining the original water balance of the basin. Because of its subalpine climate, the Tahoe basin is not suited for agriculture. Some limited golf course disposal was attempted, and a major water reclamation experiment which involved spraying effluent on stands of Jeffrey pine, *Pinus jeffreyi,* was undertaken. Many trees were killed by excessive nutrients or waterlogging of the roots. This not only damaged the forest, but the soil continued to release the added sewage nitrogen many years after the last application (Perkins et al., 1975). Since nitrogen, along with iron, limits algal growth in Tahoe, it is an important element to control. Six years after spraying ceased. Heavenly Valley Creek, which drains the sprayed area, still had average concentrations of 1 mg liter^{-1} NO$_3$-N while similar control streams showed less than 10 percent of this level. This particularly undesirable land disposal option, used for a few years on public land, was discontinued.

Because of the ultraoligotrophic nature of Lake Tahoe, sewage treatment obviously was required to a level well beyond the normal secondary level if lake disposal was to be considered. Nutrient-addition bioassays, a lake nutrient budget, and proof of holomixis showed this solution was unacceptable. Unlike lakes Washington and Shagawa, algal growth in Lake Tahoe is principally nitrogen-limited, and thus the removal of nitrogen, not phosphorus, would be required. Unfortunately, the removal of nitrogen is more difficult and even more costly than the expensive removal of phosphorus. To control the eutrophicating potential of sewage, nutrient removal to a level comparable to an unpolluted basin stream would be necessary. Such levels are not as yet feasible, as the most highly treated outflows contain at least 2 mg liter^{-1} NH$_4$-N and 500 μg liter^{-1} PO$_4$-P. Bioassays showed that this level of nutrient addition, even when diluted in the lake, would still substantially degrade the lake's water quality. It was economical, due to the clumped distribution of ski resorts, gambling casinos, and lakeshore dwellers, to sewer all but the most remote communities. Consequently, the wastewater was exported from the lake basin, although, curiously, the most advanced and costly level of treatment possible was still provided before discharge into distant rivers.

Some attempt at restricting land disturbance from construction activity was also attempted. Unlike installing sewage systems, which most people think well of, land use planning with restrictions on land development is usually unpopular. The public controversy still continues among the various agencies of two states, five counties, and the numerous federal agencies involved, with new attention to atmospheric sources.

The diversion of nutrients from Lake Tahoe was a preventative measure rather than a cure for a serious existing problem. Similar measures are now planned for Lake George, New York. Except for the growth of attached algae, the lake problems were more potential than real at the time sewage diversion was proposed, since Lake Tahoe's water quality was still extremely high. This is in contrast to lakes Washington and Shagawa where nuisance blue-green algal blooms were already present. The results of waste export are shown in Fig. 20-7. It can be seen that the increase in phytoplankton produc-

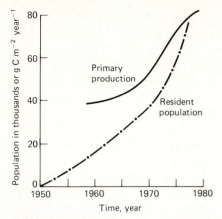

Figure 20-7 Changes in the annual primary production in the pelagic waters of Lake Tahoe in relation to the increase in resident population in the watershed. Nonresidents have increased in a similar fashion. Soil and nutrients eroded during home and recreational construction as well as leachate from old septic tanks reaches the lake and is the probable cause of the measured doubling in primary production. (*Modified from Goldman, 1981.*)

tivity since 1959 has at best only been slowed, but certainly not reversed. Since the water in Lake Tahoe has the enormous hydrologic residence time of 700 years, no measurable decline would be anticipated for many years. As with the phosphorus-limited lakes, the role of internal recycling of major nutrients is only partially known in Lake Tahoe, and thus the relation between hydrologic residence time and nutrient washout is still speculative. Because of the long residence time, it is also unlikely that the attached algae at the margins of the lake will be significantly reduced for many years. A predictive model of the productivity of Lake Tahoe which incorporates residence time indicates that the lake will make a very slow recovery from the increased nutrient inputs of the last few decades (N. J. Williams, unpublished data).

Some lakes should be preserved from the products of permanent human settlement. This level of preservation is now too late for Lake Tahoe, which might well have been preserved as a National Park like Crater Lake, Oregon.

Lake Trummen: Dredging

Lake Trummen, Sweden, is a small, shallow, northern temperate lake which was formerly oligotrophic ($A = 1.0$ km^2, $\bar{z} =$ about 1.5 m, $z_{max} = 2.2$ m). As with many lakes, municipal waste inflow for many years was probably the cause of its eutrophication. By the mid-1950s, large blooms of blue-green algae were present. Although no extensive limnological study was available, waste discharges to the lake were diverted with the hope that the lake would recover. Unfortunately, after 10 years of diversion no improvement was apparent and nuisance algae were still abundant.

We saw in Shagawa Lake that shallow dimictic lakes recycle sewage phosphorus from the sediments sufficiently to slow down nutrient flushing from the lake. Similar recycling of ammonia will also occur in polluted lakes. In summer, occasional breakdowns in complete thermal stratification in shallow lakes allows nutrients released by anoxic sediments to support large algal blooms. The solution for Lake Trummen was to dredge out 0.5 m of the bottom mud to remove the nutrients which had accumulated over the years and were preventing the lake's reversion to its original state (Bengtsson et al., 1975). Only the upper sediments contained the legacy of the previous pollution, yet this 0.5-m layer amounted to 6 million m^3. The cost was greatly reduced by using the sediment to build a park beside the lake. The interstitial waters, which are always high in nutrients (see Fig. 10-5), were cleared of phosphate by precipitation with alum in a treatment lagoon before return to the lake. The cost of dredging was $500,000 1974 U.S. dollars and the research costs were about $400,000. Releases of interstitial phosphate and sediment phosphate were sufficiently reduced so that blue-green algal blooms are no longer a nuisance (Cronberg et al., 1975). Major changes also occurred at higher trophic levels since zooplankton, especially cladocerans, were much reduced (Andersson et al., 1975). A decline was observed in the common cyprinid fish, many of which are zooplankton feeders. This reduction in overall fertility should be anticipated when cleaning up lakes where fish populations are important. Benthic invertebrate populations of chironomid midge larvae, often found in eutrophic lakes, so far remain little affected by the restoration.

In most polluted lakes, the cost of dredging and spoil disposal is prohibitive, but diversion of waste or nutrient removal should not be proposed without considering the interaction of the sediments and the overlying water. Surface area or volume alone provide insufficient information to predict the time required for lake restoration. For example, if Lake Erie's algal problems are mainly due to phosphorus, the lake manager needs to know how long it will be before the supply of phosphorus in the sediments is exhausted or covered over. This information is essential if limnologists are to make realistic predictions of lake recovery.

WHEN NUTRIENT DIVERSION IS IMPOSSIBLE

Two prominent features of eutrophication are excess nutrient levels and low oxygen content of the lake bottom waters. These can be altered by physical or chemical methods. In many lakes it is not economically feasible to divert or adequately treat inflowing nutrients. For example, in Clear Lake, California, Upper Klamath Lake, Oregon, or Lake George, Uganda, natural inflows account for almost all the nutrient input, yet these lakes are very eutrophic. In other cases, diffuse sources of nutrient inflow, such as agricultural drainage, make water treatment virtually impossible. Many lakes are easily made eutrophic by changes in the drainage basin. For such small but often important systems an approach of physical manipulation is probably the best procedure available. Two of the acceptable methods currently in use are flushing and aeration.

Flushing

Obviously, if a nutrient-rich lake is flushed with sufficient nutrient-poor water, its water quality will improve (Welch et al., 1972). In addition, the very process of the extra stirring due to inflowing water will generally reduce stagnation. Green Lake, near Seattle, Washington ($A = 104$ ha, $\bar{z} = 3.8$ m, $z_{max} = 8.8$ m), is a small eutrophic lake which had nuisance blooms of the blue-green alga *Aphanizomenon*. When flushed with a large volume of low-nutrient water, the hydrologic residence time was reduced from 14 to 3 months. In this case a decrease in nitrate occurred and phosphate levels were little changed. Nuisance algal levels were reduced and water clarity increased (Fig. 20-8). This form of lake management is expensive, since only a few cities have sufficient surplus clean water available for flushing. If the lake volume is small relative to the available flushing water, it is not necessary that the flushing water be nutrient-poor, only free of algae. If the flushing rate approaches the rate of cell division of the problem algae, a simple washout can be achieved. This would require replacing the lake water every week. Even with rapid flushing, buoyant blue-green algae may resist washout if prevailing wind direction blows them away from the lake outlet.

Aeration and Mixing

Low dissolved oxygen results in poor fish habitat, high sediment nutrient releases, and large algal crops. Mixing the photic zone with the aphotic zone redistributes photosynthetically produced oxygen, but this does not occur on windless days or in sheltered lakes. Lake quality in these situations can be enhanced by mechanical mixing (Fig. 20-9).

At present most lakes are totally destratified with compressed air (Fig. 20-10), but more-efficient giant paddles or water pumps may eventually become common. The main ef-

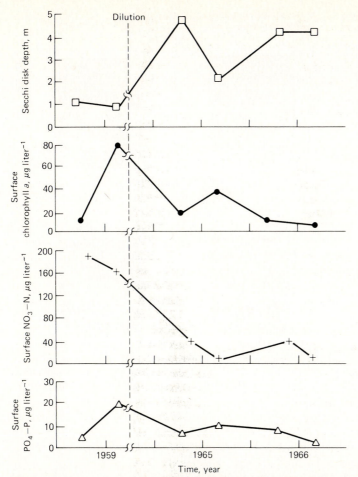

Figure 20-8 Changes in water chemistry and biology before and after dilution of small, eutrophic Green Lake (Seattle, Washington) with low-nutrient water. Values are for mid-May and early August for each year. (*Redrawn from Oglesby, 1968; Sylvester and Anderson, 1964.*)

fect is to redistribute oxygen, rather than increase its concentration. Air can be used economically to add oxygen during mixing, but it is also possible to aerate a lake without thermally destratifying it. The technique, called *hypolimnetic aeration,* is shown diagrammatically in Fig. 20-10 (Bernhardt, 1967; Pastorok et al., 1981). The great advantage is that oxygen is added to the oxygen-depleted hypolimnion without mixing nutrients into the epilimnion. As with total destratification, hypolimnetic aera-

tion removes iron and manganese which cause problems for water supplies. These metals may give a taste to the water and precipitate as brown stains on porcelain fixtures. Nutrient releases from the sediments are reduced by the restoration of the oxidized microzone (Fig. 10-5, Chap. 7), and toxic ammonia may be converted back to relatively harmless nitrate. Unpleasant taste and odor and hydrogen sulfide, which is associated with putrefaction, are also reduced (Barnett, 1975). In addition, oxy-

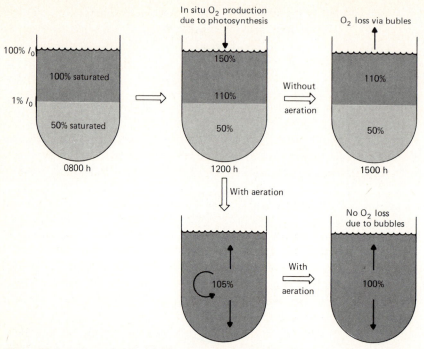

Figure 20-9   Increase in overall lake oxygen levels by simple mechanical aeration or mixing at noon. Major losses in photosynthetically generated oxygen occur when oxygen bubbles form and are vented from a stratified lake. In the stirred lake, photosynthetic oxygen is used to replenish the oxygen debt in the sediments.

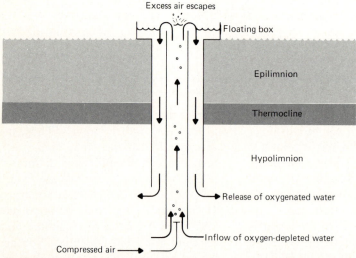

Figure 20-10   Diagrammatic representation of one type of hypolimnetic aeration system (airlift pump). The main feature common to all these systems is the rising air bubbles which oxygenate and lift hypolimnetic water followed by the release of the oxygenated water back below the thermocline.

genated sediments allow substantial increases in benthic invertebrates which serve as food for fish.

In temperate regions, hypolimnetic aeration allows trout to live in the hypolimnia of eutrophic lakes which normally would have insufficient oxygen to support them. Total destratification would not work in this case because the mixing of warm upper water with the hypolimnion would produce a temperature too warm for trout, especially in hot climates. A warm-water fishery in the epilimnion can be maintained simultaneously with a cold-water fishery in the hypolimnion by this method. Since eutrophic lakes are more productive, trout grow faster under conditions of hypolimnetic aeration than in natural, more oligotrophic lakes. In eutrophic dimictic lakes, fish kills can occur in winter from oxygen depletion. Aeration in winter can reduce or eliminate fish kills that would require expensive restocking. Artificial aeration or mixing is probably the safest and cheapest method of dealing with eutrophication problems, other than nutrient diversion.

A fairly new lake aeration system called *sidestream pumping* involves the addition of pressurized oxygen to small quantities of hypolimnial water pumped into a mixing chamber. Since pure oxygen is added, much more oxygen will dissolve than from air which contains only 20 percent oxygen. A solution supersaturated with oxygen flows back to the hypolimnion and can maintain the desired oxygen level in an otherwise anoxic region. The capital costs of this experimental system are small compared to hypolimnetic aeration, and it is easily adapted to changes in the oxygen demand of the sediment (Fast et al., 1975).

In eutrophic lakes and estuaries it is also possible to oxidize the superficial sediments with nitrate injection. This immobilizes phosphate, iron, or hydrogen sulfide, preventing its release during the otherwise reduced conditions. It has been employed successfully in Sweden and also in at least one estuary (Ripl, 1976; Horne and Roth, 1979).

PLANT HARVESTING AND CHEMICAL CONTROL

In many lakes, nutrient inflows are so diffuse and numerous that any nutrient treatment or diversion project is unlikely to succeed. Examples of these types of lakes are small mountain lakes where the phosphorus from even one dwelling is sufficient to produce eutrophication; small, shallow recreation lakes in parks or highly urbanized surroundings where street runoff is important; and large lakes where natural inflows of nitrogen and phosphorus combine with intensive recycling to produce eutrophic conditions. In many of these cases, dredging, aeration, or both, even if effective, are uneconomical.

A major problem with these lakes is the invasion of the shallow waters by rooted aquatic macrophytes. One solution is *macrophyte harvesting* by an underwater mowing device that cuts off the plants at the stem, combined with a rake (Fig. 20-11) for transferring the cut plants to a disposal barge (Livermore and Wunderlich, 1969). Other harvesters may throw the plants ashore. Many lakes are successfully "cleaned up" with harvesters which, although they have a high capital cost, may also be rented (Nichols and Cottam, 1972). It is important to move the decaying cut vegetation away from the lake edge to prevent release of nutrients back into the lake. The vegetation may be used for compost, landfill, or even methane generation.

Wetlands which occur on inflowing streams and around the lake edge naturally buffer lakes, streams, and estuaries from sediment and nutrient inflow. Using macrophyte harvesting, it may be possible to remove sufficiently large quantities of nutrients to reduce the lake's overall trophic state. Unlike phytoplankton, however, rooted macrophytes have direct, permanent access to sediment resources, and the

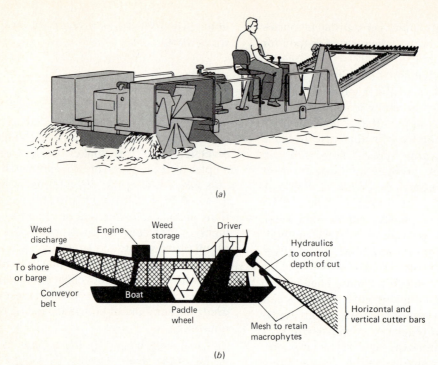

(a)

(b)

Figure 20-11 Large aquatic macrophyte harvesters. (*Redrawn from information provided by Aquamarine Co., Waukesha, Wisconsin.*)

nutrient content of the water may not regulate their growth. Nevertheless, macrophytes frequently remove sufficient nutrients from the water to prevent phytoplankton blooms, and there is speculation that some higher plants may even release algal antimetabolites. Higher plant control can lead to the paradoxical situation where the clearance of weed-choked lake waters gives rise to algal blooms which cloud previously clear water. Harvesting must be carried out frequently enough to balance nutrient inputs. It might even be possible to harvest enough macrophytes to restore the lake sediments to their original state.

In very large lakes, particularly those in equatorial zones, floating macrophytes can be a great nuisance by clogging outlets, harboring insect and snail hosts, and increasing water loss through transpiration. These plants are usually

quite large and have descriptive common names like water lettuce (*Pistia stratiotes*), water fern (*Salvinia molesta*), and water hyacinth (*Eichhornia crassipes*). In general, these macrophytes have a very high water content and are virtually useless as cattle feed without extensive dewatering. In Lake Kariba, central South Africa, one-third of this enormous reservoir ($A = 4400$ km²) was covered with *Salvinia* only a few years after its formation (Mitchell, 1969). Although the area covered has diminished, it continues to impede fishing and navigation. Its decay is responsible for toxic hydrogen sulfide concentrations below the thermocline in portions of the lake. At Kainji Reservoir on the Niger River in Nigeria, Africa, water level reductions during operation of the hydroelectric plant strand the plants on the shoreline. Water level control may also be used to reduce spawning success of fish likely to overpopulate the lake (Chap. 14). Lake drawdown is currently the practice in many temperate lakes for control of both floating and attached macrophytes, but is incompatible with most recreational use. The sight of several meters of exposed mud covered with rotting weeds does not encourage tourism!

Animals such as crayfish, grass carp, tilapia, nutria, capybara, or manatee may provide a degree of macrophyte control (see Chap. 11). Control has also been achieved by introducing insects which may effectively attack and destroy specific emergent or floating plants. Introductions of this sort must of course be undertaken with caution so as not to create some new problem where none previously existed.

Algal harvesting may be used to collect concentrated surface blue-green blooms. A major difficulty with this method is the design of a harvesting screen which will collect the algae before the motion of the collecting barge dissipates the floating algae. Pilot-scale algal harvesting devices (Fig. 20-12) have successfully

Figure 20-12 An experimental algal harvester at work on surface films of *Aphanizomenon* in Clear Lake, California. The floating blue-green algae are collected by driving the submerged blade of the leading collector under them. The resulting material is dewatered by a backwashing rotating microstrainer in the rear craft.

been tested on the very dense algal blooms in Clear Lake, California (Oswald, 1976).

The most widely used method of lake algal clearance is the application of *copper sulfate*. The cupric ion is toxic and acts to prevent photosynthesis or growth by affecting the cell membranes. Copper is not a catalytic poison and is used up in the toxic action. If the copper concentration is too low, those membrane sites not blocked by the copper ion will continue to operate and the cells will recover (Steemann Nielsen et al., 1969). Repeated applications lead to algal populations with increased copper tolerance; in some cases, resistance has built up to the point where it is uneconomical to continue (Ahlgren, 1970). In such cases algal harvesting or waste treatment should be employed.

In many lakes where algae are not very abundant but float and are wind-concentrated to nuisance levels, copper treatment has been the cheapest control technique. Copper sulfate is added to surface waters to produce a concentration of 500 μg liter^{-1}. Although this rapidly kills algae, the toxic effect lasts only a day or so and fish and zooplankton are not visibly disturbed. Presumably they can avoid the high copper levels by migrating because in laboratory tests some zooplankton and fish larvae may be sensitive to ionic copper levels as low as 10 μg liter^{-1}. Nuisance blue-green algae, the prime targets of copper treatment, are seriously affected at 5 μg liter^{-1} (Horne and Goldman, 1974, Fig. 10-12).

Soluble copper persists in lake waters much longer than expected from simple precipitation theory, presumably due to chelation with dissolved organic material. After conventional copper treatment, most ionic copper is precipitated as basic copper carbonate, also called malachite. A very small quantity of ionic copper penetrates the living cell and a small quantity remains in solution in a nontoxic or sublethal chelated form. Over a period of days or weeks, this is gradually incorporated into the particulate fraction (Elder and Horne, 1978b).

Finally, copper is lost to the sediments where, under the anoxic conditions of eutrophic lakes, it often precipitates as insoluble copper sulfide. Although copper continues to be used with little regard to long-term consequences on the structure of the aquatic community, the insolubility of copper sulfide may be the reason for the apparent lack of long-term food-chain effects following copper sulfate applications.

STREAMS, RIVERS, AND ESTUARIES

Limnologists should not neglect the applied aspects of flowing waters. Too many beautiful and once-productive small streams that could be restored have been regarded as hopelessly damaged. Similarly, the political difficulties of restoring larger streams and rivers are sometimes overwhelming since parts of the river may be in different states or even countries with no common antipollution code. This is unfortunate because rivers and streams are usually easier to restore than lakes due to the continual supply of purer water from upstream.

Frequent causes of lotic pollution are too much organic nutrient or toxicant loading. Physical damage is caused by dams, dredging, artificial channelization, accelerated watershed drainage, and removal of riparian vegetation. In extreme cases the river may be reduced to a concrete channel or even be completely buried underground in a conduit.

Although desirable, it is usually uneconomical to physically restore significant portions of the river or stream. Practical considerations such as flood control, roads, docks, or housing developments which originally changed the watercourse often still apply. Over the last century, extensive building in flood plains has often placed artificial structures squarely in the course of flood destruction.

There is little doubt that dams cause serious losses in production of such valuable anadromous fish as salmon and sturgeon (Chap. 14).

The case of the Columbia River, a large river in northwest North America, is typical. Here the construction of a series of high dams for electric power generation virtually destroyed the large salmon runs. The adults were usually unable to ascend the fish ladders built for them and only a few could be caught and trucked around the obstructions. Those adults reaching the headwaters were able to spawn successfully, but expensive hatcheries had to be constructed to supply the bulk of the next generation. These fry must be released in the river to imprint them with the river "scent" which allows them to find the correct river when they return as adults from the sea. Unfortunately, the juveniles migrating downstream have great difficulty passing through the still waters of the reservoir, and the turbines at each dam kill about 10 percent of those passing through. The restoration of the original salmon runs has not been accomplished.

In the case of other rivers such as the Sacramento and Truckee in California, the spawning grounds themselves have been inundated by the construction of dams in the upper tributaries. Once again, large natural runs of salmonid fish and sturgeon were lost and costly hatcheries had to be constructed. It is interesting to note that many fishery authorities were unaware of the size of some fish runs before dam construction and were thus unable to offer convincing scientific evidence that might have prevented or altered dam construction. This is a consequence of the great difficulty in estimating fish populations, especially in larger streams or rivers (Chap. 14).

Once constructed, few dams, concrete channels, or levees have ever been removed no matter what the damage caused. Natural catastrophies occasionally destroy these obstacles, but for the most part dams must be considered permanent. The collapse of detritus-based fisheries and benthic invertebrate communities below reservoirs due to sediment retention by the reservoir is a long-term loss to the river's production (El-Zarka, 1973). In other cases productivity may rise (Fig. 14-12).

Unlike lakes where pollutants may be diverted, rivers and streams are hard to clean up by waste diversion unless a parallel sewer is constructed the whole length of the waterway. It is informative for the limnologist to walk along a stream or river and count the numerous officially undescribed and unregulated pipes, drains, ditches, and solid waste accumulation. Some pipes may have been in place for centuries and there is no longer any knowledge of what they drain. Most highways near rivers serve as drains for rainwater, and it is an astonishing fact that in the United States half the oil pollution of lakes and streams is derived from the asphalt of roads and leaks from automobiles.

Restoration of the River Thames, England

A major success for river restoration at a reasonable cost has been achieved over a period of about 20 years in the River Thames.

This river is 380 km long and starts from springs, drains fertilized farmland, and eventually flows through the center of London, which has a population of about 10 million. The estuarine tidal portion is about 100 km long. The Thames was once known for its large salmon runs. The species was so abundant 200 years ago that people grew tired of the taste. Apprentices in London even had clauses in their contracts forbidding the serving of salmon to them more than twice per week!

The River Thames has been receiving the wastes of London for almost two thousand years, and King Edward III decreed some basic environmental protection regulations in 1357 in a vain attempt to reduce its smell. However, the introduction of the water closet in 1800 and the rapid increase in population to over 2 million by 1850 destroyed the fishery which had until then survived the pollu-

tion. By 1850 there was probably an extensive stretch of low dissolved oxygen in the river centered around the major sewage discharge in the heart of the city (km 30, see Fig. 20-13). In fact the foul smells produced were so common that the British Parliament, which is situated on the riverbank, swifty enacted the world's first major sewage diversion scheme. The waste was piped farther downstream and discharged during the ebb tide. The extra dilution given by this diversion together with a precipitation treatment produced a much healthier river between 1891 and 1910 (Fig. 20-13a,b). Migratory marine fish such as smelt and flounder were once again caught in much of the river, and the dissolved oxygen at km 79 was adequate, even at the most critical summer periods (Fig. 20-14b,c). Km 79 is the traditional station used for measurements in the Thames since its oxygen content is unaffected by variation in river discharge. The dissolved-oxygen content of upstream stations is heavily influenced by river discharge. Fish upstream are able to survive pollution at high river discharge and conversely tend to suffer most during droughts.

As the population rose and industry added more toxic material to the river, the fishery once again declined until by 1950 there were only a few eels to be found in a 30-km-long zone of low oxygen (Fig. 20-13). The average dissolved-oxygen content at km 79 in autumn had fallen from 60–75 percent saturation in 1980–1910 to less than 10 percent (Figs. 20-13, 20-14). Dissolved oxygen was not the only factor involved in changes in the fisheries in the 1800–1920 period. During the summer and fall the mud must have had an anoxic surface which produced hydrogen sulfide (Fig. 10-6). This toxic gas is a health hazard for all animals and became so prevalent in the 1950s that it even blackened the lead-based paints on buildings near the river. Unfortunately, there were few biological studies in the river between 1910 and 1950.

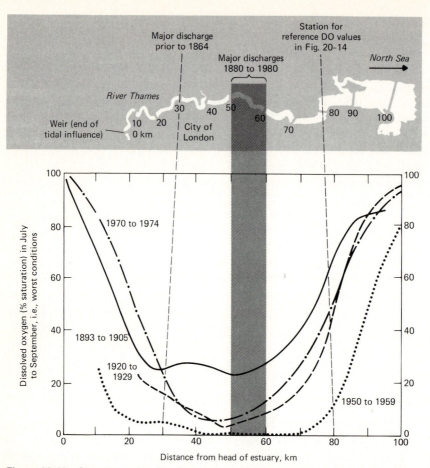

Figure 20-13 Changes in the dissolved-oxygen (DO) content in various parts of the tidal River Thames over the last century. Minimum oxygen levels always occur during low river discharge in late summer. Note that the minimum DO between 30 and 60 km was 20–30 percent saturation in 1893–1904, fell to 5–10 percent by 1920, and was zero in 1950–1959. The 1974 levels are now similar to those of 1920 and further improvement is expected. The major waste discharges up to 1864 and after 1880 are shown by lines and a shaded area. (*Redrawn from DSIR, 1964; and Gameson and Wheeler, 1977.*)

Despite the lack of biological information, a 15-year study by the government's Department of Scientific and Industrial Research (DSIR, 1964) concluded that river flow, rather than tidal mixing, was the major source of oxygen. Sewage and industrial waste rather than storm runoff were the main causes of the oxygen depletion. From this information DSIR predicted that the oxygen deficit could be eliminated and biota restored by improved waste treatment.

It was determined that upgrading sewage treatment from *primary* to *secondary* would sufficiently reduce the oxygen deficit. In addition it was calculated that some of the new sewage treatment plants would have to be expanded to include a final nitrification step. In this process ammonia is converted to nitrate

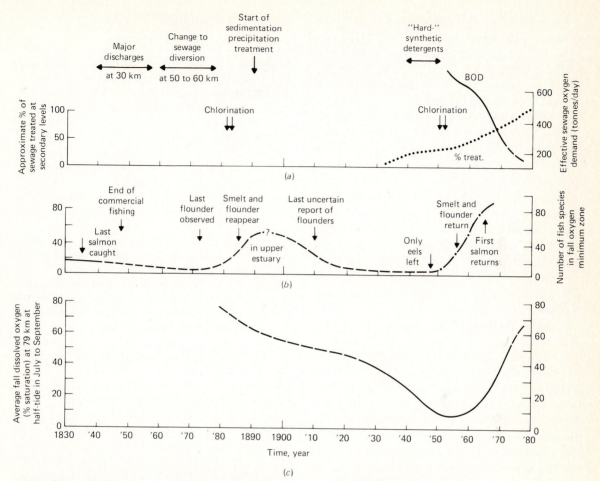

Figure 20-14 Changes in fish populations, dissolved oxygen, and waste loadings in the tidal River Thames, England, between 1830 and the present. The direct discharge of sewage in the heart of London (km 30) started about 1000 years ago. It accelerated after the invention of the water closet in 1800, and decimated the fishery by 1850. At this time a barrier of low oxygen probably existed around km 30. (*a*) Sewage diversion to km 50–60 temporarily halted the decline, and dissolved oxygen almost certainly increased between 1850 and 1900. (*b*) At this time fish in the river increased. (*c*) Further loading from various new sources then depleted the oxygen until it fell to virtually zero in the 1950s. At this time there were very few living macroscopic creatures in the river. Occasional chlorination and detergents (Fig. 16-13) probably further deterred fish and some crustaceans (Horne et al., 1982). As secondary treatment removed some of the biological oxygen demand of the wastewater in 1965–1970 [part (*a*)], the oxygen block diminished [part (*c*)] and smelt, flounder, and eventually salmon reappeared [part (*b*)]. (*Drawn from various sources including DSIR, 1964; Gameson and Wheeler 1977; and personal observation of A.J. Horne.*)

by supplying oxygen. The final outflow is thus low in both biological oxygen demand and ammonia, since ammonia requires oxygen in the river during conversion to nitrate. An additional benefit is the removal of ammonia as a significant toxicant (Table 8-4).

The results of this restoration project are shown in Figs. 20-13 and 20-14, which demonstrate the depletion and gradual increase in dissolved oxygen at km 79. The relationship between oxygen in the river and upgrading of sewage treatment is shown in Fig. 20-14*a*, *c*. The low oxygen levels have been eliminated, although oxygen is not as high as it once was. Fish have increased from one species in 1950 to at least 80 by 1974, and smelt and flounder are once again present. Dissolved oxygen at km 79 has increased from 10 to almost 60 percent at the most critical fall period (Fig. 20-14*b*). The section of the estuary with essentially zero oxygen levels has been eliminated, although the levels found from 1893–1905 have not yet been reached (Fig. 20-13). Further increases in oxygen are expected as complete waste treatment is achieved.

A condition which limits the degree of success is flow regulation from the construction of locks and weirs as well as the dredging of shoals and riffles to assist navigation. Any loss in river turbulence decreases the natural stream reaeration. In some of the remaining low oxygen areas of the Thames, this has been partially overcome by the installation of mechanical aerators which mix atmospheric oxygen into the water. In addition the large quantities of toxic chemicals such as cyanide from gas-producing plants have also been greatly reduced.

In 1966 a single salmon was caught in the Thames, and recently the upper waters were restocked with salmon fry. It is hoped that their migrations will no longer be blocked and that the sensitive fish will return as final proof of success of the Thames restoration.

FURTHER READINGS

Bates, J. M., and C. I. Weber (eds.). 1981. *Ecological Assessments of Effluent Imputs on Communities of Indigenous Aquatic Organisms.* ASTM Spec. Tech. Publ. 730. 370 pp.

Esch, G. W., and R. W. McFarlane (eds.). 1976. *Thermal Ecology II.* U.S. TIS. Conf.-750425, Springfield, Va. 404 pp.

Hart, C. W., and S. L. H. Fuller (eds.). 1974. *Pollution Ecology of Freshwater Invertebrates.* Academic, New York. 389 pp.

————, and ———— (eds.). 1979. *Pollution Ecology of Estuarine Invertebrates.* Academic, New York. 406 pp.

Hynes, H. B. N. 1966. *The Biology of Polluted Waters.* Liverpool University Press, England, 202 pp.

Jolley, R. L., H. Gorchev, and D. H. Hamilton (eds.). 1978. *Water Chlorination: Environmental Impact and Health Effects,* vol. 2. Ann Arbor Science, Ann Arbor, Mich. 909 pp.

Mitchell, R. (ed.). 1972, 1978. *Water Pollution Microbiology,* vols. I and II. Wiley, New York. 416 and 442 pp.

Warren, C. E. 1971. *Biology and Water Pollution Control.* Saunders, Philadelphia. 434 pp.

Weitzel, R. L. (ed.). 1979. *Methods and Measurement of Periphyton Communities: A Review.* Amer. Soc. Testing and Materials. Philadelphia. 183 pp. (Periphyton = attached algae &/or aufwuchs in this review.

Welch, E. B., and T. Lindell 1980. *Ecological Effects of Waste Water.* Cambridge, New York. 337 pp.

References

Agassiz, L. 1850. *Lake Superior: Its Physical Character, Vegetation, and Animals.* Gould, Kendal, and Lincon, Boston. 428 pp.

Ahlgren, I. 1970. "Limnological Studies of Lake Norrviken, a Eutrophicated Swedish Lake. II. Phytoplankton and its Production," *Schweiz. Z. Hydrol.,* **32**:353–396.

Alexander, W. B., B. A. Southgate, and R. Bassindale. 1935 (reprinted 1961). "Survey of the River Tees. Part II: The Estuary—Chemical and Biological." *Dep. Sci. Ind. Res. Water Pollut. Res., Tech. Paper,* no. 5. 171 pp.

Allen, K. R. 1951. "The Horokiwi Steam. A Study of a Trout Population." *Bull. Mar. N. Z. Fish.,* **10**:1–238.

American Fisheries Society. 1966. *A Symposium on Estuarine Fisheries.* R. F. Smith (Chairman). Special Publ. No. 3. 154 pp.

Anderson, G. C., and R. P. Zeutschel. 1970. "Release of Dissolved Organic Matter by Marine Phytoplankton in Coastal and Offshore Areas of the Northeast Pacific Ocean," *Limnol. Oceanogr.,* **15**:402–407.

Anderson, H. M., and A. J. Horne. 1975. "Remote Sensing of Water Quality in Reservoirs and Lakes in Semi-arid Climates." University of California, Berkeley, UCB-SERL Report No. 75-1. 132 pp.

Andersson, G., H. Berggren, and S. Hamrin. 1975. "Lake Trummen Restoration Project. III. Zooplankton, Macrobenthos, and Fish." *Verh. Int. Ver. Limnol.,* **19**:1097–1106.

APHA. 1981. *Standard Methods for the Examination of Water and Wastewater.* 15th ed. American Public Health Association, New York.

Armstrong, F. A. J., and D. W. Schindler. 1971. "Preliminary Chemical Characterization of Waters in the Experimental Lakes Area, Northwestern Ontario." *J. Fish. Res. Board Can.,* **28**:171–187.

Arnold, D. E. 1971. "Ingestion, Assimilation, Survival, and Reproduction by *Daphnia pulex* Fed Seven Species of Blue-Green Algae." *Limnol. Oceanogr.,* **16**:906–920.

Axler, R. P., G. W. Redfield, and C. R. Goldman. 1981. "The Importance of Regenerated Nitrogen to Phytoplankton Productivity in a Subalpine Lake." *Ecology,* **62**:345–354.

———, R. M. Gersberg, and C. R. Goldman. 1982.

"Inorganic nitrogen assimilation in a subalpine lake." *Limnol. Oceanogr.,* **27**(1):53–65.

Bachmann, R. W., and C. R. Goldman. 1965. "Hypolimnetic Heating in Castle Lake, California." *Limnol. Oceanogr.,* **10:**233–239.

Baker, A. L., and A. J. Brook. 1971. "Optical Density Profiles as an Aid to the Study of Microstratified Phytoplankton Populations in Lakes." *Arch. Hydrobiol.,* **69:**214–233.

———,———, and A. R. Klemer. 1969. "Some Photosynthetic Characteristics of a Naturally Occurring Population of *Oscillatoria agardhii* (Gomont). *Limnol. Oceanogr.,* **14:**327–333.

Baldwin, N. S. 1964. "Sea Lamprey in the Great Lakes." *Canadian Audubon Magazine,* November–December. 7 pp.

Balon, E. K., and A. G. Coche (eds.). 1974. "Lake Kariba: A Man-made Tropical Ecosystem in Central Africa." *Monogr. Biol.,* vol. 24. Junk Publishers, The Hague. 767 pp.

Barbour, C. D., and J. H. Brown. 1974. "Fish Species Diversity in Lakes." *Am. Nat.,* **108:**473–489.

Barnes, R. S. K. 1974. *Estuarine Biology.* E. Arnold, London, and Crane, Russak and Co., New York. Studies in Biology no. 49. 76 pp.

Barnett, R. H. 1975. "Case Study of Casitas Reservoir." ASCE Hydraulics Division: Symposium on Reaeration Research. Gatlinburg, Tenn., October.

Barret, E., and G. Brodin. 1955. "The Acidity of Scandinavian Precipitation." *Tellus,* **7:**251–257.

Bates, J. M., and C. I. Weber (eds.), 1981. *Ecological Assessments of Effluent Inputs on Communities of Indigenous Aquatic Organisms.* ASTM Spec. Tech. Publ. 730. 370 pp.

Beaumont, P. 1975. "Hydrology," pp. 1–38. In B. A. Whitton (ed.), *River Ecology.* University of California Press, Berkeley.

Beeton, A. M. 1969. "Changes in the Environment and Biota of the Great Lakes," pp. 150–187. In *Eutrophication: Causes, Consequences and Correctives.* National Academy of Sciences, Washington.

———, and D. C. Chandler. 1963. "The St. Lawrence Great Lakes," pp. 534–558. In D. G. Frey (ed.), *Limnology in North America.* The University of Wisconsin Press, Madison.

Bengtsson, L., S. Fleischer, G. Lindmark, and W. Ripl. 1975. "Lake Trummen Restoration Project.

I. Water and Sediment Chemistry." *Verh. Int. Ver. Limnol.,* **19:**1080–1087.

Bennett, E. B. 1978. "Characteristics of the Thermal Regime of Lake Superior." *J. Great Lakes Res.,* **4:**310–319.

Benson, B. B., and D. M. Parker. 1961. "Relations among the Solubilities of Nitrogen, Argon, and Oxygen in Distilled Water and Sea Water." *J. Phys. Chem.,* **65:**1489–1496.

Bernhardt, H. 1967. "Aeration of Wahnbach Reservoir without Changing the Temperature Profile." *J. Am. Water Works Assoc.,* **59:**943–964.

Berman, T. 1970. "Alkaline Phosphatases and Phosphorus Availability in Lake Kinneret." *Limnol. Oceanogr.,* **15:**663–674.

———, and W. Rodhe. 1971. "Distribution and Migration of *Perdinium* in Lake Kinneret." *Mitt. Int. Ver. Theor. Angew. Limnol.,* **19:**266–276.

Billaud, V. A. 1968. "Nitrogen Fixation and the Utilization of Other Inorganic Nitrogen Sources in a Sub-arctic Lake." *J. Fish. Res. Board Can.,* **25:**2101–2110.

Bindloss, M. E., A. V. Holden, A. E. Bailey-Watts, and I. R. Smith. 1972. "Phytoplankton Production, Chemical and Physical Conditions in Loch Leven," pp. 639–659. In Z. Kajak and A. Hillbricht-Ilkowska (eds.), *Productivity Problems of Freshwaters.* PWN (Polish Scientific Publishers), Warsaw–Kraków.

Birge, E. A. 1915. "The Heat Budgets of American and European Lakes." *Trans. Wis. Acad. Sci. Arts Lett.,* **18:**166–213.

———, and C. Juday. 1911. "The Inland Lakes of Wisconsin: The Dissolved Gases and Their Biological Significance." *Bull. Wis. geol. nat. Hist. Surv.,* **22.** 259 pp.

Blake, B. F. 1977. "The Effect of the Impoundment of Lake Kainji, Nigeria, on the Indigenous Species of Mormyrid Fishes." *Freshwater Biol.,* **7:**37–42.

Blanton, J. O. 1973. "Vertical Entrainment into the Epilimnia of Stratified Lakes." *Limnol. Oceanogr.,* **18:**697–704.

Blum, J. L. 1956. "The Ecology of River Algae." *Bot. Rev.,* **22:**291–341.

Bone, D. H. 1971. "Relationship between Phosphates and Alkaline Phosphatase of *Anabaena flos-aquae* in Continuous Culture." *Arch. Mikrobiol.,* **80:**147–153.

Boney, A. D. 1975. *Phytoplankton.* Institute of Biology Study 52. Crane, Russak Co., New York. 116 pp.

Boyce, F. M. 1974. "Some Aspects of Great Lakes Physics of Importance to Biological and Chemical Processes." *J. Fish. Res. Board Can.,* **31:**689–730.

Boyd, C. E., and C. P. Goodyear. 1971. "Nutritive Quality of Food in Ecological Systems." *Arch. Hydrobiol.,* **69:**256–270.

Brezonik, P. L. 1973. *Nitrogen Sources and Cycling in Natural Waters.* Ecological Research Series. EPA-660/3-73-002. 167 pp.

———, and G. F. Lee. 1968. "Denitrification as a Nitrogen Sink in Lake Mendota, Wisconsin." *Environ. Sci. Technol.,* **2:**120–125.

Bricker, O. P., and R. M. Garrels. 1967. "Minerologic Factors in Natural Water Equilibria." In S. O. Faust and J. V. Hunter (eds.), *Principles and Applications of Water Chemistry.* Wiley, New York.

Brinkhurst, R. O., and D. G. Cook (eds.) 1980. *Aquatic Oligochaetes.* Plenum, New York. 529 pp.

———, and B. G. M. Jamieson. 1971. *Aquatic Oligochaeta of the World.* Oliver & Boyd, Edinburgh. 860 pp.

Britt, N. W. 1955. "Stratification in Western Lake Erie in the Summer of 1953: Effects on the *Hexagenia* (Ephemeroptera) Population." *Ecology,* **36:**239–244.

Brock, T. D. 1967. "Life at high temperatures." *Science,* **158:**1012–1019.

Brooks, J. L., and S. I. Dodson. 1965. "Predation, Body Size, and Composition of Plankton." *Science,* **150:**28–35.

Brown, E. J., and D. K. Button. 1979. "Phosphate-limited Growth Kinetics of *Selenastrum capricornutum* (Chlorophyceae)." *J. Phycol.,* **15:**305–311.

Brown, G. W. 1970. "Predicting the Effect of Clear-cutting on Stream Temperature." *J. Soil Water Conserv.,* **25:**11–13.

Bryan, G. W., and L. G. Hummerstone. 1971. "Adaptation of the Polychaete *Nereis diversicolor* to Estuarine Sediments Containing High Concentrations of Heavy Metals. I. General Observations and Adaptations to Copper." *J. Mar. Biol. Assoc. U. K.,* **51:**845–863.

Buchanan, R. E., and N. E. Gibbons (eds.). 1974. *Bergey's Manual of Determinative Bacteriology.* 8th ed. Williams & Wilkins, Baltimore.

Burgis, M. J. 1971. "The Ecology and Production of Copepods, Particularly *Thermocyclops hyalinus,* in the Tropical Lake George, Uganda." *Freshwater Biol.,* **1:**169–192.

———. 1974. "Revised Estimates for the Biomass and Production of Zooplankton in Lake George, Uganda." *Freshwater Biol.,* **4:**535–541.

———, J. P. E. C. Darlington, I. G. Dunn, G. G. Ganf, J. J. Gwahaba, and L. M. McGowan. 1973. "The Biomass and Distribution of Organisms in Lake George, Uganda." *Proc. R. Soc. London, Ser. B,* **184:**271–298.

Burns, C. W. 1969. "Relation between Filtering Rate, Temperature, and Body Size in Four Species of *Daphnia.*" *Limnol. Oceanogr.,* **14:**693–700.

Burns, N. W., and C. Ross. 1972. "Project Hypo —An Intensive Study of the Lake Erie Central Basin Hypolimnion and Related Surface Water Phenomena." Canadian Centre for Inland Waters, Paper 6. 182 pp.

Burton, T. M., and G. E. Likens. 1973. "The Effect of Strip-Cutting on Stream Temperatures in the Hubbard Brook Experimental Forest, New Hampshire." *Bioscience,* **23:**433–435.

Butler, E. I., E. D. S. Corner, and S. M. Marshall. 1969. "On the Nutrition and Metabolism of Zooplankton. VI. Feeding Efficiency of *Calanus* in Terms of Nitrogen and Phosphorus." *J. mar. biol. Assoc. U. K.,* **49:**977–1001.

Capblancq, J., and H. Laville. 1972. "Etude de la productivité du lac de Port-Bielh, Pyrénées centrales," pp. 73–88. In Z. Kajak and A. Hillbricht-Ilkowska (eds.), *Productivity Problems of Freshwaters* PWN (Polish Scientific Publishers), Warsaw-Kraków.

Carpelan, L. H. 1958. "The Salton Sea. Physical and Chemical Characteristics." *Limnol. Oceanogr.,* **3:**373–397.

Carpenter, E. J., and R. R. L. Guillard. 1971. "Intraspecific Differences in Nitrate Half-saturation Constants for Three Species of Marine Phytoplankton." *Ecology,* **52:**183–185.

———, C. C. Remsen, and S. W. Watson. 1972. "Utilization of Urea by Some Marine Phytoplankters." *Limnol. Oceanogr.,* **17:**265–269.

Carr, J. F., and J. K. Hiltunen. 1965. "Changes in the

Bottom Fauna of Western Lake Erie from 1930 to 1961." *Limnol. Oceanogr.,* **10:**551–569.

Carson, R. 1962. *Silent Spring.* Houghton Mifflin, Boston. 368 pp.

Casey, H. 1977. "Origin and Variation of Nitrate Nitrogen in the Chalk Springs, Streams and Rivers in Dorset and Its Utilization by Higher Plants." *Prog. Water Technol.,* **8.** 14 pp.

Caspers, H. 1951. "Bodengreiferuntersuchungen über die Tierwelt in der Fahrrinne der Unterelbe und im Vormündungsgebiet der Nordsee." *Verh. Dtsch. Zool. Ges.,* **1951:**404–418.

Castenholz, R. W. 1973. "Ecology of Blue-Green Algae in Hot Springs." In N. G. Carr, and B. A. Whitton (eds.), "The Biology of Blue-green Algae." *Bot. Monogr. No. 9,* pp. 379–414.

———, and C. E. Wickstrom. 1975. "Thermal Streams." pp. 264–285. In B. A. Whitton (ed.), *River Ecology.* Univ. California Press, Berkeley.

Chapman, V. J. (ed.). 1977. *Ecosystems of the World,* vol. 1: *Wet Coastal Ecosystems.* Elsevier Press, New York. 428 pp.

Christie, W. J. 1974. "Changes in the Fish Species Composition of the Great Lakes." *J. Fish. Res. Board Can.,* **31:**827–854.

Cloern, J. E. 1979. "Phytoplankton Ecology of the San Francisco Bay System: The Status of Our Current Understanding," pp. 247–264. In T. J. Conomos (ed.), *San Francisco Bay.* American Association for the Advancement of Science, San Francisco.

Cole, G. A. 1968. "Desert Limnology," pp. 423–486. In G. W. Brown, Jr. (ed.), *Desert Biology.* Academic, New York.

———. 1979. *Textbook of Limnology.* 2d ed. Mosby, St. Louis. 426 pp.

Confer, J. L. 1972. "Interrelations among Plankton, Attached Algae, and the Phosphorus Cycle in Artificial Open Systems," *Ecol. Monogr.,* **42:**1–23.

Cook, D. G., and M. G. Johnson. 1974. "Benthic Macroinvertebrates of the St. Lawrence Great Lakes," *J. Fish. Res. Board Can.* **31:**763–782.

Cordone, A., S. Nicola, P. Baker, and T. Frantz. 1971. "The Kokanee Salmon in Lake Tahoe." *Calif. Fish Game,* **57:**28–43.

Corner, E. D. S., and A. G. Davies. 1971. "Plankton as a Factor in the Nitrogen and Phosphorus Cycles in the Sea." *Adv. mar. Biol.,* **9:**101–204.

Coull, B. C. (ed.). 1977. *Ecology of Marine Benthos.* The University of South Carolina Press. Columbia. 467 pp.

Cowgill, U. M., and G. E. Hutchinson. 1970. "Chemistry and Mineralogy of the Sediments and Their Source Materials. In "Ianula: An Account of the History and Development of the Lago di Monterosi, Latium, Italy." *Trans. Am. Philos. Soc.,* **60:**37–101.

Craig, P. C., and P. J. McCart. 1975. "Classification of Stream Types in Beaufort Sea Drainages between Prudhoe Bay, Alaska and the Mackenzie Delta, N. W. T. Canada." *Arctic Alpine Res.,* **7:**183–198.

Craik, A. D. D., and S. Leibovich. 1976. "A Rational Model for Langmuir Circulations." *J. Fluid Mech.,* **73:**401–426.

Cronberg, G., C. Gelin, and K. Larsson. 1975. "Lake Trummen Restoration Project. II. Bacteria, Phytoplankton and Phytoplankton Productivity." *Verh. Int. Ver. Limnol.,* **19:**1088–1096.

Csanady, G. T. 1969. "Dispersal of Effluents in the Great Lakes." *Water Res.,* **3:**835–972.

———. 1975. "Hydrodynamics of Large Lakes." *Ann. Rev. Fluid Mech.,* **7:**357–385.

Cummins, K. W. 1973. "Trophic Relations of Aquatic Insects." *Ann. Rev. Entomol.,* **18:**183–206.

———. 1974. "Structure and Function of Stream Ecosystems." *Bioscience,* **24:**631–641.

———, R. C. Petersen, F. O. Howard, J. C. Wuycheck, and V. I. Holt. 1973. "The Utilization of Leaf Litter by Stream Detritivores." *Ecology,* **54:**336–345.

Cushing, E. J., and H. E. Wright, Jr. (eds.). 1967. *Quaternary Paleoecology.* Yale University Press, New Haven, Conn. 433 pp.

Darley, W. M. 1974. "Silicification and Calcification," pp. 655–675. In W. D. P. Stewart (ed.), "Algal Physiology and Biochemistry." *Bot. Monogr. No. 10.*

de Bernardi, R., G. Giussani, and E. L. Pedretti. 1981. "The Significance of Blue-green Algae as Food for Filterfeeding Zooplankton: Experimental Studies on *Daphnia spp.* Fed by *Microcystis aeruginosa.*" *Verh. Int. Ver. Limnol.,* **21:**477–483.

Denison, P. J., and F. C. Elder. 1970. "Thermal Inputs to the Great Lakes 1968–2000." *Proc. 13th Conf. Great Lakes Res.,* pp. 811–828.

Dillon, T. M., and T. M. Powell. 1979. "Observations of a Surface Mixed Layer." *Deep-Sea Res.*, **26A:**915–932.

———, ———, and L. O. Myrup. 1975. "Low Frequency Turbulence and Vertical Temperature Microstructure in Lake Tahoe, California–Nevada." *Verh. Int. Ver. Limnol.*, **19:**110–115.

Dobson, H. F. H., M. Gilbertson, and P. G. Sly. 1974. "A Summary and Comparison of Nutrients and Related Water Quality in Lakes Erie, Ontario, Huron, and Superior." *J. Fish. Res. Board Can.*, **31:**731–738.

Dodson, S. I. 1974. "Adaptive Change in Plankton Morphology in Response to Size-selective Predation: A New Hypothesis of Cyclomorphosis." *Limnol. Oceanogr.*, **19:**721–729.

Dowd, J. E., and D. S. Riggs. 1965. "A Comparison of Estimates of Michaelis-Menten Kinetic Constants from Various Linear Transformations." *J. Biol. Chem.*, **240:**863–869.

Droop, M. 1957. "Auxotrophy and Organic Compounds in Nutrition of Marine Phytoplankton." *J. Gen. Microbiol.*, **16:**286–293.

———. 1974. "The Nutrient Status of Algal Cells in Continuous Culture." *J. Mar. Biol. Assoc. U. K.*, **54:**825–855.

DSIR. 1964. "Effects of Polluting Discharges on the Thames Estuary." *Dept. Sci. Ind. Res. Water Pollut. Res. Tech. Paper No. 11.* London. 201 pp.

Dumont, H. J. 1972. "A Competition-based Approach of the Reverse Vertical Migration in Zooplankton and Its Implications, Chiefly Based on a Study of the Interactions of the Rotifer *Asplanchna priodonta* (Gosse) with Several Crustacea Entomostraca." *Int. Rev. ges. Hydrobiol.*, **57:**1–38.

Dunst, R. C., S. M. Born, P. D. U. Hormark, S. A. Smith, S. A. Nichols, J. O. Peterson, D. R. Knauer, S. L. Sens, D. R. Winter, and T. L. Wirth. 1974. "Survey of Lake Rehabilitation Techniques and Experiences." *Wis. Dep. Nat. Resour. Tech. Bull. No. 75.* 177 pp.

Dussart, B. 1966. *Limnologie. L'étude des eux continentales.* Gauthier-Villars, Paris. 677 pp.

Edmondson, W. T. 1955. "Seasonal Life History of *Daphnia* in an Arctic Lake." *Ecology*, **36:**439–455.

———. 1965. "Reproductive Rate of Planktonic Rotifers as Related to Food and Temperature in Nature." *Ecol. Monogr.*, **35:**61–111.

———. 1972a. "The Present Condition of Lake Washington." *Verh. Int. Ver. Limnol.*, **18:**284–291.

———. 1972b. "Nutrients and Phytoplankton in Lake Washington." *Limnol. Oceanogr. Special Symp. Vol. 1*, pp. 172–193.

———. 1974a. "Secondary Production." *Mitt. Int. Ver. Theor. Angew. Limnol.*, **20:**229–272.

———. 1974b. "The Sedimentary Record of the Eutrophication of Lake Washington." *Proc. Nat. Acad. Sci. U.S.A.*, **71:**5093–5095.

———. 1977. "Population Dynamics and Secondary Production." *Ergeb. Limnol.*, **8:**56–64.

———, and D. E. Allison. 1970. "Recording Densitometry of X-Radiographs for the Study of Cryptic Laminations in the Sediment of Lake Washington." *Limnol. Oceanogr.*, **15:**138–144.

———, and A. Litt. 1982. "*Daphnia* in Lake Washington." *Limnol. Oceanogr.*, **27:**272–293.

Egeratt, A. W. S. M., and J. L. M. Huntjens (eds.). 1975. "The Sulphur Cycle." *Plant Soil*, **48:**228 pp.

Eisma, D., and A. J. van Bennekom. 1978. "The Zaire River and Estuary and the Zaire Outflow in the Atlantic Ocean." *Neth. J. Sea Res.*, **12:**255–272.

Elder, J. F. 1974. "Trace Metals from Ward Creek and their Influence upon Phytoplankton Growth in Lake Tahoe." Ph.D. thesis, University of California, Davis. 144 pp.

———, R. H. Fuller, and A. J. Horne. 1979. "Physiochemical Factors and Their Effects on Algal Growth in a New Southern California Reservoir." *Water Resour. Bull.*, **15:**1608–1617.

———, and A. J. Horne. 1977. "Biostimulatory Capacity of Dissolved Iron for Cyanophycean Blooms in a Nitrogen-rich Reservoir." *Chemosphere, No. 9*, pp. 525–530.

———, and ———. 1978a. "Ephemeral Cyanophycean Blooms and Their Relationships to Micronutrient Chemistry in a Southern California Reservoir." University of California, Berkeley, Sanitary Engineering Research Lab. UCB-SERL Report No. 78–1. 182 pp.

———, and ———. 1978b. "Copper Cycles and $CuSO_4$ Algicidal Capacity in Two California Lakes." *Environ. Manage.*, **2:**17–30.

Elliot, J. M., and P. A. Tullett. 1977. "The Downstream Drifting of Larvae of *Dixa* (Diptera:Dixidae) in Two Stony Streams." *Freshwater Biol.,* **7:**403–407.

Elster, H-J. 1974. "History of Limnology." *Mitt. Int. Ver. Theor. Angew. Limnol.,* **20:**7–30.

Elwood, J. W., J. D. Newbold, R. V. O'Neill, R. W. Stark, and P. T. Singley. 1981. "The Role of Microbes Associated with Organic and Inorganic Substrates in Phosphorus Spiralling in a Woodland Stream." *Verh. Int. Ver. Limnol.,* **21:**850–856.

El-Zarka, S. E. D. 1973. "Kainji Lake, Nigeria," pp. 197–219. In W. C. Akerman, G. F. White, and E. B. Worthington (eds.), *Man-made Lakes: Their Problems and Environmental Effects.* American Geophysical Union, Washington.

Emery, K. O., and G. T. Csanady. 1973. "Surface Circulation of Lakes and Nearby Land-locked Seas." *Proc. Nat. Acad. Sci. U.S.A.,* **70:**93–97.

Eppley, R. W., J. N. Rogers, and J. J. McCarthy. 1969. "Half-saturation Constants for Uptake of Nitrate and Ammonium by Marine Phytoplankton." *Limnol. Oceanogr.,* **14:**912–920.

———, and W. H. Thomas. 1969. "Comparison of Half-saturation Constants for Growth and Nitrate Uptake of Marine Phytoplankton." *J. Phycol.,* **5:**375–379.

Erman, D. C., and N. A. Erman. 1975. "Macroinvertebrate Composition and Production in Some Sierra Nevada Minerotrophic Peatlands." *Ecology,* **56:**591–603.

———, J. D. Newbold, and K. B. Roby. 1977. "Evaluation of Streamside Bufferstrips for Protecting Aquatic Organisms." University of California, Davis. Water Resources Center, Publ. No. 165. 47 pp.

Esch, G. W., and R. W. McFarlane (eds.). 1976. *Thermal Ecology II.* U.S. TIS. Conf.-750425. Springfield, Va. 404 pp.

Falter, C. M. 1977. "Early Limnology of Dworshak Reservoir, N. Idaho," pp. 285–294. In R. D. Andrews et al. (eds.), *Proceedings of the Symposium on Terrestrial and Aquatic Studies in the Northwest,* EWSC Press, Washington.

Fast, A. W. 1973. "Effects of Artificial Destratification on Primary Production and Zoobenthos of El Capitan Reservoir, California." *Water Resour. Res.,* **9:**607–623.

———, W. J. Overholtz, and R. A. Tubb. 1975. "Hypolimnetic Oxygenation Using Liquid Oxygen." *Water Resour. Res.,* **11:**294–299.

Fee, E. J. 1976. "The Vertical and Seasonal Distribution of Chlorophyll in Lakes of the Experimental Lakes Area, Northwestern Ontario: Implications for Primary Production Estimates." *Limnol. Oceanogr.,* **21:**767–783.

———. 1980. "Reply to Comments by Patalas and Schindler." *Limnol. Oceanogr.,* **25:**1152–1153.

Fittkau, E. J. 1964. "Remarks on Limnology of Central-Amazon Rain-forest Streams." *Verh. Int. Ver. Limnol.,* **15:**1092–1096.

Fitzgerald, G. P. 1969. "Field and Laboratory Evaluations of Bioassays for Nitrogen and Phosphorus with Algae and Aquatic Weeds." *Limnol. Oceanogr.,* **14:**206–212.

———, and T. C. Nelson. 1966. "Extractive and Enzymatic Analyses for Limiting or Surplus Phosphorus in Algae." *J. Phycol.,* **2:**32–37.

Fogg, G. E. 1969. "The Leewenhoek Lecture, 1968. The Physiology of an Algal Nuisance." *Proc. R. Soc. London, Ser. B,* **173:**175–189.

———. 1971. "Extracellular Products of Algae in Freshwater." *Arch. Hydrobiol. Beih.,* **5:**1–25.

———. 1975. *Algal Cultures and Phytoplankton Ecology.* 2d ed. The University of Wisconsin Press, Madison. 175 pp.

———, and A. J. Horne. 1967. "The Determination of Nitrogen Fixation in Aquatic Environments," pp. 115–120. In H. L. Golterman and R. C. Clymo (eds.), *Chemical Environment in the Aquatic Habitat.* Noord-Hollandsche Uitgevers-Mij., Amsterdam.

———, and ———. 1970. "The Physiology of Antarctic Freshwater Algae," pp. 632–638. In M. W. Holdgate (ed.), *Antarctic Ecology,* Academic, New York.

Folt, C., and C. R. Goldman. 1981. "Allelopathy between Zooplankton: A Mechanism for Interference Competition." *Science,* **213:**1133–1135.

Forbes, S. A. 1887. "The Lake as a Microcosm." *Bull. Peoria (Ill.) Sci. Assoc.* Reprinted 1925 in *Bull. Ill. Nat. Hist. Surv.,* **15:**537–550.

Foree, E. J., W. J. Jewell, and P. L. McCarty. 1971. "The Extent of Nitrogen and Phosphorus Regeneration from Decomposing Algae." *Fifth International Water Pollution Research Conference.* Pergamon, London.

Forel, F. A. 1869. "Introduction à l'étude de la faune profonde du Lac Léman." *Bull. Soc. Vaud. Sci. Nat. (Lausanne),* **10:**217.

———. 1892, 1895, 1904. *Le Léman: monographie*

limnologique, tome I (1892): *Geographie, hydrographie, géologie, climatologie, hydrologie,* 543 pp.; tome II (1895): *Mecanique, hydraulique, chimie, thermique, optique, acoustique,* 651 pp.; tome III (1904): *Biologie, histoire, navagation, pêche,* 715 pp. Lausanne, F. Rouge, reprinted Genève, Slatkine Reprints, 1969.

———. 1901. *Handbuch der Seenkunde: allgemeine Limnologie.* Bibliothek geographische Handbücher, Stuttgart.

Frank, P. W., C. D. Boll, and R. W. Kelley, 1957. "Vital Statistics of Laboratory Cultures of *Daphnia pulex* deGeer as Related to Density." *Physiol. Zool.,* 30:276–305.

Frey, D. G. 1963. *Limnology in North America.* University of Wisconsin Press, Madison. 734 p.

———. 1964. "Remains of Animals in Quaternary Lake and Bog Sediments and Their Interpretation." *Ergeb. Limnol.,* 2:1–116.

Fryer, G., and T. D. Iles. 1972. *The Cichlid Fishes of the Great Lakes of Africa: Their Biology and Evolution.* Oliver & Boyd. Edinburgh. 641 pp.

Fuji, N. 1978. "Pollen Analysis of a 200-m Core Sample from Lake Biwa." *Verh. Int. Ver. Limnol.,* 20:2663–2665.

Galat, D. L., E. L. Lider, S. Vigg, and S. R. Robertson. 1981. "Limnology of a Large Deep, North American Terminal Lake, Pyramid Lake, Nevada, U.S.A." *Hydrobiologia,* 82:281–317.

Gameson, A. L. H., and A. Wheeler. 1977. "Restoration and Recovery of the Thames Estuary," pp. 72-101. In S. J. Cairns, K. L. Dikson, and E. E. Herricks (eds.), *Recovery and Restoration of Damaged Ecosystems.* University Press of Virginia, Charlottesville.

Gammon, J. R., A. Spacie, G. Hamelink, and R. L. Kaessler. 1981. "Role of Electrofishing in Assessing Environmental Quality of the Wabash River," pp. 307–324. In J. M. Bates and C. I. Weber (eds.), *Ecological Assessment of Effluent Impacts on Communities of Indigenous Aquatic Organisms.* ASTM Spec. Tech. Publ. 730.

Ganf, G. G. 1974. "Phytoplankton Biomass and Distribution in a Shallow, Eutrophic Lake (Lake George, Uganda)." *Oecologia,* 16:9–29.

———, and P. Blažka. 1974. "Oxygen Uptake, Ammonia and Phosphate Excretion by Zooplankton of a Shallow Equatorial Lake (Lake George, Uganda)." *Limnol. Oceanogr.,* 19:313–325.

———, and A. J. Horne. 1975. "Diurnal Stratification, Photosynthesis and Nitrogen-Fixation in a Shallow, Equatorial Lake (Lake George, Uganda)." *Freshwater Biol.,* 5:13–39.

———, and A. B. Viner. 1973. "Ecological Stability in a Shallow, Equatorial Lake (Lake George, Uganda)." *Proc. R. Soc. London, Ser. B,* 184:321–346.

Gerloff, G. C., and F. Skoog. 1954. "Cell Contents of Nitrogen and Phosphorus as a Measure of Their Availability for Growth of *Microcystis aeruginosa.*" *Ecology,* 35:348–353.

———, and ———. 1957. "Availability of Iron and Manganese in Southern Wisconsin Lakes for the Growth of *Microcystis aeruginosa.*" *Ecology,* 38:551–556.

Gersberg, R., K. Krohn, N. Peek, and C. R. Goldman. 1976. "Denitrification Studies with ^{13}N-labelled Nitrate." *Science,* 192:1229–1231.

Gessner, F. 1955. *Hydrobotanik,* vol. I: *Energiehaushalt.* VEB Deutsche Verlag Wissenschaft, Berlin. 517 pp.

———. 1959. *Hydrobotanik,* vol. II: *Stoffhaushalt.* VEB Deutsche Verlag Wissenschaft. Berlin. 701 pp.

Giesy, J. P. 1980. "Microcosms in Ecological Research." Technical Information Center. U.S. Department of Energy. Springfield, Va. (DOE Symposium Series 52). Conf.-781101. 1110 pp.

Gilbert, J. J., and G. A. Thompson. 1968. "Alpha Tocopherol Control of Sexuality and Polymorphism in the Rotifer *Asplanchna.*" *Science,* 159:734–738.

Gold, K. 1964. "Aspects of Marine Dinoflagellate Nutrition Measured by [14]C Assimilation." *J. Protozool.* 11:85–89.

Goldman, C. R. 1960. "Primary Productivity and Limiting Factors in Three Lakes of the Alaskan Peninsula." *Ecol. Monogr.,* 30:207–270.

———. 1961. "The contribution of alder trees (*Alnus tenuifolia*) to the primary productivity of Castle Lake, California." *Ecology,* 42:282–288.

———. 1963. "The Measurement of Primary Productivity and Limiting Factors in Freshwater with Carbon-14," pp. 103–113. In M. S. Doty (ed.), *Conference on Primary Productivity.* U.S. Atomic Energy Commission. TID-7633.

———. 1965. "Micronutrient Limiting Factors and Their Detection in Natural Phytoplankton Populations," pp. 121–135. In C. R. Goldman (ed.),

"Primary Productivity in Aquatic Environments." *Mem. Ist. Ital. Idrobiol.*, **18**(suppl.).

————. 1968*a*. "The Use of Absolute Activity for Eliminating Serious Errors in the Measurement of Primary Productivity with ^{14}C." *J. Cons. Perm. Int. Explor. Mer.*, **32**:172–179.

————. 1968*b*. "Aquatic Primary Production." *Am. Zool.*, **8**:31–42.

————. 1969. "Photosynthetic Efficiency and Diversity of a Natural Phytoplankton Population in Castle Lake, California. Proceedings IBP/PP Technical Meeting, *Prediction and Measurement of Photosynthetic Productivity*, pp. 507–517.

————. 1972. "The Role of Minor Nutrients in Limiting the Productivity of Aquatic Ecosystems." *Symposium on Nutrients and Eutrophication*, vol. 1, pp. 21–33. American Society of Limnology and Oceanography.

————. 1981. "Lake Tahoe: Two Decades of Change in a Nitrogen Deficient Oligotrophic Lake." *Verh. Int. Ver. Limnol.*, **21**:45–70.

————, and R. C. Carter. 1965. "An Investigation by Rapid Carbon-14 Bioassay of Factors Affecting the Cultural Eutrophication of Lake Tahoe, California–Nevada." *J. Water Pollut. Control Fed.*, **37**:1044–1059.

————, and R. W. Hoffman. 1977. "Environmental Aspects of the Purari River Scheme," pp. 325–341. In J. H. Winslow (ed.), *The Melanesian Environment*. Australian National University Press, Canberra.

————, and D. T. Mason. 1962. "Inorganic Precipitation of Carbon in Productivity Experiments Utilizing Carbon-14." *Science*, **136**:1049–1050.

————, ————, and J. E. Hobbie. 1967. "Two Antarctic Desert Lakes." *Limnol. Oceanogr.*, **12**:295–310.

————, ————, and B. J. B. Wood. 1963. "Light Injury and Inhibition in Antarctic Freshwater Phytoplankton." *Limnol. Oceanogr.*, **8**:313–322.

————, ————, and ————. 1972. "Comparative Study of the Limnology of Two Small Lakes on Ross Island, Antarctica." *Antarct. Res. Ser.*, **20**:1–50.

————, M. D. Morgan, S. T. Threlkeld, and N. Angeli. 1979. "A Population Dynamics Analysis of the Cladoceran Disappearance from Lake Tahoe, California–Nevada." *Limnol. Oceanogr.*, **24**:289–297.

————, E. A. Stull, and E. deAmezaga. 1973. "Vertical Patterns of Primary Productivity in Castle Lake, California." *Verh. Int. Ver. Limnol.*, **18**:1760–1767.

————, M. G. Tunzi, and R. Armstrong. 1969. "^{14}C Uptake as a Sensitive Measure of the Growth of Algal Cultures." *Proc. Eutrophication-Biostimulation Assessment Workshop*. Berkeley. pp. 158–170.

————, and R. G. Wetzel. 1963. "A Study of the Primary Productivity of Clear Lake, Lake County, California." *Ecology*, **44**:283–294.

————, N. Williams, and A. Horne. 1975. "Prospects for Micronutrient Control of Algal Populations," pp. 97–105. In P. L. Brezonik and J. L. Fox (eds.), *Water Quality Management through Biological Control*. University of Florida, Gainesville, Engineering Science Report ENV-07-75-1.

Golterman, H. L. 1975*a*. *Physiological Limnology*. Elsevier Scientific Publishing Co., New York. 489 pp.

————. 1975*b*. "Chemistry of Running Waters," pp. 39–80. In B. A. Whitton (ed.), *River Ecology*. Blackwell, Oxford.

————, and R. S. Clymo (eds.). 1967. *Chemical Environment in the Aquatic Habitat*. Noord-Hollandsche Uitgevers-Mij., Amsterdam. 322 pp.

————, ————, and M. A. M. Ohnstad. 1978. *Methods for Physical and Chemical Analysis of Fresh Waters*. 2d ed. Blackwell Scientific Publishers, Oxford. IBP Handbook, No. 8. 213 pp.

Gorham. E. 1955. "On the Acidity and Salinity of Rain." *Geochim. Cosmochim. Acta*, **7**:231–239.

————. 1964. "Morphometric Control of Annual Heat Budgets in Temperate Lakes." *Limnol. Oceangr.*, **9**:525–529.

————, J. W. G. Lund, J. E. Sanger, and W. E. Dean, Jr. 1974. "Some Relationships between Algal Standing Crop, Water Chemistry, and Sediment Chemistry in the English Lakes." *Limnol. Oceanogr.*, **19**:601–617.

Granhall, U., and A. Lundgren. 1971. "Nitrogen Fixation in Lake Erken." *Limnol. Oceanogr.*, **16**:711–719.

Grant, J. W. G., and I. A. E. Bayly. 1981. "Predator Induction of Crests in Morphs of the *Daphnia Carinata* King Complex." *Limnol. Oceanogr.*, **26**:201–218.

Griffith, E. J., A. Beeton, J. M. Sponser, and D. T. Mitchell (eds). 1973. *Handbook of Environmental Phosphorus*. Wiley, New York. (See especially chapters by H. L. Golterman, "Vertical Movement of Phosphate in Freshwater," and F. H. Rigler, "A Dynamic View of the Phosphorus Cycle in Lakes.")

Griffiths, B. M. 1939. "Early References to Waterblooms in British Lakes." *Proc. Limn. Soc. London,* **151:**12–19.

Griffiths, M., and W. T. Edmondson. 1975. "Burial of Oscillaxanthin in the Sediment of Lake Washington." *Limnol. Oceanogr.,* **20:**945–952.

Gross, M. G. 1977. *Oceanography*. 2d ed. Prentice-Hall, Englewood Cliffs, N. J. 497 pp.

Groth, P. 1971. "Untersuchungen über einige Spuren-elemente in Seen" ("Investigations of Some Trace Elements in Lakes"). *Arch. Hydrobiol.,* **68:**305–375 (summary in English).

Gudmundsson, F. 1979. "The Past Status and Exploitation of the Mývatn Waterfowl Populations." In P. M. Jónasson (ed.), "Lake Mývatn." *Oikos,* **32:**232–249.

Guillard, R. R. L., and P. Kilham. 1977. "The Ecology of Marine Planktonic Diatoms," pp. 372–469. In D. Werner (ed.), *The Biology of Diatoms*. University of California Press, Berkeley.

Gunter, G. 1961. "Some Relations of Estuarine Organisms to Salinity." *Limnol. Oceanogr.,* **6:**182–190.

———. 1967. "Some Relationships of Estuaries to the Fisheries of the Gulf of Mexico," pp. 621–638. In G. H. Lauff (ed.), *Estuaries*. American Association for the Advancement of Science, publ. no. 83, Washington.

Guseva, K. A. 1939. "Bloom on the Ucha Reservoir." *Biul. Moskov. Obshch. Ispytat. Prirody, Otdel. Biol.,* **48:**30. Reprinted in *Nat. Res. Counc. Can.,* technical translation 879 by G. Bekov, 1960.

Håkanson, L. 1981. *A Manual of Lake Morphometry*. Springer-Verlag, New York. 78 pp.

Hall, C. A. S. 1977. "Models and the Decision-making Process: The Hudson River Power Plant Case." In C. A. S. Hall and J. W. Day (eds.), *Ecosystem Modeling in Theory and Practice*. Wiley, New York. 684 pp.

Hall, D. J., W. E. Cooper, and E. E. Werner. 1970. "An Experimental Approach to the Production Dynamics and Structure of Freshwater Animal Communities." *Limnol. Oceanogr.,* **15:**839–928.

———, S. T. Threlkeld, C. W. Burns, and P. H. Crowley. 1976. "The Size Efficiency Hypothesis and the Size Structure of Zooplankton Communities." *Ann. Rev. Ecol. Syst.,* **7:**177–208.

Haney, J. F. 1973. "An in situ Examination of the Grazing Activities of Natural Zooplankton Communities." *Arch. Hydrobiol.,* **72:**87–132.

Harbeck, G. E. 1955. "The Effect of Salinity on Evaporation." *U.S. Geol. Surv. Prof. Pap.* 272A. 6 pp.

Harding, D. 1966. "Lake Kariba. The Hydrology and Development of Fisheries." In "Man-made Lakes." *Symp. Inst. Biol.,* **15:**7–20.

Hargrave, B. T., and G. H. Geen. 1968. "Phosphorus Excretion by Zooplankton." *Limnol. Oceanogr.,* **13:**332–342.

Harkness, W. J. K., and J. R. Dymond. 1961. "The Lake Sturgeon. The History of Its Fishery and Problems of Conservation." *Ont. Dep. Lands For., Fish Wildl. Br.* 121 pp.

Harrison, M. J., R. E. Pacha, and R. Y. Morita. 1972. "Solubilization of Inorganic Phosphates by Bacteria Isolated from Upper Klamath Lake Sediment." *Limnol. Oceanogr.,* **17:**50–57.

Hart, C. W., and S. L. H. Fuller (eds.). 1974. *Pollution Ecology of Freshwater Invertebrates*. Academic, New York. 389 pp.

———, and ———, (eds.). 1979. *Pollution Ecology of Estuarine Invertebrates*. Academic, New York. 406 pp.

Harvey, H. W. 1937. *The Chemistry and Fertility of Sea Waters*. Cambridge, England. 240 pp.

Hasler, A. D. 1966. *Underwater Guideposts: Homing of Salmon*. The University of Wisconsin Press, Madison.

Hayes, F. R., and J. E. Phillips. 1958. "Lake Water and Sediment. IV. Radiophosphorus Equilibrium with Mud, Plants and Bacteria under Oxidized and Reducing Conditions." *Limnol. Oceanogr.,* **3:**459–475.

Healy, F. P. 1977. "Ammonia Uptake by Some Freshwater Algae." *Can. J. Bot.,* **55:**61–69.

Hecky, R. E., E. J. Fee, H. Kling, and J. W. M. Rudd. 1978. "Studies on the Planktonic Ecology of Lake Tanganyika." Canadian Fisheries and Marine Services. Technical Report 816, Winnipeg. 51 pp.

Hedgepeth, J. W. 1957. "Estuaries and Lagoons. II. Biological Aspects." In "Treatise on Marine Ecology and Paleoecology." *Geol. Soc. Am. Mem. 67,* 1:693–729.

———. 1977. "Models and Muddles. Some Philosophical Observations." *Helgol. wiss. Meeresunters.,* **30:**92–104.

Hensen, V. 1887. "Über die Bestimmung des Planktons oder des in Meere treibenden Materials on Pflanzen und Thieren." *Ber. Kommn Wiss. Unters. dt. Meere.,* **5:**1–109.

Heron, J. 1961. "The Seasonal Variation of Phosphate, Silicate, and Nitrate in the Waters of the English Lake District." *Limnol. Oceanogr.,* **6:**338–346.

Heywood, R. B. 1972. "Antarctic Limnology: A Review." *Br. Antarct. Surv. Bull.,* No. 29, pp. 35–65.

Hildebrand, S. F., and W. C. Schroeder. 1928. Reprinted 1972. "Fishes of Chesapeake Bay." *U.S. Fish Wildl. Serv., Fish. Bull.,* ser. IV, vol. 53. pt. I. Smithsonian Institution, Washington. 388 pp.

Holm, L. G., L. W. Weldon, and R. D. Blackburn. 1969. "Aquatic Weeds." *Science,* **166:**699–709.

Holm, N. P., and D. E. Armstrong. 1981. "Role of Nutrient Limitation and Competition in Controlling the Populations of *Asterionella formosa* and *Microcystis aeruginosa* in Semicontinuous Culture." *Limnol. Oceanogr.,* **26:**622–634.

Horie, S. (ed.). 1974. *Paleolimnology of Lake Biwa and the Japanese Pleistocene.* Otsu Hydrobiology Station, Kyoto University, Japan. 288 pp.

———. 1981. "On the Significance of Paleolimnological Study of Ancient Lakes—Lake Biwa and Other Relic Lakes." *Verh. Int. Ver. Limnol.,* **21:**13–44.

Horne, A. J. 1972. "The Ecology of Nitrogen Fixation on Signy Island, South Orkney Islands." *Br. Antarct. Surv. Bull.,* No. 27, pp. 1–18.

———. 1975. "The Ecology of Clear Lake Phytoplankton." Special report of the Clear Lake Algal Research Unit, Lakeport, Calif. 116 pp.

———. 1977. "Nitrogen Fixation—A Review of This Phenomenon as a Polluting Process." *Prog. Water Technol.,* **8:**357–372.

———. 1978. "Nitrogen Fixation in Eutrophic Lakes," pp. 1–30. In R. Mitchell (ed.), *Water Pollution Microbiology,* vol. 2. Wiley, New York.

———. 1979a. "Management of Lakes Containing N₂-fixing Blue-green Algae." *Arch. Hydrobiol. Beih.,* **13:**133–144.

———. 1979b. "Nitrogen Fixation in Clear Lake, California. IV. Diel Studies on *Aphanizomenon* and *Anabaena* Blooms." *Limnol. Oceanogr.,* **24:**329–341.

———, M. Bennett, R. Valentine, P. P. Russell, R. E. Selleck, and P. W. Wild. 1982. "The Effects of Chlorination of Wastewater on Juvenile Dungeness Crabs in San Francisco Bay Waters." *Calif. Dep. Fish Game Fish Bull.,* **172** (in press).

———, and C. J. W. Carmiggelt. 1975. "Algal Nitrogen Fixation in California Streams: Seasonal Cycles." *Freshwater Biol.,* **5:**461–470.

———, and G. E. Fogg. 1970. "Nitrogen Fixation in Some English Lakes." *Proc. R. Soc. London, Ser. B,* **175:**351–366.

———, ———, and D. J. Eagle. 1969. "Studies in situ of the Primary Production of an Area of Inshore Antarctic Sea." *J. Mar. Biol. Assoc. U.K.,* **49:**393–405.

———, and C. R. Goldman. 1972. "Nitrogen Fixation in Clear Lake, California. I. Seasonal Variation and the Role of Heterocysts." *Limnol. Oceanogr.,* **17:**678–692.

———, and ———. 1974. "Supression of Nitrogen Fixation by Blue-green Algae in a Eutrophic Lake with Trace Additions of Copper." *Science,* **83:**409–411.

———, P. Javornicky, and C. R. Goldman. 1971. "A Freshwater 'Red Tide' on Clear Lake, California." *Limnol. Oceanogr.,* **16:**684–689.

———, and W. J. Kaufman. 1974. "Biological Effects of Ammonia Salts and Dilute, Treated Petroleum Refinery Effluent on Estuarine Aufwuchs, Phytoplankton and Fish Communities." Sanitary Engineering Research Laboratory, University of California, Berkeley. UCB-SERL Rep. No. 74–5. 112 pp.

———, and J. C. Roth. 1979. "Nitrate Ploughing to Eliminate Hydrogen Sulfide Production in the Tillo Mudflat." Report to City of South San Francisco, Public Works Department, February. 23 pp.

———, ———, D. Kelley, and F. McLaren. 1979. "A Biological Survey of Martis Creek, California." McLaren Environmental Engineering, Sacramento, Calif.

———, and A. B. Viner. 1971. "Nitrogen Fixation

and Its Significance in Tropical Lake George, Uganda." *Nature*, **232**:417–418.

———, and R. C. Wrigley. 1975. "The Use of Remote Sensing to Detect How Wind Influences Planktonic Blue-green Algal Distribution." *Verh. Int. Ver. Limnol.*, **19**:784–791.

Hrbáček, J. 1962. "Species Composition and the Amount of Zooplankton in Relation to the Fish Stock." *Rozpr. Cesk. Akad. Ved. Rada Mat. Prir. Ved.*, **72**:1–116.

Hunt, E. G., and A. I. Bischoff. 1960. "Inimical Effects on Wildlife of Periodic DDT Applications to Clear Lake." *Cal. Fish Game*, **46**:91–106.

Hutchinson, G. E. 1957, 1967, 1975. *A Treatise on Limnology*, vol. I (1957): *Geography, Physics and Chemistry*, 1015 pp.; vol. II (1967): *Introduction to Lake Biology and the Limnoplankton*, 1115 pp.; vol. III (1975): *Limnological Botany*, 660 pp. Wiley, New York.

———, and U. M. Cowgill. 1973. "The Waters of Merom: A Study of Lake Huleh. III. The Major Chemical Constituents of a 54 m Core." *Arch. Hydrobiol.*, **72**:145–185.

Hutchinson, G. L., and F. G. Viets. 1969. "Nitrogen Enrichment of Surface Water by Absorption of Ammonia Volatilized from Cattle Feed-lots." *Science*, **166**:514–515.

Hynes, H. B. N. 1966. *The Biology of Polluted Waters*. University of Liverpool Press, England. 202 pp.

———. 1972. *The Ecology of Running Waters*. University of Toronto Press, Toronto. 555 pp.

———. 1975. "The Stream and Its Valley." *Verh. Int. Ver. Limnol.*, **19**:1–15.

International Association of Theoretical and Applied Limnology. 1971. "Factors That Regulate the Wax and Wane of Algal Populations." Symposium No. 19. *Mitt. Int. Ver. Theor. Angew. Limnol.* 318 pp.

Itasaka, O., and M. Koyama. 1980. "Elementary Components in Water," p. 28. In S. Mori (ed.), *An Introduction to the Limnology of Lake Biwa*. Shizuoka Women's University, Japan.

Ivlev, V.S. 1939. "Transformation of Energy by Aquatic Animals. Coefficient of Energy Consumption by *Tubifex tubifex* (Oligochaeta)." *Int. Rev. ges. Hydrobiol. Hydrograph.*, **38**:449–458.

———. 1963. "On the Utilization of Food by Plankton-Eating Fishes." *Fish. Res. Board Can.* (Translation Series No. 447, 17 pp.)

Jackson, P. B. M. 1962. "Why Do Nile Crocodiles Attack Boats?" *Copeia*, **1**:204–206.

James, H. R., and E. A. Birge. 1938. "A Laboratory Study of the Adsorbtion of Light by Lake Waters." *Trans. Wis. Acad. Sci. Arts Lett.*, **31**:1–154.

Jansson, B-O., and F. Wulff. 1977. "Baltic Ecosystem Modeling," pp. 323–343. In C. A. S. Hall and J. W. Day (eds.), *Ecosystem Modeling in Theory and Practice*. Wiley, New York. 684 pp.

Jassby, A. D., and C. R. Goldman. 1974. "Loss Rates from a Lake Phytoplankton Community." *Limnol. Oceanogr.*, 19:618–627.

Jefferies, R. L., and A. J. Davy (eds.). 1979. *Ecological Processes in Coastal Environments*. Blackwell Press, Oxford. 684 pp.

Jenkin, P. M. 1936. "Reports on the Percy Staden Flamingos." *Phil. Trans. Roy. Soc. Lond. B.* **240**:401–493.

———. 1942. "Seasonal Changes in the Temperature of Windermere (English Lake District)." *J. Anim. Ecol.*, **11**:248–269.

Jenkins, D. 1975. "The Analysis of Nitrogen Forms in Waters and Wastewaters." *Proc. Conf. on Nitrogen as a Water Pollutant*. IAWPR. vol. 1. 30 pp.

Jerlov, N. G. 1968. *Optical Oceanography*. Elsevier Press, London. 194 pp.

Johannes, R. E. 1968. "Nutrient Regeneration in Lakes and Oceans." *Adv. Microbiol. Sea*, **1**:203–213.

Jolley, R. L., H. Gorchev, D. H. Hamilton (eds.). 1978. *Water Chlorination: Environmental Impact and Health Effects*, vol. 2. Ann Arbor Science, Ann Arbor, Mich. 909 pp.

Jónasson, P. M. 1978. "Zoobenthos of Lakes." *Verh. Int. Ver. Limnol.*, **20**:13–37.

———, E. Lastein, and A. Rebsdorf. 1974. "Production, Insulation and Nutrient Budget of Lake Esrom." *Oikos*, **25**:255–277.

Jones, J. G. 1971. "Studies on Freshwater Bacteria: Factors Which Influence the Population and Its Activity." *J. Ecol.*, **59**:593–613.

———. 1972. "Studies on Freshwater Microorganisms. Phosphatase Activity in Lakes of Differing Degrees of Eutrophication." *J. Ecol.*, **60**:777–791.

———. 1977. "Variation in Bacterial Populations in

Time and Space." *Freshwater Biol. Assoc. Ann. Rep.* No. 45, pp. 55–61.

Jones, K., and W. D. P. Stewart. 1969. "Nitrogen Turnover in Marine and Brackish Habitats. III. The Production of Extracellular Nitrogen by *Calothrix scopulorum*." *J. mar. biol. Assoc. U.K.,* **49:**475–488.

Jørgensen, C. B. 1966. *Biology of Suspension Feeding*. Pergamon. New York. 357 pp.

Kalff, J. 1967. "Phytoplankton Dynamics in an Arctic Lake." *J. Fish. Res. Board Can.,* **24:**1861–1871.

———. 1971. "Nutrient Limiting Factors in an Arctic Tundra Pond." *Ecology,* **52:**655–659.

Kaushik, N. K., and H. B. N. Hynes. 1971. "The Fate of the Dead Leaves That Fall into Streams." *Arch. Hydrobiol.,* **68:**465–515.

Kawanabe, H. 1978. "Some Biological Problems." *Verh. Int. Ver. Limnol.,* **20:**2674–2677.

Keeney, D. R. 1972. "The Fate of Nitrogen in Aquatic Ecosystems." *Univ. Wis. Water Resour. Cent. Lit. Rev.* 3. 59 pp.

———, R. L. Chen, and D. A. Graetz. 1971. "Importance of Denitrification and Nitrate Reduction in Sediments to the Nitrogen Budget of Lakes." *Nature,* **233:**66–67.

Keirn, M. A., and P. L. Brezonik. 1971. "Nitrogen Fixation by Bacteria in Lake Mize, Florida, and in Some Lacustrine Sediments." *Limnol. Oceanogr.,* **16:**720–731.

Kellogg, W. W., R. D. Cadle, E. R. Allen, A. L. Lazrus, and E. A. Martell. 1972. "The Sulfur Cycle." *Science.,* **175:**587–596.

Kelts, K., and K. J. Hsü. 1978. "Freshwater Carbonate Sedimentation," pp. 295–323. In A. Lerman (ed.), *Lakes: Chemistry, Geology, Physics.* Springer-Verlag, New York.

Kerfoot, W. C. (ed.). 1980. *Evolution and Ecology of Zooplankton Communities.* Special Symposium Vol. 3, American Society of Limnology and Oceanography. University of New England Press, New Hampshire. 793 pp.

Kilham, P., and D. Tilman. 1979. "The Importance of Resource Competition and Nutrient Gradients for Phytoplankton Ecology." *Arch. Hydrobiol. Beih.,* **13:**110–119.

King, C. E. 1967. "Food, Age and the Dynamics of a Laboratory Population of Rotifers." *Ecology,* **48:**111–128.

——— (ed.). 1977. *Proceedings of the First International Rotifer Symposium.* Springer-Verlag, New York.

Kipling, C. 1976. "Year Class Strengths of Perch and Pike in Windermere." *Freshwater Biol. Assoc. Ann. Rep.* No. 44, pp. 68–75.

———, and W. E. Frost. 1970. "A Study of the Mortality, Population Numbers, Year Class Strengths, Production and Food Consumption of Pike, *Esox lucius,* in Windermere from 1944 to 1962." *J. Anim. Ecol.,* **39:**115–157.

Kirchner, W. B. 1975. "An Examination of the Relationship between Drainage Basin Morphology and the Export of Phosphorus." *Limnol. Oceanogr.,* **20:**267–270.

Kittel, T., and P. J. Richerson. 1978. "The Heat Budget of a Large Tropical Lake, Lake Titicaca (Peru-Bolivia)." *Verh. Int. Ver. Limnol.,* **20:**1203–1209.

Knighton, A. D. 1976. "Stream Adjustment in a Small Rocky Mountain Basin." *Arctic Alpine Res.* **8:**197–212.

Koehl, M. A. R., and J. R. Strickler. 1981. "Copepod Feeding Currents: Food Capture at Low Reynolds Number." *Limnol. Oceanogr.,* **26:**1062–1073.

Kozhov, M. 1963. *Lake Baikal and Its Life.* Junk Publishers, The Hague. 344 pp.

Krueger, D. A., and S. I. Dodson. 1981. "Embryological Induction and Predation Ecology in *Daphnia Pulex*." *Limnol. Oceanogr.,* **26:**219–223.

Kusnezow, S. I. 1959. *Die Rolle der Mikroorganismen im Stoffkreislauf der Seen.* Deutsches Verlag der Wissenschaften, Berlin.

Lagler, K. F., J. E. Bardach, R. R. Miller, and D. R. M. Passino. 1977. *Ichthyology.* 2d ed. Wiley, New York. 506 pp.

Lallatin, R. D. 1972. "Alternative Eel River Projects and Conveyance Routes. Appendix C. Clear Lake Water Quality." California Department of Water Resources, Red Bluff. 145 pp.

Lange, W. 1971. "Limiting Nutrient Elements in Filtered Lake Erie Water." *Water Res.,* **5:**1031–1048.

Larimore, R. W., L. Durham, and G. W. Bennett. 1950. "A Modification of the Electric Fish Shocker for Lake Work." *J. Wildl. Manag.,* **14:**320–323.

Larsen, D. P., D. W. Schults, and K. W. Malueg.

1981. "Summer Internal Phosphorus Supplies in Shagawa Lake, Minnesota." *Limnol. Oceanogr.,* **26:**740–753.

Lauff, G. (ed.). 1967. *Estuaries.* American Association for the Advancement of Science, Washington. Publication 83. 757 pp.

Lean, D. R. S. 1973. "Phosphorus Dynamics in Lake Water." *Science,* **179:**678–680.

———, and M. N. Charlton. 1976. "A Study of Phosphorus Kinetics in a Lake Ecosystem," pp. 283–294. In J. O. Nriagu (ed.), *Environmental Biochemistry.* Ann Arbor Science, Ann Arbor, Mich.

LeCren, E. D. 1958. "Observations on the Growth of Perch (*Perca fluviatilis*) over Twenty-two Years with Special Reference to the Effects of Temperature and Changes in Population Density." *J. Anim. Ecol.,* **27:**287–334.

Lee, D. H. K. 1970. "Nitrates, Nitrites, and Methemoglobinemia." *Environ. Res.,* **3:**484–511.

Lefevre, M., H. Jakob, and M. Nisbet. 1952. "Auto-et hetero-antagonisme chez les algues d'eau douce." *Ann. Stat. Cent. Hydrobiol. Appl.,* **4:**5–197.

Lemmin, U., J. T. Scott, and U. H. Czapski. 1974. "The Development from Two-dimensional to Three-dimensional Turbulence Generated by Breaking Waves." *J. Geophys. Res.,* **79:**3442–3448.

Leonard, R. L., L. A. Kaplan, J. F. Elder, R. N. Coats, and C. R. Goldman. 1979. "Nutrient Transport in Surface Runoff from a Subalpine Watershed, Lake Tahoe Basin, California." *Ecol. Monogr.,* **49:**281–310.

Leopold, L. B., M. G. Wolman, and J. P. Miller. 1964. *Fluvial Processes in Geomorphology.* Freeman, San Francisco. 522 pp.

Lerman, A. (ed.). 1978. *Lakes—Chemistry, Geology, Physics.* Springer-Verlag, New York. 363 pp.

Leslie, J. 1838. "Treatise on Various Subjects of Natural and Chemical Philosophy." Cited in Murray and Pullar (1910), pp. 91–92.

Lewin, J. C. 1962. "Silicification," pp. 445–455. In R. A. Lewin (ed.), *Physiology and Biochemistry of Algae.* Academic, London.

Lewis, W. M. 1973. "The Thermal Regime of Lake Lanao (Philippines) and Its Theoretical Implications for Tropical Lakes." *Limnol. Oceanogr.,* **18:**200–217.

———. 1979. *Zooplankton Community Analysis—Studies on a Tropical System.* Springer-Verlag, New York. 163 pp.

Likens, G. E., F. H. Borman, N. M. Johnson, D. W. Fisher, and R. S. Pierce. 1970. "Effects of Forest Cutting and Herbicide Treatment on Nutrient Budgets in the Hubbard Brook Watershed Ecosystem." *Ecol. Monogr.,* **40:**23–47.

———, ———, R. S. Pierce, J. S. Eaton, and N. M. Johnson. 1977. *Bio-geochemistry of a Forested Ecosystem.* Springer-Verlag, New York. 146 pp.

———, and N. M. Johnson. 1969. "Measurement and Analysis of the Annual Heat Budget for the Sediments in Two Wisconsin Lakes. *Limnol. Oceanogr.,* **14:**115–135.

Lindegaard, C., and P. M. Jónasson. 1979. "Abundance, Population Dynamics and Production of Zoobenthos in Lake Mývatn, Iceland." In P. M. Jónasson (ed.), "Lake Mývatn." *Oikos,* **32:**202–227.

Lindeman, R. L. 1942. "The Trophic-dynamic Aspect of Ecology." *Ecology,* **23:**399–418.

Lindström, K. and W. Rodhe. 1977. "Selenium as a Micronutrient for the Dinoflagellate *Peridinium cinctum* fa. *westii.*" *Verh. Int. Ver. Limnol.,* **21:**168–173.

Livermore, D. F., and W. E. Wunderlich. 1969. "Mechanical Removal of Organic Production from Waterways." pp. 494–519. In *Eutrophication: Causes, Consequences, Correctives.* National Academy of Sciences, Washington.

Livingstone, D. A. 1963. "Mean Composition of World River Water. Chemical Composition of Rivers and Lakes. Data of goechemistry 6th ed. Chap. G." *U.S. Geol. Surv. Prof. Pap.* 400-G pp. 1–64.

Livingstone, R. J. (ed.). 1979. *Ecological Processes in Coastal and Marine Systems. Marine Science,* vol. 10. Plenum, New York. 548 pp.

Lorenz, J. R. 1863. "Brackwasserstudien an der Elbemündung." *Sitzungsber. math–naturwiss. Kl. Kaiser Akad. Wiss. Wein.,* **48**(2):602–613.

Lund, J. W. G. 1949, 1950. "Studies on *Asterionella.* I. The Origin and Nature of the Cells Producing the Spring Maximum." *J. Ecol.,* **37:**389–419 (1949); and II. "Nutrient Depletion and the Spring Maximum." *J. Ecol.,* **38:**1–35 (1950).

———. 1954. "The Seasonal Cycle of the Plankton

Diatom, *Melosira italica* (Ehr.) Kütz. subsp. *subarctica* O. Müll." *J. Ecol.,* **42**:151–179.

———. 1964."Primary Production and Periodicity of Phytoplankton." *Verh. Int. Ver. Limnol.,* **15**:37–56.

———. 1965. "The Ecology of Freshwater Phytoplankton." *Biol. Rev.,* **40**:231–293.

———. 1971. "An Artificial Alteration of the Seasonal Cycle of the Plankton Diatom *Melosira italica* subsp. *subarctica* in an English Lake." *J. Ecol.,* **59**:521–533.

———. 1972. "Eutrophication." *Proc. R. Soc. London, Ser. B,* **180**:371–382.

———, C. Kipling, and E. D. Le Cren. 1958. "The Inverted Microscope Method of Estimating Algal Numbers and the Statistical Basis of Estimations by Counting." *Hydrobiologia,* **11**:143–170.

Macan, T. T. 1970. *Biological Studies of the English Lakes.* American Elsevier, New York. 260 pp.

———. 1974. "Running Water." *Mitt. Int. Ver. Theor. Angew. Limnol.,* **20**:301–321.

McCarthy, J. J. 1972. "The Uptake of Urea by Marine Phytoplankton." *J. Phycol.,* **8**:216–222.

McHarg, I. L., and M. G. Clarke. 1973. "Skippack Watershed and the Evansburg Project: A Case Study for Water Resources Planning," pp. 299–330. In C. R. Goldman, J. McEvoy, and P. J. Richerson (eds.), *Environmental Quality and Water Development.* Freeman, San Francisco.

MacIsaac, J. J., and R. C. Dugdale. 1969. "The Kinetics of Nitrate and Ammonia Uptake by Natural Populations of Marine Phytoplankton." *Deep-Sea Res.,* **16**:45–57.

MacKenthun, K. M., et al. 1964. "Limnological Aspects of Recreational Lakes." *U.S. Public Health. Serv. Publ.* **1167.** 176 pp.

McLaren, F. R. 1977. "Water Quality Studies of the Truckee River." McLaren Environmental Engineering, Sacramento, Calif.

———, A. J. Horne, D. Kelley, and J. C. Roth. 1979. "Studies of Martis Creek, 1979." McLaren Environmental Engineering, Sacramento, Calif.

McMahon, J. W. 1969. "The Annual and Diurnal Variation in the Vertical Distribution of Acid-soluble Ferrous and Total Iron in a Small Dimictic Lake." *Limnol. Oceanogr.,* **14**:357–367.

Macnae, W. 1967. "Zonation within Mangroves Associated with Estuaries in Queensland," pp. 432–441. In G. H. Lauff (ed.), *Estuaries.* Ameri-can Association for the Advancement of Science, publ. no. 83, Washington.

McRoy, C. P., and R. J. Barsdate. 1970. "Phosphate Absorption in Eelgrass." *Limnol. Oceanogr.,* **15**:6–13.

———, ———, and N. Nebert. 1972. "Phosphorus Cycling in an Eelgrass (*Zostera marina* L.) Ecosystem." *Limnol. Oceanogr.,* **17**:58–67.

Magnuson, J. J. 1962. "An Analysis of Aggressive Behavior, Growth and Competition for Food and Space in Medaka, *Oryzias latipes Pisces (Cyprinodontidae)*." *Can. J. Zool.,* **40**:313–363.

Maitland, P. S. 1974. "The Conservation of Freshwater Fishes in the British Isles." *Biol. Conserv.,* **6**:7–14.

Malueg, K. W., R. M. Brice, D. W. Schults, and D. P. Larson. 1973. "The Shagawa Lake Project." U.S. Environmental Protection Agency, EPA R3-73-026. 49 pp.

Mangelsdorf, P. C. 1967. "Salinity Measurements in Estuaries," pp. 71–79. In G. H. Lauff (ed.), *Estuaries.* American Association for the Advancement of Science, publ. no. 83, Washington.

Mann, K. H., R. H. Britton, A. Kowalczewski, T. J. Lack, C. P. Mathews, and I. McDonald. 1972. "Productivity and Energy Flow at All Trophic Levels in the River Thames, England," pp. 579–596. In Z. Kajak and A. Hillbricht-Ilkowska (eds.), *Productivity Problems of Freshwater.* PWN (Polish Scientific Publishers), Warsaw-Kraków.

Manny, B. A. 1972. "Seasonal Changes in Dissolved Organic Nitrogen in Six Michigan Lakes." *Verh. Int. Ver. Limnol.,* **18**:147–156.

———, R. G. Wetzel, and R. E. Bailey. 1978. "Paleolimnological Sedimentation of Organic Carbon, Nitrogen, Phosphorus, Fossil Pigments, Pollen and Diatoms in a Hypereutrophic, Hardwater Lake: A Case History of Eutrophication." *Pol. Arch. Hydrobiol.,* **25**:243–267.

Margalef, R. 1965. "Ecological Correlations and the Relationship between Primary Productivity and Community Structure," pp. 355–364. In C. R. Goldman (ed.), "Primary Productivity in Aquatic Environments." *Mem. 1st. Ital. Idrobiol.,* **18** (suppl.).

Marker, A. F. H. 1976*a*. "The Benthic Algae of Some Streams in Southern England I. Biomass of the Epilithon in Some Small Streams." *J. Ecol.,* **64**:343–358.

———. 1976*b*. "The Benthic Algae of Some Streams in Southern England. II. The Primary Production of the Epilithon in a Small Chalk-stream." *J. Ecol.,* **64**:359–373.

Marshall, N. B. 1966. *The Life of Fishes.* Universe Books, New York. 402 pp.

Mason, D. T. 1967. "Limnology of Mono Lake, California." *Univ. Calif. Berkeley Publ. Zool.,* **83.** 110 pp.

Mathews, C. P., and A. Kowalcezwski. 1969. "The Disappearance of Leaf Litter and Its Contribution to Production in the River Thames." *J. Ecol.,* **57**:543–552.

Melack, J. M. 1976. "Primary Productivity and Fish Yields in Tropical Lakes." *Trans. Am. Fish Soc.,* **105**:575–580.

Meyer, H., and K. Mobius. 1865, 1872. *Fauna der Kieler,* bucht 1 (1865), pp. 1–88; bucht 2 (1872), pp. 1–139.

Mill, H. R. 1895. "Bathymetric Survey of the English Lakes." *Geogr. J.,* **6**:46–73.

Miller, R. G. 1951. "The Natural History of Lake Tahoe Fishes." Ph.D. thesis, Stanford University, Stanford, Calif. 160 pp.

Milliman, J. D., and E. Boyle. 1975. "Biological Uptake of Dissolved Silica in the Amazon River Estuary." *Science,* **189**:995–997.

Minshall, W. 1978. "Autotrophy in Stream Ecosystems." *Bioscience,* **28**:767–771.

Mitchell, D. S. 1969. "The Ecology of Vascular Hydrophytes on Lake Kariba." *Hydrobiologia,* **34**:448–464.

Mitchell, G. H. 1956. "The Geological History of the Lake District." *Proc. Yorks. Geol. Soc.,* **30.**

Mitchell, R. (ed.). 1972, 1978. *Water Pollution Microbiology,* vols. I and II. Wiley, New York. 416 and 442 pp.

Moore, J. W. 1976. "Seasonal Succession of Algae in Rivers. I. Examples from the Avon, a Large Slow-flowing River." *J. Phycol.,* **12**:342–349.

Mordukai-Boltovskoi, Ph.D (ed.). 1979. "The River Volga and Its Life." *Monogr. Biol.,* vol. 33. 473 pp.

Morgan, M. D. 1980. "Life History Characteristics of Two Introduced Populations of *Mysis relicta.*" *Ecology,* **61**:551–561.

Mori, S. 1974. "Diatom Succession in a Core from Lake Biwa," pp. 247–254. In S. Hori (ed.), "Paleolimnology of Lake Biwa and the Japanese Pleistocene." *Contrib. Paleolimnol. L. Biwa,* No. 43. Otsu, Japan.

———, and T. Miura. 1980. "List of Plant and Animal Species Living in Lake Biwa." *Memoirs Fac. Sci. Kyoto Univ. Ser. Biol.,* **8**:1–33.

Moriarty, D. J. W. 1973. "The Physiology of Digestion of Blue-green Algae in the Cichlid Fish, *Talapia nilotica.*" *J. Zool. London,* **171**:25–39.

———, J. P. E. C. Darlington, I. G. Dunn, C. M. Moriarty, and M. P. Tevlin. 1973. "Feeding and Grazing in Lake George, Uganda." *Proc. R. Soc. London, Ser. B,* **184**:299–319.

Morisawa, M. 1968. *Streams, Their Dynamics and Morphology.* McGraw-Hill, New York. 175 pp.

Morris, I. 1974. "Nitrogen Assimilation and Protein Synthesis," pp. 583–609. In W. D. P. Stewart (ed.), "Algal Physiology and Biochemistry." *Bot. Monograph.,* **10.**

Morris, R. H., D. P. Abbott, and E. C. Haderlie. 1980. *Intertidal Invertebrates of California.* Stanford Univ. Press. 690 pp.

Mortimer, C. H. 1941, 1942. "The Exchange of Dissolved Substances between Mud and Water in Lakes." *J. Ecol.,* **29**:280–329, (1941); **30**:147–201 (1942).

———. 1971. "Chemical Exchanges between Sediments and Water in the Great Lakes—Speculations on Probable Regulatory Mechanisms." *Limnol. Oceanogr.,* **16**:387–404.

———. 1974. "Lake Hydrodynamics." *Mitt. Int. Ver. Theor. Angew. Limnol.,* **20**:124–197.

———. 1981. "The Oxygen Content of Air-saturated Freshwaters over Ranges of Temperature and Atmospheric Pressure of Limnological Interest." *Mitt. Int. Ver. Theor. Angew. Limnol.* No. 22. 22 pp.

———, and E. B. Worthington. 1942. "Morphometric Data for Windermere." *J. Anim. Ecol.,* **11**:245–247.

Moss, B. 1972. "The Influence of Environmental Factors on the Distribution of Freshwater Algae: An Experimental Study. I. Introduction and the Influence of Calcium Concentration." *J. Ecol.,* **60**:917–932.

Mosser, J. L., and T. D. Brock. 1976. "Temperature Optima for Algae Inhabiting Cold Mountain Streams." *Arctic Alpine Res.,* **8**:111–114.

Moyle, P. B. 1976. *Inland Fishes of California*. University of California Press, Berkeley. 405 pp.

Mullin, M. M., and P. M. Evans. 1974. "The Use of a Deep Tank in Plankton Ecology. II. Efficiency of a Planktonic Food Chain." *Limnol. Oceanogr.*, **19:**902–911.

Murphy, C. R. 1972. "An Investigation of Diffusion Characteristics of the Hypolimnion of Lake Erie," pp. 39–44. In N. M. Burns and C. Ross (eds.), "Project Hypo—An Intensive Study of the Lake Erie Central Basin Hypolimnion and Related Surface Water Phenomena." Canadian Centre for Inland Waters, Paper 6.; US-EPA Tech. Rep. TS-05-71-208-24.

Murphy, T. P., D. R. S. Lean, and C. Nalewajko. 1976. "Blue-green Algae: Their Excretion of Ion Selective Chelators Enables Them to Dominate Other Algae." *Science*, **192:**900–902.

Murray, J., and L. Pullar. 1910. *Bathymetric Survey of the Scottish Freshwater Lochs*. Vol. 1. Challenger Office, Edinburgh. 785 pp.

Myrup, L. O., T. M. Powell, D. A. Godden, and C. R. Goldman. 1979. "Climatological Estimate of the Average Monthly Energy and Water Budgets of Lake Tahoe, California-Nevada." *Water Resour. Res.*, **15:**1499–1508.

National Academy of Sciences. 1969. *Eutrophication: Causes, Consequences, Correctives*. Washington. 611 pp. Includes lakes, rivers, and estuaries.

———. *Nitrates: An Environmental Assessment*. Environmental Studies Board. National Research Council. Washington. 750 pp.

Nauwerck, A. 1963. "Die Beziehungen Zwischen Zooplankton und Phytoplankton im See Erken." *Symb. Bot. Ups.*, **17:**1–163.

Needham, P. R., and A. C. Jones. 1959. "Flow, Temperature, Solar Radiation and Ice in Relation to Activities of Fishes in Sagehen Creek, California." *Ecology*, **40:**465–474.

Neess, J. C., R. C. Dugdale, V. A. Dugdale, and J. J. Goering. 1962. "Nitrogen Metabolism in Lakes. I. Measurement of Nitrogen Fixation with N^{15}." *Limnol. Oceanogr.*, **7:**163–169.

Nelson, D. M., J. J. Goering, S. S. Kilham, and R. R. L. Guillard. 1976. "Kinetics of Silicic Acid Uptake and Rates of Silica Dissolution in the Marine Diatom *Thalassiosira pseudonana*." *J. Phycol.*, **12:**246–252.

Newell, R. C. 1970. *Biology of Intertidal Animals*. American Elsevier, New York. 555 pp.

Nichols, F. H. 1979. "Natural and Anthropogenic Influences on Benthic Community Structure in San Francisco Bay," pp. 409–426. In T. J. Conomos (ed.), *San Francisco Bay*. American Association for the Advancement of Science, San Francisco.

Nichols, S. A., and G. Cottam. 1972. "Harvesting as a Control for Aquatic Plants." *Water Res. Bull.*, **8:**1205–1210.

Odum, E. P. 1971. *Fundamentals of Ecology*. 3d ed. Saunders, Philadelphia. 574 pp.

———, and A. de la Cruz. 1967. "Particulate Organic Detritus in a Georgia Salt Marsh-Estuarine Ecosystem," pp. 383–388. In G. H. Lauff (ed.), *Estuaries*. American Association for the Advancement of Science, publ. no. 83. Washington.

Odum, H. T., and R. F. Pigeon (eds.). 1970. "A Tropical Rain Forest. A Study of Irradiation and Ecology." Office of Information Services, U.S. Atomic Energy Commission, Washington. 1684 pp.

Officer, C. B. 1976. *Physical Oceanography of Estuaries*. Wiley-Interscience, New York. 465 pp.

Oglesby, R. T. 1968. "Effects of Controlled Nutrient Dilution of a Eutrophic Lake." *Water Res.*, **2:**106–108.

———. 1978. "The Limnology of Cayuga Lake," pp. 1–120. In J. A. Bloomfield (ed.), *Lakes of New York State*, vol. I: *Ecology of the Finger Lakes*. Academic, New York.

———, C. A. Carlson, and J. A. McCann (eds.). 1972. *River Ecology and Man*. Academic, New York. 465 pp.

Ohle, W. 1952. "Die Hypolimnishe Kohlendioxyd-Akkumulation als produktionsbiologischer Indikator." *Arch. Hydrobiol*, **46:**153–285.

Ólafsson, J. 1979. "The Chemistry of Lake Mývatn." *Oikos*, **32:**82–112.

Olausson, E., and I. Cato. 1980. *Chemistry and Biogeochemistry of Estuaries*. Wiley-Interscience. New York. 452 pp.

Omernik, J. M. 1976. "The Influence of Land Use on Stream Nutrient Levels." U.S. EPA-600/3-76-014. Corvallis, Ore. 68 pp. plus appendix.

Oswald, W. J. 1976. "Removal of Algae in Natural Bodies of Water." University of California, Berkeley. Sanitary Engineering Research Lab. Rep. UCB-SERL No. 76-1. 140 pp.

Overbeck, J. and H. D. Babenzien. 1964. "Bakterien und Phytoplankton eines Kleingewässers im Jahreszyklus." *Z. Allg. Mikrobiol.*, **4**:59–76.

Paasche, E. 1973. "Silicon and the Ecology of Marine Planktonic Diatoms. II. Silicate-uptake Kinetics in Five Diatom Species." *Mar. Biol.*, **19**:262–269.

Paerl, H. 1973. "Detritus in Lake Tahoe: Structural Modification by Attached Microflora." *Science,* **180**:496–498.

———, R. C. Richards, R. L. Leonard, and C. R. Goldman. 1975. "Seasonal Nitrate Cycling as Evidence for Complete Vertical Mixing in Lake Tahoe, California-Nevada." *Limnol. Oceanogr.,* **20**:1–8.

Pamatmat, M. M., and K. Banse. 1969. "Oxygen Measurements by the Seabed. II. *In Situ* Measurements to a Depth of 180 m." *Limnol. Oceanogr.,* **14**:250–259.

Parma, S. 1980. "The History of the Eutrophication Concept and the Eutrophication in the Netherlands." *Hydrobiol. Bull.* (Amsterdam). **14**:5–11.

Pasciak, W. J., and J. Gavis. 1974. "Transport Limitation of Nutrient Uptake in Phytoplankton." *Limnol. Oceanogr.,* **19**:881–888.

Pastorok, R. A., M. W. Lorenzen, and T. C. Ginn. 1981. "Artificial Aeration and Oxygenation of Reservoirs: A Review of Theory, Techniques, and Experiences." Tetra-Tech. Co. Final Rep. TC-3400. Waterways Experimental Station U.S. Army Corps of Engineers. 192 pp. + appendix.

Patalas, K. 1969. "Composition and Horizontal Distribution of Crustacean Zooplankton in Lake Ontario." *J. Fish. Res. Board Can.,* **26**:2135–2164.

———. 1971. "Crustacean Plankton and the Eutrophication of St. Lawrence Great Lakes." *J. Fish. Res. Board Can.,* **29**:1451–1462.

Patrick, R., B. Crum, and J. Coles. 1969. "Temperature and Manganese as Determining Factors in the Presence of Diatom or Blue-green Algal Floras in Streams." *Proc. Nat. Acad. Sci. U.S.A.,* **64**:472–478.

Patterson, C. C., and J. D. Salvia. 1968. "Lead in the Modern Environment, How Much Is Natural?" *Scientist and Citizen,* **10**:66–79.

Paulson, L. J. 1980. "Models of Ammonia Excretion for Brook Trout (*Salvelinus fontinalis*) and Rain-

bow Trout (*Salmo gairdneri*)." *Can. J. Fish. Aquatic Sci.,* **37**:1421–1425.

Pearsall, W. H. 1932. "Phytoplankton in English Lakes. II. The Composition of the Phytoplankton in Relation to Dissolved Substances." *J. Ecol.,* **20**:24–262.

———. 1950. *Mountains and Moorlands.* Collins, London. 312 pp.

———, and W. Pennington. 1973. *The Lake District.* Collins, London. 320 pp.

Pearson, W. D., and R. H. Kramer. 1972. "Drift and Production of Two Aquatic Insects in a Mountain Stream." *Ecol. Monogr.,* **24**:365–385.

Pechlaner, R., G. Bretschko, P. Gollmann, H. Pfeifer, M. Tilzer, and H. P. Weissenbach. 1972. "The Production Processes in Two High Mountain Lakes (Vorder and Hinterer Finstertaler See, Kuhtai, Austria," pp. 239–269. In Z. Kajak and A. Hillbricht-Ilkowska (eds.), *Productivity Problems of Freshwaters,* PWN (Polish Scientific Publishers), Warsaw-Kraków.

Pennak, R. W. 1971. "Towards a Classification of Lotic Habitats." *Hydrobiologia,* **38**:321–334.

———. 1978. *Freshwater Invertebrates of the United States.* 2d ed. Wiley, New York. 803 pp.

Perkins, E. J. 1974. *Biology of Estuaries and Coastal Waters.* Academic, New York. 678 pp.

Perkins, M. A., C. R. Goldman, and R. L. Leonard. 1975. "Residual Nutrient Discharge in Streamwaters Influenced by Sewage Effluent Spraying." *Ecology,* **56**:453–460.

Pfeifer, R. F., and W. F. McDiffett, 1975. "Some Factors Affecting Primary Productivity of Stream Riffle Communities." *Arch. Hydrobiol.,* **75**:306–317.

Polmiluyko, V. P., and M. V. Ochkivskaya. 1970. "Comparison of Nitrate Reductase Activity in *Microcystis aeruginosa* and *Chlorella vulgaris* in Culture." *Hydrobiol. J. (USSR),* **6**:77–80.

Pomeroy, L. R., H. M. Matthews, and H. S. Min. 1963. "Excretion of Soluble Organic Compounds by Zooplankton." *Limnol. Oceanogr.,* **8**:50–55.

Pond, S. and G. L. Pickard. 1978. *Introductory Dynamic Oceanography.* Pergamon, New York. 241 pp.

Porter, K. G. 1976. "Enhancement of Algal Growth and Productivity by Grazing Zooplankton." *Science,* **192**:1332–1334.

———. 1977. "The Plant-Animal Interface in

Freshwater Ecosystems." *Am. Sci.,* **65:**159–170.

Porter, R., and D. W. Fitzimons (eds.) 1978. *Phosphorus in the Environment.* Ciba Foundation, no. 57 (new series). 330 pp. (See especially chapter by C. S. Reynolds, "Phosphorus and the Eutrophication of Lakes," pp. 201–228.)

Powell, T., and A. Jassby. 1974. "The Estimation of Vertical Eddy Diffusivities below the Thermocline in Lakes." *Water Resour. Res.,* **10:**191–198.

Powers, C. F., D. W. Schults, K. W. Malueg, R. M. Brice, and M. D. Schuldt. 1972. "Algal Responses to Nutrient Additions in Natural Waters. II. Field Experiments." *Limnol. Oceanogr. Special Symp.* vol. I., pp. 141–154.

Prepas, E., and F. H. Rigler. 1978. "The Enigma of *Daphnia* Death Rates." *Limnol. Oceanogr.,* **23:**970–988.

Preston, T., W. D. P. Stewart, and C. S. Reynolds. 1980. "Bloom-forming Cyanobacterium *Microcystis aeruginosa* overwinters on Sediment Surface" *Nature,* **288:**365–367.

Prichard, D. W. 1967. "What Is an Estuary? Physical Viewpoint," pp. 3–5. In G. H. Lauff (ed.), *Estuaries.* American Association for the Advancement of Science, Washington.

Priscu, J. C. 1982. "Physiological ecology of Castle Lake phytoplankton: a comparison of shallow and deep-water communities." Ph.D. thesis. Univ. California, Davis.

Provasoli, L. 1963. "Organic Regulation of Phytoplankton Fertility." In M. N. Hill (ed.), *The Sea,* **2:**165–219.

Ragotzkie, R. A. 1974. "The Great Lakes Rediscovered." *Am. Sci.,* **62:**454–464.

———. 1978. "Heat Budgets of Lakes," pp. 1–19. In A. Lerman (ed.), *Lakes—Chemistry, Geology, Physics.* Springer-Verlag, New York.

Ranwell, D. S. 1972. *Ecology of Salt Marshes and Dunes.* Chapman & Hall, London. 258 pp.

Rasmussen, R. A., J. Krasnec, and D. Pierotti. 1976. "N_2O analysis in the atmosphere via electron capture-gas chromatography." *Geophys. Res. Letters,* **3:**615–618.

Ravera, O. 1953. "Gli Stadi di Sviluppo dei Copepodi Pelagici del Lago Maggiore." *Mem. Ist. Ital. Idrobiol.,* **7:**129–151.

Rawson, D. S. 1952. "Mean Depth and the Fish Production of Large Lakes." *Ecology,* **33:**513–521.

———. 1955. "Morphometry as a Dominant Factor in the Productivity of Large Lakes." *Verh. Int. Ver. Limnol.,* **12:**164–175.

Regier, H. A., and K. H. Loftus. 1972. "Effects of Fisheries Exploitation on Salmonid Communities in Oligotrophic Lakes." *J. Fish. Res. Board Can.,* **29:**959–968.

Reid, G. K., and R. D. Wood. 1976. *Ecology of Inland Waters and Estuaries.* 2d ed. Van Nostrand, New York. 485 pp.

Reid, P. C. 1975. "Large Scale Changes in North Sea Phytoplankton." *Nature,* **257:**217–219.

Remane, A., and C. Schlieper. 1971. *Biology of Brackish Water.* 2d ed. Schweizerbart' sche, Stuttgart.

Reuter, J. E. 1982. "Nitrogen assimilation of the epilithic periphyton community in Lake Tahoe, California-Nevada." Ph.D. thesis. Univ. California, Davis.

Reynolds, C. S. 1971. "The Ecology of the Planktonic Blue-green Algae in the North Shropshire Meres." *Field Stud.,* **3:**409–432.

———. 1973. "Growth and Buoyancy of *Microcystis aeruginosa,* Kütz amend. Elenkin, in a Shallow Eutrophic Lake." *Proc. R. Soc. London, Ser. B,* **184:**29–50.

Richardson, J., and D. Livingstone. 1962. "An Attack by a Nile Crocodile on a Small Boat." *Copeia,* **1:**203–204.

Richerson, P. J., C. Widmer, and T. Kittel. 1977. "The Limnology of Lake Titicaca (Peru-Bolivia), a Large, High-Altitude Tropical Lake." Institute of Ecology, publ. no. 14, University of California, Davis. 78 pp.

Richter, D. D., C. W. Ralston, and W. R. Harms. 1982. "Prescribed Fire: Effects on Water Quality and Forest Nutrient Cycling." *Science,* **215:**661–663.

Ricker, W. E. 1975. *Computation and Interpretation of Biological Statistics of Fish Populations.* Department of Environment Fisheries and Marine Service, Ottawa, Canada. 382 pp.

Rigler, F. H. 1956. "A Tracer Study of the Phosphorus Cycle in Lake Water." *Ecology* **37:**550–562.

Ringler, N. H., and J. D. Hall. 1975. "Effects of Logging on Water Temperature and Dissolved Oxygen in Spawning Beds." *Trans. Am. Fish. Soc.,* **104:**111–121.

Ripl, W. 1976. "Biochemical Oxidation of Polluted

Lake Sediment with Nitrate—A New Lake Restoration Method." *Ambio,* **5:**132–135.

Rodhe, W. 1948. "Environmental Requirements of Freshwater Plankton Algae. VII. Iron as a Limiting Factor for Growth." *Symb. Bot. Ups.,* **10:**104–117.

———. 1965. "Standard Correlations between Pelagic Photosynthesis and Light," pp. 365–382. In C. R. Goldman (ed.), "Primary Productivity in Aquatic Environments." *Mem. Ist. Ital. Idrobiol.* **18** (suppl).

———. 1979. "The Life of Lakes." *Arch. Hydrobiol. Beih.,* **3:**5–9.

Roth, J. C. 1968. "Benthic and Limnetic Distribution of Three *Chaoborus* Species in a Southern Michigan Lake (Diptera, *Chaoboridae*)." *Limnol. Oceanogr.,* **13:**242–249.

———, and A. J. Horne. 1981. "Algal Nitrogen Fixation and Microcrustacean Abundance: An Unregarded Interrelationship between Zoo- and Phyto-plankton." *Verh. Int. Ver. Limnol.,* **21:**333–338.

———, and S. E. Neff. 1964. "Studies of Physical Limnology and Profundal Bottom Fauna, Mountain Lake, Virginia." *Va. Agric. Exp. Stn. Tech. Bull. 169.* 44 pp.

———, and J. A. Stewart. 1973. "Nearshore Zooplankton of Southeastern Lake Michigan, 1972." *Proc. 16th Conf. Great Lakes Res.,* pp. 132–142.

Round, F. E. 1971. "The Growth and Succession of Algal Populations in Freshwaters." *Mitt. Int. Ver. Theor. Angew. Limnol.,* **19:**70–99.

Rudd, R. L. 1964. *Pesticides in the Living Landscape.* The University of Wisconsin Press, Madison. 320 pp.

Russell, F. S., A. J. Southward, G. T. Boalch, and E. I. Butler. 1971. "Changes in Biological Conditions in the English Channel of Plymouth during the Last Half-century." *Nature,* **234:**468–470.

Russell-Hunter, W. D. 1968. *A Biology of Lower Invertebrates.* 181 pp. Macmillan, New York.

———. 1969. *A Biology of Higher Invertebrates.* Macmillan, New York. 224 pp.

Ruttner, F. 1963. *Fundamentals of Limnology.* English translation by D. G. Frey and F. E. J. Fry. Toronto University Press, Canada. 307 pp. (First German edition in 1940.)

Ruttner-Kolisko, A. 1972. "Das Zooplankton der Binnengewässer. I. Tiel. Rotatoria." *Die Binnengewässer,* **26:**99–234.

Saunders, G. W. 1971. "Carbon Flow in the Aquatic System." In J. Cairns (ed.), *Structure and Function of Freshwater Microbial Communities.* Research Division Monograph 3, pp. 31–45. Virginia Polytechnic Institute, Blacksburg.

Schelske, C. L. 1962. "Iron, Organic Matter and Other Factors Limiting Primary Productivity in a Marl Lake." *Science,* **136:**45–46.

———, and J. C. Roth. 1973. "Limnological Survey of Lakes Michigan, Superior, Huron, and Erie." *Great Lakes Res. Div., Univ. Mich., Publ. 17.* 108 pp.

———, and E. F. Stoermer. 1972. "Phosphorus, Silica, and Eutrophication of Lake Michigan." *Limnol. Oceanogr. Special Symp.* vol. 1, pp. 157–170.

Schindler, D. W. 1980. "The Effect of Fertilization with Phosphorus and Nitrogen versus Phosphorus Alone on Eutrophication of Experimental Lakes." *Limnol. Oceanogr.,* **25:**1149–1152.

Schmidt, G. W. 1969. "Vertical Distribution of Bacteria and Algae in a Tropical Lake." *Int. Rev. ges. Hydrobiol.,* **54:**791–797.

Seliger, H. H., J. H. Carpenter, M. Loftus, and W. D. McElroy. 1970. "Mechanisms for the Accumulation of High Concentrations of Dinoflagellates in a Bioluminescent Bay." *Limnol. Oceanogr.,* **15:**234–245.

Sellers, C. M., A. G. Heath, and M. L. Bass. 1975. "The Effect of Sublethal Concentrations of Copper and Zinc on Ventilatory Activity, Blood Oxygen, and pH in Rainbow Trout (*Salmo gairdneri*)." *Water Res.,* **9:**401–408.

Sellery, G. C. 1956. *E. A. Birge, a Memoir.* The University of Wisconsin Press, Madison. 221 pp. [With an appraisal of Birge, the limnologist, "An Explorer of Lakes," by C. H. Mortimer, pp. 165–211.]

Serruya, C., and U. Pollingher. 1971. "An Attempt at Forecasting the *Peridinium* Bloom in Lake Kinneret (Lake Tiberias)." *Mitt. Int. Ver. Theor. Angew. Limnol.,* **19:**277–291.

Shapiro, J. 1960. "The Cause of a Metalimnetic Minimum of Dissolved Oxygen." *Limnol. Oceanogr.,* **5:**216–227.

———. 1973. "Blue-green Algae: Why They Become Dominant." *Science,* **179:**382–384.

———, W. T. Edmondson, and D. E. Allison. 1971. "Changes in the Chemical Composition of Sediments in Lake Washington, 1958–1970." *Limnol. Oceanogr.*, **16:**437–452.

Simpson, D. G., and G. Gunther. 1956. "Notes on Habits, Systematic Characters, and Life Histories of Texas Saltwater Cyprinodonts." *Tulane Stud. Zool. Bot.*, **4:**115–134.

Skabitchewsky, A. P. 1929. "Über die biologie von *Melosira baicalensis* (K. Meyer)." *Wisl. Russ. Gidrobiol. Zh.*, **8:**93–114.

Smayda, T. J. 1957. "Phytoplankton Studies in Lower Narragansett Bay." *Limnol. Oceanogr.*, **2:**342–358.

Smith, I. R. 1975. "Turbulence in Lakes and Rivers." *Freshwater Biol. Assoc., U.K. Publ. 29.* 79 pp.

Smith, M. S., M. K. Firestone, and J. M. Tiedje. 1978. "The acetylene inhibition method for short-term measurement of soil denitrification and its evaluation using ^{13}N." *Soil. Sci. Soc. Amer. J.*

Smith, R. A. 1852. "Relationship of Sooty Skies and Acid Precipitation in Manchester, England." Referred to in *National Geographic*, November 1981, p. 661.

Smith, R. C. 1968. "The Optical Characterization of Natural Waters by Means of an 'Extinction Coefficient'." *Limnol. Oceanogr.*, **13:**423–429.

———, J. E. Tyler, and C. R. Goldman. 1973. "Optical Properties and Color of Lake Tahoe and Crater Lake." *Limnol. Oceanogr.*, **18:**189–199.

Smith, S. H. 1968. "Species Succession and Fishery Exploitation in the Great Lakes." *J. Fish. Res. Board Can.*, **25:**667–693.

Smyly, W. J. P. 1978. "Strategies for Co-existence in Two Limnetic Cyclopoid Copepods." *Verh. Int. Ver. Limnol.*, **20:**2501–2504.

Solórzano, L. 1969. "Determination of Ammonia in Natural Waters by the Phenolhypochlorite Method." *Limnol. Oceanogr.*, **14:**799–801.

Sørensen, J. 1978. "Capacity for denitrification and reduction of nitrate to ammonia in a coastal marine sediment." *Appl. Environ. Microbiol.* **35:**301–305.

Steele, J. H. 1974. *The Structure of Marine Ecosystems.* Blackwell Scientific Publications, Ltd., Oxford. 128 pp.

Steeman Nielsen, E. L., L. Kamp-Nielsen, S. Wium-Anderson. 1969. "The Effect of Deleterious Concentrations of Copper on the Photosynthesis of *Chlorella pyrenoidosa*." *Physiol. Plant. Pathol.*, **22:**1121–1131.

———, and S. Wium-Anderson. 1971. "The Influ-ence of Cu on Photosynthesis and Growth in Diatoms." *Physiol. Plant. Pathol.*, **24:**408–414.

Stewart, K. M. 1972. "Isotherms under Ice." *Verh. Int. Ver. Limnol.*, **18:**303–311

———. 1973. "Detailed Time Variations in Mean Temperature and Heat Content of Some Madison Lakes." *Limnol. Oceanogr.*, **18:**218–226.

Stewart, W. D. P. 1967. "Transfer of Biologically Fixed Nitrogen in a Sand Dune Slack Region." *Nature*, **214:**603–604

———. (ed.) 1974. *Algal Physiology and Biochemistry.* Botanical Monographs, vol. 10. University of California Press, Berkeley.

———, and G. Alexander. 1971. "Phosphorus Availability and Nitrogenase Activity in Aquatic Blue-green Algae." *Freshwater Biol.*, **1:**389–404.

———, G. P. Fitzgerald, and R. H. Burris. 1967. "In situ Studies on N_2-fixation Using the Acetylene Reduction Technique." *Proc. Nat. Acad. Sci. U.S.A.*, **58:**2071–2078.

Stillinger, F. H. 1980. "Water Revisited." *Science*, **209:**451–457.

Stone, R. W., W. J. Kaufman, and A. J. Horne. 1973. "Long-term Effects of Toxicants and Biostimulants on the Waters of Central San Francisco Bay." University of California, Berkeley, Sanitary Engineering Research Lab. UCB-SERL Rep. No. 73-1. 111 pp.

Storrs, P. N., E. A. Pearson, and R. E. Selleck. 1966. "A Comprehensive Study of San Francisco Bay, a Final Report. Vol. V. Summary of Physical, Chemical, and Biological Water and Sediment Data." University of California, Berkeley, Sanitary Engineering Research Lab. UCB-SERL Rep. No. 67-2.

Strickland, J. D. H., and T. R. Parsons. 1972. "A Practical Handbook of Seawater Analysis." *Fish. Res. Board Can. Bull.* 167 (2d ed.). 310 pp.

———, O. Holm-Hansen, R. W. Eppley, and R. J. Linn. 1969. "The Use of a Deep Tank in Plankton Ecology. I. Studies of the Growth and Composition of Phytoplankton Crops at Low Nutrient Levels." *Limnol. Oceanogr.*, **14:**23–34.

Stumm, W. and J. O. Leckie. 1971. "Phosphate Exchange with Sediments: Its Role in the Productivity of Surface Waters." *Proceedings of Fifth International Water Pollution Conference.* Pergamon, London.

———, and E. S. Zollinger. 1972. "The Role of

Phosphorus in Eutrophication," pp. 11–42. In R. Mitchell (ed.), *Water Pollution Microbiology*. Wiley, New York.

Sugawara, K. 1939. "Chemical Studies in Lake Metabolism." *Bull. Chem. Soc. Japan*, **14**:375–451.

Swale, E. M. F. 1969. "Phytoplankton in Two English Rivers." *J. Ecol.*, **57**:1–23.

Swift, M. C. 1970. "A Qualitative and Quantitative Study of Trout Food in Castle Lake, California." *Calif. Fish Game*, **56**:109–120.

Sylvester, R. O., and G. C. Anderson. 1964. "A Lake's Response to Its Environment." *J. Sanit. Eng. Div. Am. Soc. Civ. Eng.*, **90**.

Talling, J. F. 1969. "The Incidence of Vertical Mixing, and Some Biological and Chemical Consequences, in Tropical African Lakes." *Verh. Int. Ver. Limnol.*, **17**:998–1012.

———. 1970. "Generalized and Specialized Features of Phytoplankton as a Form of Photosynthetic Cover," pp. 431–446. In *Prediction and Measurement of Photosynthetic Productivity*. Centre for Documentation, Wageningen, Netherlands.

———. 1971. "The Underwater Light Climate as a Controlling Factor in the Production Ecology of Freshwater Phytoplankton." *Mitt. Int. Ver. Theor. Angew. Limnol.*, **19**:214–243.

———. 1976. "The Depletion of Carbon Dioxide from Lake Water by Phytoplankton." *J. Ecol.*, **64**:79–121.

Teal, J. M. 1957. "Community Metabolism in a Temperate Cold Spring." *Ecol. Monogr.*, **27**:283–302.

Thienemann, A. 1922. "Die beiden *Chironomus* Arten Tiefenfauna der norddeutschen Seen. Ein hydrobiologisches Problem." *Arch. Hydrobiol.*, **13**:609–646.

Thorpe, S. A. 1971. "Experiments on Instability of Stratified Shear Flows: Miscible Fluids." *J. Fluid Mech.*, **46**:299–319.

———. 1977. "Turbulence and Mixing in a Scottish Loch." *Philos. Trans. R. Soc. London, Ser. A*, **286**:125–181.

Tilman, D. 1978. "The Role of Nutrient Competition in a Predictive Theory of Phytoplankton Population Dynamics." *Mitt. Int. Ver. Theor. Angew. Limnol.*, **21**:585–592.

———, and S. S. Kilham. 1976. "Phosphate and Silicate Growth and Uptake Kinetics of the Diatoms *Asterionella formosa* and *Cyclotella meneghiniana* in Batch and Semi-continuous Culture." *J. Phycol.*, **12**:375–383.

Tilzer, M. M., C. R. Goldman, and E. de Amezaga. 1975. "The Efficiency of Photosynthetic Light Energy Utilization by Lake Phytoplankton." *Verh. Int. Ver. Limnol.*, **19**:800–807.

———, and A. J. Horne. 1979. "Diel Patterns of Phytoplankton Productivity and Extracellular Release in Ultra-Oligotrophic Lake Tahoe." *Int. Rev. ges. Hydrobiol.*, **64**:157–176.

Timm, T. 1980. "Distribution of Aquatic Oligochaetes," pp. 55–77. In R. O. Brinkhurst and D. G. Cook (eds.), *Aquatic Oligochaete Biology*. Plenum, New York.

Toetz, D. W. 1971. "Diurnal Uptake of NO_3 and NH_4 by a *Ceratophyllum*-Periphyton Community." *Limnol. Oceanogr.*, **16**:819–822.

Tomlinson, T. E. 1970. "Trends in Nitrate Concentration in English Rivers in Relation to Fertilizer Use." *Water Treat. Exam.*, **19**:277–293.

Toms, R. G. 1975. "Management of Water Quality," pp. 538–564. In B. A. Whitton (ed.), *River Ecology*. University of California Press, Berkeley.

Torrey, M. S., and G. F. Lee. 1976. "Nitrogen Fixation in Lake Mendota, Madison, Wisconsin." *Limnol. Oceanogr.*, **21**:365–378.

Truesdale, G. A., A. L. Downing, and G. F. Lowden. 1955. "The Solubility of Oxygen in Pure Water and Seawater." *J. Appl. Chem.*, **5**:53–62.

Trussell, R. P. 1972. "The Percent Un-ionized Ammonia in Aqueous Ammonia Solutions at Different pH Levels and Temperatures." *J. Fish. Res. Board Can.*, **29**:1505–1507.

Tuite, C. H. 1981. "Standing Crop Densities and Distribution of *Spirulina* and Benthic Diatoms in East African Alkaline Saline Lakes." *Freshwat. Biol.*, **11**:345–360.

Tyler, J. 1968. "The Secchi Disc." *Limnol. Oceanogr.*, **13**:1–6.

Tyler, J. E., and R. W. Priesendorfer. 1962. "Transmission of Energy within the Sea." In M. N. Hill (ed.), *The Sea*, **1**:397–451.

Tyler, M. A., and H. H. Seliger. 1978. "Annual Subsurface Transport of a Red Tide Dinoflagellate to Its Bloom Area: Water Circulation Patterns and Organism Distributions in the Chesapeake Bay." *Limnol. Oceanogr.*, **23**:227–246.

Vallentyne, J. R. 1960. "Fossil Pigments," pp. 83–105. In M. B. Allen (ed.), *Comparative Bio-*

chemistry of Photoreactive Systems. Academic, New York.

Various authors. 1973. "Physical, Chemical, Phytoplankton, Zooplankton, Zoobenthos and Fisheries in Lake George, Uganda." *Proc. R. Soc. London, Ser. B,* **184**:235–346.

———. 1974. *J. Fish. Res. Board Can.,* **31**:689–854. (On Laurentian Great Lakes physics, chemistry, biology.)

Visser, S. A., and J. P. Villeneuve. 1975. "Similarities and Differences in the Chemical Composition of Waters from West, Central and East Africa." *Verh. Int. Ver. Limnol.,* **19**:1416–1425.

Viner, A. B. 1969. "The Chemistry of the Water of Lake George, Uganda." *Verh. Int. Ver. Limnol.,* **17**:289–296.

———. 1972. "Responses of a Mixed Phytoplankton Population to Nutrient Enrichments of Ammonia and Phosphate, and Some Associated Ecological Implications." *Proc. R. Soc. London, Ser. B,* **183**:351–370.

———, and I. R. Smith. 1973. "Geographical, Historical and Physical Aspects of Lake George." *Proc. R. Soc. London, Ser. B,* **184**:235–270.

Vollenweider, R. A. 1961. "Photometric Studies in Inland Waters. 1. Relations Existing in the Spectoral Extinction of Light in Water." *Mem. Ist. Ital. Idrobiol.,* **13**:87–113.

———. 1965. "Calculation Models of Photosynthesis Depth Curves and Some Implications Regarding Day Rate Estimates in Primary Production Measurements," pp. 425–457. In C. R. Goldman (ed.), "Primary Productivity in Aquatic Environments." *Mem. Ist. Ital. Idrobiol.,* **18** (suppl.).

———. 1968. *Scientific Fundamentals of the Eutrophication of Lakes and Flowing Waters, with Particular Reference to Nitrogen and Phosphorus as Factors in Eutrophication.* Tech. Rept. No. DAS/CSJ/68.27. Organization for Economic Cooperation and Development, Paris. 159 pp.

———. 1969. *A Manual on Methods for Measuring Primary Production in Aquatic Environments.* IBP Handbook No. 12. Davis, Philadelphia. 213 pp.

Waananen, A. O., D. D. Harris, and R. C. Williams. 1970. "Floods of December 1964 and January 1965 in the Four Western States. Part 2. Stream Flow and Sediment Data." *U.S. Geol Surv. Water-Supply Pap.* 18665. 861 pp.

Walker, W. W. 1979. "Use of Hypolimnetic Oxygen Depletion Rates as a Trophic State Indicator for Lakes." *Water Resour. Res.,* **15**:1463–1470.

Walsby, A. E. 1972. "Structure and Function of Gas Vacuoles." *Bacteriol. Rev.,* **36**:1–32.

———. 1974. "The Extracellular Products of *Anabaena cylindrica* Lemm. I. Isolation of a Macromolecular Pigment-peptide Complex and Other Components." *Br. Phycol. J.,* **9**:371–381.

———, and C. S. Reynolds. 1979. "Sinking and Floating in Phytoplankton Ecology." In I. Morris (ed.), *The Ecology of Phytoplankton.* Blackwell Scientific Publications, Ltd., Oxford.

———, and A. Xypolyta. 1977. "The Form Resistance of Chitin Fibres Attached to the Cells of *Thalassiosira fluviatilis* Hustedt." *Br. Phycol. J.,* **12**:215–223.

Ward, A. K., and R. G. Wetzel. 1975. "Sodium: Some Effects in Blue-green Algal Growth." *J. Phycol.,* **4**:357–363.

Ward, H. B., and G. C. Whipple. 1959. In W. T. Edmondson (ed.), *Freshwater Biology.* 2d ed. Wiley, New York. 1248 pp.

Ward, J. V., and J. A. Stanford (eds.). 1979. *The Ecology of Regulated Streams.* Plenum, New York. 398 pp.

Warren, C. E. 1971. *Biology and Water Pollution Control.* Saunders, Philadelphia. 434 pp.

Watt, W. D. 1966. "Release of Dissolved Organic Material from the Cells of Phytoplankton Populations." *Proc. R. Soc. London, Ser. B,* **164**:521–551.

Weatherley, A. H., and P. Dawson. 1972. "Zinc Pollution in a Freshwater System: Analysis and Proposed Solutions." *Search,* **4**:471–476.

Weitzel, R. L. (ed.). 1979. *Methods and Measurement of Periphyton Communities: A Review.* Amer. Soc. Testing and Materials. Philadelphia. 183 pp.

Welch, E. B. (ed.). 1980. *Ecological Effects of Waste Water.* Cambridge, New York. 337 pp.

———, J. A. Buckley, and R. M. Bush. 1972. "Dilution as an Algal Bloom Control." *J. Water Pollut. Control Fed.,* **44**:2245–2265.

Welch, H. 1967. "Energy Flow through the Major Macroscopic Components of an Aquatic Ecosystem." Ph.D. thesis, University of Georgia, Athens.

Welch, P. S. 1935. *Limnology.* McGraw-Hill, New York. 472 pp. (2d ed., 1952, 536 pp.)

Westlake, D. F., H. Casey, F. H. Dawson, M. Ladle, R. H. K. Mann, and A. F. H. Marker. 1972. "The Chalk-stream Ecosystem," pp. 613–635. In Z. Kajak and A. Hillbricht-Ikowska (eds.), *Productivity Problems of Freshwaters*. PWN (Polish Scientific Publishers), Warsaw-Kraków.

Wetzel, R. G. 1960. "Marl Encrustation on Hydrophytes in Several Michigan Lakes." *Oikos*, **11**:223–236.

———. 1964. "A Comparative Study of the Primary Productivity of Higher Aquatic Plants, Periphyton and Phytoplankton in a Large, Shallow Lake." *Int. Rev. ges. Hydrobiol. Hydrogr.*, **49**:1–61.

———. 1975. *Limnology*, Saunders, Philadelphia. 743 pp.

———. 1979. "The Role of the Littoral Zone and Detritus in Lake Metabolism." *Arch. Hydrobiol. Beih.*, **13**:145–161.

———, and A. Otsuki. 1974. "Allochthonous Organic Carbon of a Marl Lake." *Arch. Hydrobiol.*, **73**:31–56.

Whitton, B. A. 1970. "Toxicity of Heavy Metals to Freshwater Algae: A Review." *Phykos*, **9**:116–125.

———. (ed.). 1975. *Studies in Ecology*, vol. 2: *River Ecology*. University of California Press, Berkeley. 725 pp.

Williams, L. R. 1971. "A Possible Role of Heteroinhibition in the Production of *Anabaena flosaquae* Waterblooms." Ph.D. thesis, Rutgers University, New Brunswick, N.J. 88 pp.

Williams, W. D. 1972. "The Uniqueness of Salt Lake Ecosystems," pp. 349–361. In Z. Kajak and A. Hillbricht-Ilkowska (eds.), *Productivity Problems of Freshwaters*. PWN (Polish Scientific Publishers), Warsaw-Kraków.

———. 1978. "Limnology of Victorian Salt Lakes, Australia." *Verh. Int. Ver. Limnol.*, **20**:1165–1174.

———. 1981. "Inland Salt Lakes: An Introduction." In W. D. Williams (ed.), *Salt Lakes: Proceedings of an International Symposium on Athalassic (Inland) Salt Lakes*. Junk Publishers, The Hague.

Willoughby, L. G. 1969. A Study of the Aquatic Actinomycetes of Bleham Tarn." *Acta Hydrobiol. Hydrographia Protistol.*, **34**:465–483.

Winberg, G. G. 1970. "Energy Flow in the Aquatic Ecological System." *Pol. Arch. Hydrobiol.*, **17**:11–19.

Wright, H. E., T. C. Winter, and H. L. Patten. 1963. "Two Pollen Diagrams from Southeastern Minnesota; Problems in the Regional Late-glacial and Post-glacial Vegetation History." *Geol. Soc. Am. Bull.*, **74**:1371–1396.

Wright, J. C. 1965. "The Population Dynamics and Production of *Daphnia* in Canyon Ferry Reservoir, Montana." *Limnol. Oceanogr.*, **10**:583–590.

Wright, R. T., and J. E. Hobbie. 1966. "Use of Glucose and Acetate by Bacteria and Algae in Aquatic Ecosystems." *Ecology*, **47**:447–464.

Wrigley, R. C., and A. J. Horne. 1974. "Remote Sensing and Lake Eutrophication." *Nature*, **250**:213–214.

———, and ———. 1975. "Surface Algal Circulation Patterns in Clear Lake by Remote Sensing." *NASA Tech. Memo. X-62, 451.* 11 pp.

Wurtsbaugh, W. A., R. W. Brocksen, and C. R. Goldman. 1975. "Food and Distribution of Underyearling Brook and Rainbow Trout in Castle Lake, California." *Trans. Amer. Fish. Soc.*, **104**:88–95.

Young, J. O. 1975. "Preliminary Field and Laboratory Studies on the Survival and Spawning of Several Species of Gastropoda in Calcium-poor and Calcium-rich Waters." *Proc. Malac. Soc. London*, **41**:429–437.

Zaret, T. M., and R. T. Paine. 1973. "Species Introduction in a Tropical Lake." *Science*, **182**:449–455.

Zenkevitch, L. 1963. *Biology of the Seas of the U.S.S.R.* G. Allen, London. 955 pp.

Zevenboom, W., and L. R. Mur. 1978. "On Nitrate Uptake by *Oscillatoria agardhii*." *Verh. Int. Ver. Limnol.*, **20**:2302–2307.

———, G. J. DeGroot, and L. R. Mur. 1980. "Effects of Light on Nitrate-limited *Oscillatoria agardhii* in Chemostat Culture." *Arch. Mikrobiol.*, **125**:59–65.

———, and L. R. Mur. 1981. "Ammonia-limited Growth and Uptake by *Oscillatoria agardhii* in Chemostat Culture." *Arch. Mikrobiol.*, **129**:61–66.

Indexes

Name Index

Name Index

Abbott, D. P., 337
Agassiz, L., 3–4
Ahlgren, I., 405
Akerman, W. C., 407
Alexander, G., 135
Alexander, V., 119
Alexander, W. B., 320
Allen, E. R., 150, 152, 163
Allen, K. R., 271
Allen, M. B., 358
Allison, D. E., 155, 359
American Fisheries Society, 338
Anderson, G. C., 401
Anderson, H. M., 37, 175, 202
Andersson, G., 399–400
Andrews, R. D., 268
Angeli, N., 233, 241
Anton, H., 238
APHA (American Public Health Association), 88, 132
Aquamarine Company, 404
Armstrong, D. E., 285
Armstrong, F. A. J., 145
Armstrong, R., 290
Arnold, D. E., 232
Axler, R. P., 125, 226

Babenzien, H. D., 169

Bachmann, R. W., 47
Bailey, R. E., 361
Bailey-Watts, A. E., 373
Baker, A. L., 199, 203, 218
Baker, P., 257
Baldwin, N. S., 262
Balon, E. K., 34, 251, 269, 271
Banse, K., 112
Barbour, C. D., 271
Bardach, J. E., 188–192, 253, 271
Barnes, R. S. K., 320–321, 323, 334, 338
Barnett, R. H., 401
Barret, E., 7
Barsdate, R. J., 139, 395
Bass, M. L., 159
Bassindale, R., 320
Bates, J. M., 268, 410
Bayly, I. A. E., 94, 225
Beadle, L. C., 363
Beaumont, P., 299
Beeton, A. M., 140, 261–263, 370–371
Bengtsson, L., 399
Bennett, E. B., 133, 145, 369
Bennett, G. W., 253
Bennett, M., 409
Benson, B. B., 116
Bergey, 167

Berggren, H., 399–400
Berman, T., 137, 140, 219
Bernhardt, H., 401
Billaud, V. A., 119, 121
Bindloss, M. E., 373
Birge, E. A., 4, 37, 54, 56, 96, 112, 363
Bischoff, A. I., 267
Blackburn, R. D., 196
Blake, B. F., 269
Blanton, J. O., 50
Blažka, P., 130, 137, 140
Bloomfield, J. A., 123, 133, 138, 145, 153
Blum, J. L., 316
Boalch, G. T., 236
Boll, C. D., 241
Bone, D. H., 137
Boney, A. D., 220
Borman, F. H., 93, 123, 133, 299–300, 304, 316
Born, S. M., 390
Boyce, F. M., 44, 50–51, 56, 64, 68, 78, 84, 86
Boyd, C. E., 281
Boyle, E., 146
Bretschko, G., 245, 379–380
Brezonik, P. L., 118–119, 122, 130, 161

Brice, R. M., 138, 393, 395–396
Bricker, O. P., 144
Brinkhurst, R. O., 196, 242, 246
Britt, N. W., 241
Britton, R. H., 279, 290
Brock, T. D., 384–385
Brocksen, R. W., 257
Brodin, G., 7
Brook, A. J., 199, 203, 218
Brooks, J. L., 225
Brown, E. J., 137
Brown, G. W., 301
Brown, G. W., Jr., 385
Brown, J. H., 271
Bryan, G. W., 159
Buchanan, R. E., 167
Buckley, J. A., 400
Burgis, M. J., 73, 373–374
Burns, C. W., 225, 234
Burns, N. W., 49–50, 67, 86
Burris, R. H., 116
Burton, T. M., 300–301
Bush, R. M., 400
Butler, E. I., 236, 290
Button, D. K., 137

Cadle, R. D., 150, 152, 163
Cairns, J., 94
Cairns, S. J., 408
Cantor-Lund, H., 173, 181, 203, 209
Capblancq, J., 246
Carlson, C. A., 306
Carmiggelt, C. J. W., 8, 118, 121
Carpelan, L. H., 385
Carpenter, E. J., 125
Carpenter, J. H., 330
Carr, J. F., 241
Carr, N. G., 203
Carson, R., 267
Carter, R. C., 397
Casey, H., 303–304, 308
Caspers, H., 335
Castenholz, R. W., 203, 385
Cato, I., 340
Chandler, D. C., 370
Chapman, V. J., 5, 328, 338
Charlton, M. N., 141
Chen, R. L., 116
Christie, W. J., 262–264, 271
Clarke, M. G., 7
Cloern, J. E., 325
Clymo, R. S., 94, 116, 132, 273
Coats, R. N., 302
Coche, A. G., 34, 251, 269, 271
Cole, G. A., 5, 381, 385
Coles, J., 159
Commins, M. L., 182, 184, 223
Confer, J. L., 140
Conomos, T. J., 325, 334
Cook, D. G., 196, 242, 246

Cooper, W. E., 9
Cordone, A., 257
Corner, E. D. S., 130, 137, 290
Cottam, G., 403
Coull, B. C., 338
Cowgill, U. M., 354, 361
Craig, P. C., 384
Craik, A. D. D., 73
Cronberg, G., 399
Crowley, P. H., 225
Crum, B., 159
Csanady, G. T., 68, 71, 83, 86
Cummins, K. W., 275, 310, 316
Cushing, E. J., 360
Czapski, U. H., 77

Dalton, J., 97
Darley, W. M., 144, 164
Darlington, J. P. E. C., 255, 374
Davis, A. G., 130, 137
Davy, A. J., 338
Dawson, F. H., 308
Dawson, P., 159
Day, J. W., 321, 325
de Amezaga, E., 82, 290
Dean, W. E., Jr., 360
de Bernardi, R., 232
de Duillier, F., 3
DeGroot, G. J., 125
de la Cruz, A., 329
Denison, P. J., 56
Dikson, K. L., 408
Dillon, T. M., 76, 79
Dobson, H. F. H., 123, 133, 145, 369
Dodson, S. I., 94, 225
Doty, M. S., 288
Dowd, J. E., 283
Downing, A. L., 104
Droop, M. N., 94, 137
DSIR (Department of Scientific and Industrial Research, England), 408–409
Dugdale, R. C., 116, 125
Dugdale, V. A., 116
Dumont, H. J., 232
Duncan, C. P., 80
Dunn, I. G., 255, 374
Dunst, R. C., 390
Durham, L., 253
Dussart, B., 5
Dymond, J. R., 190

Eagle, D. J., 217
Eaton, J. S., 123, 133, 299–300, 304
Edmondson, W. T., 155, 196, 230–231, 237–238, 240–241, 247, 355, 359, 391–394
Egeratt, A. W. S. M., 163
Eggleton, F. E., 363

Eisma, D., 338
Elder, F. C., 56
Elder, J. F., 50, 154–155, 157–158, 160, 302, 405
Elliot, J. M., 312
Elster, H-J., 11
Elwood, J. W., 20
El-Zarka, S. E. D., 407
Emery, K. O., 71
Eppley, R. W., 125, 282, 291
Erman, D. C., 241, 302
Erman, N. A., 241
Esch, G. W., 410
Evans, P. M., 290

Falter, C. M., 268
Fast, A. W., 241, 403
Faust, S. O., 144
Fee, E. J., 9, 82, 145
Fisher, D. W., 93, 316
Fittkau, E. J., 311
Fitzgerald, G. P., 116, 135, 137
Fitzimons, D. W., 141
Fleischer, S., 399
Fogg, G. E., 2–3, 33, 94, 116, 119, 121, 123, 129, 201, 204, 217, 220, 377
Folt, C., 94
Forbes, S. A., 4
Foree, E. J., 139
Forel, F. A., 2, 5, 19, 96
Fox, J. L., 161
Frank, P. W., 241
Frantz, T., 257
Frey, D. G., 4–5, 359, 370
Fritsch, A., 5
Frost, W. E., 259
Fry, F. E. J., 4–5
Fryer, G., 271
Fuji, N., 357
Fuller, R. H., 158
Fuller, S. L. H., 410

Galat, D. L., 119, 382
Gameson, A. L. H., 408
Gammon, J. R., 268
Ganf, G. G., 73, 99, 107, 130, 137, 140, 145, 287, 289–290, 353, 373–374, 379
Garrels, R. M., 144
Gavis, J., 200
Gelin, C., 399
Gerloff, G. C., 115, 152
Gersberg, R. M., 116
Gessner, F., 218, 296
Gibbons, N. E., 167
Giesy, J. P., 11
Gilbert, J. J., 237
Gilbertson, M., 123, 133, 145, 369

Ginn, T. C., 401
Giussani, G., 232
Godden, D. A., 53
Goering, J. J., 116, 144, 146
Gold, K., 94
Goldman, C. R., 5, 7, 33–34, 36, 38,
    47, 53, 74, 82, 94, 116,
    118–120, 123, 133, 145, 149,
    152–153, 159–161, 163, 219,
    226–227, 233, 241, 257, 288–
    290, 302, 324, 353, 374, 376–
    377, 379–380, 382, 385, 390,
    395, 397–399, 405
Gollman, P., 245, 379–380
Golterman, H. L., 5, 94, 116,
    132–133, 141, 273
Goodyear, C. P., 281
Gorchev, H., 410
Gorham, E., 7, 54, 56, 360
Graetz, D. A., 116
Granhall, U., 119
Grant, J. W. G., 94, 225
Green, G. H., 226
Griffith, E. J., 140
Griffiths, B. M., 201
Griffiths, M., 355
Gross, M. G., 72
Groth, P., 153, 155, 159, 163
Gudmundsson, F., 246
Guillard, R. R. L., 125, 144, 146, 201
Gunter, G., 102, 336, 338, 382
Guseva, K. A., 159
Gwahaba, J. J., 374

Haderlie, E. C., 337
Haeckel, E., 3, 222
Håkanson, L., 13, 23
Hall, C. A. S., 321, 325
Hall, D. J., 9, 225
Hall, J. D., 300
Hamelink, G., 268
Hamilton, D. H., 410
Hamrin, S., 399–400
Haney, J. F., 234
Harbeck, G. E., 381
Harding, D., 269
Hargrave, B. T., 226
Harkness, W. J. K., 190
Harms, W. R., 92
Harris, D. D., 299
Harrison, M. J., 137
Hart, C. W., 410
Harvey, H. W., 235
Hasler, A. D., 4, 89
Hayes, F. R., 133
Healy, F. P., 125
Heath, A. G., 159
Hecky, R. E., 145
Hedgepeth, J. W., 5, 10, 338
Henry, W., 97

Hensen, V., 3
Heron, J., 123, 129, 133, 138, 207,
    365
Herricks, E. E., 408
Heywood, R. B., 376, 385
Hildebrand, S. F., 338
Hill, M. N., 94
Hillbricht-Ilkowska, A., 245–246, 279,
    290, 308, 373, 379–380, 386
Hiltunen, J. K., 241
Hobbie, J. E., 36, 282, 382
Hoffman, R. W., 7
Holden, A. V., 373
Holdgate, M. W., 33, 377
Holm, L. G., 196
Holm, N. P., 285
Holm-Hansen, O., 291
Holt, V. I., 310, 316
Horie, S., 357, 359–360
Hormark, P. D. U., 390
Horne, A. J., 7, 33, 37, 42, 50,
    61–62, 99, 107–108, 116, 118–
    121, 123, 125, 127, 129–130,
    133, 138, 145, 151, 153–158,
    160–161, 195, 199, 213, 215–
    217, 219–220, 232, 277, 287,
    289–290, 300, 306, 315, 326,
    353, 373–377, 379, 394, 403,
    405, 409
Howard, F. O., 310, 316
Hrbáček, J., 225
Hsü, K. J., 102
Hummerstone, L. G., 159
Hunt, E. G., 267
Hunter, J. V., 144
Huntjens, J. L. M., 163
Hutchinson, G. E., 4–5, 23, 31, 37,
    56, 181, 196, 210, 220, 343,
    354–355, 360–361, 381
Hutchinson, G. L., 128
Hynes, H. B. N., 5, 94, 297, 307,
    309, 311, 316, 410

Iles, T. D., 271
Itasaka, O., 145, 153
Ivlev, V. S., 243, 290

Jackson, P. B. M., 194
Jakob, H., 155
James, H. R., 37
Jamieson, B. G. M., 246
Jansson, B-O., 321
Jardin, J., 3
Jassby, A., 82
Javornicky, P., 219
Jefferies, R. L., 338
Jenkin, P. M., 42, 194
Jenkins, D., 115
Jerlov, N. G., 37

Jewell, W. J., 139
Johannes, R. E., 226
Johnson, M. G., 246
Johnson, N. M., 56, 93, 123, 133,
    145, 299–300, 304, 316
Johnson, P. W., 277
Jolley, R. L., 410
Jónasson, P. M., 109, 243–245, 247
Jones, A. C., 384
Jones, J. G., 137, 171, 196
Jones, K., 121, 128
Jørgensen, C. B., 340
Joule, J. P., 67
Juday, C., 4, 96, 112, 363

Kaessler, R. L., 268
Kajak, Z., 245–246, 279, 290, 308,
    373, 379–380, 385
Kalff, J., 380
Kamp-Nielsen, L., 405
Kaplan, L. A., 302
Kashimura, 177
Kata, 177
Kaufman, W. J., 125, 151, 217
Kaushik, N. K., 94, 307, 316
Kawanabe, H., 357
Keeney, D. R., 116, 130
Keirn, M. A., 119
Kelley, D., 300, 315
Kelley, R. W., 241
Kellogg, W. W., 150, 152, 163
Kelts, K., 102
Kelvin, W. T., 67
Kerfoot, W. C., 234, 247
Kilham, P., 201, 286
Kilham, S. S., 137, 144, 146
King, C. E., 237, 241
Kipling, C., 200, 259–261
Kirchner, W. B., 140
Kittel, T., 53, 123, 133, 145, 153
Klemer, A. R., 203
Kling, H., 145
Knauer, D. R., 390
Koehl, M. A. R., 235
Kofoid, C. A., 4
Kowalczewski, A., 279, 290, 310
Koyama, M., 145, 153
Kozhov, M., 107, 110, 123, 133, 145,
    210, 244
Kramer, R. H., 312
Krohn, K., 116
Krueger, D. A., 94
Kuznetsov, S. I., 6, 119, 130

Lack, T. J., 279, 290
Ladle, M., 308
Lagler, K. F., 188–192, 253, 271
Lallatin, R. D., 138, 145
Lange, W., 155, 160

Larimore, R. W., 253
Larsen, D. P., 138, 393, 395–396
Larsson, K., 399
Lastein, E., 109
Lauff, G. H., 5, 324, 329, 336,
    339–340, 348
Laville, H., 246
Lazrus, A. L., 150, 152, 164
Lean, D. R. S., 134, 141, 155–156
Leckie, J. O., 135
LeCren, E. D., 200, 259–261
Lee, D. H. K., 124
Lee, G. F., 118–119, 122
Leeuwenhoek, A. van, 2–3
Lefevre, M., 155
Leibovich, S., 73
Lemmin, U., 77
Leonard, R. L., 302, 398
Leopold, L. B., 297
Lerman, A., 5, 53, 102
Leslie, J., 3
Lewin, J. C., 144
Lewin, R. A., 144
Lewis, W. M., 56
Lider, E. L., 382
Likens, G. E., 56, 93, 123, 133, 145,
    299–301, 304, 316
Lindegaard, C., 245
Lindell, T., 410
Lindeman, R. L., 4
Lindmark, G., 399
Lindström, K., 163
Linn, R. J., 291
Litt, A., 230
Livermore, D. F., 403
Livingstone, D. A., 145, 153, 193–194
Livingstone, R. J., 340
Loftus, K. H., 264
Loftus, M., 330
Lopez, M., 217
Lorenz, J. R., 3–4
Lorenzen, M. W., 401
Lowden, G. F., 104
Lund, J. W. G., 10, 149, 200, 204,
    206–211, 220, 354, 360–361,
    363, 379
Lundgren, A., 119

Macan, T. T., 145, 153, 311, 363,
    365–366, 385
McCann, J. A., 306
McCart, P. J., 384
McCarthy, J. J., 127–128
McCarty, P. L., 139
McDiffett, W. F., 316
McDonald, I., 279, 290
McElroy, W. D., 330
McEvoy, J., 7
McFarlane, R. W., 410
McGowan, L. M., 374

McHarg, I. L., 7
MacIsaac, J. J., 125
MacKenthun, K. M., 140
McLaren, F. R., 112, 123, 133, 145,
    153, 300, 315
McMahon, J. W., 156–157
Macnae, W., 339
MacRobbie, E. A. C., 164
McRoy, C. P., 139, 395
Magnuson, J. J., 271
Maitland, P. S., 267
Malueg, K. W., 138, 393, 395–396
Mangelsdorf, P. C., 324
Mann, K. H., 279, 290, 308
Manny, B. A., 128, 361
Margalef, R., 290
Marker, A. F. H., 177, 308
Marshall, N. B., 271
Marshall, S. M., 290
Martell, E. A., 150, 152, 164
Mason, D. T., 33, 36, 56, 74, 145,
    152, 376–377, 379, 381–382
Mathews, C. P., 279, 290, 310
Matthews, H. M., 141
Melack, J. M., 250
Meyer, H., 3
Mill, H. R., 54
Miller, J. P., 297
Miller, R. G., 257
Miller, R. R., 188–192, 253, 271
Milliman, J. D., 146
Min, H. S., 141
Minshall, G. W., 308, 316
Mitchell, D. S., 196, 375, 404
Mitchell, D. T., 140
Mitchell, G. H., 364
Mitchell, R. B., 125, 130, 410
Miura, T., 210
Mobius, K., 3
Moore, J. W., 316
Mordukai-Boltovskoi, Ph.D., 316
Morgan, M. D., 232–233, 241
Mori, S., 145, 153, 210, 357
Moriarty, C. M., 255
Moriarty, D. J. W., 255, 375
Morisawa, M., 297, 299
Morita, R. Y., 137
Morris, I., 126, 200
Morris, R. H., 337
Mortimer, C. H., 4, 54, 68, 81, 86,
    104–105, 151, 153
Moss, B., 148
Mosser, J. L., 384
Moyle, P. B., 257, 271
Müller, J., 3
Müller, P. E., 3
Mullin, M. M., 290
Mur, L. R., 125
Murphy, C. R., 67
Murphy, T. P., 155–156
Murray, J., 3, 5

Myrup, L. O., 53, 79

Nalewajko, C., 155–156
Nanomura, A., 330
NASA (National Aeronautics and Space
    Administration), 322
National Academy of Sciences, 130,
    361
Nauwerck, A., 228
Nebert, N., 139, 395
Needham, J. G., 4
Needham, P. R., 384
Neess, J. C., 116
Neff, S. E., 42
Nelson, D. M., 144, 146
Nelson, T. C., 135, 137
Newbold, J. D., 20, 302
Newell, R. C., 334
Nichols, F. H., 334
Nichols, S. A., 390, 403
Nicola, S., 257
Nisbet, M., 155
Nriagu, J. O., 141

Ochkivskaya, M. V., 124
Odum, E. P., 278, 324, 326, 328–329
Odum, H. T., 374
Oglesby, R. T., 123, 133, 138, 145,
    153, 306, 401
Ohle, W., 112, 363
Ohnstad, M. A. M., 273
O'Kelley, J. C., 164
Ólafsson, J., 138
Olausson, E., 340
Omernik, J. M., 316
O'Neill, R. V., 20
Oswald, W. J., 405
Otsuki, A., 102
Overbeck, J., 169
Overholtz, W. J., 403

Paasche, E., 146
Pacha, R. E., 137
Paerl, H. W., 93–94, 147, 170, 202,
    397, 398
Paine, R. T., 265–266
Pamatmat, M. M., 112
Parker, D. M., 116
Parma, S., 361
Parsons, T. R., 88, 115, 132, 143,
    153, 273, 288
Pasciak, W. J., 200
Passino, D. R. M., 188–192, 253, 271
Pastorok, R. A., 401
Patalas, K., 225, 370
Patrick, R., 159
Patten, H. L., 358
Patterson, C. C., 388

Paulson, L. J., 251, 254
Pearsall, W. H., 129, 177, 360, 366–367
Pearson, E. A., 331, 333
Pearson, W. D., 312
Pechlaner, R., 245, 379–380
Pedretti, E. L., 232
Peek, C. A., 156
Peek, N., 116
Pennak, R. W., 181, 194, 294, 316
Pennington, W., 177
Perkins, E. J., 5, 340
Perkins, M. A., 398
Petersen, R. C., 310, 316
Peterson, J. O., 390
Pfeifer, H., 245, 379–380
Pfeifer, R. F., 316
Phillips, J. E., 133
Phinney, H. K., 156
Pickard, G. L., 68
Pierce, R. S., 93, 123, 133, 299–300, 304, 316
Pigeon, R. F., 374
Pollingher, U., 220
Polmiluyko, V. P., 124
Pomeroy, L. R., 141
Pond, S., 68
Porter, K. G., 236, 247
Porter, R., 141
Powell, T. M., 53, 76, 79, 82
Powers, C. F., 393
Prepas, E., 241
Preston, T., 286
Prichard, D. W., 348
Priesendorfer, R. W., 38
Priscu, J. C., 125
Provasoli, L., 94
Pullar, L., 3, 5

Ragotzkie, R. A., 53, 133, 145, 369
Ralston, C. W., 92
Ranwell, D. S., 340
Ravera, O., 239
Rawson, D. S., 23, 271, 363
Rebsdorf, A., 109
Redfield, G. W., 226
Regier, H. A., 264
Reid, G. K., 5
Reid, P. C., 236
Remane, A., 320–321
Remsen, C. C., 128
Reuter, J. E., 125
Reynolds, C. S., 141, 200–201, 214, 218, 286
Richards, R. C., 224, 398
Richardson, J., 193–194
Richerson, P. J., 7, 53, 123, 133, 145, 153, 217
Richter, D. D., 92
Ricker, W. E., 271

Riggs, D. S., 283
Rigler, F. H., 141, 241
Ringler, N. H., 300
Ripl, W., 399, 403
Robertson, S. R., 382
Roby, K. B., 302
Rodhe, W., 11, 152, 163, 219, 290
Rogers, J. N., 125
Romanenko, V. I., 6
Ross, C., 49–50, 67, 86
Roth, J. C., 8, 42, 133, 145, 227–228, 232, 246, 300, 315, 369–370, 385, 403
Round, F. E., 373
Rudd, J. W. M., 145
Rudd, R. L., 267
Russell, F. S., 236
Russell, P. P., 409
Russell-Hunter, W. D., 196, 235
Ruttner, F., 4–5, 363
Ruttner-Koliska, A., 181

Salt, G. W., 182, 184, 223
Salvia, J. D., 388
Sanger, J. E., 360
Saunders, G. W., 94
Schelske, C. L., 133, 145–146, 152, 369–370, 385
Schindler, D. W., 9, 145
Schladow, G., 80
Schlieper, C., 320–321
Schmidt, G. W., 169
Schroeder, W. C., 338
Schuldt, M. D., 393
Schults, D. W., 138, 393, 395–396
Scott, J. T., 77
Secchi, P. A., 30
Seliger, H. H., 330, 332
Selleck, R. E., 331, 333, 409
Sellers, C. M., 159
Sellery, G. C., 4
Sens, S. L., 390
Serruya, C., 220
Shapiro, J., 113, 155, 226, 285
S.I.L. (Societas Internationalis Limnologiae), 220
Simony, F., 5
Simpson, D. G., 102, 382
Singley, P. T., 20
Skabitchewsky, A. P., 210
Skoog, F., 115, 152
Sly, P. G., 123, 133, 145, 369
Smayda, T. J., 331
Smith, I. R., 60, 63, 68, 86, 123, 373
Smith, R. A., 7
Smith, R. C., 34, 38
Smith, R. F., 338
Smith, S. A., 390
Smith, S. H., 262
Smyly, W. J. P., 230

Solórzano, L., 115
Sørensen, J., 116, 130
Southgate, B. A., 320
Southward, A. J., 236
Spacie, A., 268
Spencer, J. M., 140
Stanford, J. A., 316
Stark, R. W., 20
Steele, J. H., 286, 291
Steemann-Nielsen, E. L., 160, 405
Stewart, J. A., 228
Stewart, K. M., 54–56
Stewart, W. D. P., 116, 121, 126, 128, 135, 144, 164, 286
Stillinger, F. H., 27
Stoermer, E. F., 146
Stone, R. W., 151
Storrs, P. N., 331, 333
Strickland, J. D. H., 88, 115, 132, 143, 153, 273, 288, 291
Strickler, J. R., 235
Stuart, T. J., 85
Stull, E. A., 82
Stumm, W., 135–136
Sugawara, K., 113, 116
Swale, E. M. F., 316
Swift, M. C., 256
Sylvester, R. O., 401

Talling, J. F., 38, 56, 101, 113, 290, 363
Tansley, A. G., 4
Teal, J. M., 291
Tevlin, M. P., 255
Thienemann, A., 4, 241
Thomas, W. H., 125, 282
Thompson, G. A., 237
Thorpe, S. A., 60, 70, 83–84, 86
Threlkeld, S. T., 225, 233, 241
Tilman, D., 137, 146, 286
Tilzer, M. M., 99, 245, 287, 289–290, 353, 379–380
Timm, T., 242
Toetz, D. W., 126
Tomlinson, T. E., 305
Toms, R. G., 306
Tonolli, L., 6
Tonolli, V., 6
Torrey, M. S., 119
Truesdale, G. A., 104
Trussell, R. P., 127
Tubb, R. A., 403
Tuite, C. H., 194
Tullett, P. A., 312
Tunzi, M. G., 290
Tyler, J. E., 30, 34, 38
Tyler, M. A., 332

Valentine, R., 409

Vallentyne, J. R., 358
van Bennekom, A. J., 338
Viets, F. G., 128
Vigg, S., 382
Villeneuve, J. P., 145
Vinberg, G. G., 6
Viner, A. B., 119, 123, 133, 145, 195, 373–375
Vinogradskiy, S. N., 6
Visser, S. A., 145
Vollenweider, R. A., 38, 199, 273, 288, 290, 361

Waananen, A. O., 299
Walker, W. W., 113
Walsby, A. E., 121, 128, 200–201
Ward, A. K., 150
Ward, H. B., 196
Ward, J. V., 316
Warren, C. E., 410
Watson, S. W., 128
Watt, W. D., 74
Weatherley, A. H., 159
Weber, C. I., 268, 410
Weissenbach, H. P., 245, 379–380
Weitzel, R. L., 410
Welch, E. B., 400, 410

Welch, H., 276, 278
Welch, P. S., 4, 19–20, 363
Weldon, L. W., 196
Werner, D., 201
Werner, E. E., 9
Westlake, D. F., 308
Wetzel, R. G., 5, 94, 102, 113, 150, 153, 324, 353, 361, 385
Wheeler, A., 408
Whipple, G. C., 196
White, G. F., 407
Whitton, B. A., 5, 133, 159, 164, 203, 299, 306, 316, 385
Wickstrom, C. E., 385
Widmer, C., 123, 133, 145, 153
Wild, P. W., 409
Williams, L. R., 155
Williams, N. J., 161, 399
Williams, R. C., 299
Williams, W. D., 375, 381, 385–386
Willoughby, L. G., 196
Winberg, G. G., 291
Winslow, J. H., 7
Winter, D. R., 390
Winter, T. C., 358
Wirth, T. L., 390
Wium-Anderson, S., 160, 405
Wolman, M. G., 297

Wood, B. J. B., 33, 74, 376–377, 379
Wood, R. D., 5
Worthington, E. B., 54, 407
Wright, H. E., 358, 360
Wright, J. C., 240
Wright, R. T., 282
Wrigley, R. C., 61–62, 85, 199, 216, 220
Wulff, F., 321
Wunderlich, W. E., 403
Wurtsbaugh, W. A., 156–157, 161, 257
Wuycheck, J. C., 310, 316

Xypolyta, A., 200

Yamamoto, 177
Young, J. O., 307

Zaret, T. M., 265–266
Zenkevitch, L., 333
Zeutschel, R. P., 290
Zevenboom, W., 125
Zollinger, E. S., 136

Subject Index

Subject Index

Abert Lake (Oreg.), 343

Acartia, 331

Acid rain, 7–8, 99, 388

Acidity (*see* pH)

Acipenser fulvescens (sturgeon), 187–190, 249, 255, 406

Actinomycetales, 170

Advective transport, 66–69

Aechmophorus occidentalis (Western grebe), 267

Aeration and mixing, 97, 387, 400–403

Agricultural practices (*see* Cultural eutrophication; Pollution)

Air narcosis (bends), 118, 268

Albedo (reflection of light), 25, 36–37

Albert, Lake (East Africa), 343

Algae:
    estuarine, 318, 319, 333
    lake, 2, 9, 19, 30, 71, 88, 222, 254, 255, 277
        as base of aquatic food chain, 166
        calcium and, 147, 149
        chlorine and, 151
        classification of, 171–177
        copper and, 88, 160, 163
        copper sulfate to control, 387, 405–406
        dissolved organic nitrogen and, 128, 129

Algae, lake *(Cont.)*:
    in eutrophic lakes (*see* Eutrophic lakes, algae in)
    excretion by, 89
    harvesting of, 387, 405
    iron and (*see* Iron, algae and)
    in kettle lakes, 78
    lake color and, 36
    in Lake Tahoe, 398
    in Lake Washington, 391, 398
    light and (*see* Light, algae and)
    macrophytes and concentrations of, 179
    magnesium and, 150
    manganese and, 160
    measurement of, 389, 390
    molybdenum and, 163
    nitrates and, 124, 210, 212, 215, 216
    nitrogen and, 115
    nitrogen fixation and (*see* Nitrogen fixation, by algae)
    oxygen and carbon dioxide production and, 97–98, 107
    pH and, 148
    phosphorus and, 131, 132, 135–137, 139–140, 393
    photosynthesis by, 101, 107
    place of, in biological pyramid,

Algae, lake *(Cont.)*:
    275
    in polar lakes, 377
    potassium needs of, 142
    protozoans and, 180
    in Shagawa Lake, 392–393, 398
    silica needs of, 143
    in temperate lakes, 364
    thermal bar and, 51–52
    in tropical lakes, 372
    viral attacks on, 166
    water movement and, 59, 71, 73, 75
    watershed and, 19
    in windrows, 73
    zonation and, 15
    zoobenthos feeding on, 222
    zooplankton grazing on (*see* Grazing by zooplankton)
    red tides and, 33, 89, 198, 218–220, 330
    stream and river, 20–21, 292, 293, 312, 385
    (*See also* Attached algae; Blue-green algae; Chlorophyceae; Phytoplankton; *specific algae*)

Alkaline phosphatases, 89, 131, 133, 136–137, 349

Alkalinity (*see* pH)

Allelochemistry, defined, 89
Alligators, 193, 249, 388
Allochthonous materials in streams and rivers, 292, 293
Alnus tenuifolia (mountain alder), 90
Alosa pseudoharengus (alewife), 249, 262–265
Alpine lakes (*see* Polar and alpine lakes)
Aluminum, 134
Amazon River (South America), 9, 179, 302, 319
Amictic lakes, 43, 44, 381
  (*See also* Polar and alpine lakes)
Amids, 190
Ammonia and ammonium ion (NH_3 and NH_4), 87, 91, 126–128, 144, 169, 399
  in lakes: blue-green algae and, 107, 125, 212, 216
    denitrification of, 121–122
    effects of aeration on, 401
    enzyme uptake kinetics of, 123–125
    in eutrophic lakes, 122, 126, 128
    excretion of, 97, 126, 226
    measurement of, 115
    nitrates and, 114, 115, 121, 122, 126, 408–409
    nitrogen fixation and, 120
    nitrogenase and, 118
    pH and, 100, 127, 128
    rainfall and, 127
  in streams and rivers, 127, 128
Ammonification, defined, 122
Amphibians, 192–196
  (*See also specific amphibians*)
Amphipod crustaceans, 224, 242, 309, 335, 385
  (*See also specific amphipods*)
Amplitude of waves, 63, 78
Anabaena (a blue-green alga), 118–120, 198, 212, 213, 215, 216, 271, 382, 392
Anabaena circinalis, 218
Anaconda, 193
Anadromous fish, 191, 192, 248
  (*See also specific anadromous fish*)
Analog experiments, 9
Anchor ice, 301, 383
Angiosperms, 177–179
  (*See also* Macrophytes)
Anguillids (eel family), 191–192, 199, 252
Animal life (*see specific types of animal life*)
Animal plankton:
  discovered, 3
  (*See also* Zooplankton)
Annelids, 222
  (*See also specific annelids*)
Annual heat budget (maximum heat content), 54–55

Annual variations in zooplankton, 226–232
Anostracans, 381
Antimetabolite, defined, 89
Aphanizomenon (a blue-green alga), 37, 118, 119, 201, 203, 212–216, 219, 258, 371, 392
Aphanizomenon flos-aquae, 367
Aphotic zone (*see* Hypolimnion)
Apparent color of lakes, 25, 33, 35–36
Applied limnology, 387–410
  case studies of, 391–400
  diversion used in (*see* Diversion)
  harvesting and chemical control in, 387, 403–406
  measurements in, 389–390
Arctic and Antarctic lakes (*see* Polar and alpine lakes)
Arid lakes (*see* Tropical lakes)
Arrhythmic (nonperiodic) currents, 57
Arsenic, 163
Artemia salina (brine shrimp), 231, 381, 382
Arthropods, 29, 182
  (*See also specific arthropods*)
Artificial reoxygenation (aeration), 97, 387, 400–403
Asplanchna (a rotifer), 181, 182, 233–234, 236, 238
Asplanchna priodonta, 231, 232
Asterionella (a diatom), 55, 146, 197, 200, 210, 211, 258, 364, 371, 391, 392
Asterionella formosa, 173, 203–210, 285, 286, 387
Aswan High Dam (Egypt), 298
Atitlan, Lake (Guatemala), 265
Atlantic salmon, 191
Atmosphere:
  and chemical composition of water, 90–92
  as source of ammonia, 128
  as source of iron, 157–158
  as source of oxygen and carbon dioxide, 96–97
Atmospheric light, 31–32
Attached algae (periphyton), 18, 166, 176–177
  in estuaries, 326
  in lakes, as measure of lake restoration, 391
  in streams and rivers, 307, 308, 354
  (*See also specific attached algae*)
Aufwuchs, 18, 278, 294, 327, 376
Autochthonous materials in rivers, 292
Avicennia (a mangrove), 337
Azolla (a water fern), 121, 179

Bacteria, 18, 30, 74, 89, 94, 170
  in distilled water, 88
  in estuaries, 169, 319, 320
  in lakes, 75, 203, 277

Bacteria, in lakes (*Cont.*):
  classification of, 167, 168
  denitrification by, 82, 115, 116, 121–122, 167
  effects of thermal pollution on, 56
  general characteristics of, 166–169
  iron and, 155
  lake color and, 36
  in Lake Tahoe, 397
  light and, 30
  phosphorus and, 132, 137
  protozoans and, 180
  role of, in oxidation-reduction reactions, 105
  temperature and activity of, 46
  viral attacks on, 166
  zoobenthos feeding on, 222
  zooplankton grazing on, 221, 232
  in streams and rivers, 21, 169, 308, 310–313, 385
Baikal, Lake (U.S.S.R.), 106, 195, 350, 351, 368
  fish in, 261, 269–270
  heat in, 48
  origin of, 343
  phytoplankton in, 210
  zooplankton in, 242
Bar-built estuaries, 349
Basin, lake, 12, 90, 342, 343, 345, 346
Bass, 276, 356
  (*See also specific species of bass*)
Batch culture, defined, 282
Beavers, 195, 297, 348
Beetles, 184
Belostomatidae (giant water bug), 184
Bends (nitrogen and air narcosis), 118, 268
Benthic invertebrates (*see* Benthic organisms)
Benthic organisms, 9, 184, 400
  defined, 18
  in estuaries, 320, 322
  in lakes: classification of, 186–187
    effects of aeration on, 401, 403
    effects of dams on, 406–407
    feeding patterns of benthic fish, 254
    light and, 29
    respiration of, 106
  in streams and rivers, 20–21
  (*See also* Benthos; Zoobenthos; *specific benthic organisms*)
Benthos:
  defined, 18
  of Laurentian Great Lakes, 373
  (*See also* Zoobenthos)
Big Cedar Lake (Wash.), 347
Big Soda Lake (Nev.), 345
Bilharzia (schistosomiasis), 187, 372
Biological pyramid, 272, 274–275
Biological structure, 18–22
  of lakes, 18–19

Biological structure (Cont.):
  of streams and rivers, 20–22
  (See also specific life forms)
Birds:
  in estuaries, 333, 336
  in lakes, 52, 192–196, 327, 377,
    382, 383
  (See also specific birds)
Birth rate (natality) of zooplankton,
  238, 239
Bison, 195
Bivalve (lamellibranch) mollusks, 185
  (See also specific bivalves)
Biwa, Lake (Japan), 210, 343, 359
Black bass, 191, 267
Black crappie, 265
Blue baby, 124
Blue-green algae, 33, 36, 60, 188
  in estuaries, 327
  in lakes, 89, 179, 220, 254
    ammonia and, 107, 125, 212, 216
    copper and, 160
    dissolved organic nitrogen and,
      129
    in eutrophic lakes, 201, 212, 215,
      390
    harvesting of, 405
    iron and, 155, 157, 213, 215, 216
    lake color and, 36
    in Lake Trummen, 399
    in Lake Washington, 391, 398
    in Laurentian Great Lakes, 371
    movement of, 61, 175, 201–203
    nitrogen and, 115, 125, 215
    nitrogen fixation and, 116, 121
    photosynthesis by, 101
    in polar and alpine lakes, 376,
      377, 379
    reflection and, 36–37
    in salt lakes, 381–383
    seasonal cycles of, 204, 212–216
    in Shagawa Lake, 392, 394, 398
    silica shortage and, 146
    sodium and, 150
    spatial variations in, 217–218
    in tropical lakes, 373–375
    viruses and, 166
    zooplankton grazing on, 232, 236
  in streams and rivers, 307
  (See also specific species of blue-
    green algae)
Bog (dystrophic) stage of lakes, 89,
  350
Bogs, color of, 89
Bonneville Salt Flats (U.S.), 19
Bonney, Lake (Antarctica), 43, 111,
  377
Bosmina (a cladoceran), 228, 231, 233,
  359, 371, 373
Bosmina longirostris, 232, 233
Botryococcus braunii (a green alga),
  201
Bottom color of lakes, 33

Bottom-dwelling organisms (see In-
    fauna; Zoobenthos)
Bowfin, 190
Brachionus (a rotifer), 233
Brackish waters, 2, 319
Branta bernicla (brant goose), 336
Breaking waves, 58, 69, 74–75, 79
Brevoortia (menhaden), 336
Brine flies, 382
Brine midges, 382
Bromine (Br), 90
Brooks Lake (Alaska), 150, 298
Brown trout, 270
Bruguiera (a mangrove), 337
Brunt-Vaisala frequency (N), 79
Buffalos, 195, 372

Caddisflies (see specific species of
    caddisflies)
Caimans, 193
Calanoid copepods, 182, 221, 229,
  233, 236, 238
Calciphile, defined, 149
Calciphobe, defined, 149
Calcium (Ca), 96, 101, 134, 135, 142,
  144, 147–149, 160, 318
Caldera lakes, 344, 345
Callinectes sapidus (blue crab), 336
Cancer magister (Dungeness crab),
  334, 336
Cape buffalo, 195
Capybara, 405
Carbon (C):
  as base of food pyramid, 286
  eutrophication and increase in, 349
  paleolimnologic measurement of, 358
  in tropical lakes, 373
Carbon dioxide (CO_2), 87–88
  in lakes: in Lake Tahoe, 397
    mosses and, 178
    primary-productivity measurements
      and, 287, 288
    as source of global pollution, 92
  (See also Oxygen, and carbon
    dioxide)
Carbonic anhydrase, 100, 160
Carex (sedge), 178, 326
Carnivores, 166, 311–312
  (See also specific carnivorous
    organisms)
Carotenes, 171
Carp, 190–191, 308, 316, 356, 388,
  405
  pollution and, 191, 267, 268
  stocking with, 249, 270
  uses of, 265
Castle Lake (Calif.), 9, 43, 90, 108
  fish in, 256
  nutrients in, 159, 163
  origins of, 346, 351, 359
  water movement in, 82
  zooplankton in, 231

Catadromous fishes, 191–192, 248,
  317–318
Caterpillars, 311
Catfish, 252, 255, 256, 267, 268
Catostomids (sucker family), 191, 250,
  255, 313–314
  (See also specific catostomids)
Ceanothus bushes, 116
Cell shape of phytoplankton, 200–201
Cellulase, 310
Centrarchids (sunfish family), 191,
  250, 268, 270, 276
Cephalaspidomorphi, 187
  (See also specific
    cephalaspidomorphs)
Ceratium (a dinoflagellate), 163, 175,
  176, 180, 198
Ceratophyllum (coontail), 179, 281
Ceriodaphnia (a cladoceran), 228, 233
Chad, Lake (Africa), 19, 20, 343
Chaetognaths, 184
Chalk (hard-water) streams, 293,
  306–307
Chaoborus astictopus (lake gnat), 267
Chaoborus and Chaoborus larvae
    (phantom midge), 184, 221,
    232, 242, 246, 374
Chaotic (turbulent) eddies, 65, 276
Chara (skunk weed), 178, 376
Charophyceae group of algae, 177
Chelating agents, 88, 89
Chelator, defined, 89
Chemical control of lakes, 403–406
Chemical structure:
  of lakes, 17–19, 87–94
  of streams and rivers, 20, 293
Chemicals (see specific chemicals, for
    example: Copper sulfate; spe-
    cific entries, for example: Inor-
    ganic chemical compounds;
    Organic chemical compounds)
Chemocline, defined, 17
Chemolithotrophic bacteria, 169
Chemostat (continuous-flow culture),
  282
Chemotroph, defined, 166
Chesapeake Bay (U.S.), 319, 330, 334,
  354
Chilwa, Lake (Malawi), 382
Chironomids, 243, 245, 280, 358, 381
  larvae of, 184, 241, 242, 277, 312,
    379, 400
Chironomus (a midge), 246
Chironomus anthracinus, 243, 245, 246
Chironomus islandicus, 245
Chironomus lakes, 241
Chishi, Lake (Zambia),
  193–194
Chlorella (a green alga), 124, 180, 236
Chloride, 88, 381
Chlorine (Cl), 90, 142, 145, 150–151
Chlorophyceae (green algae family),
  198, 330

Chlorophyceae *(Cont.)*:
  estuarine, 327
  in lakes, 89, 146, 175–177, 254, 356
    high-rate photosynthesis and, 98
    in polar lakes, 376
    reflection and, 37
    in Shagawa Lake, 392
    spatial variations in, 216
  in streams and rivers, 307, 354
  (See also specific green algae)
Chlorophyll, 89, 149, 163, 167, 316,
    358, 390
Chlorophyll *a*, 142, 171, 175, 199,
    282, 287, 372, 389, 394
Chondrichthyes, 187
Chromatic adaptation, defined, 33
Chrysophycean flagellates, 143
  (See also specific flagellates)
Chrysophyte algae, 163, 198
Chytrid (a fungus), 169, 198, 207,
    210–212
Cichla ocellaris (peacock bass), 265
Cichlids, 191, 250, 269
Ciliates, 203, 208, 330, 381
Cirque lakes, 346
Cisco fish, 262–265
Cladocerans:
  in estuaries, 333
  in lakes, 163, 182–184, 221, 275
    feeding by, 233
    fossils of, 358
    in Lake Trummen, 399
    in Laurentian Great Lakes, 371
    measurement of, 222
    paleolimnological investigation of,
      359
    in polar lakes, 378
    population dynamics of, 238, 240
    reproduction of, 236
    in salt lakes, 381
    seasonal variations in, 226,
      228–232
  in tropical lakes, 375
Cladophora (blanket weed), 177, 245,
    354
Clams, 222, 322, 330, 333, 334
Clay, 134, 135, 140, 144
Clear Lake (Calif.), 367, 371, 374
  algal control in, 405
  fish in, 258, 267
  nutrients in, 119
  origins of, 343, 346, 351, 359
  oxygen and carbon dioxide in, 98
  phytoplankton in, 201, 212–216,
    219, 220
  water movement in, 71
  zonation in, 15
  zooplankton in, 246
Climate:
  cycles of, 389
  effects of, on available iron, 152
  effects of water presence on, 28

Climate *(Cont.)*:
  influence of, on chemicals, 90
  and lake succession, 350
  lake watershed and, 19
  (See also Temperature)
Clinograde distribution of oxygen,
    109–111
Clupea harengus harengus (Atlantic
    herring), 336
Clupea harengus pallasii (Pacific her-
    ring), 336
Clupeids, 190, 269
  (See also specific clupeids)
Clyde River (Scotland), 267
Coarse (rough) fish, 251–252
Cobalt (Co), 143, 149, 155, 159, 163
Cocconeis (a diatom), 275
Coho salmon, 267
Cohort of fish, 258, 259
Cold-water streams, 383–385
Coliform bacteria, 169
Color:
  of bogs, 89
  of estuaries, 33
  of lakes: apparent, 25, 33, 35–36
    of dystrophic lakes, 89, 350
    light and, 33–36
    phytoplankton and, 199
Colorado River (U.S.), 382
Columbia River (U.S.), 112, 406
Communication, chemical, 88–90
Comparative limnology, 7, 9, 363–386
  defined, 362
  measurement methods used in,
    363–364
  of polar and alpine lakes, 376–380
  of salt lakes, 380–382
  of streams and rivers,
    383–385
  of temperate lakes, 364–372
  of tropical lakes, 372–376
Compensation depth, 97–98
Configurational close packing, effects
  of, on water molecule structure,
    27
Congo River (Africa), 302
Constance, Lake (Germany), 6
Continuous-flow culture (chemostat),
    282
Convective mixing, epilimnion stirred
  by, 58
Cooling:
  of epilimnion, 47–48
  (See also Evaporation)
Copepods:
  in estuaries, 333
  in lakes, 18, 182–184, 222, 239, 275
    in polar and alpine lakes, 379
    reproduction of, 236, 238
    in salt lakes, 381
    seasonal variations in, 226, 228,
      229

Copepods, in lakes *(Cont.)*:
    in tropical lakes, 374–375
  (See also specific copepods)
Copper (Cu), 88, 123, 143, 153,
    159–163
Copper sulfate, algal control with, 387,
    405–406
Coral reefs, 147, 232
Coregonus (whitefish), 254
Coregonus artedii (lake herring), 262,
    263, 265
Coregonus clupeaformis
    (whitefish), 190, 249, 254,
    262–265
Coregonus hoyi (bloater), 265
Coregonus nigripinnis (blackfin cisco),
    265
Coriolis effect, 69
Coriolis force, 71, 75, 80
Cormorants, 194
Corophium (a crustacean), 333, 334
Corvina, 382
Cottids (sculpin family), 255, 256, 316
Coypu, 195
Crabs, 181, 320, 333, 335, 337
  (See also specific crabs)
Crangon (a shrimp), 333
Crappies, 191, 265
Crassostrea virginica (common Ameri-
    can oyster), 334, 336
Crater Lake (Oreg.), 345, 349, 350,
    399
Crater lakes, 345
Crayfish, 18, 29, 181, 243, 248, 249,
    253, 280, 309, 353, 405
Crickets, 311
Crocodiles, 193–195, 249, 338, 372,
    388
Crocodilus niloticus (Nile crocodile),
    193–194
Crose Mere (England), 218
Crustaceans:
  in estuaries, 317, 323–324, 333
  in lakes, 134, 181–184, 194, 221,
    229, 371, 378, 381
  in streams and rivers, 307, 311
  (See also specific crustaceans)
Cryogenic (thaw; thermokarst) lakes,
    347
Cryptomonads, 217
Cryptophyte algae, 198
Cultural eutrophication, 98, 318, 349,
    354–358
    blue-green algae as first sign of, 212
    (See also Blue-green algae)
    management of, 388
    silica shortage and, 146
Cultural oligotrophication, 356
Culture of rice, 121, 179, 249, 301
Currents:
  lake: in epilimnion, 69
    kinetic energy spectrum of, 65

Currents, lake *(Cont.)*:
  and long-period internal waves, 79
  measurements of, 59–61
  overall view of patterns of, 82–83
  types of, 57, 58
  in streams and rivers, 20, 294, 297
Cutophages, 167
Cuyahoga River (U.S.), 306
Cyanophage viruses, 198
Cyclomorphosis, defined, 225
Cyclopoid copepods, 182, 229–230,
      236, 371, 374
Cyclops (a copepod), 364, 367, 371
Cyclops abyssorum, 229–231, 368
Cyclostomes, 187
Cyclotella meneghiniana (a diatom),
      285, 286
Cyperus papyrus (papyrus), 178, 312,
      375–376
Cyprinids (minnow family), 190–192,
      194, 248, 250, 254, 313–314,
      400
  (See also specific cyprinids)
Cyprinodon (pupfish), 382

Dace, 180, 256
Dalton's law, 97
Dams and reservoirs, 1, 7, 97, 118,
      261, 268–270, 298, 319, 348,
      404–406
Daphnia (water flea), 66, 73, 127, 182,
      222, 229–231, 233, 235–240,
      254, 371, 375
Daphnia pulicaria, 232, 233
Daphnia rosea, 225, 231–233
Data sets, 7
DDD, 267
DDT, 92, 267
Death rate (mortality) of zooplankton,
      238, 241
Deep scattering layers, 225
Deforestation and logging,
      92–94, 261, 269, 354, 356, 391
Denitrification, 82, 115, 116, 121–122,
      167
Density:
  of phytoplankton, 201
  of water, 27, 28, 44, 45
Depth:
  of estuaries, 324, 330
  of lakes, 15
    chemical structure and, 17, 18
    color and, 33, 36
    compensation depth, 97–98
    diversion and, 391
    heat and, 39, 42, 43, 53–55
    of Lake Tahoe, 395, 397
    of Lake Washington, 391
    light and, 25
    macrophytes and, 178, 179

Depth, of lakes *(Cont.)*:
    management and, 395
    nitrogen, nitrate and nitrite distri-
      bution and, 116, 117, 122
    of oligotrophic lakes, 341
    oxygen distribution and, 108–111
    phosphorus and, 139
    of polar lakes, 377
    Secchi depth, 30, 36, 386–390,
      392, 394
    thermal pollution and, 56
    thermocline and, 47
    of tropical lakes, 372
    water movement and, 59, 71
    zoobenthos and, 242, 243,
      245–246
Derbyshire neck (goiter), 90–91
Desmids, 232
Detergents *(see* Pollution)
Detritivore, defined, 134
Detritivorous fishes, 248, 249, 254–255
Detritus, 75
  estuarine, 317, 320, 322, 333–334
  lake, 277, 280
    algal growth on, 176
    contributed by macrophytes, 179
    conversion of, 185
    decomposition of, 103
    fungi and, 170
    zoobenthos and, 222, 232, 243,
      245
  stream and river, 20, 21, 293,
      308–311, 313
Diamesa (a chironomid), 385
Diapause, defined, 226
Diaptomus (a copepod), 226, 228–231,
      371
Diaptomus ashlandi, 229
Diaptomus nauplii, 226
Diaptomus novamexicanus, 231
Diaptomus shoshone, 233
Diatomaceous earth, 172
Diatoms, 37, 389
  estuarine, 326, 330, 331, 333
  frustules of, 142, 143, 146–147,
      172, 359
  lake, 198, 277, 280
    attached, 176
    bacteria and, 169
    characteristics of, 175
    fossils of, 358
    in Lake Washington, 391
    movements of, 175, 200, 203
    nitrate and ammonia uptake by,
      125
    photosynthesis by, 127
    reflection and, 37
    in Shagawa Lake, 392
    seasonal cycles of, 204–210, 283,
      285
    silica and, 142–144, 146–147
    spatial variations in, 216, 217

Diatoms, lake *(Cont.)*:
    thermocline formation and, 43
  in streams and rivers, 307, 308
Diel variations:
  in chemical composition of rivers, 293
  in extracellular products of photosyn-
      thesis, 289
  iron and, 50, 157–158
  oxygen and carbon dioxide and,
      105–107, 111, 112
  in streams and rivers, 307, 308
  in zooplankton population, 225
Diffuse (indirect) sunlight on lake sur-
      face, 32
Diffusive transport, 66–69
Dimictic lakes, 52, 54, 55, 392
  defined, 39, 43
  phytoplankton in, 210, 211
  temperate, 364
  (See also specific dimictic lakes)
Dinobryon sertularia (an alga), 163
Dinoflagellates:
  in estuaries, 332, 340
  in lakes, 32, 89, 173, 198, 232
    lake color and, 36
    movement of, 203
    nitrate and ammonia uptake by,
      125
    in polar and alpine lakes, 379
    red tides and, 89, 198, 218–220,
      330
    seasonal variations in, 204
  in streams and rivers, 308
Discharge:
  defined, 294
  river, 20, 297–299, 303
  waste heat, 55–56, 267
Dissolved organic nitrogen (DON),
      128–130
Dissolved organic phosphorus (DOP),
      132, 133
Dissolved oxygen, 389
  in Thames River, 96, 407, 408, 410
  trace elements, iron and, 158
Dissolved phosphate, 132
Dissolved total phosphorus, 132
Distilled water, 88
Diversion:
  by aeration and mixing, 97, 387,
      400–403
  case study of, 391–392
  by flushing, 387, 400
  importance of, 390–391
  from Lake Tahoe, 397, 398
Dobsonflies, 184
Dolomite, 149
Dolphins, 61
DON (dissolved organic nitrogen),
      128–130
Donk Lake (Belgium), 231
DOP (dissolved organic phosphorus),
      132, 133

Downwelling, 52, 58, 73
Dragonfly larvae, 18
Dredging of Lake Trummen, 399–400
Drift, 20–21, 58, 69–71, 293, 311–316
Drift-basin lakes, 347
Drifters, current measurements with, 60
Dunaliella (a phytoplankter), 381
Dynamic food web, 272–273, 278, 279
Dystrophic (bog) stage of lakes, 89, 350

Earthworms, 311
ECPP (extracellular products of photo-synthesis), 288–290
Eddies, 61, 65–66, 276
Eddy diffusion, 67–68
Eddy diffusion constant *(K)*, 67
    of silica, 146
    (*See also* Michaelis-Menten kinetics)
Edward, Lake (Africa), 243
Egrets, 194
Eichhornia crassipes (water hyacinth), 179, 404
Ekman drift, 71
Ekman spiral, 58, 72
Electric eels, 252
Electrofishing, 252–253, 264
Elephants, 195
Emerald Bay (Lake Tahoe), 232
Emergent macrophytes, 18, 178–179
    (*See also* Macrophytes)
Energetic eddies, 65
Energy:
    regulation of flow of, 273–281
    (*See also* Heat; Light; Wind)
Energy cascades, 65
Energy spectrum, defined, 57, 65
English Lake District, 9, 101, 120, 126, 169, 364–367, 379, 391
    comparative limnological study of, 364–366
    fish in, 258, 259
    heat in, 43, 54, 55
    nitrogen levels in, 120
    oxygen and carbon dioxide in, 106
    paleolimnology of, 359–360
    phytoplankton in, 207–212
    zooplankton in, 229
Entrainment of hypolimnion, 48–50, 139
Ephemeroptera (mayfly family), 184, 241, 311, 313
Ephippium, defined, 222
Ephydrid flies, 381
Epibenthic organisms, 18, 29
Epilimnion (mixed layer), 16, 17
    defined, 39, 41
    heat and, 45–48, 55, 56
    hypolimnetic entrainment and, 50
    iron in, 158

Epilimnion (*Cont.*):
    oxygen and carbon dioxide in, 106
    pollution effects on, 55
    of temperate lakes, 364
    thermal bar and, 52
    water movement in, 58, 59, 69–75, 82
    zooplankton in, 226
Epiphytic algae, 177, 178, 363, 379
Epischura (a copepod), 226, 236
Equations of motion, 67–68
Equilibrium carbon dioxide, defined, 100
Erie, Lake (*see* Laurentian Great Lakes)
Erken, Lake (Sweden), 229, 231
Erosion, 19
    chemical composition and, 87, 90
    chemical pollution and logging and, 92–94
    cultural eutrophication and, 356
    and nitrate concentration, 115
        (*See also* Nitrates)
    and overpopulation of hippopotami, 195
    phosphorus and, 114, 140, 304
    silica and silicon and, 142
Esocids (pike family), 190, 250, 254, 258–260
    (*See also specific esocids*)
Esox lucius (pike), 258, 259
Esrom, Lake (Denmark), 243–245
Esthwaite Water (England), 169, 207, 210
Estuaries, 3, 87, 88, 187, 317–341, 403
    bacteria in, 169, 319, 320
    birds in, 194
    chlorine in, 143, 151
    color of, 33
    crustaceans in, 181, 184
    eutrophic, 322–326, 354–356, 403
    fish in, 22, 248, 249, 265, 267, 317–319, 323–324, 326, 333, 336–338, 354
    food-chain dynamics in, 290
    heat in, 39, 41
    macrophytes in, 177
    mammals in, 195
    management of, 406–410
    measurement of, 318
    nitrogen in, 115
    origin of, 342, 348–349
    oxygen and carbon dioxide in, 96, 103
    phosphorus in, 139
    phytoplankton in, 210, 318, 326–331
    potassium in, 150
    red tides in, 218
    salinity of, 28, 317–322, 333
    structure of, 22–23
    sulfur in, 151

Estuaries (*Cont.*):
    tropical, 337–339
    water movement in, 58, 61
    zooplankton in, 222, 318, 327, 331–336
    (*See also* Swamps and marshes; *specific estuaries*)
Euglena (an alga), 175
Euglenophyte algae, 198
Euphotic (photic) zone, 15, 97
    depth of, 47
    zooplankton in, 226
    (*See also* Photosynthesis)
Euryhaline organisms in estuaries, 319
Eutrophic estuaries, 322–326, 354–356, 403
Eutrophic lakes, 18, 25, 55, 88, 147, 212, 367, 371, 372, 377
    algae in, 351, 353
        Asterionella, 210
        blue-green, 201, 212, 215, 390
    ammonia in, 122, 126, 128
    bacteria in, 169
    cleaning of, 153
    defined, 341
    diversion and, 390
    fisheries in, 265, 270, 351
    insects in, 184
    iron-phosphorus-sulfur interaction in, 151
    lake succession and, 349–352
    light and, 24
    measurement of, 390
    nitrogen in, 116, 118, 122, 125, 126, 349
    nitrogen fixation in, 118, 119, 349
    oxygen and carbon dioxide in, 95, 99, 103
    pH value in, 99
    phosphorus and phosphate in, 131, 132, 151, 349
    photosynthesis in, 98, 101
    pollution and (*see* Cultural eutrophication; Pollution)
    redox potential in, 105
    surface area/watershed ratio of, 19
    trace elements in, 160
    water movement in, 71, 82
    zoobenthos of, 222, 241–243, 245, 246
    zooplankton of, 229
Eutrophic streams and rivers, 353–354
Evaporation, 52, 87, 88
    and epilimnion temperature, 47–48
    of fresh water, salt lake formation and, 380, 381
Evaporative cooling, 47, 48, 52
Excretion, 18, 222
    algal, 89
    of alkaline phosphatases, 131
    of ammonia, 87, 126, 226
    of carbon dioxide, 87

Excretion *(Cont.)*:
  by hippopotami, 374
  of iron, 158
  by macrophytes, 178
  of nitrogen, 254
  phosphate, 87, 89, 133, 137, 139, 226
  phosphorus, 89, 254
  by zooplankton, 56, 89, 139, 226, 282
Exotic predators, 265–268, 270
Extracellular products of photosynthesis (ECPP), 288–290

Facultative forms of bacteria, 167
Fahlensee (Switzerland), 345
Fall overturn, 17, 39
  of oxygen and carbon dioxide, 106–108
Fall variations *(see Seasonal variations)*
Feldspars, 142, 144
Ferns, 121, 177, 375, 404
Fertilizers *(see Cultural eutrophication; Pollution)*
Fetch of wind, 63, 78
Field measurements, 9
Field studies, laboratory studies integrated with, 7–9
Filinia (a rotifer), 233
Filtration (grazing) rate, 234
Fish eagles, 194
Fish and fisheries, 1
  in estuaries, 22, 248, 249, 265, 267, 317–319, 323–324, 326, 333, 336–338, 354
  in lakes, 18, 47, 97, 126, 187–192, 225, 243, 248–271
    changes in population of, 258–260
    classification of, 187–191
    cultural eutrophication and destruction of, 356
    effects of copper sulfate on, 405
    effects of nitrogen supersaturation on, 118
    effects of pollution on, 190, 249, 253, 265–268
    in eutrophic lakes, 265, 270, 351
    feeding by, 212, 221, 232, 248, 253–259
    and fishery conservation and management, 260–271
    as food source, 193–195
    fossils of, 358
    hypolimnetic entrainment and migration of, 48
    iron and, 158
    laminar flow utilization by, 61
    macrophytes and, 179
    as measure of lake restoration, 390
    measurement of, 250–252
    molybdenum and, 163

Fish and fisheries, in lakes *(Cont.)*:
    phosphorus and, 137
    poikilothermal, 56, 275
    in polar lakes, 378
    regulation of energy flow in, 276
    in salt lakes, 363, 382, 383
    stocking of, 249, 265–268, 270, 388
    and sustainable yield of fisheries, 286
    in tropical lakes, 372, 375
    zooplankton and, 221, 235, 248, 254
  in streams and rivers, 21, 248–250, 260, 265, 267, 270, 298, 300, 301, 306, 313, 383–385
  in Thames River, 407, 410
  (See also Fish kills; Larvae; specific fishes)
Fish hawks, 194
Fish kills, 265–268, 297, 389
  summer, 106
  winter, 55, 108, 265, 403
Fjords, 322–323, 347, 348, 354
Flagellates, 48, 125, 203, 204, 216, 283
  (See also specific flagellates)
Flamingos, 194, 232, 383
Flatworms and flatworm cocoons, 185, 359
Flies and fly larvae *(see specific flies)*
Flocculation in estuaries, 320, 333
Floods (spates), 293, 298
  river drift and, 313
Flounder, 407
Fluorescent yield, defined, 199
Fluorine, 90
Flushing, 387, 400
Fluviatile-dam lakes, 347–348
Foaming or wetting agents (organic surfactants), 29
Fontinalis moss, 178
Food chain:
  aquatic, algae as base of, 166
    (See also Algae)
  lake, dynamics of, 272–291
    kinetics of nutrient uptake and, 282–286
    measurements of, 273
    primary productivity and, 286–290
    regulation of nutrients and energy flow and, 273–281
  management of, 388
  in rivers and streams, 307–311
    dynamics of, 277, 280, 282, 290
Food web, 272–273, 275, 276, 278, 279, 286, 312
Foraminiferans, 381
Fragilaria (a diatom), 197, 206, 207, 359, 371
Frazil ice, 363, 383
Free-floating (pelagic) macrophytes, 177, 179

Free-floating macrophytes *(Cont.)*:
  (See also Macrophytes)
Fresh water, world distribution of, 25
Freshwater fish:
  hypotonicity of, 192, 319
Frogs, 192, 248, 249
Frozen lakes *(see Dimictic lakes)*
Frozen water, world distribution of, 25
Frustules (diatom), 142, 143, 146–147, 172, 359
Fungi:
  in distilled water, 88
  in estuaries, 319
  in lakes, 8, 21, 88, 89, 115, 169–172, 180
  in streams and rivers, 21, 308, 310–312
Fungi Imperfecti, 171

Galilee, Sea of (Lake Tiberias, Israel), 219
Gambusia (mosquito fish), 265, 270
Gambusia affinis (archer fish), 254
Gammarus (an amphipod), 270
Gammarus pulex, 242
Garda, Lago di (Italy), 357
Gas bubble disease, 112
Gastropods, 185, 187
  (See also specific gastropods)
Geese, 336
Genetic engineering of bacteria, 167
Geneva, Lake (Lac Léman, Switzerland), 2, 96
George, Lake (Uganda), 372–376, 400
  fish in, 254
  mammals in, 195
  movement in, 60, 71
  phosphorus in, 137
  photosynthesis in, 98
  as polymictic lake, 43
  succession of, 351
Geosmin, 89
Geothermal heating, 47
Gerrids (water strider family), 29
Gigartina (an alga), 327
Gizzard shad, 267
Glacial flour, 36
Glacial lakes, 341–343, 346–347, 358
Glacial rock basin, 346–347
Glenodinium (a dinoflagellate), 379
Gliders, 52
Gliding of phytoplankton single filaments, 203
Gloeotrichia (a blue-green alga), 118
Goiter (Derbyshire neck), 90–91
Gongrosira (a green alga), 178
Graben lakes, 343, 351
Gram-negative (and gram-positive) bacteria, 167
Grass carp (white amur), 255, 270, 308, 405

Gravel riffles, 292
Gravity waves (*see* Internal gravity
    waves)
Grazing (filtration) rate, 234
Grazing by zooplankton, 212, 221,
    231–236, 312
    on blue-green algae, 213, 220
    excretion while, 139
    on phytoplankton, 197, 198, 204,
        209
Great Lakes (*see* Laurentian Great
    Lakes)
Great Salt Lake (Utah), 100, 381
Grebes, 194, 267
Green algae (*see* Chlorophyceae)
Green Bay of Lake Michigan, 281, 371
Green Lake (Wis.), 347, 400
Green plants (*see* Algae; Macrophytes)
Ground moraine, 347
Groundwater, 92, 126, 300–301
Growth kinetics, 282–283
Guinea, Gulf of, 319
Gulf Coast, 318, 338
Gymnodinium (a dinoflagellate), 379
Gyre, defined, 71

Hagfish, 187
Halobacteria, 381
Haplochromis nigripinnis, 254, 275
Hard water, 147–149
Hard-water (chalk) streams, 293,
    306–307
Harvesting, 387, 403–406
Heat, 39–56
    flow of, 45–46
    light and, 30–31
    measurements of, 40
    and thermal stratification, 39, 41–43
        (*See also* Thermocline)
Heat cycles, 52–55
Height of waves, 63
Hemocyanin, 160
Henry's law, 97
Hepatitis, 167
Herbivores, 166, 311–312
    (*See also specific herbivores*)
Heron, Lake (Minn.), 347
Herons, 194
Herring, 190, 254
    (*See also specific herring species*)
Heterocysts, 117–118, 215–216
Heterograde distribution of oxygen,
    110, 111
Heterotrophy, defined, 128
Heterotrophs (*see specific heterotrophs,
    for example*: Bacteria; Fungi)
Hexagenia (a mayfly), 242, 247
Hippopotami, 195, 372, 374, 388
Holomictic lake, defined, 39, 43
Holoplankton, 197
Homothermous lakes, 17

Horizontal component of lake chemical
    structure, 18
Hornindalsvatn (fjord lake, Norway),
    347
Hubbard Brook (N.H.), 302
Humic acids, 88, 89, 115, 350
Huron, Lake (*see* Laurentian Great
    Lakes)
Hyalella azteca (an amphipod), 270
Hydraulic retention time, 13, 14
Hydrogen bonding, 24, 26–27
Hydrogen-ion concentration (*see* pH)
Hydrogen sulfide, 38, 381, 401, 403,
    404, 407
Hydropsyche (a caddisfly), 275, 311, 313
Hydropyrus (brine fly), 382–383
Hypertonicity of saltwater fish entering
    fresh water, 192, 319
Hyphomycetes, 310
Hypolimnetic aeration, 97, 401–403
Hypolimnetic entrainment, 48–50, 139
Hypolimnion (aphotic zone), 15, 17, 98
    defined, 39, 41
    effects of eutrophication on, 265
    heat and temperature of, 45–47, 55
    iron in, 155, 158
    nitrogen in, 114, 118
    oxygen and carbon dioxide in, 106
    phosphorus in, 139
    thermal bar and, 52
    trace elements in, 159
    upwelling and, 48
    water movement in, 58, 59, 79,
        81–82
    zooplankton in, 226
Hypotonicity of freshwater fish entering
    salt water, 192, 319

Ice:
    anchor, 301, 303
    formation of, on rivers and streams,
        383–385
    frazil, 363, 383
Ichthyophthirius (a protozoan), 181
Ictalurids, 255
Illinois River (U.S.), 4
Indirect (diffuse) sunlight on lake sur-
    face, 32
Industrial pollution (*see* Pollution)
Inertial period of long-period internal
    waves, 80
Infauna:
    in estuaries, 333, 335
    in lakes, 18
    (*See also* Benthic organisms; Zoo-
        benthos; *specific bottom-dwell-
        ing organisms*)
Inland water shrimp, 231
Inorganic chemical compounds, 88–90
Insecticides and pesticides (*see*
    Pollution)

Insects, 405
    in lakes, 184–185
        fish feeding on, 256
        and food-chain dynamics, 277,
            280
        ingestion of diatoms by, 277
        in salt lakes, 382
    in streams and rivers, 182, 184–185,
        311, 383–385
    (*See also* Larvae; *specific insects*)
Intermittency index, 82
Internal gravity waves, 57, 58, 69,
    76–78, 80
    in thermocline, 75–82
        long-period gravity waves, 79–82
        short-period gravity waves, 79
Internal loading, phosphate and phos-
    phorus, 394
Internal seiches, 58, 80
Interstitial spaces between sediment
    particles, 135
Inverse lake stratification, 17
Invertebrates:
    in estuaries, 319
    in lakes, food-chain dynamics and,
        273, 277
    in streams and rivers, 292, 293,
        306–316
    (*See also* Benthic organisms; *specific
        invertebrates*)
Iodine, 90–91
Iron (Fe), 17, 89, 132, 142, 149,
    151–159, 163
    algae and, 143, 153, 155–158
        blue-green algae, 155, 157, 213,
            215, 216
    diel variations due to transfer of, 50
    effects of aeration on, 401, 403
    epilimnetic, 48
    eutrophication and increase in, 349
    in Lake Tahoe, 398
    measurement of, 153
    in nitrite reductase, 123
    nitrogen fixation and, 118, 119, 143
    in nitrogenase, 118
    pH value and, 100
    phosphate and, 134, 135, 143, 153,
        155
    redox potential and, 105, 159
Iron cycle, 143, 153–156
Ischadium (a bivalve), 336–337
Isoetes (quillwort), 178
Isopods, 381
Isotopic tracers, 285–286

Jellyfish, 184
Jesus Lizard, 193
Jordan River (Middle East), 219
Josephine Lake (Minn.), 217
Journals, limnology, 6
Juncus (reed), 327

Kafue River (Africa), 343
Kainji Reservoir (Nigeria), 404–405
Kaolin, 135
Kariba, Lake (Africa), 179, 375
Kariba Dam (Africa), 269
KE (kinetic energy) spectrum, 63–67
Kellacottia (a rotifer), 182
Kelvin-Helmholtz instability (mixing), 70, 79
Kelvin waves, 75, 79–82
Keratella (a rotifer), 228, 229, 231, 233
Keratella hiemalis, 379
Kettle lakes, 78, 347
Kimberly Lake (South Africa), 348
Kinetic energy (KE) spectrum, 63–67
Kinetics:
  enzyme, 124–127, 136, 137
  growth, 282–283
Kingfishers, 32, 194
King's River (U.S.), 348
Kinneret, Lake (Israel), 163, 219–220
Kioga, Lake (Africa), 343
Kivu, Lake (Africa), 346

Laboratory studies, field studies integrated with, 7–9
Lake basin, 12, 90, 342, 343, 345, 346
Lake district, 342–343, 349, 363
  (*See also* English Lake District)
Lake succession, cultural eutrophication and, 349–352, 356
Lake trout (mackinaw), 270
Lake whitening, 96, 98, 103
Lamellibranch (*see* Bivalve mollusks)
Laminar flow, 61–63
Land erosion (*see* Erosion)
Langmuir spirals, 58, 63, 73–74, 259
Large lakes (*see* Size, of lakes)
Large-mouth bass, 265
Larvae, 134
  in estuaries, 320, 334
  in lakes, 18, 19, 184, 221, 249
    effects of chlorine on, 151
    feeding mechanisms of, 222
    phosphorus recycling by, 134
  in streams and rivers, 292, 313
  (*See also specific insects and fishes*)
Lateral lakes, 348
Laurentian Great Lakes, 9, 13, 28, 146, 242, 367–372
  ammonia in, 126
  comparative limnological study of, 365
  fish and fisheries in, 187, 190, 249, 261–265, 267, 270
  food-chain dynamics in, 285
  heat in, 43, 48, 49, 54
  management of, 394, 400
  movement in, 69, 78
  origin of, 342–343

Laurentian Great Lakes (*Cont.*):
  oxygen and carbon dioxide in, 102, 107
  succession of, 349, 350, 356
Lead pollution, 388
Leeches, 185, 280
Leibig's law, 282
Léman, Lac (Lake Geneva, Switzerland), 2, 96
Lemna (duckweed), 121, 179
Length of waves, 63, 65, 66, 69, 78
Lentic (standing) waters, 2, 12, 292
  (*See also* entries beginning with the term: Lake; *the subentry* lake *under specific headings, for example:* Algae, lake; *specific entries, for example:* Apparent color of lakes)
Lepisosteidae (gars), 190
Leptodora (a cladoceran), 233
Leven, Loch (Scotland), 373
Lichen, 385
Light:
  algae and, 30, 36–37
    phytoplankton, 197, 204, 208, 209, 215
  in eutrophic lakes, 351
  heat and, 30–31
  oxygen and carbon dioxide distribution and, 95, 97, 98
  in polar lakes, 376
  (*See also* Photosynthesis; Water-light interaction)
Light inhibition of photosynthesis, 377, 379
Light zonation, 15
Lignin, 170
Limnetic zone (*see* Pelagic zone)
Limnocalanus (a copepod), 236, 373
Limnology, 1–11
  defined, 2
  early, 2–6
  future of, 7–10
  introduction to, 1–6
  present-day, 6–7
  (*See also* Applied limnology; Comparative limnology; *specific aspects of limnology*)
Linsey Pond (Conn.), 347, 354
Lions, 372
Liquid crystal, 24, 26
Liquid water, world distribution of, 25
Lithosphere, defined, 166
Lithotrophs, defined, 166
Littoral avoidance by zooplankton, 226
Littoral fish, 254
Littoral zone:
  of estuaries, 22
  of lakes, 14–15
Liverworts, 177
Logging and deforestation, 92–94, 261, 269, 354, 356, 391

Long-period internal gravity waves, 75, 79–82
Lota lota (burbot), 265
Lotic (running) waters (*see* Streams and rivers)
Luxury consumption of phosphate, 131, 135, 159, 207
Lymnaea (a snail), 367
Lyngbya (a blue-green alga), 371

Mackinaw (lake trout), 270
Macoma (vacuum-cleaner clams), 334
Macroalgae (seaweeds), 324, 326–331
Macrophytes:
  in lakes, 18, 176–180, 278, 280, 281
    classification of, 177–178
    control of, with grass carp, 270
    harvesting of, 387, 403–404
    light reflection and, 36–37
    phosphorus and, 139–140, 178
    in polar lakes, 363
    in salt lakes, 381
    snail growth and, 187
    zoobenthos and, 242
  in ponds, 20
  in streams and rivers, 177–179, 293, 307–308, 312, 316
  (*See also specific macrophytes*)
Maggiore, Lago (Italy), 347
Magnesium (Mg), 142, 145, 147, 149–150, 163, 318
Mahega, Lake (Uganda), 345
Malaria, 372
Malawi, Lake (Africa), 246, 343
Mallomonas (a flagellate), 175
Mammals, 192–196, 275
  (*See also specific mammals*)
Management:
  fishery, 260–271
  food-chain, 388
  of lakes and estuaries (*see* Applied limnology)
Managua, Lake (Nicaragua), 47, 48
Manatees, 195–196, 308, 405
Manganese (Mn), 143, 153, 159–160, 401
Mangrove swamps, 337–338
Marchantia aquatica (liverwort), 178
Marion, Lake (Canada), 83, 133
Maritime lakes, 377
Marl lakes, 96, 101–102, 148
Marshes (*see* Swamps and marshes)
Mathematical modeling, 9–10
Mathematical submodels, 10
Maximum heat content (annual heat budget), 54–55
Mean flow, turbulent vs., 66–67
Meanders of streams and rivers, 20, 21
Medusae, 222
Melosira (a diatom), 146, 175, 198,

Melosira (Cont.):
    200, 226, 359, 371
Melosira italica, 210–211
Melosira moniliformis, 330–331
Melosira solida, 359
Mendota, Lake (Wis.), 43, 54, 55,
    118, 347
Meromictic fjords, 322–323
Meromictic lakes, 39, 43
Meroplankton, 198
    (*See also specific meroplankters*)
Mesocyclops (a copepod), 364
Mesocyclops leuckarti, 229, 230, 237
Mesophytic algae, 177, 178, 307, 308
Mesotrophic lakes:
    ammonia in, 127–128
    lake succession and, 349, 351
    nitrogen in, 120–121, 125, 126
    paleolimnological investigation of,
        359
    temperate, 367
    zoobenthos of, 242
Metabolite, defined, 89
Metalimnion (*see* Thermocline)
Metals (*see* Trace elements)
Methane and methane bubbles, 118
Methyl mercury, 267
Mexico, Gulf of, 319, 349
Michigan, Lake (*see* Laurentian Great
    Lakes)
Microcystis (a blue-green alga), 124,
    198, 201, 212, 213, 215, 216,
    219, 236, 371, 375
Microcystis aeruginosa, 173, 254,
    373–374
Micronutrients (*see specific
    micronutrients*)
Microphytes:
    lake, 177
    stream and river, 307–308
    (*See also specific microphytes*)
Midges (*see specific midge species*)
Mining operations, 88, 160
    (*See also* Pollution)
Mink, 195
Minnesota River (U.S.), 348
Mississippi River (U.S.), 319, 348,
    349
Mixed layer (*see* Epilimnion)
Mixing:
    and aeration, 387, 400–403
    epilimnion stirred by convection, 58
    Kelvin-Helmholtz, 70, 79
Modeling, mathematical, 9–10
Molecular structure of water, 24, 26–27
Mollusks, 147, 149, 160, 221, 222,
    307, 311, 336
    (*See also specific mollusks*)
Molybdenum (Mo), 9, 118, 143, 155,
    159, 161–163, 231, 286
Monimolimnion, defined, 43
Mono Lake (Calif.), 19, 28, 102, 381,

Mono Lake *(Cont.):*
    382
Monod (Michaelis-Menten) curve,
    282–283, 285
Monoma, Lake (Wis.), 55
Monomictic lakes, 54
    blue-green algae in, 212
    defined, 39, 43
    phytoplankton in, 210–212
    temperate, 364, 367
Moraine, 347
Morone (Roccus) chrysops (white
    bass), 254
Morone saxatilis (striped bass), 336
Morphometry of lakes, 12–13
    (*See also* Depth, of lakes)
Mortality (death rate) of zooplankton,
    238, 241
Mosquitoes, 381
Mosses, 178, 292, 293, 316, 377, 379,
    385
Movement:
    of blue-green algae, 61, 175,
        201–203
    of dinoflagellates, 203
    of phytoplankton, 200–203
    of plankton, 65–66, 71, 155
    of zooplankton, 48
    (*See also* Water movement)
Mud flats (*see* Estuaries)
Mugil cephalus (grey mullet), 336
Mullet, 336, 388
Multivoltine copepods, 236
Muskrat, 195
Mussels, 187, 330, 335
Myriophyllum (water milfoil), 179, 281
Mysids, 236
Mysis (a crustacean), 270, 371
Mysis relicta (opossum shrimp), 221,
    226, 231–233, 242, 368
Mytilus edulis (a mussel), 330, 335
Mývtan, Lake (Iceland), 245, 345
Myxinidae (hagfishes), 187

Natality (birth rate) of zooplankton,
    238, 239
Natrix (water snake; grass snake), 193
Necturus (mud puppy), 192
Negative heterograde distribution of ox-
    ygen, 110, 111
Nekton, estuarine, 18, 318, 333
Nematodes, 311, 379
Neomysis (opossum shrimp), 231, 333
Nereis (a polychaete), 333
Ness, Loch (Scotland), 54, 82, 211
Neuston, defined, 18
New Hope Creek (N.C.), 309
Newts, 192
Nicaragua, Lake (Nicaragua), 187
Niger River, 302, 304
Nile River, 9, 269, 298, 302, 349
Nile River Delta, 178

Nitella (an alga), 178
Nitracline of Lake Tahoe, 397, 398
Nitrate reductase, 115, 123, 124, 163
Nitrates (NO_3), 19, 82, 94, 122–126,
    140, 169
  in lakes: algae and, 124, 210, 212,
    215, 216
    ammonia and, 114, 115, 121, 122,
      126, 408–409
    copper and transformation of, 160
    hypolimnetic, 82
    measurement of, 115–116
    nitrogen fixation and, 118, 120
    removal of, 391, 394
    seasonal cycles of, 126
    transport of, 19
  in streams and rivers, 293, 302–304
Nitrification, defined, 122
Nitrite reductase, 123
Nitrites (NO_2), 115, 122–126, 169
Nitrogen (N):
  in lakes, 17, 19, 84, 88, 112,
    114–130, 150
    and *Asterionella* growth, 207, 209
    from atmospheric sources, 91
    and blue-green algae growth, 115,
      125, 215
    climate and, 90
    denitrification process, 82, 115,
      116, 121–122, 167
    dispersion of, 84
    dissolved organic nitrogen,
      128–130
    in eutrophic lakes, 116, 118, 122,
      125, 126, 349
    excretion of, 254
    forms of, 117–118
    in Lake Tahoe, 397–398
    in Laurentian Great Lakes, 368,
      371
    measurement of, 115–116
    paleolimnologic measurement of,
      358
    partial pressure of, 97
    phytoplankton and, 198, 254
    plant leaves as source of, 179
    in polar and alpine lakes, 380
    in refractory compounds, 88
    removal of, 392, 403
    in tropical lakes, 374, 375
  narcosis and, 118, 268
  in streams and rivers, 114, 115, 123,
    124, 129, 293, 304, 307
  as tracer, 285, 286
Nitrogen cycle, 116–117, 143, 163
Nitrogen fixation, 115, 116, 118–121,
    131
  by alder trees, 90
  by algae, 118–119, 286
    by blue-green algae, 212, 216
  bacteria and, 167
  in eutrophic lakes, 118, 119, 349

Nitrogen fixation *(Cont.)*:
  iron and, 143, 157
  measurement of, 116
  by meroplankton, 198
  molybdenum and, 163
  as source of ammonia, 128
  in tropical lakes, 373
Nitrogenase, 118, 157, 163
Nitzschia (a diatom), 275, 307
Nodularia (a blue-green alga), 118, 382
Nonperiodic (arrhythmic) currents, 57
Nostoc (a blue-green alga), 112, 118,
    121, 286
Nuphar (water lilies), 179, 281
Nutria, 405
Nutrients:
  in estuaries, 22, 317, 318, 320, 322,
      323, 354
  in lakes: and blue-green algae
      growth, 212, 215
    determining concentrations of, 88
    effects of algal movements on,
        200
    excess *(see* Cultural eutro-
        phication)
    heat cycles and, 55
    hydraulic retention time and, 13
    hypolimnetic, 82
    hypolimnetic entrainment and, 50
    for macrophytes, 178
    and *Melosira* growth, 211
    oxygen production and, 95
    and phytoplankton growth, 197,
        198, 204, 208, 209
    in polar lakes, 376
    regulation of, 273–281
    river dispersion of, 83, 84
    thermal stratification and, 17
    thermocline formation and, 43
    trapped by thermal bar, 51–52
    in tropical lakes, 372
    uptake kinetics of, 282–286
    water movement and release of,
        59, 82
  in streams and rivers, 20, 302–307
  (See also Food chain; Recycling of
      nutrients; *specific nutrients)*
Nymphaea (water lily), 179, 281

Oceans:
  chlorine in, 143
  Ekman spiral in, 71
  frustules in, 147
  protozoans in, 180
  salinity-induced density effects in, 28
  upwelling in, 48
Oconto River (U.S.), 281
Odonata and *Odonata* larvae (dragon-
    flies), 18, 184, 311
Odoriferous compounds, 88, 89
Oil-degrading bacteria, 167

Oil pollution, 167, 407
Okeechobee, Lake (Fla.), 343
Oligochaete worms, 185, 242, 311,
    379, 381
Oligotrophic estuaries, 354
Oligotrophic lakes:
  algae in, 390
  ammonia in, 122, 126–128
  bacteria in, 169
  defined, 341
  fish in, 259
  iron in, 153, 154
  lake succession and, 349, 351–353
  nitrates and nitrites in, 122, 124–126
  nitrogen in, 116, 128
  nitrogen fixation in, 119
  oxygen and carbon dioxide in, 95
  polar lakes as, 377
    (See also Polar and alpine lakes)
  temperate, 367
  trace elements in, 143, 158–159, 163
  water movement in, 82
  zooplankton in, 225
Omnivores:
  defined, 166
  (See also specific omnivores)
Omnivorous fishes, 256
  (See also specific omnivorous fishes)
Onchocerciasis (river blindness), 184,
    372
Oncorhynchus nerka (kokanee salmon),
    187, 232
Ontario, Lake *(see* Laurentian Great
    Lakes)
Opossum shrimps, 221
  (See also specific species)
Optical structure, thermal stratification
    and, 16
Optically deep lakes, 15
Optically shallow lakes, 15
Organic chemical compounds, 88–90
Organic surfactants (foaming or wetting
    agents), 29
Organisms:
  effects of presence, on oxygen and
      carbon dioxide concentrations,
      102–104
  functional classification of, 166
  *(See also specific groups of organ-
      isms and specific organisms)*
Organotrophs, defined, 166
Origin of lakes, 341–348
Orthodon microlepidotus (blackfish),
    254
Orthograde distribution of oxygen, 109,
    110
Oscillatoria (a blue-green alga),
    206–209, 371, 391
Oscillatoria rubescens, 79, 203, 217,
    231
Osmerus mordax (rainbow smelt), 264
Osmoregulation of salt balance, 192

Osprey, 194
Osteichthyes, 187, 189
Ostracods, 379
Ouzels, 194
Oxbow lakes, 20, 348, 351
Oxidation-reduction (redox) potential
    (E_h), 104–105, 135, 152, 154,
    155, 159
Oxygen (O_2), 52, 75, 87, 88, 211
  air narcosis and, 268
  and bacteria classification, 167
  and carbon dioxide, 95–113, 217
    diel and seasonal variations in,
        105–111
    effects of temperature, salinity,
        and organisms on, 102–103
    measurement of, 96
    redox potential and, 104–105
    sources of, 96–102
  depletion *(see* Cultural
      eutrophication)
  in lakes: effects of zooplankton res-
      piration on, 226
    fish kills and depletion of, 265, 267
    in Lake Tahoe, 397, 398
    as measure of lake restoration, 390
    nitrogen and, 116–118
    nitrogen fixation and, 116
    phosphorus released from sedi-
        ments and, 139
    in primary-productivity measure-
        ment and, 287, 288
    in tropical lakes, 375
    and trophic state of lakes, 351
    zoobenthos distribution and, 222,
        242–243
  redox potential, 104–105, 135, 152,
      154, 155, 159
  in streams and rivers, 309, 407–410
    carbon dioxide and, 94, 97, 102,
        103, 111–112
Oxygen deficit, defined, 111
Oxygen sag, defined, 111
Oysters, 320, 330, 333, 334, 336

Pacific salmon, 191, 249, 265, 267
Paleolimnology, 358–360
  defined, 359
  and function of paleolimnologists,
      342
Palmella (an alga), 385
Paralytic shellfish poisoning (PSP), 330
Paramecium (a ciliate), 180
Paraná River (Argentina), 179, 255,
    256, 348
Parasitism, 198, 204, 207, 209, 213
  and phytoplankton growth, 197, 198
  (See also Bacteria; Fungi; Proto-
      zoans; Viruses)*
Parent rock, chemical composition and,
    90

Particulate phosphorus, 132
Patches:
  phytoplankton, 220
  plankton, 65
  zooplankton, 330
Paternoster lakes, 346
PCBs (polychlorinated biphenyls), 267
Pelagic fish, 254
Pelagic macrophytes (*see* Free-floating
  macrophytes)
Pelagic (limnetic) zone:
  of estuaries, 22
  of lakes, 15–16
  phosphorus in, 133–135
Penaeus setiferus (harvest shrimp), 336
Penguins, 194
Pennate diatoms:
  in estuaries, 324, 331
  in Laurentian Great Lakes, 371
Peptidoglycans, 167
Perca fluviatilis (perch), 248, 258–260,
  262
Percids (perch family), 190, 191
  (*See also specific percids*)
Peridinium (a dinoflagellate), 175, 176,
  198, 213–214, 218
Peridinium pernardii, 219
Periodic (rhythmic) waves, 57
Periphyton (*see* Attached algae)
Perris, Lake (Calif.), 48
Pesticides and insecticides (*see*
  Pollution)
Petromyzon marinus (sea lamprey),
  261–263
Petromyzontids (lamprey family), 187
  (*See also specific petromyzontids*)
pH (puissance d'Hydrogène; hydrogen-
  ion concentration), 95–96
  ammonia concentration and, 100,
    127, 128
  calcium and, 147–149
  phosphate and, 100, 135–136, 395
  phosphorus and, 134
  and photosynthesis, 98
  primary-production measurement
    and, 287
  as source of oxygen and carbon
    dioxide, 98–102
  trace elements and, 155, 158, 160
  of tropical lakes, 375
Phoca (Pusa) sibirica (freshwater seal),
  195
Phoenicopterus minor (lesser flamingo),
  194
Phormidium (a blue-green alga), 307
Phosphate (PO_4):
  in lakes, 87, 94, 144
    blue-green algae and, 216
    effects of aeration on, 403
    in eutrophic lakes, 131, 132, 151,
      349
    excretion of, 87, 89, 133, 137,

Phosphate (PO_4), in lakes (*Cont.*):
    139, 226
    iron and, 134, 135, 143, 153, 155
    in Laurentian Great Lakes, 368
    luxury consumption of, 131, 135,
      159, 207
    nitrates and, 114, 115, 122
    pH and, 100, 135, 136, 395
    in phosphorus cycle, 132–135,
    pollution due to, 147
    reduction of level of, by *Asteri-*
      onella, 207
    removal of, 391–395
    transport of, 19
    uptake of, 135, 136
  in streams and rivers, 297, 298, 302,
    305
Phosphorus (P):
  in lakes, 17, 84, 114, 131–144, 150,
    280
    algae and, 131, 132, 135–137,
      139–140, 393
    Asterionella and, 206–207, 209,
      210
    and blue-green algae, 213, 215
    climate and, 90
    cultural eutrophication and, 98,
      354–356, 358
    dispersion of, 84
    in eutrophic lakes, 131, 132, 151,
      349
    excretion of, 89, 254
    induction of phosphatases and up-
      take of, 135–137
    iron and, 151, 153
    in Lake Tahoe, 398
    macrophytes and, 139–140, 178
    measurement of, 132
    nitrogen and, 114
    paleolimnologic measurement of,
      358
    phytoplankton and, 131, 136, 137,
      139, 198
    in polar and alpine lakes, 380
    recycling of, 137–140
    in refractory compounds, 88
    removal of, 387, 391–394, 399
    requirements for, 129, 130
    sources of, 91, 140
    sulfur and, 151
    transport of, 19
  in streams and rivers, 293, 298, 302,
    304
  as tracer, 285, 286
Phosphorus cycle, 132–135, 143
Photic zone (*see* Euphotic zone)
Photoautotrophs, 176
  defined, 166
  (*See also specific photoautotrophs*)
Photosynthesis:
  in estuaries, 317
  extracellular products of, 288–290

Photosynthesis (*Cont.*):
  in lakes, 15, 29, 30, 148
    algae and, 101, 171
    ammonium and, 127
    bacteria and, 169
    calcium and, 101, 142
    copper and, 160, 161
    effects of, on blue-green algae,
      216
    in eutrophic lakes, 98, 101
    iron and, 151
    light inhibition of, 377, 379
    manganese and, 159
    measurement of, 285
    oxygen and carbon dioxide and,
      95–101
    in polar and alpine lakes, 362,
      377, 379, 380
    respiration and, 103
    seasonal variations in, 107–108
    water as electron source in, 25
    zonation and, 15
Phototrophs:
  defined, 166
  (*See also specific phototrophs*)
Phragmites australis (a macrophyte),
  178
Phragmites beds, 139
Phycobilins, 175
Phycomycetes, 170
Physical structure:
  of lakes, 15–17
  of rivers and estuaries, 20, 22
Phytoplankton:
  in estuaries, 210, 318,
    326–331
  in lakes, 18, 56, 89, 173, 197–220,
    389
    ammonia and, 115, 126
    bacteria on, 169
    calcium and, 149
    carbon dioxide and growth of, 98
    copper and, 160
    distribution of, 198–199
    effects of macrophytes on, 404
    in eutrophic lakes, 351
    factors affecting growth of,
      197–198
    in food chain, 280,
      282–283
    iron and, 155
    in kettle lakes, 78
    lake color and, 36
    in Lake Tahoe, 397, 398
    in Lake Washington, 392
    in Laurentian Great Lakes, 368,
      371
    measurement of, 199–200
    molybdenum and, 161, 163
    movements and, 200–203
    nitrogen and, 198, 254
    in oligotrophic lakes, 351

Phytoplankton, in lakes *(Cont.)*:
  paleolimnologic investigation of, 359
  patchiness of, 220
  phosphorus and, 131, 136, 137, 139, 198
  polar and alpine lakes and, 363, 377, 379–380
  protozoans and, 181
  in salt lakes, 381, 382
  seasonal variation in, 204–216, 219, 283–285
  silica and, 142
  spatial variations of, 216–219
  thermal pollution effect on, 56
  trace element uptake by, 159
  transport of, 58
  in tropical lakes, 362, 372–374
  viruses and, 166
  zinc and, 160
  zooplankton excretion and growth, 56
  in rivers and streams, 296, 307
  (See also specific types of phytoplankton)
Piedmont lakes, 347
Pinus jeffreyi (Jeffrey pines), 398
Piscivorous fishes, 248, 254
  (See also specific piscivorous fishes)
Pistia (water lettuce), 18, 375
Pistia stratiotes (common water lettuce), 179, 404
Pitot tube, 294
Planktivorous fishes, 248, 254
  (See also specific planktivorous fishes)
Plankton, 4
  animal: discovered, 3
    (See also Zooplankton)
  defined, 3, 18
  in estuaries, 320, 323–326, 333
  in lakes, 52, 73
    iron and, 155
    movement of, 29, 65–66, 71
    in polar and alpine lakes, 376
    in tropical lakes, 373
    viscous drag effect on, 29
  (See also Phytoplankton)
Plants:
  in estuaries, 326–331
  in lakes: deep-dwelling, chromatic adaptation of, 33
    effects of pH value on, 99–100
    respiration of, 106
    solar radiation and distribution of, 29
    succession patterns in, 350
    water movement and, 58, 59
  pollens of, 358, 360
  (See also Macrophytes; Photosynthesis; Phytoplankton; Surround-

Plants *(Cont.)*:
    ing vegetation)
Plecoptera and *Plecoptera* larvae (stonefly), 184, 309, 311, 382
Pleuston, defined, 18
Plumes of sediment, 83–85
Pneumatophores of mangroves, 337
Poikilothermal (cold-blooded) fish, 56, 275
Poincaré waves, 75, 79–82
Poisoning:
  paralytic shellfish, 330
  (See also Toxicity)
Polar and alpine lakes:
  insects in, 184
  light in, 29–30, 33, 36
  management of, 362–363, 376–380
  origins of, 346
  oxygen and carbon dioxide in, 107–108, 111
  zooplankton in, 243, 245
  (See also specific polar and alpine lakes)
Polarized light, 29
Pollens of plants, 358, 360
Pollution:
  control of (*see* Applied limnology)
  of estuaries, 22
  in lakes: abundance of *Sphaerotilus* due to, 167
    ammonia and, 127
    carp and, 191, 267, 268
    chlorine and, 150–151
    deforestation and logging and, 92–94, 261, 269, 354, 356, 391
    effects of: on fish life, 190, 249, 253, 265–268
      on heat cycle, 55
      on mollusks, 187
    erosion and (*see* Erosion)
    interfering with chemical communication, 89
    of Laurentian Great Lakes, 368, 371
    nitrates and nitrites and (*see* Nitrates; Nitrites)
    nitrogen and, 146
    oxygen depletion and, 95
    phosphate and phosphorus and, 114, 132, 136, 140, 146
    presence of coliform bacteria as indicative of, 169
    reflection patterns and, 37
    rivers and, 83, 294, 306
    silica and silicon and, 116, 146
    sulfur and, 150
    thermal, 55–56, 267
    water movement and, 71
  in streams and rivers, 111, 294, 306
  (See also Acid rain; Applied limnology; Blue-green algae; Cultural eutrophication)

Polyarthra (a rotifer), 231
Polyarthra dolichoptera, 379
Polychaete worms, 333, 334, 336
Polychlorinated biphenyls (PCBs), 367
Polymictic lakes:
  blue-green algae in, 212
  defined, 40, 43
  heat in, 47
  Laurentian Great Lakes as, 371
    (See also Laurentian Great Lakes)
  phytoplankton in, 210, 212
  temperate, 364
  tropical, 372, 373
Polyodon (paddlefish), 190
Polyodon spathula (paddlefish), 254
Polyphemus (a cladoceran), 233
Polyphosphate granules, 135
Ponds, 19–20
Pontoporeia (an amphipod), 371
Pontoporeia affinis, 242
Pools in streams and rivers, 20, 292, 295, 296
Posidonia (a sea grass), 327
Positive heterograde distribution of oxygen, 110, 111
Potamogeton (a weed), 179
Potassium (K), 88, 142, 145, 150, 280
Predators:
  exotic, 265–268
  (See also specific organisms)
Prevention, management by, 395–399
Prochilodus platensis (sabalo), 255, 256
Procladius (a midge), 242
Protozoans:
  growth of, in distilled water, 88
  in lakes, 18, 88, 89, 180–182, 221, 222
    light and, 30
    in polar lakes, 378
    on river detritus, 21
    in salt lakes, 382
  (See also specific protozoans)
Prymnesium (a dinoflagellate), 330
Psammolittoral zone, water flow in, 18
Pseudospora (a protozoan), 181
PSP (paralytic shellfish poisoning), 330
Purari River (New Guinea), 350
Pure culture studies, 9
Pusa (Phoca) sibirica (freshwater seal), 195
Pycnoclines (salt density gradients), 320
Pyramid Lake (Nev.), 102, 147, 380, 382
Pyrenoids, 175
Pyrheliometer measurement:
  of epilimnion temperature, 47
  of solar radiation, 30

Quanta of light, 30–31
  defined, 30

Rae Lakes (Calif.), 270
Rainfall:
  acid, 7–8, 99, 388
  in estuaries, 319, 327
  in lakes, 87, 90, 94, 147
    ammonia from, 114, 128
    carbon dioxide from, 144
    iron and, 157–158
    and lake succession, 350
    nitrates from, 123, 126
    nitrogen from, 114
    phosphorus from, 140
    in salt lakes, 381
    in temperate lakes, 364
    in tropical lakes, 376
    watershed and, 19
  in streams and rivers, 297
  (*See also* Erosion)
Rallus longirostris (clapper rail), 337
Random water movement, 74–75
Ranunculus (water crowfoot), 308
Reactive phosphorus, defined, 132
Real color of lakes, 33, 35
Recycling of nutrients:
  in estuaries, 322
  in lakes, 18
    bacterial, 167
    by fish, 252, 254
    hypolimnetic temperature and, 46
    of phosphate, 131, 136
    of phosphorus, 134
    of silica, 144
    trace elements and, 285
    in tropical lakes, 374
    by zoobenthos, 222
  in streams and rivers, 20, 311
Red midge larvae, 247
Red tides, 33, 89, 198, 218–220,
    330
Redhorse, 267, 268
Redox (oxidation-reduction) potential
    (E_h), 104–105, 135, 152, 154,
    155, 159
Reduced microzone, redox potential in,
    105
Reeds, 178, 327, 372, 375–376
  (*See also specific reeds*)
Reelfoot Lake (Tenn.), 343
Reflection of light (albedo), 25, 36–37
Refractory chemical compounds, 88–89
Refractory phosphorus, 140
Relic lakes, 343
Reproduction of zooplankton, 236–238
Reptiles, 192–196, 337, 383
  (*See also specific reptiles*)
Reservoirs and dams, 1, 7, 97, 118,
    261, 268–270, 298, 319, 348,
    404–406
Resource competition, 285
Respiration:
  in aphotic zone, 15, 98
  bacterial, sulfur and, 151

Respiration (*Cont.*):
  dependent on temperature, 103
  iron and, 143
  photosynthesis and, 103
  zooplankton, 226
  (*See also* Carbon dioxide; Oxygen)
Restoration of lakes, 356, 367, 387,
    390
  (*See also* Applied limnology)
Reversal of cultural eutrophication, 359
  (*See also* Applied limnology)
Reynolds number (R_e), 62
Rhine River (Germany), 9, 306
Rhizophora (a mangrove), 337
Rhythmic (periodic) waves, 57
Rice culture, 121, 179, 249, 301
Richardson's number (R_i), 44–45, 62
Riffles of streams and rivers, 20, 292,
    295
Riparian, 300
River blindness (onchocerciasis), 184,
    372
River valleys, drowned, 348
Rivers (*see* Streams and rivers)
Roccus (Morone) chrysops (white
    bass), 254
Rocks:
  chemical composition and parent, 90
  glacial rock basin, 346–347
  river vegetation and types of, 293
  surrounding pH and, 99
Rooted macrophytes, 177–179, 350
  (*See also* Macrophytes)
Rorippa nasturtium-aquaticum (water
    cress), 179, 249
Roseate flamingos, 383
Rotenone, 250–251
Rotifers:
  in lakes, 18, 181–183, 221
    feeding by, 233–234, 254
    in Laurentian Great Lakes, 371
    measurement of, 222, 223
    in polar and alpine lakes, 378, 379
    population dynamics of, 239
    in salt lakes, 382
    seasonal variations in, 226, 228,
      229, 231
  reproduction of, 236, 238
  in streams and rivers, 21, 181
Rough (coarse) fish, 251–252
Running (lotic) water (*see* Streams and
    rivers)
Runoff waters:
  nitrates from, 123
  trace elements in, 159
Runs of rivers and streams, 295, 296
Ruppia (widgeon grass), 178, 327, 381
Rutilus rutilus (roach), 250, 255–256

Sacramento River (Calif.), 261, 406
Saginaw Bay, 371

Salamanders, 192
Salicornia (a marsh plant), 327
Saline water, world distribution of, 25
Salinity:
  effects of, on oxygen and carbon
    dioxide distribution, 102–104
  of estuaries, 28, 317–322, 333
  (*See also* Oceans)
Salinity-induced density of water, 28
Salminus maxillosus (dorado), 256
Salmo gairdnerii (rainbow trout), 127,
    249, 256, 270, 275
Salmon, 112, 169, 171, 194, 298, 301,
    306, 307, 310, 319
  (*See also specific salmons*)
Salmon disease, 167
Salmonids, 89–90, 97, 159, 190, 191,
    249, 261–264, 278, 379
  (*See also specific salmonids*)
Salt density gradients (pycnoclines),
    320
Salt Fork River (Ill.), 297
Salt lakes, 48, 100, 231, 380–382
Salt wedge, 4, 22, 318–323
Saltwater fish:
  hypotonicity of, 192, 319
  (*See also specific saltwater fish*)
Salvelinus fontinalis (brook trout), 256
Salvelinus namaycush (lake trout),
    262–264
Salvinia (water fern), 179, 375
Salvinia molesta (water fern), 404
San Francisco Bay (Calif.), 22, 190,
    330–331, 333, 334, 337, 349
Saprolegnia (a fungus), 169
Saprophytic fungi, 169
Sauger, 267, 268
Savonius rotor, 59
Schistosomiasis (bilharzia), 187, 372
Schizothrix (a blue-green alga), 382
Scirpus (American bullrush), 178, 327
Scirpus robustus (alkali bullrush), 178
Scroll lakes, 20, 348, 351
Sea grasses, 336
  (*See also specific sea grasses*)
Seals, 195, 377
Seasonal variations:
  in ammonia concentration, 127–128
  in diatom population, 204–210, 283,
    285
  in iron concentration, 155–157
  in lake color, 35
  in nitrate concentration, 124, 126
  in nitrogen concentration, 114, 117
  in oxygen and carbon dioxide con-
    centration, 106–111
  in phytoplankton population,
    204–216, 219, 283–285
  in polar and alpine lakes, 376,
    379–380
  in stream and river discharge,
    297–298

Seasonal variations *(Cont.)*:
  in stream and river nutrients, 302
  in trace element concentration, 158
  in tropical lakes, 373
  in zooplankton populations, 226–232
Seaweeds (macroalgae), 324, 326–331
Secchi disk measurement of lake transparency, 30, 36, 389–390, 392, 394
Sediment plumes, 83–85
Sediments:
  in estuaries, 322–323, 327, 331, 333
  in lakes, 18, 87
    adsorption of nutrients by, 94
    cultural eutrophication and, 354–358
    effects of, on zoobenthos, 242
    in English Lake District, 367
    fertilizers and production of, 93
    frustules in, 142, 147
    heat cycles and, 55
    in Lake Tahoe, 397
    macrophytes and, 403–404
    measurement of, 389
    Melosira in, 210–212
    nitrogen released by, 114
    paleolimnologic investigation of, 342, 358–359
    phosphate from, 216
    phosphorus and, 134, 135, 139, 178
    redox potential in, 159
    sulfur released by, 150
    water movement and, 59, 62, 82
    in watershed, 19
    zoobenthos in, 222, 225
Seiches, 3, 5
  defined, 71–72
  Forel's theory of, 5
  internal, 58, 80
  surface, 58, 71–73
Selenga River (U.S.S.R.), 269
Selenium (Se), 163
Serrasalmo (piranha), 256
Severn Estuary (England), 22
Sewage *(see* Pollution)
Shad, 190, 191, 254, 267
Shagawa, Lake (Minn.), 356, 392–395
Shallow lakes *(see* Depth, of lakes)
Sharks, 187, 192, 193
Shasta Dam system (Calif.), 261
Shellfish, 22, 89, 337
  (See also specific shellfish)
Shellfish poisoning, paralytic, 330
Shore birds *(see* Birds)
Shoreline of lakes, 13
Short-period internal gravity waves, 79
Shredding organisms, 312
Shrimp, 181, 221, 326, 333, 335, 338
  (See also specific shrimps)
Siderochromes, 89, 143, 155
Sidestream pumping, 403

Silica (SiO₂), 19, 90, 132, 142–147, 198, 210, 293, 302, 368
Silica cycle, 143–145
Silicon (Si), 142–147, 172, 207, 209–210, 302
Simuliidae and *Simuliidae* larvae (blackflies), 184, 311
Size of lakes:
  heat flow and, 54
  lake whitening and, 102
  light and, 25
  of tectonically formed lakes, 343
  water movement and, 69, 78–81
  zooplankton population and, 225
Skeletonema (a diatom), 210
Slicks (windrows), 73–74
Slide Lake (Wyo.), 345
Small lakes *(see* Size of lakes)
Smelt, 191, 264, 407
Smith Lake (Alaska), 119, 215
Snag Lake (Calif.), 346
Snails, 185, 187, 194, 311, 327, 334, 381
Snakes, 193
Soda lakes, 99
Sodium (Na), 88, 90, 142, 145, 147, 150
Sodium chloride (NaCl), 318
Soft-water streams, 306–307
Soil erosion *(see* Erosion)
Soil water, 147
Solar radiation *(see* Heat; Light)
Solution lakes, 347
Spartina (cord grass), 326, 327
Spates *(see* Floods)
Spatial variations in phytoplankton, 216–218
Species diversity:
  in estuaries, 319–321
  in salt lakes, 381, 382
  in tropical lakes, 375
Specific heat of water, defined, 27
Sphaerotilus (sewage fungus), 167, 311, 356
Sphagnum moss, 178, 293
Spicules of sponges, 143
Spirochaetes, 381
Spirogyra (a green alga), 2
Spirulina (a blue-green alga), 381
Sponges, 143
Spring overturn, 17
Spring variations *(see* Seasonal variations)
Standing surface gravity waves (surface seiches), 58, 71–73
Standing waters *(see* Lentic waters)
Staurastrum (a green alga), 359, 367
Stephanodiscus (a diatom), 210
Stocking of fish, 249, 265–270, 388
Stoke's law, 201
Streams and rivers, 2, 20, 87, 90, 292–316, 341, 363, 403

Streams and rivers *(Cont.)*:
  algae in, 20–21, 292, 293, 312, 315
  ammonia in, 127, 128
  bacteria in, 21, 169, 308, 310–313, 385
  carnivorous and herbivorous invertebrates in, 311–312
  characteristics of, 295–297
  comparative limnology of, 383–385
  crustaceans in, 181
  discharge into, 297–300
  effects of, on movement in lakes, 83–84
  eutrophication in, 353–354
  fish in, 21, 248–250, 260, 261, 265, 267, 270, 298, 300, 301, 306, 313, 383–385
  food chain in, 307–311
    dynamics of, 277, 280, 282, 290
  heat in, 39–41
  insects in, 182, 184–185, 311, 383–385
  iron in, 158
  macrophytes in, 177–179, 293, 307–308, 312, 316
  mammals in, 195
  management of, 406–410
  measurement of, 294–295
  movement in, 58, 61, 292
  nitrogen in, 114, 115, 123, 124, 129, 293, 304, 307
  nitrogen fixation in, 121
  nutrients in, 20, 302–307
  origin of, 342
  oxygen in, 309, 407–410
    and carbon dioxide, 94, 97, 102, 103, 111–112
  phosphorus in, 133, 140
  rotifers in, 181
  silica in, 144
  structure of, 20–22, 293
  temperature of, 298–302
  tropical, 375, 376
  watershed of, 19
  worms and mollusks in, 187
  (See also specific rivers)
Structure of lakes, 12–20
  biological, 18–19
  chemical, 17–18
  physical, 15–17
Sublittoral zone of lakes, 15
Substrate-growth curve, 123–127, 136, 137, 282–283, 285
Suisun Marsh ducks, 329
Sulfate (SO₄), 318, 381
Sulfur (S), 142, 145, 150–152, 286, 388
Sulfur cycle, 152
Summer kills, 106
Summer overturn of oxygen and carbon dioxide, 106
Summer variations *(see* Seasonal

Summer variations *(Cont.)*:
  variations)
Superior, Lake (*see* Laurentian Great
  Lakes)
Surber sampler, 294
Surface area of lakes, 12–13
Surface drifts, 58, 69–71
Surface light, measurement of, 30
Surface seiches (standing surface grav-
  ity waves), 58, 71–73
Surface tension of water, 28–29
Surface waves, 63, 64, 69, 71–73
Surface wind shear, 58
Surfactants, organic, 29
Surrounding vegetation:
  of lakes: chemical composition and,
    90
    lake color and, 35
    light and, 32
    thermal stratification and, 40
  of rivers and streams, 293, 298, 300,
    302, 309
Sustainable yield of fisheries, 286
Swamps and marshes, 22
  fish and fisheries in, 260, 261
  iron in, 158
  macrophytes in, 177, 178
  mammals in, 195
  mangrove swamps, 337–338
  sulfur released by, 151
  (*See also* Estuaries)
Swan mussels, 187
Swans, 336
Swimmer's itch, 187
Synchaeta (a rotifer), 233
Synedra (a diatom), 371, 292

Tabellaria (a diatom), 197, 206, 207,
  371
Tahoe, Lake (Calif.):
  ammonia in, 126
  fish in, 256
  heat in, 43, 48
  light reaching, 36
  management of, 395–399
  minerals in, 158–159
  movement in, 79
  origins of, 343
  succession in, 351, 356
  trace elements in, 160
Tanganyika, Lake (Africa), 43, 80,
  111, 210, 343, 351
Tanytarsus (an insect), 247
Tanytarsus gracilentus, 245
Tectonic processes:
  estuary formation by, 348
  lake formation by, 341, 343
Temperate lakes, 19, 54–55, 364–372,
  405
  (*See also specific temperate lakes*)

Temperature:
  lake: *Asterionella* growth and, 209
    calcium solubility and, 147
    depth and, 5, 7
    lake succession and, 350
    nitrogen concentration and, 114,
      117
    optimum, for denitrification, 121
    oxygen and carbon dioxide content
      and, 96, 97, 102–104
    in polar lakes, 378, 379
    trace elements and, 158
    in tropical lakes, 374
    and zooplankton population dy-
      namics, 241
  in streams and rivers, 293, 295,
    298–302, 383–384
  and water density, 28
  (*See also* Heat; Light; Thermal
    stratification)
Terns, 194
Tethys Sea, 343
Thalassia (a sea grass), 327
Thames, River (England), 3, 96, 356
  estuary of, 319, 349
  food-chain dynamics in, 277–280
  oxygen and carbon dioxide in, 111
  restoration of, 387, 407–410
Thaw (thermokarst; cryogenic) lakes,
  347
Thermal bar, 51–52
  defined, 51
Thermal pollution (waste-heat discharge), 55–56, 267
Thermal stratification, 2, 5, 39, 41–43
  in estuaries, salt wedge and, 22
  in lakes: *Melosira* growth and, 210,
    211
    phytoplankton growth and, 197
    in salt lakes, 381
  in ponds, 20
  in streams and rivers, 20
  (*See also* Epilimnion; Hypolimnion;
    Thermocline)
Thermocline:
  defined, 39, 41
  depth of, 47
  establishment and destruction of,
    50–51
  formation of, 43–44
  and hypolimnetic entrainment, 48–50
    phosphorus leaking around edges
      of, 139
  in metalimnion, 16
  phytoplankton near, 203, 217
  upwelling, and exposure of, to wind,
    48
  water movement in, 58, 69, 71, 75–
    82
  zooplankton in, 226
Thermocyclops hyalinus (a
  zooplankter), 236, 374, 375

Thermokarst (thaw; cryogenic) lakes,
  347
Thurston Lake (Calif.), 343
Tiberias, Lake (Sea of Galilee, Israel),
  219
Tidal marshes (*see* Swamps and
  marshes)
Tides:
  effects of, on estuaries, 22, 318,
    322, 326, 327, 353, 354
  red, 33, 89, 198, 218–220, 330
Tilapia (a fish), 191, 269, 356, 372,
  405
Tilapia nilotica, 254, 375
Titicaca, Lake (Peru), 345
Total mixing, 52
Total phosphorus, defined, 132
Toxicity:
  of ammonium, 127
  of arsenic and selenium, 163
  of chlorine, 143, 150–151
  of copper, 88, 160
  of magnesium, 149
  of manganese, 159–160
  of zinc, 88
  (*See also* Pollution)
Toxotes jaculator (archer fish), 254
Trace elements, 17
  from atmospheric sources, 91–92
  pH values and, 100
  phytoplankton growth and, 198
  in polar and alpine lakes, 380
  watershed and, 19
  (*See also specific trace elements*)
Transition metals, defined, 118
Transparency of lakes:
  of Lake Tahoe, 396
  management and, 394
  measurement of, 30, 36, 389–390,
    392, 394
Transport (*see* Water movement)
Trichoptera (caddisfly), 184, 311
Trophi of rotifers, 181
Tropical estuaries, 337–339
Tropical (arid) lakes, 169, 192–194,
  262, 372–376
  heat in, 48
  oxygen and carbon dioxide in, 111
  structure of, 19
  (*See also specific tropical lakes*)
Trout:
  in lakes, 160, 162, 171, 190, 248,
    388, 403
    as cold-water fish, 249
    in English Lake District, 367
    feeding patterns of, 254, 256
    overfishing and decimation of, 249
  in streams and rivers, 301, 313, 316
  (*See also specific trouts*)
Truckee River (Calif.), 275, 406
True color, apparent color modified by,
  25

Trummen, Lake (Sweden), dredging of, 387, 399–400
Tubifex worms, 243, 247
Tulare Lake (Calif.), 347–348
Tunicates, 222
Turbulence, 82–84
Turbulent billows, 58
Turbulent cascade of energy, 57
Turbulent (chaotic) eddies, 65, 276
Turbulent flow, 61–63, 66–68
  mean vs., 66–67
Turbulent mixing, 84
Turkana, Lake (Africa), 343, 345
Turtles, 249
Typha (cattail; reed mace), 178, 281

Ulothrix (a green alga), 175
Ulva (a seaweed), 326
Unchelated metals, toxicity of, 143
Underwater light, 30, 32
Universal solvent, water as, 87
Univoltine copepods, 236
Upper Klamath Lake (Oreg.), 215, 216, 400
Upwelling, 48, 58, 73
Urea, 115, 128, 192
Urease, 128
Utricularia (bladderwort), 179, 281

Valves of diatoms, 175
Vampyrella (a protozoan), 181
Vanda, Lake (Antarctica), 36, 43, 47, 111, 377, 381
Varves, 342, 358
Vegetation (*see* Logging and deforestation; Plants; Surrounding vegetation)
Vertical component of lake chemical structure, 18
Vibration motion, effects of, on water molecule structure, 27
Victoria (giant water lily), 179
Victoria, Lake (Africa), 343
Viruses, 18, 166
Viscosity of water, 28–29, 380, 381
Viscous drag, defined, 29
Volcanic lakes, 341–343, 345–346
  (*See also specific volcanic lakes*)
Vorderer Finstertaler See (Europe), 378–380
Vorticella (a protozoan), 180

Wabash River (U.S.), 267–268
Wading birds, 336
Wakatipu, Lake (New Zealand), 71, 347
Waldo Lake (Calif.), 349
Walden Pond (Mass.), 347
Walker Lake (Nev.), 380, 382

Walleye pike, 267
Warburg system of oxygen measurement, 96
Ward Creek (Calif.), 302
Washington, Lake (Wash.):
  succession in, 350, 356
  waste diversion from, 391–392, 394–396, 398
  zooplankton in, 229, 231, 235
Waste discharge (*see* Pollution)
Waste heat discharge (thermal pollution), 55–56, 267
Wastewater treatment (*see* Water treatment plants)
Water, chemical structure of, 25–27
Water-light interaction, 24–38
  properties of water and, 25–29
  and role of light, 29–37
Water lilies, 179, 376
Water movement:
  in lakes, 57–86
    advective and diffusive transport and, 66–69
    effects of rivers on, 83–84
    in epilimnion, 59, 69–75, 82
    in hypolimnion, 59, 79, 81, 82
    kinetic energy spectrum in, 63–67
    laminar and turbulent flow and, 61–63, 66–68
    measurement of, 59–61
    in thermocline, 58, 69, 71, 75–82
    and zoobenthos distribution, 242
  in streams and rivers, 58, 61, 292
Water temperature at depth, 5, 7
Water treatment plants, 355, 393–394
  Lake Tahoe and, 396
  Shagawa Lake and, 392–395, 398, 399
Waterfowl (*see* Birds)
Watershed:
  lake: and chemical composition of water, 87, 90
    management of (*see* Diversion)
    in tropical lakes, 374
  of streams and rivers, 292, 293, 304
Waves, 29
  breaking, 58, 69, 74–75, 79
  dissolved gasses in, 97
  kinetic energy in, 63, 64
  length of, 63, 65, 66, 69, 78
  types of, 57, 69
  upwelling and, 48, 58, 73
  (*See also* Wind; *specific types of waves*)
Wetlands (*see* Estuaries)
Wetting or foaming agents (organic surfactants), 29
White amur (grass carp), 255, 270, 308, 405
Whole-lake manipulation, 7, 9
Wind:
  and distribution of gases, 97

Wind (*Cont.*):
  effects of, on thermocline, 50–51
  heat and, 53–55
  and Langmuir spirals, 58, 63, 73–74, 259
  speed of, 50–51, 53, 55
  and wave propagation, 29
  (*See also* Waves)
  (*See also* Water movement)
Wind mixing, 20
Windermere (England) (*see* English Lake District)
Windrows (slicks), 73–74
Winkler method of oxygen measurement, 96
Winnebago, Lake (Wis.), 347
Winnipeg, Lake (Canada), 19, 20
Winter kills, 55, 108, 265, 403
Winter overturn of oxygen and carbon dioxide, 106–108
Worms, 18, 134, 185–187, 222

Xanthidium (a green alga), 173
Xanthophylls, 175

Yeasts, 170, 180
Yellowstone Lake (Mont.), 346

Zambesi River (Africa), 269
Zinc (Zn), 88, 143, 153, 159, 160
Zizania aquatica (wild rice), 179
Zonation of lakes, 13–15
Zoobenthos:
  estuarine, 318, 331–337
  in lakes, 221–224, 241–247
    feeding patterns of, 243, 244
    fossils of, 359
    general characteristics of, 221–222
    life cycles of, 243–246
    as measure of lake restoration, 390
    measurement of, 224
    in tropical lakes, 372
  (*See also specific zoobenthic organisms*)
Zooplankton:
  discovered, 3
  estuarine, 222, 318, 327, 331–336
  in lakes, 9, 18, 221–241, 280
    chlorine and, 151
    copper and, 160
    effects of copper sulfate on, 405
    epilimnetic temperature and, 47
    excretion by, 56, 89, 139, 226, 282
    fish and, 221, 236, 248, 254
    fossils of, 358, 359
    general characteristics of, 221
    grazing by (*see* Grazing by zooplankton)

Zooplankton, in lakes *(Cont.)*:
  hypolimnetic entrainment and migration of, 48
  iron and, 153, 158
  lake color and, 171
  in Lake Trummen, 399, 400
  in Laurentian Great Lakes, 371
  as measure of lake restoration, 390
  measurement of, 222–225
  molybdenum and, 162

Zooplankton, in lakes *(Cont.)*:
  phosphorus and, 131, 137
  in polar lakes, 377
  population structure, 225–241
  requirements of energy in, 275, 276
  in salt lakes, 382, 383
  thermal pollution effects on, 56
  in tropical lakes, 372–374
  water movement and, 59, 73, 75

Zooplankton, in lakes *(Cont.)*:
  in windrows, 73–74
  in streams and rivers, 307, 311
  (See also specific groups of zooplankton)
Zostera (eel grass), 178, 324, 327, 336
Zostera marina, 336
Zurich, Lake (Switzerland), 79, 203, 217, 392